Metallurgy
Theory
and
Practice

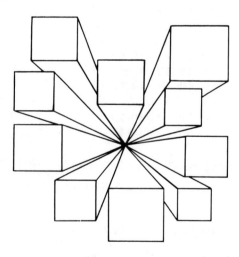

Dell K. Allen

Associate Professor of Industrial Technology
Brigham Young University
Member American Society for Metals

 American Technical Publishers, Inc.
Chicago, Illinois 60637

Copyright © 1969, by American Technical Publishers, Inc.

1st Printing 1969
2nd Printing 1971
3rd Printing 1973
4th Printing 1975
5th Printing 1976
6th Printing 1977
7th Printing 1978
8th Printing 1979
9th Printing 1981

Library of Congress Catalog Card Number: 70-82378

ISBN: 0-8269-3500-1

PRINTED IN THE UNITED STATES OF AMERICA

To CURTIS, MARK, MELODY, HEIDI, PAUL, and DON

Preface

Rapid technological developments have placed increased demands on graduates of industrial and engineering technology programs. One of the important demands is that graduates must have knowledge regarding the proper design, selection, processing, and application of metals and alloys for the production of useful products. Formerly much of this work was done by the engineer. Today, however, the engineering curriculum has few courses dealing with the selection or processing of metals and alloys.

The shift of the engineering curriculum toward more science courses at the expense of applied courses has resulted in the rapid growth of technology training programs at both the associate and baccalaureate degree level.

The technician or the technologist who graduates from these programs may work with both engineering people and production people. Consequently, he must understand the language and techniques of each group. This book has been written to introduce students in technical programs to both theory and application of metallurgical principles as they are applied to the design, selection, heat treating, and processing of metals and alloys.

This text introduces the student to terminology related to the mechanical, physical, and chemical properties of metals. The principles of materials testing are introduced, followed by a brief introduction to the atomic and crystal structure of metals and theories regarding deformation of metals. Tools of the metallurgist are presented, including microscopic and macroscopic examination and temperature measurement. The various types of alloys are discussed and related to an explanation of the iron-carbon diagram. Principles of heat treatment and surface treatments are covered in some detail and interpreted in terms of the basic metallurgical principles previously given. A few chapters deal with specific ferrous and non-ferrous metals and alloys and provide the student with information regarding the selection and treatment of these materials. The final chapters discuss the metallurgical aspects of metal forming, molding, casting, welding, and machining. The book acquaints the student with the basic principles of metallurgy and helps him apply these principles in the design, selection, processing, and testing of metal products.

It is hoped that laboratory work will accompany the study of this book in order to give the student an opportunity to apply and test the principles discussed.

The text may also be used as a reference book as the student becomes involved in design and drafting, foundry processes, welding, metal finishing, quality control, and metal machining.

The instructor of the course should become aware of the vast amount of resource material available for use, including laboratory apparatus, slides, films, transparencies, booklets, samples, models, suggested course outlines and programmed materials.

The student who uses this book will find it much more meaningful if he tries to relate or apply the major concepts and principles to real-life problems. There are a great many facts and figures presented. These are primarily intended for the learning of broader concepts and principles. The facts and figures may soon be forgotten, but it is hoped that the student will retain the terminology and major concepts so he can intelligently discuss heat treatment and processing of metals in terms of these concepts. In addition to the objectives specified for the course by the instructor, it is further hoped that the student will be given the opportunity to become acquainted with various sources of metallurgical literature related to topics of his own special interest.

The author is deeply indebted to Professors Carl G. Johnson and William R. Weeks for the use of certain text material and illustrations from their Metallurgy, 4th Edition, which has served the educational community for many years.

The excellent cooperation of many individuals and companies has made available a large number of illustrations, many of which were prepared especially for this book. Although the contributors are too numerous to mention individually, the author is indebted for their assistance.

The author is also indebted to Mrs. Helen Stillman, Mrs. Linda Williams, Mrs. Carol Lake, and Miss Bonnie Duell who typed the manuscript, and most of all, to his wife, Wanda, who typed the draft copies and who rendered valuable assistance and encouragement during the many hours required for the preparation of this book.

Finally, the author wishes to thank his students and laboratory assistants who have thoughtfully and critically commented on the course and content in which this material has been used.

DELL K. ALLEN

Contents

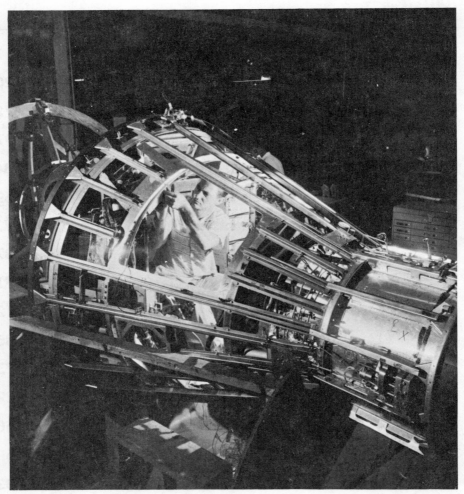

Space capsule constructed of high strength, temperature resistant alloys. Frame is titanium, heat shields beryllium, and 'Rene' 41 outside. (Titanium Metals Corp. of America)

Properties of Metals

1.1 Introduction. The statement "Metallurgy is one of the oldest of the arts and one of the newest of the sciences" pretty well sums up the long and interesting history of the field of metallurgy.

The art of metallurgy dates back to early Biblical times when Tubal Cain, a descendant of Cain, was reputed as an "instructor of every artificer in brass and iron." (Genesis 4:22).

Early Egyptians called iron *ba-en-pet*, the metal from heaven. The Assyrians, Babylonians, Chaldeans, and Hebrews likewise had names for the iron-nickel meteors which would fall from the sky, and from which they would fashion ornaments and tools.

It is reported that gold was used for ornaments, plates, and utensils as early as 3500 B.C. Gold objects showing a high degree of culture have been excavated at the ruins of the ancient city of Ur in Mesopotamia.

Silver was used as early as 2400 B.C. and tin as early as 3000 B.C. Native copper was used in the Near East soon after 4000 B.C. for tools and utensils. Egypt had great copper mines on the island of Crete. Lead pipes were made by the Romans for their water distribution system by extracting the metal from its ore.

The art of hardening and tempering of steel swords during the Middle Ages was a jealously guarded secret, passed on from father to son by the famous swordmakers of Damascus, and of Toledo in Spain.

The art and science of separating metals from their ores and preparing them for use is called metallurgy. Because of its wide scope, the field of metallurgy may be divided into two branches. One is called extractive metallurgy and the other, physical metallurgy.

1.1.1. Extractive metallurgy is the branch dealing with taking metals from their ores and refining them

1

to the desired state. It includes a wide variety of specialized commercial processes, including mineral dressing, roasting, sintering, smelting, leaching, electrolysis, amalgamation, and pyrometallurgy (refining molten metal).

1.1.2. Physical metallurgy is the branch covering adapting metals to human use. It includes operations and processes to convert a refined commercial metal into a useful finished product. This involves combining metals into alloys to obtain special properties such as combining chromium or nickel with iron to make rust resistant stainless steel. It involves the improvement of properties by treatments such as hardening to improve strength. The forming of metal into final shape and the surface treatment of the finished product are also classed as physical metallurgy. This book deals primarily with physical metallurgy and includes recent developments in metallurgical processes and materials.

During the last fifty years men have tried diligently to find explanations for the behavior of metals during solidification, forming, heat treatment, and actual service. The results of this research will be examined beginning with consideration of the properties and uses of metals, testing of materials, and methods of metallurgical examination. The various explanations for metallic behavior during alloying, thermal treatment, and deformation will be considered. Both ferrous and nonferrous metals and alloys are discussed with special emphasis on their properties and applications. Metallurgical factors related to the shaping of metals, powder metallurgy, casting, welding, and machining, are reported in some detail.

1.2 Properties of Metals and Their Uses. Why have metals come to play so large a part in man's activities? Wood and stone are both older in use, yet to a considerable extent they have been supplanted by metals. The reason for the increased use of metals is found in their characteristic properties. Most important of these properties is strength, or ability to support weight without bending or breaking, combined with toughness, or the ability to bend rather than break under a sudden blow. Resistance to atmospheric destruction, plasticity, and the ability to be formed into desired shapes add to the remarkable combination of properties possessed by metals. Some metals are especially noted for their ability to conduct electric current and to be magnetized.

Metals can be cast into varied and intricate shapes weighing from a few ounces to many tons. Their plasticity, or ability to deform without rupture, makes them safe to use in all types of structures and also allows their formation into required

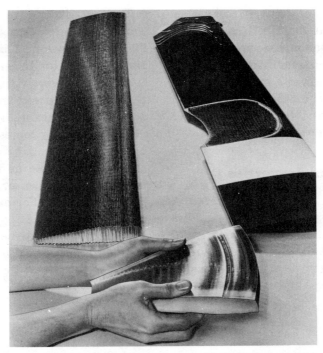

Fig. 1–1. Laminated honeycomb materials have highest rigid-ity-to-weight ratio. Aluminum helicopter tail rotor shows honeycomb before expansion, expanded core, and finished blade with bonded aluminum skin. (Hexcel Products)

shapes through forging and other processes. Metals also possess the important property of being weld-able and repairable. Other materials used in engineering construc-tion, including glass, stone, wood, and concrete, usually are discarded when the structure is no longer use-able. On the other hand, an obsolete or worn-out bridge, ship, automo-bile, or aircraft made of metal is usually cut into easily handled sec-tions, put into a furnace, remelted, cast, and finally worked into the making of a new ship, bridge, auto-mobile, or aircraft.

A knowledge of the properties of specific metals or alloys enables us to determine whether or not such materials are suitable for certain definite uses, and also to modify the thermal and mechanical treat-ments of such materials in order to obtain them in the most desirable

3

form. New processing techniques such as that described in Fig. 1–1 further enhance the properties of metals. Testing methods, in general, enable us to take a small portion of a material, and from it, predict with some certainty how the material will behave in actual service.

Mechanical Properties

1.3 Stress, Strain, and Elasticity. Mechanical properties of metals are those which, upon the application of force, reveal the elastic and inelastic reaction, or involve the relationship between stress and strain.

Before discussing mechanical properties, three terms will be defined which are most important for the student to understand. The terms are: *stress*, *strain*, and *elasticity*. Although stress and strain are used interchangeably by many people, they actually represent different quantities.

Stress is defined as the load per unit area and is measured in pounds per square inch. For example: If a 1,000 pound load (Fig. 1–2) represented by P, is applied to a wire whose cross-sectional area A, is .05 square inches, then the stress, S, is given by:

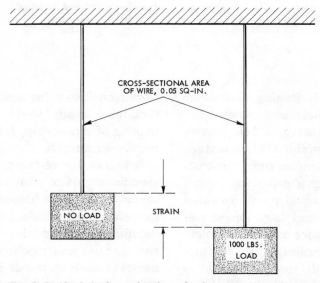

CROSS-SECTIONAL AREA OF WIRE, 0.05 SQ-IN.

NO LOAD

STRAIN

1000 LBS. LOAD

Fig. 1–2. Hooke's Law showing elastic properties of metals.

$$S = P/A = \frac{1,000 \text{ pounds}}{.05 \text{ square inch}}$$
$$= 20,000 \text{ lbs/sq inch} \quad (\text{psi})$$

This stress represents a pressure, which may be tending to pull the material apart, compress it, twist it or shear it, depending on the direction and nature of the applied load.

The stress which causes a material to stretch is called tensile stress. The stress which causes a material to get shorter is called compressive stress, and that stress which causes a material to divide into layers is called shear stress. Flexure (bending) loads and torsional (twisting) loads produce combinations of the above three stresses.

Strain, on the other hand, is the percent change in unit length during elongation or contraction of a specimen and is a measure of deformation under load. For example; if a 100-inch long wire is suspended vertically and loaded with weights, the wire will stretch or elongate under load. If the wire stretches two inches, then the strain, e, is represented by

$$e = \frac{(L_f - L_o) \times 100}{L_o}$$
$$= \frac{(102 - 100) \times 100}{100}$$
$$= 2\% \text{ strain}$$

Where:

L_o represents the original length.
L_f the final length.

Elasticity was expressed as a theory in the year 1678 by Robert Hooke, an English experimental scientist, on the basis of experiments similar to the one described above. His theory is now known as *Hooke's Law*. This law states that the degree to which an elastic body bends or stretches out of shape (strain) is in direct proportion to the force (stress) acting upon it. This law applies only within a certain range of stresses, it was later discovered. Beyond this stress is a point called the *elastic limit*. If loading is increased beyond this point, the body is permanently deformed. In reality, we find that metals are not entirely elastic even under slight loads; therefore an arbitrary method of determining the commercial elastic limit must be used. Fig. 1–3 shows a typical stress-strain diagram for a specimen which has been tested under increasing tensile loads. Elastic limit, yield strength and tensile strength values are shown in Fig. 1–3 for commercially pure iron, structural steel, and high carbon steel.

1.4 Strength. Strength with plasticity is perhaps the most important combination of properties a metal can possess. Strength is the ability of a material to *resist* deformation; plasticity is the ability to *take* deformation without breaking. A number of strength values for a metal must be known to fully understand

5

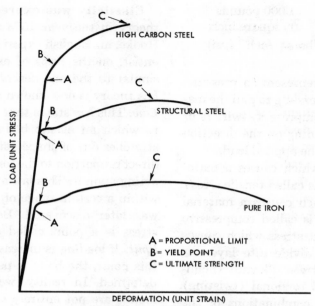

Fig. 1–3. Stress strain curves for three metals.

A = PROPORTIONAL LIMIT
B = YIELD POINT
C = ULTIMATE STRENGTH

the strength characteristics of the metal. Among these strength values are tensile strength, compression strength, fatigue strength, and yield strength. Tests for determining strength values are described in Chapter 2.

1.4.1. Tensile strength is defined as the maximum load in tension which a material will withstand prior to fracture. This is the value most commonly given for the strength of a material and is given in pounds per square inch. Tensile strength of metals may be increased by alloying, cold working, and sometimes, by heat treating.

1.4.2. Compressive strength is the maximum load in compression which a material will withstand prior to a predetermined amount of deformation. The compressive strengths of both cast iron and concrete are greater than their tensile strengths, whereas for most materials, the reverse is true.

1.4.3. Fatigue strength is the maximum load that a material can withstand without failure during a large number of reversals of load. For example: A rotating shaft which supports a weight has tensile forces on the top portion of the shaft and compressive forces on the bottom.

As the shaft is rotated there is a repeated cyclic change in tensile and compressive stresses. Fatigue strength values are used in the design of aircraft wings and other structures subject to rapidly fluctuating loads. Fatigue strength is influenced by microstructure, surface condition, corrosive environment, cold work, and other factors.

1.4.4. Yield strength is the maximum load at which a material exhibits a specified deformation. A total elongation of 0.2 percent is used in determining yield strength for many metals such as aluminum and magnesium alloys while 0.5 percent deformation under load is frequently used for copper alloys. Most engineering calculations for structures are based on yield strength values rather than tensile strength values. Yield strength values vary from about 50 percent of the tensile strength for copper to about 85 percent for cold drawn mild steel. The strength of a metal is related to its internal structure—defects, composition, heat-treatment and degree of cold work. In general, the strength values found by testing is of the order of 100 to 1000 times lower than theoretical strength values. This large difference between theoretical and observed values for strength will be discussed in Chapter 3.

1.5 Hardness. Hardness is not a fundamental property of a material but is related to its elastic and plastic properties. Generally, hardness is defined as the resistance to indentation. The greater the hardness the greater the resistance to penetration. The hardness test is widely used because of its simplicity and because it can be closely correlated with the tensile and yield strength of steels. In general, it can be assumed that a strong steel is also a hard one and one resistant to wear. Scratch or abrasion hardness tests are sometimes used for special applications, as are elastic or rebound hardness tests.

1.6 Toughness. Although there is no direct and accurate method of measuring the toughness of metals, toughness involves both ductility and strength and may be defined as the ability of a metal to absorb energy without failure.

Toughness may be expressed as the total area under a stress-strain curve as shown in Fig. 1–3. Often the impact resistance or shock resistance of a material is taken as an indication of its inherent toughness.

1.7 Plasticity. A very important mechanical property of metals is that of plasticity. Plasticity is the ability of a metal to be deformed extensively without rupture.

1.7.1. Ductility is the plasticity exhibited by a material under tension loading. It is measured by the amount the material can be permanently elongated. This ability to elongate permits a metal to be drawn from a larger size to a smaller size of wire. Copper and aluminum have high ductility.

1.7.2. Malleability, which is another form of plasticity, is the ability of a metal to deform permanently under compression without rupture. It is this property which allows the hammering and rolling of metals into thin sheets. Gold, silver, tin, and lead are examples of metals exhibiting high malleability. Gold has exceptional malleability and can be rolled into sheets thin enough to transmit light.

1.7.3. Brittleness is the property opposite to plasticity. A brittle metal is one that cannot be visibly deformed permanently; that is, it lacks plasticity. The hard metals, such as fully hardened steel, may exhibit very little plasticity and may therefore be classed as brittle; yet hardness is not a measure of plasticity. A brittle material usually develops little strength upon tensile loading but may be safely used in compression. Brittle metals show very little shock or impact strength and fail without any warning of impending failure.

On the other hand, a ductile metal may fail without any visible defor-

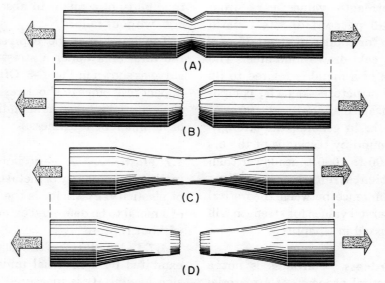

(A)

(B)

(C)

(D)

Fig. 1–4. Notching and its effects on plasticity.

Table 1–1. Mechanical properties of some common metals.

MATERIAL	TENSILE STRENGTH psi	YIELD STRENGTH psi	SHEAR STRENGTH psi	PER CENT ELONGATION	PER CENT REDUCTION IN AREA	BRINELL HARDNESS	MODULUS OF ELASTICITY	IMPACT STRENGTH FT—LBS
ALUMINUM	13,000	5,000	9,500	45	--	23	10,000,000	--
ANTIMONY	1,560	--	--	--	--	30–58	11,000,000	--
BERYLLIUM	27,000	26,400	--	0	--	97–114	36,800,000	--
BISMUTH	--	--	--	--	--	--	4,600,000	--
CADMIUM	10,300	--	--	50	--	21–23	8,000,000	--
CARBON (graphite)	--	--	--	--	--	--	700,000	--
CHROMIUM	--	--	--	--	--	110–170	--	--
COBALT	34,400	--	--	--	--	125	30,000,000	29–113 IZOD
COPPER	32,000	10,000	22,000	45	--	--	17,000,000	--
GOLD	19,000	nil	--	45	--	25	11,600,000	--
IRON (alpha)	38,000	19,000	--	43–48	70–77	67	30,000,000	--
LEAD	1,900	800	1,825	30	100	3.2–4.5	2,000,000	10.4 CHARPY
MAGNESIUM	27,000	14,000	17,000	16	--	40	6,500,000	--
MANGANESE (gamma)	72,000	35,000	--	40	--	331	--	--
MOLYBDENUM	100,000	57,000	--	--	--	160–185	50,000,000	120 IZOD
NICKEL	50–80,000	10–30,000	52,300	30–50	50–70	90–120	30,000,000	--
NIOBIUM (columbium)	50,000	--	--	30	--	125	30,000,000	--
PLATINUM	17–19,000	--	--	25–40	--	--	21,300,000	--
SILICON	--	--	--	--	--	--	16,350,000	--
SILVER	18,200	7,900	--	48	91	--	10,300,000	--
TANTALUM	50,000	--	--	40	--	--	--	--
TIN	21–3100	--	2,900	55–69	--	5.3	6,000,000	14 IZOD
TITANIUM	78,700	62,800	--	25.2	--	--	16,800,000	--
TUNGSTEN	264–590,000	--	--	--	--	--	50,000,000	--
VANADIUM	--	--	--	--	--	--	19,000,000	--
ZINC	41,000	--	31,000	10	--	82	--	43 CHARPY
ZIRCONIUM	35,900	15,900	--	31	--	--	31,000,000	--

SOURCE: COMPILED FROM DATA IN METALS HANDBOOK, 1948 ED.
PROPERTIES ARE FOR MATERIALS IN THE ANNEALED CONDITION, AND OF COMMERCIAL PURITY.

mation if the load becomes concentrated, due to a notched effect, as shown in Fig. 1–4. The test specimen A in Fig. 1–4 has a sharp reduction in cross section, and, when pulled apart, ruptures with a brittle fracture. See specimen B. Specimen C, made from the same metal, but with a gradual reduction in cross section, behaves as a plastic material, elongating noticeably before rupture; See specimen D. Sharp notches and corners act as stress raisers and may reduce the strength of a part 50 percent or more.

Mechanical properties for a number of metals are shown in Table 1–1.

Physical Properties

1.8 Coefficient of Linear Thermal Expansion. Physical properties of metals are those characteristics related to the atomic structure and include density, electrical and thermal conductivity, melting point, magnetic susceptibility, reflectivity, and coefficient of linear expansion.

With few exceptions, solids expand when they are heated and contract when cooled. They increase not only in length, but also in breadth and thickness. The increase in unit length when a solid is heated one degree is called its *coefficient of linear expansion*. For example, when steel is heated, it expands .000006 inches per inch per degree Fahrenheit. This means that a one-inch long gage

block used in precision measurement (and accurate to within four millionths of an inch) that has been calibrated at 68° Fahrenheit, would be 192 millionths of an inch longer if taken into a 100 degree Fahrenheit shop—an expansion that could not be tolerated in precision measurement.

1.9 Specific Gravity. Sometimes there is a need to compare the density of one metal with that of another. For this purpose, we need a standard. Water is the standard which physicists have selected with which to compare the densities of solids and liquids. Hence, the weight of a substance compared to the weight of an equal volume of water is called its *specific density* or *specific gravity*. Table 1–2 shows the densities of several common metals.

1.10 Melting Point. The melting point is the temperature at which a substance passes from a solid to a liquid condition. For ice this is 32°F. Pure substances have a sharp melting point. That is, they pass from a solid state to a liquid without a change in temperature; however, there is an absorption of heat during melting and a liberation of heat during freezing. The absorption or release of thermal energy when a substance changes state is called its *latent heat*. The melting points in

Table 1–2. Physical properties of some common metals.

MATERIAL	COEFF. OF LINEAR EXPANSION, MICROINCH/°F.	DENSITY gm/cm3	MELTING POINT Deg. F.	ELECTRICAL RESISTIVITY MICROHM–cm	MAGNETIC SUSCEPTIBILITY K, cgs x 10⁻⁶	REFLECTIVITY PER CENT*
ALUMINUM	13.3	2.71	1215	2.922	--	90
ANTIMONY	4.7 to 6.0	6.62	1166.9	39.0	-0.87	70
BERYLLIUM	6.9	1.82	2343	5.9	--	50-55
BISMUTH	7.4	9.80	520.3	106.8	--	--
CADMIUM	16.6	8.65	609.6	6.83	-0.18	73
CARBON (graphite)	0.3 to 2.4	2.22	6700	1375.	--	--
CHROMIUM	3.4	7.19	2370	18.9	$+3.49$	70
COBALT	6.8	8.9	2723	6.24	--	50
COPPER	9.2	8.96	1981	1.673	--	40
GOLD	7.9	19.32	1945.4	2.19	-0.15	95
IRON (alpha)	6.5	7.87	2800	9.71	--	93
LEAD	16.3	11.34	618	20.65	-0.12	62
MAGNESIUM	14	1.74	1202	4.46	$+0.55$	72
MANGANESE	12	7.43	2273	185.	$+14.0$	--
MOLYBDENUM	2.7	10.2	4760	5.17	$+0.04$	46
NICKEL	7.4	8.90	2651	6.84	--	64
NIOBIUM (columbium)	4.0	8.57	4379	13.1	$+1.5$	--
PLATINUM	4.9	21.45	3223.8	10.6	$+1.1$	67
SILICON	1.6 to 4.1	2.33	2605	105	-0.13	--
SILVER	10.49	10.49	1760.9	1.59	-0.02	92
TANTALUM	3.6	16.6	5425	12.4	$+0.93$	--
TIN	13	7.30	449.4	11.5	-0.25	75
TITANIUM	4.7	4.54	3300	80.0	$+1.25$	--
TUNGSTEN	2.4	19.3	6170	5.5	--	--
VANADIUM	4.3	6.0	3150	26.0	$+1.5$	--
ZINC	22.1	7.13	787.0	5.91	-0.15	75
ZIRCONIUM	3	6.5	3200	41.0	--	--

SOURCE: COMPILED FROM DATA IN METALS HANDBOOK, 1948 ED.

*LIGHT FROM A TUNGSTEN SOURCE

11

Table 1–2 are expressed in degrees Fahrenheit. In metallurgy temperatures are often expressed in degrees Celsius (formerly called Centigrade). The conversion from one temperature scale to the other is given by:

$$F = 9/5\ °C + 32$$
$$C = 5/9\ (°F - 32)$$

1.11 Electrical and Thermal Conductivity. The ability for a metal to easily conduct electricity and heat is one of its notable characteristics. The relative electrical resistances of metals are shown in Table 1–2. The resistance to the flow of heat is of the same order as the resistance to flow of electricity. The opposition to electric current through a wire is known as the resistance of the wire. An arbitrary resistance unit has been agreed upon internationally called the *ohm*. The resistances of metals are often expressed in millionths of an ohm or *microhms*. A microhm is the resistance in a cube of metal one centimeter on each side. Copper and aluminum are frequently used for conducting electricity since they offer little resistance to the passage of electric current.

Silver offers less resistance than either copper or aluminum but it is too expensive for commercial power transmission. Copper offers less resistance than aluminum for the same size wire, but aluminum, due to its lighter weight, offers less resistance per unit of weight.

It should be noted that there are several factors which can alter the resistance of metals, some of which are:

1. The resistance of metals to the flow of electrical and thermal energy increases with temperature.

2. Resistance to electrical flow increases with impurities and alloying.

3. Cold working (deformation) of the metal increases electrical resistance.

4. Precipitation from solid solution during age hardening increases electrical resistance. The age hardening mechanism will be explained in Chapter 8.

5. There is an inherent low resistivity for monovalent metals (Cu, Ag, Au) and the alkali metals (Li, Na, K, Rb, Cs), higher resistance for divalent alkaline earth metals (Be, Mg, Ca, Sr, Ba, Ra), and high resistance for the transition metals (Fe, Co, Ni, Ru, Rh, Pd, Os, Ir, Pt.)

During the passage of current through a conductor the resistance results in the liberation of heat; the greater the resistance, the greater the heat for passage of a given current. For electrical heating metals with high electrical resistance are needed, such as alloys of nickel and chromium.

A good conductor of heat, such as copper, is often used for heat exchangers, heating coils, and soldering irons. Utensils for cooking frequently are made of aluminum because of the combination of high heat conduction and resistance to attack by foodstuffs.

1.12 Magnetic Susceptibility. If an object is placed in a magnetic field a force is exerted on it and it is said to become magnetized. The intensity of magnetization depends upon the *susceptibility K*, which is a property of the material of which the object is composed. The value of K varies greatly from one metal to another, and it is possible to arrange three classes, according to the sign and magnitude of K.

Diamagnetic metals, for which K is small and negative, and which are therefore feebly repelled by a magnetic field. Examples are copper, silver, gold, and bismuth.

Paramagnetic metals, for which K is small and positive. Most metals are paramagnetic, examples being lithium, sodium, potassium, calcium, strontium, magnesium, molybdenum, and tantalum.

Ferromagnetic metals, for which K is large and positive, including iron, cobalt, nickel, and gadolinium. Alloys and compounds containing the above metals or manganese or chromium also are ferromagnetic. A characteristic feature of ferromagnetic substances is that they are able to retain their magnetism after the magnetizing field has been removed. Thus they are capable of becoming permanent magnets.

1.13 Reflectivity. One distinguishing characteristic of a metal is its metallic lustre. This surface color is apparently due to selective reflection

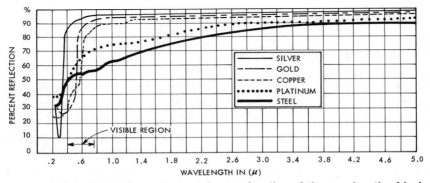

Fig. 1–5. Reflectivity of common metals as a function of the wavelength of incident light. (By permission: *College Physics,* D. C. Heath & Co.)

of the incident light at the free surface of the metal. The reflectivity (ratio of reflected to incident light) of a metal surface depends not only on the material but very much upon whether the surface is polished or rough, smooth surfaces reflecting better than rough ones. The reflectivity of polished metal surfaces varies greatly for different metals, as shown in Fig. 1–5. Also reflectivity varies greatly for different wave lengths of incident light. For infrared (heat) waves longer than about 5μ, ($\mu = 10^{-4}$ cm) the reflectivity of metals is apparently related to electrical conductivity, good conductors are good reflectors.

Excellent oxidation resistance and heat reflecting characteristics of aluminum and aluminum coated steels make them desirable for such uses as oven linings, heat deflectors, and heating element retainers.

Chemical Properties

1.14 Atomic Structure. All matter is composed of one or more essentially different substances called chemical elements, which are in turn composed of building blocks called atoms. The early Greek philosopher, Democritus, first used the word, atom, meaning not cuttable, to describe tiny particles of matter. Democritus had a theory that the atom was the smallest particle into which matter could be cut, or divided, without changing its properties. In more recent times, it has been discovered that atoms *are* the smallest particles of matter having distinct chemical characteristics. Democritus made a good guess. The size of an atom may be appreciated by the fact that 250,000,000 hydrogen atoms lined up side by side would measure one inch.

An atom resembles a miniature solar system and its chief parts are shown in Fig. 1–6. The *nucleus* or "sun" of the atom consists of protons and neutrons. *Protons* are relatively heavy (1.673 $\times$ 10^{-24}g.) and have a positive electrical charge. *Neutrons* have essentially the same weight as protons but are neutral, having no electrical charge. Revolving at high speed around the nucleus are "planets" of the atom called *electrons*. Electrons are tiny particles having negative electrical charges. The charge is so tiny that about ten billion electrons must flow each second to light the filament of an ordinary electric light bulb. Electrons are strongly attracted to the nucleus of the atom, which is positive. This attraction or binding energy holds the atom together. Each electron spins on its own axis as it travels in an elliptical path around the nucleus. Each atom has distinctive, preferred, electron paths. These paths are called *shells*. The number, arrangement, and spin of electrons in these shells determines the kind of atom and its characteristics.

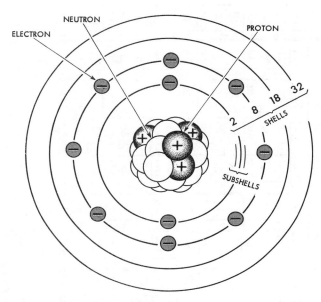

Fig. 1–6. Schematic representation of the structure of an atom.

Each shell is subdivided into energy states or levels. The number of energy levels increases with distance from the nucleus. Electrons tend to occupy the lower energy levels but may jump to higher ones when the internal energy of the atom is increased as in heating.

The first principal shell can contain a maximum of two electrons, the second, eight, and third, eighteen, and the fourth, thirty-two electrons. Electrons in the unfilled shells are known as *valence electrons* and are largely responsible for the chemical behavior of the element.

Each element is given an *atomic number*. This number is equal to the number of electrons surrounding the nucleus. The weight of an atom or its *atomic weight* is essentially equal to the weight of the neutrons and protons contained in the nucleus. Electrons weigh only about 1/2000th of the weight of a proton or neutrons, and therefore contribute only a small amount to the weight of an atom.

1.15 Periodic Table. The periodic table shown in Table 1–3 has been referred to as the "Natural Classification of the Elements." It is *natural* because it is logically based upon certain fundamental characteristics of atoms making up the known elements. It is *periodic* because ele-

Table 1–3. Periodic chart of the elements.

(From *Fundamental Chemistry*, 2nd Edition, by H. G. Deming. Published by John Wiley & Sons, Inc., Reprinted by permission)

ments with similar properties appear at regular intervals in the arrangement of the elements.

The first periodic table, or chart of the elements, was published in 1859 by Dmitri Mendeleev and was merely an arrangement of the elements in the order of their atomic weights. There were many blank spaces in the first table, but from it was predicted the existence of elements to fill these blanks and their probable properties. It was later discovered that atomic weight is variable and that an atom consists of a nucleus with a positive electrical charge, surrounded by a specific number of negative electrons. New periodic charts are based on atomic number as the distinctive characteristic of atoms rather than upon atomic weight.

The horizontal rows of elements, Table 1–3, are known as *periods* and the vertical columns as *groups*. Elements in the same group have similar electron configurations and are often substituted for one another when developing new alloys. For example: In Group VI–B we find chromium, molybdenum, and tungsten. Molybdenum and chromium have been substituted for tungsten in some high alloy steels. Sulfur in Group VI–A has been used for improving machinability of steels as have selenium and tellurium in that same group.

Chemical properties of some common metals are shown in Table 1–4. Crystal structures and lattice parameters as shown in the table are discussed in Chapter 3. In general, crystal forms, densities, melting points, specific heats, and other properties of elements are functions of the atomic number.

1.16 Metals, and Non-Metals. Chemical elements may be roughly classified into three groups; metals, non-metals, and inert gasses. Metals, in the solid state, are characterized by the following properties:

1. Crystalline structure,
2. High thermal and electrical conductivity,
3. Ability to be deformed plastically,
4. High reflectivity (metallic lustre).

Metals constitute about three-fourths of the known elements. Some non-metals, such as carbon and silicon resemble metals in some respects and are sometimes called metalloids. For example, they have a metallic lustre, crystalline form, are fair conductors of electricity, but are very brittle as are all non-metals.

1.17 Electromotive Series. The electromotive series is a list of elements arranged according to their standard electrode potential. This potential (voltage) between dissimilar metals when coupled together and immersed in an electrolyte is due to the difference in chemical activity

17

Table 1–4. Chemical properties of some common metals.

MATERIAL	SYMBOL	ATOMIC NUMBER	ATOMIC WEIGHT	CHEMICAL VALENCES	ELECTROMOTIVE POTENTIAL*	CRYSTAL STRUCTURE**	LATTICE PARAMETER*** VALUES IN kX UNITS
ALUMINUM	Al	13	26.97	3	+ 1.67	F.C.C.	4.0413
ANTIMONY	Sb	51	121.76	3,5	– 0.10	RHOMBOHEDRAL	4.4974, α = 57°6.5'
BERYLLIUM	Be	4	9.02	2	+ 1.70	C.P.H.	a= 2.2680, c = 3.5942
BISMUTH	Bi	83	209.00	3,5	– 0.20	RHOMBOHEDRAL	4.7457, 57°14.2'
CADMIUM	Cd	48	112.41	2	+ 0.40	C.P.H.	a = 2.9727, c = 5.606
CARBON (graphite)	C	6	12.01	+ 4, 2	---	HEXAGONAL	a = 2.4564, c = 6.6906
CHROMIUM	Cr	24	52.01	6, 3, 2	+ 0.71	B.C.C.(above 26°C)	2.878
COBALT	Co	27	58.93	3, 2	+ 0.28	C.P.H.	a = 2.5020, c = 4.0611
COPPER	Cu	29	63.57	2, 1	– 0.34	F.C.C.	3.6080
GOLD	Au	79	197.2	3, 1	– 1.68	F.C.C.	4.070
IRON (alpha)	Fe	26	55.85	3, 2	+ 0.44	B.C.C.	2.860
LEAD	Pb	82	207.21	2, 4	+ 0.13	F.C.C.	4.9389
MAGNESIUM	Mg	12	24.32	2	+ 2.34	C.P.H.	a = 3.2033, c = 5.1998
MANGANESE	Mn	25	54.93	7, 4, 2, 6, 3	+ 1.05	CUBIC	8.894
MOLYBDENUM	Mo	42	95.95	6, 3, 5	---	B.C.C.	3.1400
NICKEL	Ni	28	58.69	2, 3	+ 0.25	F.C.C.	3.5167
NIOBIUM (columbium)	Nb	41	92.91	5, 3	---	B.C.C.	3.294
PLATINUM	Pt	78	195.23	4, 2	1.2	F.C.C.	3.9158
SILICON	Si	14	28.06	4	---	CUBIC, DIAMOND	a = 5.417
SILVER	Ag	47	107.88	1	– 0.80	F.C.C.	4.0778
TANTALUM	Ta	73	180.95	5	---	B.C.C.	3.296
TIN	Sn	50	118.70	4, 2	+ 0.14	TETRAGONAL	a = 5.8195, c = 3.1750
TITANIUM	Ti	22	47.90	4, 3	---	C.P.H.	a = 2.95, c = 4.73
TUNGSTEN	W	74	183.92	6	---	B.C.C.	3.1585
VANADIUM	V	23	50.95	5, 4, 2	---	B.C.C.	3.033
ZINC	Zn	30	65.38	2	+ 0.76	C.P.H.	a = 2.6594, c = 4.9370
ZIRCONIUM	Zr	40	91.22	4	---	C.P.H.	a = 3.223, c = 5.123

SOURCE: COMPILED FROM DATA IN METALS HANDBOOK, 1948 ED., CHART OF THE ATOMS, W. F. MEGGINS, WELCH SCIENTIFIC CO., 1965, AND PERIODIC CHART OF THE ELEMENTS, MERCK & CO., 1964.

UNDERLINE INDICATES MOST COMMON VALENCE

*ELECTROMOTIVE POTENTIAL, VOLTS, WITH REFERENCE TO HYDROGEN AND MOST VALENCE.

**CRYSTAL STRUCTURE AT ROOM TEMPERATURE

***MULTIPY BY 1.00202 TO OBTAIN VALUES IN ANGSTROMS. (ONE ANGSTROM Å = 10^{-8} cm.)

Table 1–5. Electromotive series.

REDUCTANT		OXIDANT	POTENTIAL
Li	⇌	Li^+	+3.05
Rb	⇌	Rb^+	+2.93
K	⇌	K^+	+2.93
Cs	⇌	Cs^+	+2.92
Ba	⇌	Ba^{++}	+2.90
Sr	⇌	Sr^{++}	+2.89
Ca	⇌	Ca^{++}	+2.87
Na	⇌	Na^+	+2.71
Mg	⇌	Mg^{++}	+2.37
Be	⇌	Be^{++}	+1.85
Al	⇌	Al^{+++}	+1.66
Mn	⇌	Mn^{++}	+1.18
Zn	⇌	Zn^{++}	+0.76
Cr	⇌	Cr^{+++}	+0.74
Ga	⇌	Ga^{+++}	+0.53
$S^=$	⇌	S	+0.48
Fe	⇌	Fe^{++}	+0.44
Cd	⇌	Cd^{++}	+0.40
Tl	⇌	Tl^+	+0.34
Co	⇌	Co^{++}	+0.28
Ni	⇌	Ni^{++}	+0.25
Sn	⇌	Sn^{++}	+0.14
Pb	⇌	Pb^{++}	+0.13
H	⇌	H^+	0.00
Sn^{++}	⇌	Sn^{+4}	−0.15
Cu	⇌	Cu^{++}	−0.34
I^-	⇌	I_2	−0.54
Fe^{++}	⇌	Fe^{+++}	−0.77
Hg	⇌	Hg_2^{++}	−0.79
Ag	⇌	Ag^+	−0.80
Hg_2^{++}	⇌	Hg^{++}	−0.92
Br^-	⇌	Br_2	−1.07
Cl^-	⇌	Cl_2	−1.36
Au	⇌	Au^+	−1.68
F^-	⇌	F_2	−2.65

(Left margin: INCREASING TENDENCY TO GO TO OXIDIZED STATE — Right margin: INCREASING TENDENCY TO GO TO REDUCED STATE)

of the metals involved. The more active metals at the top of the list exhibit a stronger tendency to dissolve than those at the bottom and thus have a higher potential. A metal higher in the series than another will displace it from solution. For example, iron with a potential of +0.44 volts is higher than copper with −0.34 volts. Thus, if an iron nail is placed in a copper sulfate solution, some of the nail will be replaced with copper and the nail becomes copper plated. In compiling the list, the standard potential for hydrogen is assigned the value of zero. The electromotive series holds only for metals under the experimental conditions used in determining the potential, but it is useful in extractive metallurgy and in determining susceptibility of metals to corrosion. The potential is shown in Table 1–5 for a number of metals and nonmetals.

References

Avner, Sidney H. *An Introduction to Physical Metallurgy.* New York: McGraw-Hill, 1964.

Barrett, Charles S. *Structure of Metals, 2nd ed.* New York: McGraw-Hill, 1952.

Burton, Malcolm S. *Applied Metallurgy for Engineers.* New York: McGraw-Hill, 1956.

Cottrell, A. H. *Theoretical Structural Metallurgy.* London: Edward Arnold & Co., 1963.

Lyman, Taylor, ed. *Metals Handbook, 1944 ed.* Cleveland; American Society for Metals.

Meggers, William F. *Chart of the Atoms.* Chicago: The Welch Scientific Co., No. 4558, 1965.

Steel Founders Society of America. "Steel Casting Glossary" *Design News,* April 14, 1965. Cleveland.

Wilson, Frank W., ed. *Tool Engineer's Handbook, 2nd ed.* New York: McGraw-Hill, 1959.

Materials
Testing

2.1 Introduction. A knowledge of the properties of specific metals or alloys enables us to determine which materials are suitable for certain uses and also how to modify the thermal and mechanical treatments of such materials to obtain them in the most desirable form. Testing methods, in general, enable us to take a sample portion of a material, and from it predict with some certainty how the material will behave in actual service. Some tests involve testing the part to complete failure and are known as *destructive testing*. Examples of destructive testing are certain types of hardness tests, tensile tests, impact tests, torsion tests, and fatigue tests. Other testing methods do not harm the part nor impair it for actual service. Such tests are known as *non-destructive tests* and include x-ray inspection, magnetic particle inspection, and ultrasonic inspection. More and more emphasis is being placed on the development and use of non-destructive testing methods.

Techniques have been developed recently for experimentally determining locations of high stress in actual parts or plastic models, Fig. 2–1. These points of high stress indicate critical areas where failure may occur unless the part is redesigned or unless the loading is reduced. Experimental techniques for stress measurement involve the use of brittle lacquers, photoelastic methods, and various types of strain gages. Experimental techniques are often used by designers and engineers for testing and evaluating parts with complex shapes which render theoretical evaluation impractical.

Destructive Testing

2.2 Hardness Testing. Hardness, as the term is commonly used in metallurgy, is a measure of the resistance of a material to deformation (inden-

Fig. 2–1. Photo-elastic study of stresses on plastic gear teeth. Dark lines show areas of stress concentration. (By permission: F. D. Thompson Publications, from *Industrial Science and Engineering Magazine*, March, 1956)

tation) by an indenter of fixed geometry under a static load. The definitions for ultimate tensile strength and hardness are somewhat similar, and it has been experimentally verified that hard ferrous metals are generally also strong. No known relationship exists between the hardness and strength of non-ferrous metals, however.

A considerable amount of information can be derived from a hardness test, but intelligent appraisal of a hardness number requires understanding of the composition and condition of the metal under test as well as factors that influence accuracy of the test. A number of hardness tests have been developed each of which has advantages as well as disadvantages.

2.2.1. Scratch hardness tests are relatively quick. A simple, though crude, method of measuring relative hardness is to determine whether or not one material will scratch another. For minerals this is a commonly used method of determining hardness. One scratch hardness scale, called the *Mohs Scale*, rates very soft talc with a hardness of 1 and diamond, which will scratch all other materials, with a hardness of 10. A method similar to the scratch method is sometimes applied to hardened steel, for when the steel cannot be filed, it is pronounced "file hard" and considered to be in the hardest condition. A small fine-toothed hardness testing file is an extremely convenient tool for testing hardness, especially case-hardened objects, quenched shapes, and surface hardened pieces that should be hard. Despite the personal equation of the man handling the file and other variables such as the force applied or sharpness of the file, it is possible to make close comparisons to Rockwell hardness readings of the steels tested. Such measurements as scratch or cutting hardness are difficult to evaluate quantitatively and more refined methods have been developed for more exact measurement of hardness.

2.2.2. Penetration hardness measures hardness by resistance to pressure.

One way of measuring the hardness of a metal is to determine the depth to which a hard ball or cone will indent a metal under a given load. One such test, the *Brinell hardness test*, is made by pressing a hardened steel ball, usually 10 millimeters in diameter, into the test material by the weight of a known load. The load is usually 500 kilograms for materials such as iron and steel. The diameter of the resulting impression is measured by means of a small microscope. The Brinell hardness number is reported as the load divided by the area of the impression. Tables are available from which the hardness may be read once the diameter of the impression is obtained. Fig. 2–2 shows a Brinell hardness tester. The specimen is placed upon the anvil, which is then raised by means of a screw until the specimen is in contact with the steel ball. The load is applied by means of an air cylinder for 30 seconds, released, and the diameter of the resulting impression measured. The Brinell hardness of annealed copper is about 40, of annealed tool steel about 200, and of hardened steel about 650. When testing metals over 500 Brinell hardness, the metal ball may permanently deform, thus resulting in false hardness readings. Tungsten carbide balls are frequently employed when making Brinell hardness tests on hardened metals to avoid deformation of the indenter.

The *Vickers* hardness testing method is similar to the Brinell method. The penetrator used in the

Fig. 2–2. Brinell hardness tester. (T. Olsen Testing Machine Co.)

Fig. 2–3. Impression from Vickers Indenter through a microscope.

Vickers machine is a diamond pyramid, and the impression made by this penetrator is a dark square on a light background Fig. 2–3. This impression is more easily read than the circular impression of the Brinell method. There is also the advantage that the diamond point does not deform. In making the Vickers test a predetermined load is applied to the specimen. After removal of the load the ratio of the impressed load to the area of the resulting impression gives the hardness number. The length of the diagonal of the impression is measured with a microscope. The operation of applying and removing the load is controlled automatically. The Vickers tester is very accurate and adaptable for testing the softest and hardest materials under varying loads. It is extremely well suited to testing very hard materials. The Vickers diamond pyramid hardness number (DPH) is given by the equation:

$$DPH = \frac{1.8544\ L}{d^2}$$

L is the load in kilograms
d is the length of the diagonal of the impression in millimeters.

Fig. 2–4. Vickers Diamond Pyramid hardness tester. (Riehle Testing Machines, Div. of Ametek, Inc.)

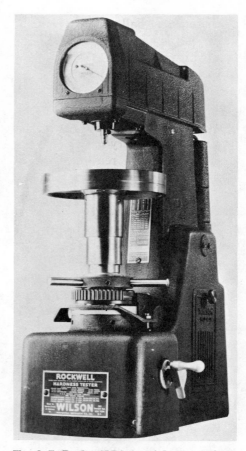

Fig. 2–5. Rockwell hardness tester. (Wilson Mechanical Instrument Div., American Chain & Cable Co.)

Hardness as determined by the Vickers method is about the same as the Brinell hardness numbers up to a value of 500. The Vickers test is primarily used for research work. The Vickers testing machine is shown in Fig. 2–4.

The *Rockwell hardness tester*, illustrated in Fig. 2–5, measures the resistance to penetration as does the Brinell test, but the depth of impression is measured instead of the diameter. Rockwell hardness is indicated directly on the scale attached to the machine. In order to minimize problems related to surface roughness of the specimen, a minor load of 10 kilograms is applied after which the major load is applied. The major load is removed, and with the minor load still acting, the hardness number is read on the dial gage. Each division on the dial gage represents a distance of .000080 inches of penetration. Because of the reversal of the order of the numbers on the dial gage, high numbers are associated with a shallow impression in hard materials, and low numbers with a deep impression in soft materials. When testing soft materials, a 1/16" diameter steel ball with a 100 kilogram load is used and the hardness number is read on the B scale. When testing hard materials a diamond cone or "brale" is used with a 150 kilogram load and the hardness is read on the C scale. Tungsten carbide and similar very hard materials are tested using a reduced load of 60 kilograms to prevent cracking of the diamond. Pressure on the diamond point may reach three or four million pounds per square inch when testing very hard materials.

The Rockwell Superficial hardness tester has been developed for testing of thin materials, and case hardened or nitrided surfaces. The test employs a light initial load of 3 kilograms and major loads of 15, 30, or 45 kilograms depending on the thickness of the hard surface. The machine does not differ in principle from the standard machine except that the graduated dial is twice as sensitive as the standard machine. The hardness number is based on the additional depth to which a test point penetrates the material under a given load beyond the depth to which the penetrator has been driven by the minor load. Because of the multiplicity of Rockwell scales, the hardness number must be designated according to the scale, indenter, and the major load employed. For example, the standard scales are Rockwell B (R_b), and Rockwell C, etc. while the Rockwell Superficial scales

are designated, 15-N70, 30-T50, and so on. With the Superficial tester, N scales are used for material of such hardness that it would be tested on the C scale if of sufficient thickness. The T scales are used for material of such hardness that it would be tested with the B scale if of sufficient thickness.

Rockwell hardness testers are widely used for production testing and process control because of their speed and the ease with which they may be used.

Another useful hardness testing machine is called the *Monotron*. This tester measures the load in kilograms required to produce a definite penetration in the material under test. The difference between the monotron and other testing machines is that the depth of impression is made constant instead of the pressure. Two dials at the top of the machine enable the operator to read depth of impression on one and load on the other. Pressure is applied until the penetrator sinks to the standard depth, 9/5000 inch. The load required to make this impression is read on the upper dial. The penetrator used is a diamond having a spherical point $3/4$ mm in diameter. One advantage for this method, based on constant depth of impression, is that the specimen is subjected to the same amount of work hardening irrespective of the initial hardness of the material. Because of this a wider range of materials can be tested without making any change in the penetrator or the methods of the test.

Very recently a new ultrasonic hardness testing device Fig. 2–6, called the *Sonodur* has appeared on the market. This device, which is an indentation type microhardness tester with an electronic readout, employs the use of a diamond tipped magnetostrictive rod which is electrically excited to its resonant fre-

quency. As the vibrating rod is brought into pressure contact with the surface of the material to be treated, the tip penetration is a function of hardness. The resonant frequency of the rod increases with penetration and the frequency change is displayed on the readout meter in terms of hardness. Thus a shift in frequency of the vibrating rod is a linear function of hardness. One other material property in addition to hardness which affects the

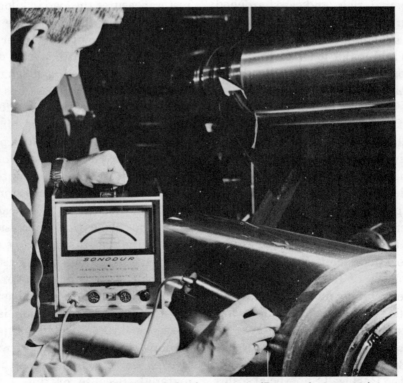

Fig. 2–6. Sonodur ultrasonic hardness tester. (Branson Instruments)

resonance of the rod is Young's Modulus of Elasticity. This means that unless care is taken the device may measure the change in Young's Modulus rather than a change in hardness, especially as materials of different composition are compared. The instrument may be calibrated to compensate for a change in Young's Modulus if the magnitude of this change is known. However, for comparative purposes with metals of a standard composition, few problems should be encountered.

The ultrasonic hardness test instrument has somewhat the same requirements for surface finish and homogeneity as conventional microhardness testing equipment. Normally, the surface finish should not generally exceed 50 microinches rms. However, rougher surfaces may be measured by averaging several readings. Since the depth of penetration is less than 7 microns in R_c65 steel and only 11 microns in R_c25 steel, surface hardness only is measured. This precludes the use of the instrument for testing heterogeneous materials such as cast iron where the measurement would be the hardness of individual constituents instead of an integrated reading over many grains, as with a Brinell or Rockwell test.

Also, brass test standards which have been cold worked on the surface will show a different value on the Sonodur instrument than they will on a Rockwell B test. The reason for this is that in the Rockwell test the ball indenter will penetrate deeply into the work and automatically integrate the hardness of the hard case with that of the soft core, whereas the Ultrasonic Microhardness Tester will respond only to changes in surface hardness. This instrument is very portable, and readings can be made in less than three seconds. The extremely small penetration presents the possibility for 100 percent non-destructive hardness testing.

2.2.3. Elastic hardness is also known as rebound hardness and may be measured with the *Shore Scleroscope*, Fig. 2–7. In this test hardness is given by the height of rebound of the diamond-pointed hammer after it has been dropped on the sample. The harder the material tested, the greater the rebound. The height of the rebound is read on a gage graduated with 100 divisions. The scleroscope can be used for large workpieces; it is portable; and the indentation made by the test is very slight, an advantage where surface finish is important. The amount of rebound is a factor more of the elastic limit of the specimen than of its tensile strength, and therefore the machine does not measure exactly the same type of hardness as do the indentation methods. The specimen must be solidly held in a horizontal position, and the hammer must fall exactly

Fig. 2–7. Shore Scleroscope model **D** shown with standard test block. (Shore Instruments & Mfg. Co.)

can be transported to the work and tests can be made on specimens too large to be taken conveniently to the other types of machines.

There are several other hardness testing machines that are satisfactory for metals. However, they all yield only comparative values for hardness. There is no definite relationship between hardness as measured by the different methods, or between hardness and strength. However, tests have been carried out and various conversion tables have been constructed which are useful when approximate conversion values are needed. See Table 2–1.

There are several factors which influence the accuracy of indentation hardness tests among which are the following:

Condition of the indenter such as flattening of a steel ball or chipping of a diamond indenter will cause inaccurate readings. The condition of the indenter should be checked frequently.

Condition of the anvil on which the specimen rests must be clean, smooth and free from burrs.

Thickness of the specimen should be enough so that no bulge or other marking appears on the surface opposite that of the indentation. On thin specimens the hardness of the anvil may be checked rather than the hardness of the specimen. This false reading is known as the "anvil effect."

vertically. The smallness of the impression made by the Shore scleroscope leads to variations in readings, thus necessitating a range of readings rather than a single one. More than one penetration should not be made on the same spot. Because of its portability the Shore scleroscope

Table 2–1. Hardness conversion table for alloy construction sheets.

Vickers or Firth Diamond Hardness Number	BRINELL		ROCKWELL HARDNESS		Shore Hardness	Tensile Strength 1000 lbs./ sq. in.
	Diameter of Impression for 3000 kg. Load and 10 mm. Ball	Hardness Number	C Scale 150 kg., 120° Diamond Cone	B Scale 100 kg., 1/16″ Ball		
	MM.					
1220	2.20	780	68	..	96	...
1114	2.25	745	67	..	94	...
1021	2.30	712	65	..	92	354
940	2.35	682	63	..	89	341
867	2.40	653	62	..	86	329
803	2.45	627	60	..	84	317
746	2.50	601	58	..	81	305
694	2.55	578	56	..	78	295
649	2.60	555	55	..	75	284
608	2.65	534	53	..	73	273
587	2.70	514	51	..	71	263
551	2.75	495	50	..	68	253
534	2.80	477	48	..	66	242
502	2.85	461	47	..	64	233
474	2.90	444	46	..	62	221
460	2.95	429	44	..	60	211
435	3.00	415	43	..	58	202
423	3.05	401	42	..	56	193
401	3.10	388	41	..	54	185
390	3.15	375	39	..	52	178
380	3.20	363	38	..	51	171
361	3.25	352	37	..	49	165
344	3.30	341	36	..	48	159
335	3.35	331	35	..	46	154
320	3.40	321	34	..	45	148
312	3.45	311	32	..	43	143
305	3.50	302	31	..	42	139
291	3.55	293	30	..	41	135
285	3.60	285	29	..	40	131
278	3.65	277	28	..	38	127
272	3.70	269	27	..	37	124
261	3.75	262	26	..	36	121
255	3.80	255	25	..	35	117
250	3.85	248	24	100	34	115
240	3.90	241	23	99	33	112
235	3.95	235	22	99	32	109
226	4.00	229	21	98	32	107
221	4.05	223	20	97	31	105
217	4.10	217	18	96	30	103
213	4.15	212	17	95	30	100
209	4.20	207	16	95	29	98
197	4.30	197	14	93	28	95
186	4.40	187	12	91	27	91
177	4.50	179	10	89	25	87
171	4.60	170	8	87	24	84
162	4.70	163	6	85	23	81
154	4.80	156	4	83	23	78
149	4.90	149	2	81	22	76
144	5.00	143	0	79	21	74
136	5.10	137	—3	77	20	71

It is recommended that the specimen thickness be at least 10 times the depth of impression.

Surface condition of the specimen to be tested should be flat, free from decarburization, scale and pits, and should be representative of the structure of the metal. Microhardness tests usually require a metallographic polished surface.

Shape of the specimen for greatest accuracy in hardness testing should be flat. Cylindrical specimens should have a flat surface prepared if at all possible, and should be supported in a V-notch anvil. If it is necessary to make a hardness test on a curved surface published correction factors should be added to the observed readings.

Location of indentations should be at least 2½ diameters from the edge of the specimen, and distance between impressions 5 diameters for ball tests and 2½ diameters for cone and pyramid tests.

Uniformity of material is needed when testing materials having structural or chemical variations in composition. Larger impressions such as the Brinell give a better average hardness reading than do hardness tests having small impressions. Hardness readings from a large number of small impressions may also be used for establishing an average value as well as for determining homogeneity of the material under test.

It should be observed, in summary, that the selection of a hardness test is usually determined by the speed, ease of performance, and accuracy desired; and that in all indentation hardness tests, the metal under test increases in hardness and resistance to penetration up to the point of rupture of the grain structure. Hardness is not a fundamental property of metals and there is no absolute standard of hardness. All hardness tests in common usage have limitations and advantages and yet have been most useful in quality control, production, and research work.

2.3 Tensile Testing. The tensile test is frequently used to obtain a great amount of information about the static mechanical properties of a material, including ductility, tensile strength, proportional limit, modulus of elasticity, elastic limit, resilience, yield point, yield strength, and breaking strength.

The tensile test is well standardized and may be carried out in a machine such as that shown in Fig. 2–8. The machine may be either mechanically or hydraulically operated but essentially consists of (1) a device for straining the specimen, (2) a device to indicate the load applied to the specimen, and (3) a gage, usually called an extensometer to measure strain over that part of the specimen in which strain is essentially uniform.

The most common types of round

Fig. 2–8. Universal testing machine with strip chart recorder for strain recording. (T. Olsen Testing Machine Co.)

and flat specimens used in tensile strength tests are shown in Fig. 2–9. Care should be taken when machining the specimens to avoid cold working the finished surface which should be smooth and free from nicks and tool marks.

The tensile test is performed by first marking the specimen gage length with prick punch marks, then measuring the cross-sectional area of the reduced section. The specimen is next locked securely in the grips of the upper and lower cross beams of the testing machine. If a stress-strain diagram is to be made, an extensometer is then fastened to the specimen. Usually a 500 lb load is

applied to seat the specimen in the grips and the extensometer is then set at zero. During the test the load is increased and both the load and elongation of the specimen are recorded.

A stress-strain diagram similar to that shown in Fig. 2–10 may then be plotted from the data. The straight line portion of the stress-strain curve shows the elastic properties of the material under test. Elasticity is the ability of a material to return to its original form after removal of the load. Theoretically, the elastic limit of a material is the limit to which it can be loaded yet recover its original form after removal of the load. In

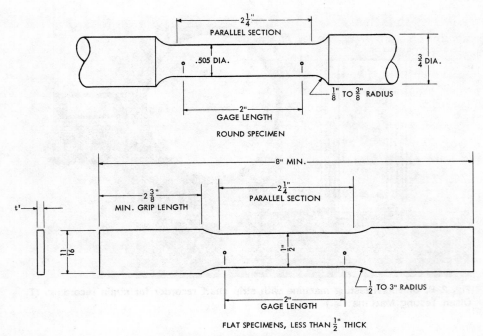

Fig. 2–9. Standard tensile test specimens.

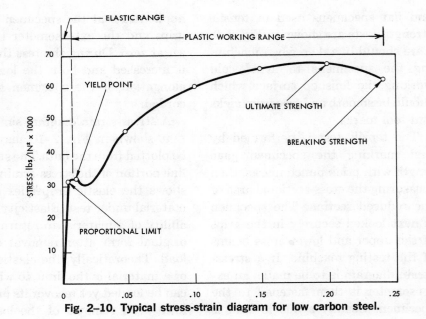

Fig. 2–10. Typical stress-strain diagram for low carbon steel.

reality, we find that metals are not entirely elastic even under slight loads. In the elastic range metals follow Hooke's Law, that is: strain is proportional to the applied stress. The stress at which this proportionality ceases to exist is known as the *proportional limit*. For practical purposes the elastic limit is the same as the proportional limit. This is an important factor in designing, as it is usually of more importance to know what load will deform a structure than what load will cause rupture.

The ratio of stress to strain or slope of the initial straight portion of the stress-strain curve within the elastic region, is known as *Young's Modulus* or the *Modulus of Elasticity* and is given the symbol *E*. This is an inherent property of considerable importance and is constant for a given material. The modulus of elasticity is a measure of stiffness and this indicates the ability of a material to resist deflection. The modulus of elasticity for steel is 30×10^6 psi and for tungsten carbide 94×10^6 psi. Because of the higher modulus of elasticity, boring bars have recently been made from tungsten carbide instead of steel in order to reduce deformation and chatter during precision metal boring operations.

The area under the stress-strain curve up to the elastic limit is known as the *modulus of resiliency* and is a measure of the ability of the material to quickly return to its original shape after being deformed.

The total area under the stress-strain diagram represents the energy required to produce fracture, or in other words, it is a measure of the toughness of the material under static loading. However, the value obtained from dynamic impact tests often do not correlate with toughness values obtained from the tensile test.

Some ductile materials, notably steel, indicate what is known as a *yield point*. This is the load at which the material will continue to elongate even though the load is not being increased. On early testing machines the yield point was indicated by a dropping of the weighing beam. The yield point is somewhat greater than the elastic limit. Many steels and non-ferrous metals do not have a well defined yield point. For such metals the yield strength is used. Yield strength is defined as the stress required to produce a given amount of permanent strain. The "offset" or permanent strain is usually 0.2 percent for aluminum and magnesium alloys, while 0.5 percent total elongation under load is frequently used for copper alloys. In using the offset method, a line is drawn parallel but offset to the right of the straight portion of the stress-strain curve. If the material under test is a copper specimen with a 2 inch gage length, the line would be offset .005 inch per inch $\times$ 2 inch, or .010 inches. The intersection of the offset line with the stress-strain curve gives the yield strength which in the case of hard, electrolytic ¼″ diameter tough pitch copper rod is 50,000 psi.

As the straining of a metal continues beyond the yield point there is a corresponding increase in the stress up to a maximum value. This maximum stress which is obtained by dividing the maximum load by the original cross sectional area and which corresponds to the highest point on the stress-strain curve is known as the *Ultimate Tensile Strength*, abbreviated UTS.

If straining is continued far enough, a tensile specimen will ultimately fracture. The stress at which fracture occurs is known as the *breaking strength, fracture strength* or *rupture strength* and is found by dividing the load at fracture by the reduced area supporting that load. Just prior to fracture, ductile materials such as low carbon steels, usually experience a localized reduction of the cross-sectional area known as necking, Fig. 2–11. Necking is usually accompanied by a reduction of the load to produce subsequent straining. The reduction in area of the necked down portion of the specimen continues until fracture occurs. Examination of the fracture which occurs in the necked down portion of the specimen usually reveals a cone shaped fracture for ductile materials, whereas brittle materials usu-

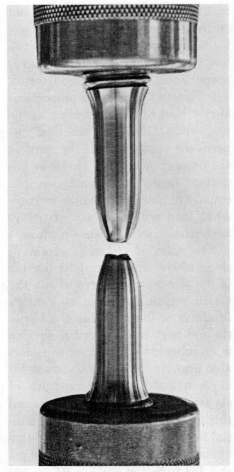

Fig. 2–11. Tensile specimen after fracture. Note necked down portion.

ally exhibit a flat fracture at right angles to the axis of the specimen. Ductility of a metal under test is often expressed in terms of percent reduction in area and percent elongation.

Percent *reduction in area* is determined by measuring the necked-down portion of the specimen with a micrometer caliper having ball ends and by using the following formula:

Reduction in area (percent) =
$$\frac{(A_o - A_f) \times 100}{A_o}$$

Where: A_o = original cross-sectional area

A_f = final cross-sectional area.

Percent *elongation* is determined by fitting together the broken halves of the specimen and measuring the distance between the original gage marks.

Elongation (percent) =
$$\frac{(L_f - L_o) \times 100}{L_o}$$

Where: L_f = final gage length

L_o = original gage length

The stress-strain diagram as shown in Fig. 2–10, can be divided into two ranges; the *elastic range*, and the *plastic range*. Most engineering calculations for structures are based on loads which will stress the members in the elastic range. Metal forming operations, on the other hand, such as forging, drawing, and spinning require high forces on the material to cause plastic deformation to occur.

2.4. Notched-Bar Impact Testing.

The behavior of metals under impact loads, or shock, may often be quite different from their behavior under loads slowly applied. The pendulum type impact testing machine, shown in Fig. 2–12, is designed to measure the notch toughness in evaluating different steels under the particular test conditions. The test may also be employed to determine the suitability of a material for low temperature applications, or to measure the resistance of a material to the propagation of a crack after it has once formed.

Two recognized tests exist in which notched specimens are fractured by a swinging pendulum. The

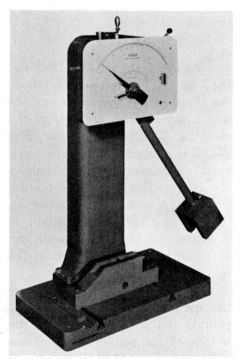

Fig. 2–12. Impact testing machine. (Riehle Testing Machines, Div. of Ametek, Inc.)

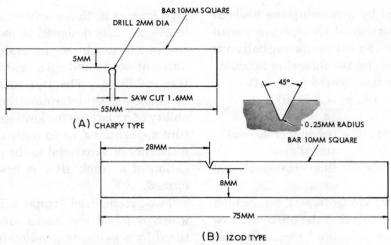

Fig. 2-13. Notched bar impact test specimens. A, Charpy type (keyhole notch), B, Izod type (v-notch).

two impact tests are the *Izod* and the *Charpy*. The tests differ only in the shape of the specimens and the way in which they are held in the machine. The Izod specimen, Fig. 2–13B is gripped vertically in a vise and the specimen is struck just above the notch. The Charpy specimen, Fig. 2–13A is supported freely by its ends in a horizontal position. It is hit by the top of the pendulum at a point behind the notch at the middle of the span.

The energy required to fracture the specimens is measured in foot-pounds and is calculated from the weight of the pendulum times the difference in height of the pendulum before and after impact. A scale on the machine indicates the impact value in foot-pounds.

It should be observed that the notched bar impact test does not test the impact behavior of metals which must be carried out at high velocity but rather the behavior of a metal with a particular notch. There is some evidence that a given geometry of notch does not impose the same stress system when used with other metals. Face centered cubic non-ferrous metals including copper, aluminum, nickel, and others remain notch tough at low temperatures whereas other metals undergo a transition from tough to brittle behavior. Austenitic stainless steels, Hadfield manganese steel, nickel steels, and tempered martensitic steels of fine grain structure have good low temperature notch toughness. For this reason the notched bar test is subject to limited interpretation. When suitable precautions are taken in specimen selection and preparation, the test can be expected to indicate only that a given metal is or is not brittle when tested at a given temperature and with a notch of given geometry.

In spite of the precautions which must be taken and the limitations of the test, notched bar impact tests are valuable in comparing a steel with a standard known to be adequate for the intended service. Notched bar tests have also been used as a sensitive index of heat treatment, especially in distinguishing between tempered martensite and tempered bainite which is difficult to do microscopically.

2.5 Torsion Testing. The torsion test is useful for testing parts such as axles, shafts, twist drills, and couplings. It has been used to good advantage in the testing of brittle materials, and at elevated temperatures as an index of forgeability. Torsion tests provide a method for determining the modulus of elasticity in shear, shearing yield strength, and the ultimate shearing strength. Torsion test data are generally valid to larger values of strain than are tension test data. ASTM (American Society for Testing and Materials) has a standardized torsion test for cast iron and one for tool steels. The test has also been applied to other ferrous and non-ferrous tubes, rods, and wires.

A torsion testing machine (Fig. 2–14) consists essentially of (1) a twisting head, in which one end of the specimen is gripped and which can apply a twisting moment to the specimen, (2) a torsion measuring head, in which the other end of the specimen can be gripped and with which the torque can be measured, and (3) a twist-measuring device, or tropometer, for measuring angular displacement of points on opposite ends of the specimen gage length.

The maximum shearing stress for a solid, round bar is given by:

$$S = \frac{16\,T}{\pi\,d_1{}^3}$$

Where: $S =$ maximum shearing stress, psi

$T =$ torque, inch-pounds

$d_1 =$ outside diameter or rod or tube

$d_2 =$ inside diameter of tube

The formula for calculating shearing stress in hollow tubes is the same as the above with $d_1{}^3$ replaced by $(d_1{}^4 - d_2{}^4)$. Both formulas are valid only as long as strain is proportional to stress.

The shearing strain in inches per inch for torsional deformation may be obtained by dividing the total angular twist by the gage length of

Fig. 2–14. Torsion testing machine, 10,000 inch-pound capacity.

the specimen and multiplying the result by half the diameter of the specimen. Stress-strain curves may be plotted from torsional data with the advantages of a uniformly applied stress and the possibility for measuring both strength and toughness very accurately even when toughness is very low.

2.6 Fatigue Testing. When a specimen is broken in a tensile testing machine, a certain definite stress is required to cause that fracture. However, a specimen of that same mate-

rial will fail under a much smaller stress when subjected to cyclic or fluctuating loads. In this way, an axle may break after months of use even though its maximum load has not increased. Such a fatigue failure is pictured in Fig. 2–15.

Metals consist of minute crystals with planes of slip oriented in many directions. Sufficient stress will cause slipping to occur on planes within the individual crystalline grains. At first this slipping may not cause harm, but continued repeated slipping causes minute cracks to form

Fig. 2–15. Fatigue fracture of aircraft crankshaft. "Clam shell" markings starting at "O" show where the fatigue crack progressed across the shaft. (National Bureau of Standards)

and spread until the cross section of a member is so reduced that it will no longer support the applied load.

The final part to fracture shows the crystalline structure of the metal with adjacent sections showing signs of having rubbed together for some time. The sometimes coarse grain of the final fracture has often led to the erroneous conclusion that the part failed due to "crystallization" in service.

All such failures are known as *fatigue failures*, and in designing parts subjected to varying stresses, the fatigue limit of a material often is more important than its tensile strength or its yield strength. The *fatigue limit* or *endurance limit* is the maximum stress that a metal will withstand without failure for a specified large number of cycles of stresses.

In fatigue testing an endurance limit of 10,000,000 cycles is common for iron and steel, but for non-ferrous light alloys, this number may be 500,000,000 cycles.

Three common fatigue tests are: the rotating-beam test, the vibrating-beam test, and the tension-compression fatigue test. The essential parts of a fatigue testing machine are: (1) a mechanical, hydraulic, or magnetic drive for applying repeated cycles of stress to the specimen, (2) a means of measuring the maximum and minimum stresses applied during a cycle, (3) a counter for indicating the number of cycles of stress applied to the specimen, and (4) a device for stopping the test automatically when the specimen breaks.

A rotating cantilever-beam fatigue machine is shown in Fig. 2–16A. The

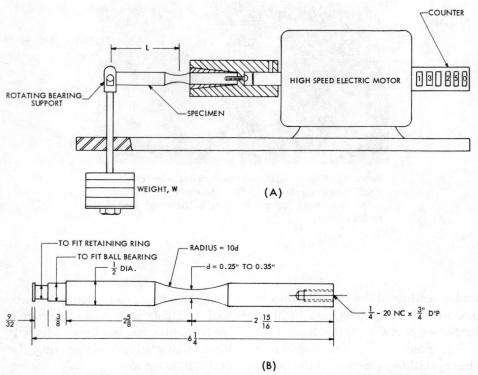

Fig. 2–16. A, Simple rotating beam fatigue testing machine, and B, unnotched fatigue specimen.

maximum bending moment at the critical section of the specimen is WL and the bending stress is equal to $WL/.098d^3$ for the specimen shown in Fig. 2–16B. Notched specimens are sometimes tested to evaluate "notch sensitivity" or sensitivity of a metal to stress raisers.

Fig. 2–17 shows typical diagrams for fatigue tests of seven common engineering metals.

Much knowledge has come to us in recent years which enables us to use metals in a safe and economical way to meet the rigid requirements in precision machines and high speed aircraft. Of prime importance for high fatigue life is good surface finish, freedom from corrosion, or from decarburization. The use of parts which have been cold rolled or shot peened to set up compressive stresses

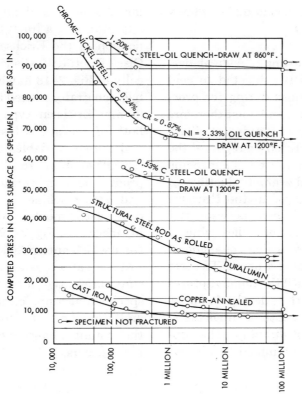

Fig. 2–17. Typical S-N (stress cycle) diagrams for fatigue tests of various metals.

near the surface of the specimen, also increases fatigue life. Roughened, notched, or threaded surfaces pronouncedly decrease the fatigue limit of metals.

It is a known fact that offsets, slots, holes, notches and sharp re-entrant angles are the locations for high stresses and possible failure in machine parts. Use of generous fil-

lets, rounding off the ends of keyways and slots, smoothing out corners and shoulders, and avoiding sharp tool cut marks will do much to eliminate stress raisers and lengthen fatigue life. Any sharp grooves or notches may change the distribution of stresses and greatly alter the usual physical properties of a material and cause the parts made

41

from it to react to loading in a wholly unexpected way.

2.7 Ductility Testing. A great deal of sheet metal is subjected to cold forming operations during the process of fabrication and must therefore be examined for "forming quality." In 1914, A. M. Erichsen reported the development of a test and a machine to measure and evaluate ductility of sheet metal strips. In the Erichsen test the sheet is held between two ring shaped clamping dies while a punch is forced against one side of the sheet, forming a dome. The test is continued until the dome ruptures at which time the "depth of cup" is measured. The depth of cup is taken as a measure of *ductility*. The rupture point is determined by observing the sheet and noting the first appearance of a fracture. One modification of the Erichsen test is the Olsen Ductility test shown schematically in Fig. 2–18 in which a ⅞ inch ball penetrator is used instead of the 20-mm Erichsen type indenter. In both of the above tests guesswork and human variables have made it difficult to obtain results that are reproducible from operator to operator and from machine to machine. Recent developments in ductility testing machines, Fig. 2–19, include features which automatically clamp the test strip at a constant, pre-set pressure, press the indenter into the specimen at a predetermined speed; and stop the test automatically at the instant a hairline crack or fracture appears.

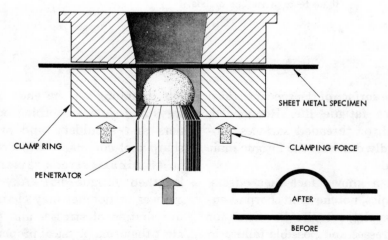

SHEET METAL SPECIMEN

CLAMP RING

CLAMPING FORCE

PENETRATOR

AFTER

BEFORE

Fig. 2–18. Schematic of Olsen ductility cup test using ⅞" ball penetrator.

Fig. 2–19. Sheet metal ductility testing machine with automatic control features. (T. Olsen Testing Machine Co.)

Fig. 2–20. Close-up of ductility tester showing deep drawn cup. Note earing tendency due to directional properties of the metal. (T. Olsen Testing Machine Co.)

In addition to Erichsen and Olsen cup tests, the machine can be used for deep drawing cup tests and hole expansion tests. Modern ductility testing machines may be used for evaluating drawing lubricant, metal formability, earing tendencies, optimum die clearance, clamping pressure and can simulate a wide range of deep drawing operations and stresses. Fig. 2–20 shows a close-up of the test equipment and a deep drawn cup.

2.8 Hardenability Testing. For many years men concerned with the heat treatment of steel parts have noted that different lots of steel of the same nominal chemical analysis and produced by the same process would sometimes exhibit marked differences in hardening characteristics. This variability in heat treatment was the cause of great concern in manufacturing plants where parts were held to close tolerances in hardness. Hence, experimental tests were developed by Jominy, Riegel, Post, Shepherd, and others to help in evaluating steels in terms of their hardenability.

Hardenability can be described as the relative ability of a steel to be hardened by quenching from a proper temperature. The difference in hardening characteristics of apparently identical steels can be traced primarily to the slight variation in chemical composition permissible within a particular grade of steel, and to difference in grain size. Steels to be heat treated may now be ordered on the basis of heat treatment response rather than chemical composition and are specified as AISI-SAE "H" steels. A method was developed in 1942 by M. A. Grossman for calculating the hardenability of steels from chemical composition and grain size. The method is based on the principle that each chemical element has a multiplying effect on the hardenability. Hardenability prediction from chemical composition has worked well with plain carbon and low alloy carbon steels. However, unexplained irregularities have been observed on steels of high hardenability and on those containing less than 0.20 percent carbon.

The current practice among steelmakers and users of steel is to experimentally determine hardenability by one of the standardized hardenability testing methods. In general, the testing method employed is designed to amplify differences in hardenability of a single type of steel. A test designed for the medium oil hardening grades of steel will not be sensitive when used with shallow water hardening or with deep air hardening grades. Only tests for medium hardening steels will be discussed here. However, detailed information for testing both shallow and deep hardening steels may be found in *Metals Handbook*, 1948 edition, American Society for Metals, pages 489-492.

2.8.1. Jominy End Quench Test is a widely used standardized test for medium hardening or common alloy steels. The test involves heating the test bar (See Fig. 2–21) to the proper temperature, placing it in a hardenability fixture, Fig. 2-22A, and then cooling the end face with a water spray, Fig. 2-22B. Two flats .015″ deep are ground along the length of the cooled specimen and Rockwell C hardness readings are taken at 1/16 inch intervals from the end of the water cooled end, using a device

Fig. 2–21. Preferred specimen of Jominy End-Quench Hardenability Test.

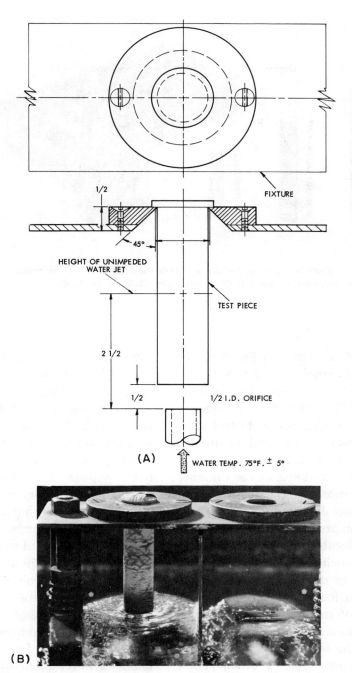

1/2

FIXTURE

45°

HEIGHT OF UNIMPEDED
WATER JET

TEST PIECE

2 1/2

1/2   1/2 I.D. ORIFICE

(A)   WATER TEMP. 75°F. ± 5°

(B)

Fig. 2–22. A, Jominy End-Quench fixture details, B, fixture in use. (Crucible Steel Co.)

45

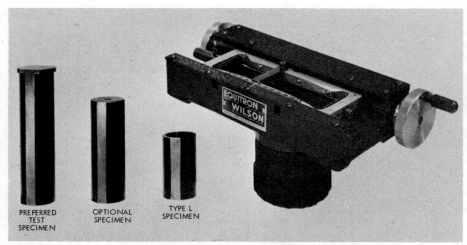

PREFERRED
TEST
SPECIMEN

OPTIONAL
SPECIMEN

TYPE L
SPECIMEN

Fig. 2–23. Fixture for spacing Jominy hardness readings at 1/16″ intervals on test bar. (Wilson Mechanical Instrument Div., American Chain & Cable Co.)

such as that shown in Fig. 2–23 for accurately spacing the intervals. Details of the test may be obtained from ASTM A255-48T, End Quench Test for Hardenability of Steel.

The results of a typical end quench test are plotted as shown in Fig. 2–24 with hardness as the ordinate and distance from the quenched end as abscissa. Jominy data can be very helpful in predicting the hardness at various locations of water quenched or oil quenched rounds. Jominy data can also be useful in selecting steels for a particular part which is later to be heat treated. Experimental investigations have shown that as the end of the water quenched Jominy bar reaches 1300°F, other points have cooling rates as shown at the top of Fig. 2–24. Cooling rates are assumed to be practically independent of chemical composition.

At the end of the bar farthest from the quench the cooling rate is very slow and is analogous to the cooling of a large section. Thus, one Jominy bar is exposed to all possible rates of cooling from very rapid to very slow. This relationship between section size and cooling rate at various positions on the end quench bar is shown in Fig. 2–25A for specimens quenched in still oil, and in Fig. 2–25B for specimens quenched in still water.

The following example will show how Jominy data for standard 1″ diameter water quenched specimens may be used for predicting the hard-

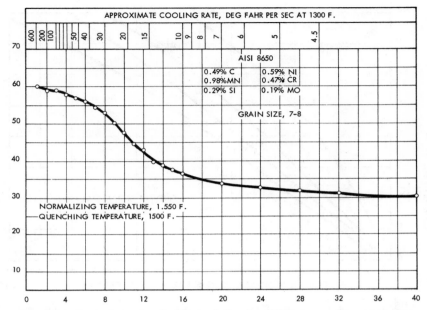

Fig. 2–24. Typical Jominy End-Quench hardenability curve. (By permission: *Metals Handbook*, 1948 ed., American Society for Metals)

ness at various positions of water or oil quenched round bars. Suppose we wish to know the hardness we can expect to achieve at the surface of an oil quenched 3.5 in. diameter bar of the steel whose Jominy curve is shown in Fig. 2–24. By looking at Fig. 2–25A, we can see that the surface hardness of a 3.5 inch specimen quenched in oil would be the same as the hardness 10/16" from the end of the Jominy bar.

By now referring back to Fig. 2–24, it can be seen that the hardness corresponding to 10/16" along the Jominy bar is R C47. Thus the sur-

face hardness of a 3.5 in. diameter oil quenched round which cooled at the same rate as this point will likewise be R C47. Now follow the same procedure as outlined above and determine the surface hardness of a similar specimen when quenched in water instead of oil.

2.8.2. Hardness Traverse Test is known as the hardness penetration test, may be used for both medium and shallow hardening steels. Samples of ½, 1, 2, 3, 4, and 5 inch round bars are quenched from the proper temperature in the desired coolant, sectioned, and hardness measure-

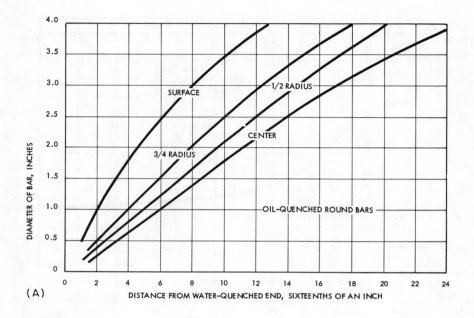

(A)

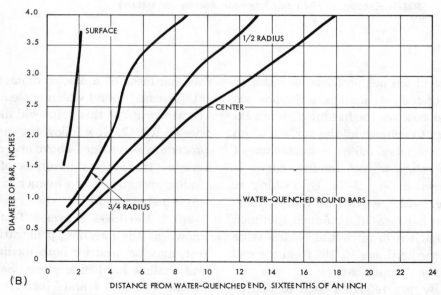

(B)

Fig. 2–25. Relationship between distance on Jominy bar and points at the surface, ¾ radius, ½ radius, and center of round bars quenched in still oil and in still water. (By permission: *Metals Handbook, 1948 ed.,* American Society for Metals)

ments made across the face of the specimen as shown in Fig. 2–26A. The hardness distribution is shown in Fig. 2–26B.

The hardness traverse test is very useful in comparing various quenching media, austenitizing temperatures, and grain size effects but has

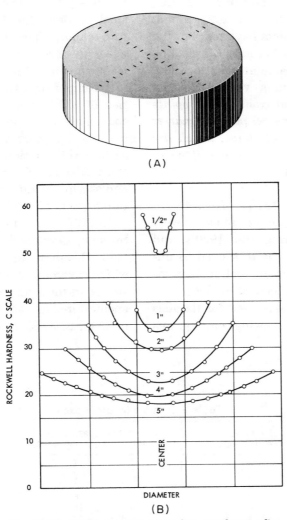

(A)

(B)

Fig. 2–26. A, hardness penetration specimen after sectioning and hardness testing. B, hardness distribution in water quenched steel rounds of SAE 1045 steel.

the disadvantage of requiring many specimens.

A study of Fig. 2–26B shows a hardness of Rockwell C30 at the surface of a 4″ round, at a point about ½″ under the surface of a 3″ round, and at the center of a 2″ round. These three points have reached the same hardness because the cooling rate was the same at each location. This represents a very important principle of heat treating. *All points on a particular steel part, no matter what its size and shape, which are cooled at the same rate will be of equal hardness.*

2.9 Grain Size Tests. Tests to determine the grain size in steels were developed during the 1920's and 1930's in response to a need to understand factors of steel quality unexplained by chemical analysis.

It had been observed for decades that a fracture showing coarse grains

accompanies excessive brittleness, and that an equally hardened steel of the same composition would require greater effort in breaking and would exhibit a "silky" appearance. Likewise, it had been observed that steel sections too large to harden throughout, harden more deeply when the grain structure is coarse than when it is fine.

Attempts to correlate the properties of a hardened steel with the coarseness or fineness of a steel were made many years ago, and as early as 1926 this relationship was used at the Uddeholm Steel Works in Sweden as a quantitative measure of certain aspects of steel quality.

In 1927, Jernkontoret and Arpi, in Sweden, developed a 10 step scale with 22 mm square steel specimens which covered the full range of grain size coarseness and fineness usually encountered in fractures. In America in 1934, B. F. Shepherd developed a

Table 2–2. A S T M grain size.

TIMKEN, ASTM NO.	GRAINS PER SQUARE INCH OF IMAGE AT 100X
1	1
2	2
3	4
4	8
5	16
6	32
7	64
8	128
9	256
10	512

set of ¾″ diameter graded standard fractures numbered serially 1 through 10. Later studies, involving microscopic examination of suitably prepared specimens, made it possible to count the number of grains per unit area. ASTM and Timken Numbers (1 through 10), which for practical purposes, are the same as Shepherd and Jernkontoret, are based on the number of grains per square inch of image at 100X magnification. The ASTM number and average number of grains per square inch at 100X magnification are shown in Table 2-2.

Austenitic grain size in ferritic steel is often determined by means of the *McQuaid-Ehn Test*. This test involves carburizing a specimen at 1700° Fahrenheit for eight hours. The carburizing treatment outlines the grain boundaries with a cementitic network. The specimen is then polished, etched, and microscopically examined at 100X magnification. Observed grain size is compared with either standard grain size charts or with engraved grain sizes on an eyepiece reticle.

The Shepherd grain size fracture standards are shown in Fig. 2–27. The fracture test for grain size consists of hardening the specimen (machined to ¾ in. diameter by 3 in. long) by quenching in 10 percent brine, notching in the middle 1/16 in. deep, and fracturing by impact.

The surface of the fracture is then compared with the Shepherd grain

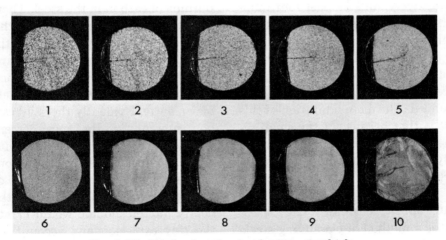

Fig. 2–27. Shepherd grain size fracture standards.

size standards to determine the proper classification of grain size.

The Shepherd Test described above has been refined to permit numerical valuation of hardening penetration as well as grain size. This test called the "P-F Test" (P = penetration; F = fracture grain size) is conducted by heating specimens to 1450, 1500, 1550, 1600°F, followed by brine quenching. The specimens are notched and impact fractured. The grain size is determined as before with one of the broken halves of the specimen. The other half is polished and then etched for three minutes in a 1 to 1 hydrochloric acid and water solution at 180°F. Hardness penetration is measured in 64ths of an inch. The P-F characteristic is recorded as 8 numbers, the first four numbers represent the penetration in 1/64th of an inch, when quenched from 1450, 1500, 1550, and 1600°F respectively, and the last four numbers, the corresponding fracture grain size. The P-F test is very useful in determining both hardenability and grain size as well as the temperature at which grain growth is pronounced. The main limitation of this test is that because of the small diameter of specimens it is adapted only to the shallow hardening plain carbon steels.

2.10 Spark Testing. The spark test is a method of classifying steels according to their composition by visual study of the sparks formed when the steel is held against a high speed grinding wheel. See Fig. 2-28. This test does not replace chemical analysis but is a very convenient and fast method of sorting mixed steels whose spark characteristics are known. One big advantage of this test is that it can be applied to metals in practically all stages of production —billets, bar stock in racks, machined forgings, or finished parts.

When any form of iron or steel is held against a grinding wheel small particles heated to incandescence are cut from the metal and hurled into the air. The trajectory of the particle, called a "carrier line," is easily followed, particularly against a dark background. Different theories have been advanced to explain the forking or bursting in the "spark picture." It seems reasonable to believe that these forks or bursts are due to rapid burning of carbon in the steel. There are some observations to support this contention, for if particles of the metal ground off are collected and examined, some are found to be hollow spheres with one side completely blown away. Evidently the solid carbon present in the incandescent fragment on coming in contact with oxygen of the air is burned to gaseous carbon dioxide. Sudden escape of this gas from the interior of the nearly molten globule would create the tiny explosion and account for the eggshell-like hollow spheres.

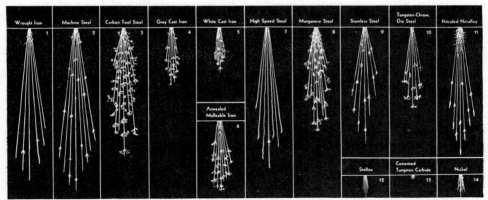

Metal	Volume of Stream	Relative Length of Stream, Inches†	Color of Stream Close to Wheel	Color of Streaks Near End of Stream	Quantity of Spurts	Nature of Spurts
1. Wrought iron	Large	65	Straw	White	Very few	Forked
2. Machine steel	Large	70	White	White	Few	Forked
3. Carbon tool steel	Moderately large	55	White	White	Very many	Fine, repeating
4. Gray cast iron	Small	25	Red	Straw	Many	Fine, repeating
5. White cast iron	Very small	20	Red	Straw	Few	Fine, repeating
6. Annealed mall. iron	Moderate	30	Red	Straw	Many	Fine, repeating
7. High speed steel	Small	60	Red	Straw	Extremely few	Forked
8. Manganese steel	Moderately large	45	White	White	Many	Fine, repeating
9. Stainless steel	Moderate	50	Straw	White	Moderate	Forked
10. Tungsten-chromium die steel	Small	35	Red	Straw*	Many	Fine, repeating*
11. Nitrided Nitralloy	Large (curved)	55	White	White	Moderate	Forked
12. Stellite	Very small	10	Orange	Orange	None	
13. Cemented tungsten carbide	Extremely small	2	Light Orange	Light Orange	None	
14. Nickel	Very small**	10	Orange	Orange	None	
15. Copper, brass, aluminum	None				None	

†Figures obtained with 12″ wheel on bench stand and are relative only. Actual length in each instance will vary with grinding wheel, pressure, etc. *Blue-white spurts. **Some wavy streaks.

Courtesy of Norton Company, Worcester, Massachusetts

Fig. 2–28. Characteristic spark patterns when grinding metals.

The spark test is best conducted by holding the steel stationary and touching a high speed portable grinder to the specimen with sufficient pressure to throw a horizontal spark stream about 12 inches long and at right angles to the line of vision. Wheel pressure against the work is important because increasing pressure will raise the temperature of the spark stream and give the appearance, to the inexperienced eye, of higher carbon content.

Observation should be made of the spark picture near and around the wheel, the middle of the spark stream, and the reaction of incandescent particles at the end of the spark stream. By experience and observation of spark streams of known samples the operator will soon be able to distinguish between steels of different chemical analyses. A 0.15 percent carbon steel shows sparks in long streaks with some tendency to burst with a sparkler effect; a carbon tool steel exhibits pronounced bursting; and a steel with 1.00 percent carbon shows brilliant and minute explosions or sparklers. As the carbon content increases, the intensity of bursting increases.

Spark bursts will vary in intensity, size, number, shape, and distance from the wheel or ends of the carrier lines. The burst is the characteristic spark of carbon. Other alloying elements may have some influence on the spark picture or the color of the spark stream. In all spark testing, the eyes of the operator should be adequately protected by safety glasses with side shields.

2.11 Corrosion Testing. The ability to resist atmospheric corrosion, or corrosion by liquids, chemicals, or gases, often is of primary importance in selecting a metal for a particular service. Accelerated corrosion tests have been devised whereby the behavior of the materials in actual service may be deduced from conditions applied for weeks, instead of for years as in actual use. Corrosion may be measured by determining the loss of tensile strength of specimens, by loss of weight in materials which dissolve in the corroding medium, or by gain in weight as when a heavy oxide coating of rust is formed.

In applying corrosion tests the test conditions should simulate the conditions of service as closely as practicable, and the specimen should be representative of the material as it would be used.

Although many corrosion tests exist which have proved to be of great practical value, only a few tests have been standardized.

One of the most generally used standardized tests is the *Salt Spray (fog) Test*. For this test specimens that have been suitably cleaned are placed in a specially prepared chamber and exposed to a fog of atomized salt water for a designated number of hours. The salt solution is prepared by dissolving 5 parts by weight of salt in 95 parts of distilled water. The exposure zone in the test chamber is maintained at 95°F. Such a test is suitable for testing ferrous and non-ferrous metals and is also used to test inorganic and organic coatings. More resistant coatings will withstand corrosive attack for some hundreds of hours.

Two types of immersion corrosion tests are common: *The Alternate Immersion Test*, and the *Total Immersion Corrosion Test*. The Alternate Immersion Test consists of a temperature controlled tank filled with the desired test solution and a suitable arrangement for alternately raising and lowering either the specimen or the solution. Test conditions should be selected and controlled in such a manner that variations in results can be attributed to variations in the material being tested rather than operator or process variations.

Duration of initial tests should be sufficient to allow removal of at least 30 mg per square inch when corrosion is uniform, to help in determining which specimens should be tested for longer periods.

Total Immersion Corrosion Tests have been tentatively accepted by the American Society for Testing and Materials (ASTM Designation: B185-43T and A279-44T) for both non-ferrous and ferrous metals. The tests may be used for quality control purposes or to assist in the selecting of a material for a specific use. During the tests specimens should be moved at uniform and constant velocity. Temperature of the corroding solution should be maintained at 95°F ± 2; and the corroding solution should be treated to maintain air saturation. Details of the method of specimen preparation, cleaning test solution, and duration of the test should be described when reporting the results of a test to facilitate interpretation of test results and their duplication by others.

The *Boiling Nitric Acid Test* has been found to be a very useful and reliable test for measuring the influence of heat treatment and welding on corrosion-resisting steels. The test involves placing a suitably prepared specimen in a one liter Erlenmeyer flask equipped with a condenser. The specimen is supported by a hook or cradle and is submerged in a 65 percent nitric acid solution. The testing period ordinarily consists of five boiling periods of 48 hours each.

The effect of the acid on the metal is determined by measuring weight loss of the specimen following each test period. Corrosion rates are reported as inches of penetration per month. An *Electrolytic Oxalic Acid Etching Test* has been developed to quickly classify steels and determine which should be subjected to the nitric acid test discussed above.

Occasionally, metals are susceptible to a type of electrochemical attack called *intergranular corrosion* which progresses along grain boundaries of a metal and may cause a metal to "fall apart," so to speak. One test for intergranular corrosion is conducted by boiling a metal for 24 hours in the so-called Strauss Solution. The solution usually consists of 47 milliliters of concentrated sulphuric acid and 13 grams of copper sulfate crystals per liter. If the metal is susceptible to intergranular corrosion the attacking reagent will cause the metal to lose its metallic resonance, cause a considerable decrease in its electrical conductivity, or cause it to develop cracks during a bend test.

One other corrosion test is the standard *Mercurous Nitrate Test*. Wrought or cast copper or copper-base alloys such as highly stressed brass and bronze may be sensitive to failure by cracking in service under conditions of a corrosive nature. This failure is spontaneous and may occur without any added stress from service. It constitutes one of the most serious defects or failures in cold worked brass.

The Mercurous Nitrate Test is an accelerated test for detecting presence of residual internal stresses that might bring about failure. The test is performed by first cleaning the specimen until all oxides and oils are removed, then totally immersing the specimen for a period of 30 minutes in an aqueous mercurous nitrate solution containing 10 ml of nitric acid (sp gr 1.42) and 10 grains of mercurous nitrate per liter of solution. The specimen is next washed in running water and excess mercury wiped from its surface. The specimen is immediately examined visually for cracks.

Non-Destructive Testing

2.12 Radiography. A radiograph is a shadow picture of a metal more or less transparent to radiation such as x-rays or Gamma rays. Various types of metals possess different absorption properties. Absorption of radiation by a material increases with its atomic number, its density, and also the wave length of the x-ray beam. Consequently, a sound section of steel will absorb more radiation than a similar piece of aluminum. It is also possible to locate regions in a particular material which differ appreciably in density from surrounding regions. These relations are the basis for radiographic non-destructive testing of materials using high energy radiation.

The experimental procedure consists of passing radiation through the specimen to be examined and detecting or measuring the varying intensity of an emergent beam over the surface of the specimen. Common detection devices for the emergent beam are x-ray film, fluorescent screens, gaseous ionization, and solid state radiation devices. Recent advances in TV x-ray systems also show great promise as high speed, sensitive detection devices.

Fig. 2–29 (A) shows an inexpensive x-ray machine recently introduced on the market, and Fig. 2–29 (B) shows an x-ray of an aluminum casting using a photographic method of recording the x-ray image.

X-Rays are produced when a target is bombarded by a rapidly moving stream of electrons. When the electrons are suddenly stopped by the target, their kinetic energy is converted to energy of radiation called x-rays. The essential elements for the generation of x-rays are: (1) a heated filament (cathode) to provide a source of electrons proceeding toward a target, (2) a target (anode) located in the path of the electrons, (3) a controllable voltage difference between the filament and target to control the velocity of electrons and

Fig. 2–29A. Desk model x-ray machine capable of penetrating 2″ of aluminum. (Field Emission Corp.)

Fig. 2–29B. An x-ray radiograph of aluminum casting made by the machine. (Conductron Corp.)

Table 2–3. Typical industrial radiation sources and their applications.

KILOVOLTAGES PEAK (KVP)	APPROXIMATE PRACTICAL THICKNESS LIMITS
X–RAYS	
50	MICRORADIOGRAPHY, WOODS, PLASTIC
100	2" ALUMINUM, 3" MAGNESIUM
150	4-1/2" ALUMINUM, 1" STEEL
250	2" STEEL OR EQUIVALENT
400	3" STEEL OR EQUIVALENT
1000	5" STEEL OR EQUIVALENT
GAMMA RAYS	
RADIUM.	8" STEEL OR EQUIVALENT
COBALT 60	APPROXIMATELY THE SAME AS RADIUM

thus the wavelength of x-rays produced, and (4) a means of regulating filament current to control the number of electrons striking the target.

Table 2–3 shows typical industrial radiation sources and their applications.

Gamma rays are also used in radiography. Gamma rays are emitted from radioactive radium or cobalt 60 and have a much shorter wave length than x-rays and consequently greater penetrating capability. Gamma radiation is somewhat limited because the wavelength is characteristic of the source and cannot be regulated for contrast or variable thickness. However, as shown in Table 2–3, radioactive sources such as cobalt 60 and radium have penetrating power equivalent to rather large x-ray units.

Safety precautions and adequate protection is extremely important when working with x-rays or radioactive substances. Overexposure to the rays by local areas of the body may result in skin burns and ulcers which may later develop into cancer. Overexposure of the entire body may lead to severe anemia, leukemia, and sterility. Necessary safeguards include shielding of all radiation sources. It should be emphasized that shielding of the direct beam alone is not sufficient because of dangerous secondary radiation from objects in the path of the direct beam.

Radiography applications have primarily been casting and weld inspection, but may also include inspection of metal forgings, and assemblies, plus plastic and other non-metallic parts.

2.13 Magnetic Particle Inspection. Magnetic particle testing is used for detecting surface and subsurface

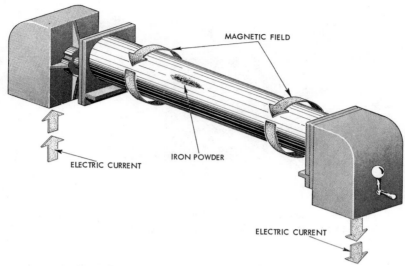

Fig. 2–30. Circular magnetization of parts for detection of longitudinal cracks by magnetic particle inspection.

flaws in ferromagnetic materials, principally iron and steel. The test is generally conducted by magnetizing the part to be inspected by passing a low-voltage (4-18 volts) high amperage electric current through it. Any discontinuities in the part, such as cracks, will produce magnetic north and south poles at opposite edges of the crack. If finely divided ferromagnetic particles are applied in the vicinity of the crack, the particles will be attracted by the magnetic flux between the two poles. The visible accumulation produces a readily discernible indication on the surface of the test part.

Fig. 2–30 shows the circular magnetization of solid parts. Passage of current longitudinally through the test part creates a circular magnetic field around the part. Magnetic lines of force are always at right angles to the direction of current which induces the magnetic field (right-hand rule). A crack at right angles to the magnetic field is visible. Cracks at right angles to the axis of the workpiece may be made visible by longitudinal magnetization as shown in Fig. 2–31.

Prior to making the test, parts should be free from oil, grease, loose scale, and dirt. Magnetizing current

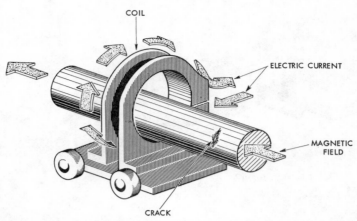

COIL

ELECTRIC CURRENT

MAGNETIC FIELD

CRACK

Fig. 2–31. Longitudinal or coil magnetization of solid parts for detection of transverse cracks by magnetic particle inspection (Magnaflux Corp.)

may vary from 400 to 800 amp per lineal inch of section thickness; optimum values should be determined by experimentation. Overly strong magnetic fields may mask defects and may also produce strong magnetic leakage at projections, corners, or angles. Insufficient magnetizing current will result in no observable pattern when flaws are actually present.

Magnetic particles are available as *dry powders* which are dusted or brushed onto the surface to be tested, and as *magnetic paste* for use in either oil or water suspensions. Magnetic paste is available in black, red, and fluorescent particle coatings. Dry particles are most sensitive for use on very rough surfaces, such as weldments and castings,

and for detecting subsurface defects. In the wet method, the liquid is pumped over the part and as the liquid flow is cut off, the part is immediately magnetized. In the dry method the current flows continuously while the powder is being applied. The wet method is used for detection of fine discontinuities such as fatigue cracks or grinding cracks on bright machined surfaces.

Magnetic particle patterns outline discontinuities and make them visible to the inspector. Surface defects provide a high, tight and closely packed formation. Subsurface discontinuities have a pattern which is broad, fuzzy, and loosely packed, becoming more so as depth increases.

A surface indication is shown in

Fig. 2–32. Indications of subsurface defects are shown in Fig. 2–33 in which holes were drilled at increasing depths from the surface of the specimen.

An inexpensive, portable. hand-held yoke kit for magnetic particle inspection is shown in Fig. 2–34. This equipment is designed for short duty cycle rather than for continu-

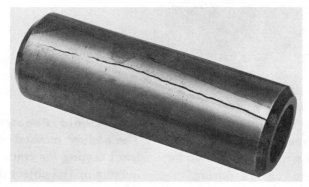

Fig. 2–32. Wet magnetic particle inspection of a seam in a diesel wrist pin. (Magnaflux Corp.)

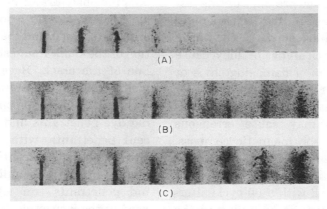

(A)

(B)

(C)

Fig. 2–33. Dry method indications of subsurface defects at varying depths. Continuous magnetization at, A, 500 amps., B, 1000 amps., and C, 2000 amps. (Magnaflux Corp.)

Fig. 2–34. Hand held yoke for magnetic particle inspection. Operated from 12-volt battery. (Sperry Products Div., Automation Industries)

ous operation. It can be used to detect surface cracks in grey iron castings, and welds, and may be used for locating fatigue cracks during the overhaul of large assemblies.

It should be pointed out that some parts which have undergone magnetic particle testing will retain their magnetism. The presence of this residual magnetism may hamper removal of chips or particles during subsequent cleaning, or may cause the attraction of damaging chips or particles to rotating parts, especially bearings. Residual magnetism in aircraft parts may cause erroneous readings in the magnetic compass. Demagnetization of parts may be accomplished by slowly moving them through a coil with a 60-cycle alternating current, or by reducing the current in the coil while the part remains within the coil.

2.14 Liquid Penetrant Testing. The earliest method of liquid penetrant testing for cracks consisted of immersing the object to be tested in a bath of light oil, removing the excess oil, and then covering the surface of the specimen with chalk. When the specimen was warmed, the oil, previously drawn into the cracks by capillary action would seep out onto the surface, and the blotting action of the chalk would show the outline of the crack. Recent improvements in liquid penetrant testing has included the development of (a) visible dye penetrants and (b) fluorescent penetrants, with greater sensitivity to fine discontinuities and greater intensity of indication. Liquid penetrants can effectively be used in the inspection of non-porous metallic materials, both ferrous and non-ferrous, and non-porous non-

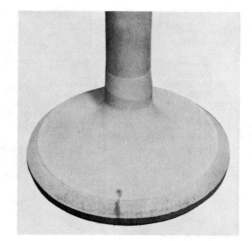

Fig. 2–35. Dye penetrant indication of crack in diesel engine valve.

metallic materials such as ceramics, plastics, and glass. Surface discontinuities such as cracks, seams, laps, or cold shuts are indicated by this method. Fig. 2–35 illustrates this method in use.

The success of any penetrant inspection is greatly dependent upon the cleanliness of the surface being examined. It should be dry and free from rust, scale, welding flux, grease, paint, oily films, and dirt. Detergent cleaners, solvents, vapor degreasing, acid or alkaline descaling solutions, ultrasonic cleaning, or abrasive blasting may be employed for cleaning the parts prior to examination.

Since penetrant indications result from a liquid which has physically entered a discontinuity it follows that the parts to be inspected should be clean and dry. Any cleaning agent which remains in or over the discontinuities will hinder entrance of the penetrant.

Visible dye penetrants are usually deep red in color and contained in a highly penetrating solvent that seeks out and fills surface cracks, porosities, and discontinuities. They may be applied by spray, brushing, or dipping. The penetrant should cover the part completely with a thin coating of liquid. Penetration time usually ranges from 5 to 60 minutes, depending upon the type of penetrant, material being inspected, and temperature. Table 2–4 shows suggested penetration time for water washable penetrants. After a suitable penetration time, the surface film of penetrant should be removed. Some penertants are water washable, while others must be removed with sol-

Table 2–4. Suggested penetration time for water washable penetrants.

MATERIAL	FORM	TYPE OF DISCONTINUITY	PENETRA-TION TIME, MIN.
ALUMINUM, MAGNESIUM, BRASS, AND BRONZE	CASTINGS	POROSITY COLD SHUTS	5 TO 15 5 TO 15
	EXTRUSIONS, FORGINGS, WELDS	LAPS	30
		LACK OF FUSION POROSITY	30* 30*
	ALL FORMS	CRACKS	30
STEEL	CASTINGS	POROSITY COLD SHUTS	30 30
	EXTRUSION FORGINGS	LAPS	60
	WELDS	LACK OF FUSION POROSITY	60* 60
	ALL FORMS	CRACKS	30
PLASTICS GLASS	ALL FORMS	CRACKS	5 TO 30

*15 MINUTES FOR BRAZED BRASS OR BRONZE PARTS.

vents. Care should be taken to minimize the possibility of removing the penetrant from the voids during rinsing, and at the same time, excess penetrant should not be left on the surface to interfere with observation of the indications.

After excess penetrant has been removed and the parts dried, a fine film of developer is applied to the part. The white coating of the developer serves as a contrasting background for the visible dye penetrant. The liquid vehicle of the developer draws the penetrant from the discontinuities to cover the developer film and leave the telltale "indication." While some discontinuities can be detected almost immediately after the developer is applied, sufficient time should be allowed for all discontinuities to be revealed. Generally, the developing time should be equal to the penetration time.

Fluorescent penetrant inspection, Fig. 2–36, makes use of a liquid vehicle that carries a dye which fluoresces brilliantly under black (ultraviolet) light. The penetrant is applied in much the same manner as visible dye penetrants. After a suitable penetration time the surface film of the penetrant is removed by a water spray. Spray rinsing is done under black light to insure that all surface penetrant has been removed.

Fig. 2–36. Fluorescent penetrant indication of crack in stainless steel weldment. (Magnaflux Corp.)

Dry developer powders are applied after the part has been dried. The developing powder is not in itself fluorescent, but is dark when viewed under black light. It acts to subdue background fluorescence and causes discontinuities to show up with a high degree of contrast. Development time should be at least one half of the penetration time to allow the developer to draw back to the surface penetrant which may be in the discontinuities. Excessively long development may cause the penetrant in large or deep discontinuities to bleed back, making broad, smudgy indications.

2.15 Ultrasonic Inspection. Nondestructive testing of parts by ultrasonic waves was first recognized in the 1930's by German and Russian investigators. In 1942, a U. S. invention by Firestone utilizing pulsed electronic wave trains, led to the development and marketing of the first practical ultrasonic flaw de-

tection equipment. The instrument used in ultrasonic flaw detection electronically generates high frequency vibrations and sends them in a pulsed beam through the part to be inspected. Any discontinuity such as blow holes, inclusions, cracks, laminations, or seams will reflect the vibrations back to the instrument. Discontinuities are determined by the elapsed time between the initial pulse and the reflection and the indications are shown by a trace on a cathode ray tube.

Ultrasonic vibrations for ultrasonic testing are generally in the 500 kilocycle to 10 megacycle frequency range. High frequency ultrasonic waves behave somewhat similar to light waves in that (1) they tend to travel in a straight line; (2) they may be reflected by a mechanical discontinuity, and (3) they may be refracted (bent) at boundaries of materials of different elastic or mechanical properties.

There are several wave forms and

modes of ultrasonic vibrations in metals, the most important of which include the following types: (1) *Longitudinal* (compression) waves in which particle vibration is parallel to the direction of propagation (similar to audible sound waves), (2) *Shear* (transverse) waves with particle vibration transverse to the direction of wave travel (similar to a cork bobbing on waves) in which the velocity of shear waves is approximately half of that of longitudinal waves, (3) *Surface (Rayleigh) waves* in which particle motion follows an elliptical path in a narrow region at, on, or just below the surface of the specimen being inspected, and (4) *Lamb waves* which may be generated in thin sheets or thin walled tubing for detecting laminar defects lying very close to the surface, and re-

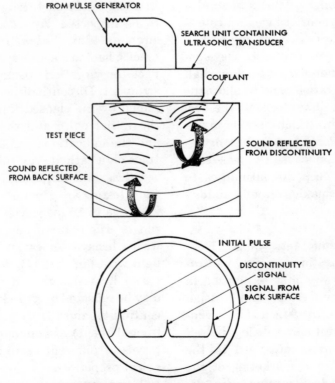

FROM PULSE GENERATOR

SEARCH UNIT CONTAINING ULTRASONIC TRANSDUCER

COUPLANT

TEST PIECE

SOUND REFLECTED FROM DISCONTINUITY

SOUND REFLECTED FROM BACK SURFACE

INITIAL PULSE

DISCONTINUITY SIGNAL

SIGNAL FROM BACK SURFACE

Fig. 2–37. Schematic drawing of principle of ultrasonic testing.

cently for testing the bond between laminations of composite materials.

During the test a short pulse of ultrasonic vibrations is transmitted into the piece to be inspected from a search unit. The pulsed beam travels until it meets a discontinuity and then is reflected back to the receiver and is subsequently displayed on the oscilloscope screen.

The search unit generally contains a quartz crystal transducer which converts high frequency electrical signals into mechanical vibrations. The pulsed signals may be repeated up to 600 times per second. A schematic drawing illustrating the principle of operation is shown in Fig. 2–37.

A liquid or paste with good wetting properties is used as an acoustic couplant between the search unit and the work piece. This permits transmission of mechanical vibrations with little loss even though the work surface may be slightly rough (up to 125 RMS). Oil, glycerine, and water are among the more commonly used couplants.

Reflected pulses are converted, amplified, and displayed on a cathode ray tube in such a manner as to indicate time difference between transmitted and reflected pulses. Time intervals are indicated as distances between the transducer and discontinuity on the cathode ray trace.

Fig. 2–38 shows typical subsurface

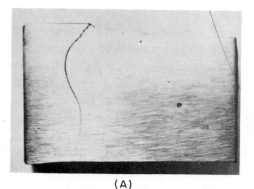

(A)

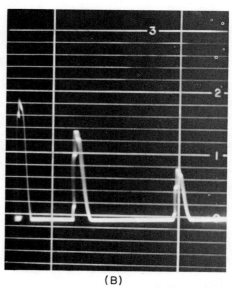

(B)

Fig. 2–38. Defect in 2½ in. thick tool steel die block determined by ultrasonic inspection. A, section through defect, B, scan indication of flaw.

flaws that may be detected using ultrasonic inspection methods.

In addition to inspection for flaws special ultrasonic techniques have been used for thickness measurement, determination of elastic moduli, study of metallurgical structure, and evaluation of the influence of processing variables on the specimen.

Advantages of ultrasonic inspection include detection of minute subsurface defects and great penetrating power allowing examination of sections up to 40 feet long. Accuracy, fast response permitting rapid and automatic inspection, and the need to test on only one surface of the specimen are also important features.

Limitations of ultrasonic inspection usually are related to unfavorable sample geometry, such as nonparallel reflecting surfaces and to undesirable interior structure such as large grain size, porosity, or dispersed precipitates.

In general, ultrasonic inspection shows great promise as one of the prominent non-destructive testing methods.

2.16 Eddy Current Testing. Nondestructive testing of ferrous and non-ferrous metals by electromagnetic induction is widely used in Europe and has been introduced into the United States. The test offers a great many possibilities for rapid, automatic, non-destructive testing and for research work.

Eddy current testing is performed by placing a test specimen within the magnetic field of a coil carrying alternating electric current. The alternating current field induces so-called

"Eddy Currents" within the test object. The induced currents in turn produce a magnetic field which is superimposed upon the original magnetic field. Accurate measurement and interpretation of the resulting impedance may be correlated with certain changes in mechanical, physical, and chemical properties of the test specimen. This field modification has exactly the same effect as would be obtained if the characteristics of the test coil itself had been changed.

Eddy current testing has been successfully used in the following applications:

1. Sorting of alloys of different composition or hardness,
2. Dimensional control of diameter, length.
3. Detection of cracks, porosity, shrinkage, internal stress due to cold working and corrosion attack,
4. Evaluation of heat treatment, depth of case and decarburization,
5. Classification of hardness, and tensile strength,
6. Measurement of plating thickness or thickness of tubing, foil, sheets, and strip,
7. Measurement of permeability and conductivity.

The change in impedance of the magnetic induction coil under the influence of the test object is a function of the properties of the test object and characteristics of the instrumentation. Significant properties of the test object include:

1. Electrical conductivity,
2. Dimensions (such as diameter of rods),
3. Magnetic permeability,
4. Presence of discontinuities, such as cracks or cavities.

Significant instrumentation characteristics include:

1. Frequency of the alternating current field of the test coil.
2. Size and shape of the test coil,
3. Distance of the test coil from the test object.

Since there are a number of variables which must be controlled during Eddy Current testing the equipment is quite sophisticated and testing procedures must be carefully planned if the test is to be meaningful. For example, special techniques are required to differentiate between effect of diameter change and surface cracks within a test specimen. A great deal of work has been done to develop theoretical and practical aspects of Eddy Current testing, including selection of the proper testing method, optimum test frequencies, design of test coils, and interpretation of test indications.

Advantages of Eddy Current nondestructive testing may be summarized by saying that it is extremely rapid, does not require contact with the test object, and lends itself well to automatic inspection and statistical quality control.

Limitations and difficulties of several kinds are associated with Eddy Current testing, and because of these inherent difficulties, various test methods have been developed. These tests include the impedance magnitude test, reactance magnitude test, impedance vector analysis, cathode ray tube ellipse tests, linear time base test, and others.

When sorting alloys by variations in electrical conductivity with the Impedance Magnitude Test the response to a 1 percent change in diameter may be 20 times as strong as a 1 percent conductivity variation. A 1 percent change in diameter is well within commercial tolerance and is insignificant whereas a 1 percent variation in conductivity may be highly significant. A reactance magnitude test with the previous example would yield only a 1 percent variation in conductivity due to diameter change instead of a 20 percent variation. Thus, when diameter variations are present a reactance magnitude test would be a better test than the impedance test.

Another limitation of Eddy Current testing is its inability to determine discontinuities at the center of cylinders or rods. This is due to the fact that Eddy currents are present only at the surface of metallic specimens.

An Eddy Current Conductivity meter is shown in Fig. 2–39. The portable instrument may be used on

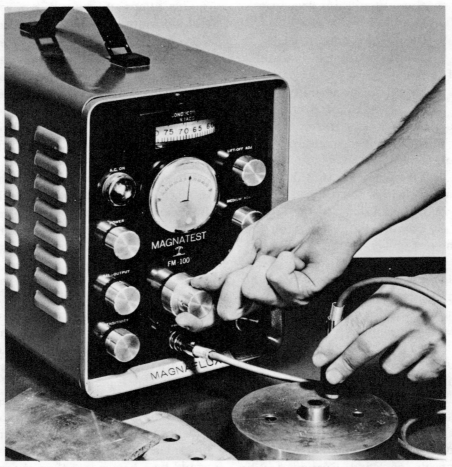

Fig. 2–39. Eddy current conductivity meter used for evaluating material, sorting non-magnetic materials, measuring thermal and electrical conductivity, and comparing tensile strength of aluminum alloys. (Magnaflux Corp.)

non-magnetic conductive material such as aluminum, magnesium, copper, stainless steel, brass, bronze, lead, or silver. The conductivity meter has broad application because so many properties such as hard-ness, thermal conductivity, tensile strength, and alloy content are related to electrical conductivity and may be quickly evaluated by comparison with known specimens.

In summary, a number of destructive and non-destructive tests have been discussed along with typical applications, advantages, and limitations. Recent trends in industry toward automation and continuous production favor non-destructive tests, especially since 100 percent inspection can be accomplished rapidly and economically. Developments in non-destructive testing and in specialized techniques are coming rapidly. These tests require sophisticated equipment and well trained technicians to properly apply and evaluate them.

References

American Society for Testing and Materials. *ASTM Standards, 1961, Part 3, Metals Test Methods.* Philadelphia.

Enos, George M., and Fontaine, W. E. *Elements of Heat Treatment.* New York: John Wiley & Son, 1955.

Lyman, Taylor, ed. *Metals Handbook, 1948 ed.* Cleveland: American Society for Metals.

Lysaght, Vincent E. *Indentation Hardness Testing.* New York: Wilson Mechanical Instrument Div., American Chain & Cable Co., 1949.

McMaster, Robert C., ed. *Non-Destructive Testing Handbook.* New York: The Ronald Press, 1959.

Simmons, Mark W. *New Ultrasonic Indentation Hardness Test Method.* Dearborn, Mich: American Society of Tool and Manufacturing Engineers, No. 1Q66-188, 1966.

Physical Metallurgy

3.1 Introduction. Much has been done in recent years to attempt to determine properties of metals. A number of theories have been proposed to help explain observed phenomena of formation, deformation, diffusion, hardening, alloy systems, fracture, and creep of metals. A few of the theories related to composition, properties, and metal fabrication will be treated briefly in this chapter. Although a great deal of our present metallurgical practice is built upon experimentation it is well to have some understanding of the theoretical advances that are being made toward explaining the simple physical laws which govern metal behavior.

3.2 Atomic Structure. Atoms are the smallest particles of matter which exhibit properties characteristic of the elements (discussed in Chapter 1). Atoms are composed of three basic parts; protrons, neutrons, and electrons. Protrons and neutrons comprise the nucleus of the atom which is positively charged. Negatively charged electrons rotate about the nucleus spinning on their own axes as they move around the nucleus in orbital paths. No more than two electrons may be in the same orbit, and to do so, they must be of opposite spin. Orbits are grouped in shells, within which are subdivisions called energy levels. The maximum number of electrons that can fit in each shell is $2N^2$, in which N is the shell number. (See Fig. 1–6). Thus the first shell may contain two electrons, the second shell eight, the third shell eighteen, and so on. Filled shells are stable whereas unfilled ones are unstable, and they tend to become filled by sharing of electrons with other elements. Shells which are not completely filled account for most of the properties of the element. Partially filled shells may give up or accept extra electrons, or they may share them with other atoms.

The force of attraction which holds solids together may be due to one of the following four types of bonds:

1. Ionic,
2. Covalent,
3. Metallic,
4. Van der Waals forces.

Ionic bond occurs when an element, such as lithium, with one excess electron in its second shell combines with fluorine which lacks one electron to complete its second shell. The lithium in giving up its extra electron as shown in Fig. 3–1 is left with a net ($+1$) positive charge and is called a positive ion. On the other hand, the fluorine which accepted the electron is also left with a net

negative (-1) charge and is known as a negative ion. The resulting strong, electrostatic attraction accounts for the bond in many common liquids and solids, including NaCl, AgCl, MgO, and others.

Covalent bond results from the sharing of electrons with like adjacent atoms. This strong bond is due to the attraction of shared electrons by the positive nuclei. In diamond, for example, each carbon atom is surrounded by four neighbors in a tetrahedral configuration. Each carbon atom shares two electrons with each of its neighbors. The diamond bond is shown in Fig. 3–2. Solids formed by covalent bonding are usually hard and have high melting points. Valency crystals generally solidify with complicated structures in order to attain the eight electrons needed for a closed-shell configuration.

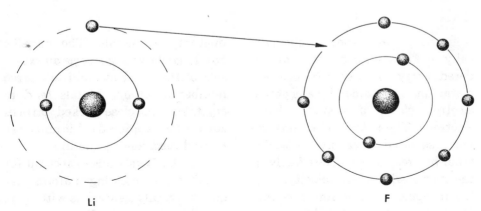

Li   F

Fig. 3–1. Electron transfer in lithium fluoride (LiF) formation.

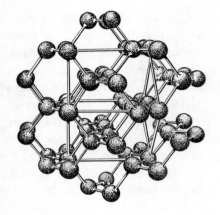

Fig. 3–2. Covalent bond in the formation of diamond. By permission: *Principles of Chemistry*, 5th ed., MacMillan Co.)

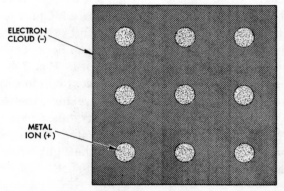

ELECTRON
CLOUD (–)

METAL
ION (+)

Fig. 3–3. Schematic representation of metallic bonding.

Metallic bond, shown schematically in Fig. 3–3, consists of an ordered array of positively charged metal ions surrounded by a negative electron cloud of shared valence electrons. The valence electrons are not associated with any particular atom but may move about freely in the metal with high velocity. The free movement of electrons accounts for the high electrical and heat conductivity of metals. The metallic bond may be thought of as an extension of the covalent bond to a large number of atoms. Metals tend to crystallize in close-packed lattices such as face-centered cubic and close packed hexagonal structures.

Van Der Waals forces account for a rather weak bonding of atoms that are electrically neutral as with inert gases and inert gas solids.

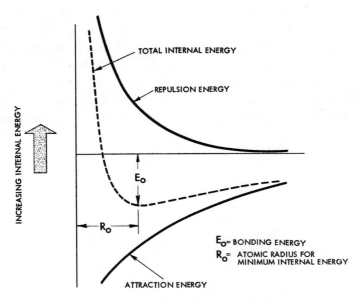

Fig. 3–4. Internal energy as a function of distance between atoms.

Cohesive energy in inert gas solids, such as neon and argon, is of the order of 1/100th as much as that of ionic crystals. Inert gases solidify in the face-centered cubic system. This type of bonding is of importance only at low temperatures when weak attractive forces can overcome the thermal agitation of the atoms. Many molecules, in addition to atoms, form crystals which are held together by Van der Waals forces, including N_2, H_2, and CH_4. As atoms approach one another (Fig. 3–4) an interaction between them occurs, affecting their internal energy.

A force of attraction between the atoms tends to bring them together whereas a repulsion arising from interference of their outer shells tends to hold them a certain distance apart. This distance of closest approach which is characteristic for each element is known as its atomic diameter. The atomic diameters for metals range from about 2 to 6 angstrom units (1 angstrom $= 10^{-8}$ cm). Atomic diameters for a number of metals are shown in Table 1–4.

3.3 Crystal Structure of Metals. During the process of solidification, the atoms of liquid metal arrange or group themselves into an orderly pattern. This orderly arrangement of atoms is known as the *space lattice* and is made up of a series of points in space representing locations of atoms interconnected by a network of imaginary lines. The smallest unit having the full symmetry of the crystal is called the *unit cell* and is but a building block of the crystal. Al-

though the atoms are too small to be seen by ordinary means, the space lattice of metals in the solid crystalline state has been determined through x-ray diffraction methods. By means of these x-rays the internal atomic arrangement of the crystal

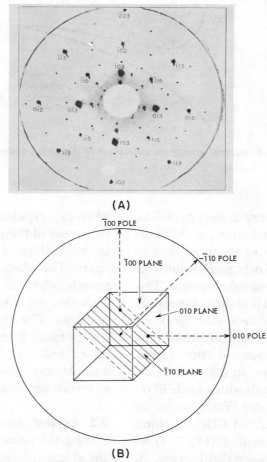

(A)

(B)

Fig. 3–5. A, Laue x-ray diffraction pattern of face-centered cubic crystal with the (100) plane perpendicular to the beam. Spots (called poles) result from reflection of x-ray beams from parallel crystal planes onto the film. The numbers are indexes of the poles. Fig. 3–5. B, poles of crystal faces. (By permission: *Engineering Metallurgy*, McGraw-Hill Book Co.)

state may be determined as shown in Fig. 3–5 A and a picture of the space lattice may be drawn as shown in Fig. 3–5 B.

All of the known types of crystals in nature can be classified into seven systems: (1) cubic, (2) hexagonal, (3) tetragonal, (4) orthorhombic (5) rhombohedral, (6) triclinic, and (7) monoclinic. Fortunately, most of the important metals crystallize in either the cubic or hexagonal system. The cubic system as shown in Figure 3–6 may be either of the face-centered cubic, F.C.C. or the body-centered cubic, B.C.C. type. The hexagonal system is often referred to as hexagonal close packed, H.C.P. (or sometimes close packed hexagonal, C.P.H.) because of closeness of the atoms. Crystal patterns for typical metals are shown in Table 1–4.

Although the number of atoms which constitute even a small crystal may be billions or trillions, the unit cell is composed of only a few atoms.

In the cubic unit cell, for example, an atom at a corner position belongs to eight adjoining atoms so that only one-eighth of each atom can be counted in one cell. In this way the body-centered cubic cell has

8 atoms at the corners $\times$ ⅛ = 1
1 atom at the center   = 1
Total 2

Typical metals which have the body-centered cubic structure at room temperature are iron, chromium, vanadium, molybdenum, niobium, tantalum, and lithium. The face-centered cubic unit cell has

8 atoms at the corners $\times$ ⅛ = 1
6 atoms in the face $\times$ ½   = 3
Total 4

Metals which are of the face-centered cubic type at room temperature include aluminum, nickel, copper, silver, platinum, gold, and lead.

The hexagonal close packed unit cell, one-third of which is shown in bold lines in Fig. 3–6, has:

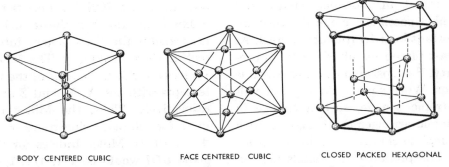

BODY CENTERED CUBIC   FACE CENTERED CUBIC   CLOSED PACKED HEXAGONAL

Fig. 3–6. The three unit cell types most common in metals.

77

8 atoms at the corners $\times$ ⅛ = 1
1 atom inside the cell   = 1
                        Total 2

Typical metals which solidify with close packed hexagonal structure include beryllium, magnesium, zinc, titanium, cobalt, cadmium, antimony, and bismuth.

Unit cells are often defined by specifying their types and parameters. The cubic cell, in which all faces have the same dimensions, requires only one parameter a; the hexagonal cell requires two parameters, one which is the width of the hexagon a, the other the height of the side face, c, (or sometimes the axial ratio c/a is given). Dimensions of the parameters are usually expressed in angstrom units.

Some metals have the capacity of crystallizing in one form of lattice structure, but upon cooling may change to another form, iron changes from F.C.C. to B.C.C. on cooling below 1313°F. This reversible change from one crystal pattern to another is called an *allotropic* change and makes possible some of the very important heat treatments of steel. At least fifteen metals show this property, the most important of which is iron. When a metal undergoes an allotropic change, such a change in atomic arrangement brings about a complete change of characteristics.

The properties of metals are dependent, in a large measure, upon the type of space lattice formed during solidification. In general, metals with the face-centered lattice lend themselves to a ductile, plastic, workable state; whereas metals with the close packed hexagonal lattice exhibit, in general, a lack of plasticity and lose their plastic nature rapidly upon shaping, such as in a cold forming operation. Metals with a body-centered lattice are not quite so plastic as the F.C.C. metals and tend to work harden quite rapidly as do those of the H.C.P. type.

Atoms are arranged in uniform layers on planes referred to as atomic or crystallographic planes. The relationship between sets of parallel planes to the axes of the unit cell is designated by *Miller Indices*. One corner of the unit cell is taken as the origin of the space coordinates, and a given set of parallel planes is identified by the reciprocals of its intersections with the coordinates of the unit cell. The lattice parameter is the basic dimension of the unit cell. For example, in a cubic crystal (Fig. 3–7) the plane *BCHG* intersects the Y axis at one unit from the origin but is parallel to the X and Z axes, intersecting them at infinity. The Miller Indices are the reciprocals of the intercepts with the X, Y, and Z axes respectively. Thus, the Miller Indices for the plane shown are $\frac{1}{\infty}$, $\frac{1}{1}$, $\frac{1}{\infty}$, or (010). Miller Indices for the plane *BDI* would be 1/1, 1/1, 1/½ or (112).

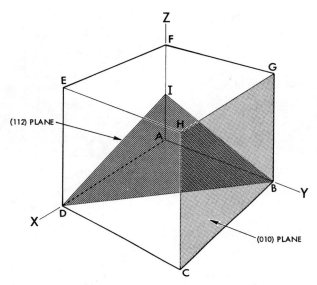

Fig. 3–7. Unit cell showing crystal planes and Miller indexes.

If a plane cuts any axis on the negative side of the origin, the index will be negative and is indicated by placing a minus sign above the index, as $(\bar{h}, k, z)$. All parallel planes have the same indices. A specific plane or set of parallel planes have parentheses placed around the Miller Indices. Braces [ ] signify a family of planes of equivalence in the crystal such as on six faces of a cubic crystal. The most important planes related to plastic deformation in metals are those of high atomic density and the greatest interplanar spacing. In the BCC structure these are the (110) planes as shown in Fig. 3–8 A and in the FCC structure are the (111)

planes as shown in Fig. 3–8 B. When metals are permanently deformed the atom groups slide over one another along planes known as slip planes. These slip planes are parallel to planes of highest atomic density indicated above. The action of plastic flow in metals will be discussed further in paragraph 3.6.

3.4 Solidification of Metals. Metals are often composed of a great number of irregularly shaped crystals and may be visualized as forming from small nuclei in molten metal. In the molten state atoms do not have a fixed arrangement, but move continuously with random mo-

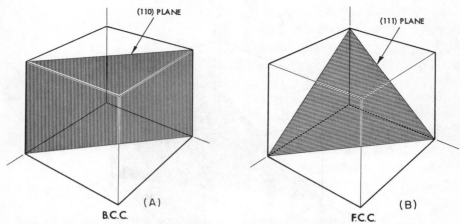

Fig. 3–8. Important crystal slip planes in the cubic system, B.C.C. and F.C.C.

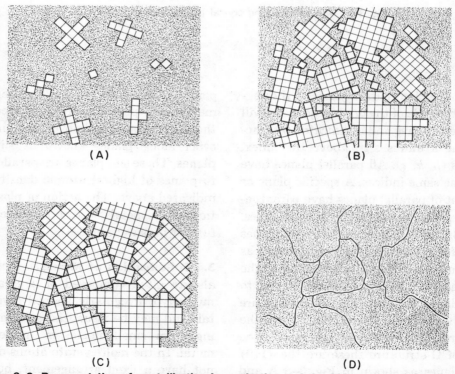

Fig. 3–9. Representation of crystallization by nucleation and dendritic grain growth. A, nucleation, B, grain growth, C, completion of crystallization, and D, grain boundaries.

tion. Of the enormous number of atoms in a metal some atoms at any given instant are in positions corresponding to the ordered arrangement they normally assume in the solid state. These chance aggregates form the nuclei or centers for further crystallization. A schematic representation of crystallization is shown in Fig. 3–9 and is composed of two stages, (a) nucleation, and (b) crystal growth.

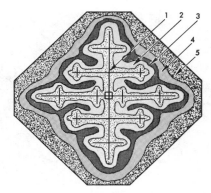

Fig. 3–10. Steps in the formation of a dendritic structure.

As each nucleus is a growing crystal, and the atoms within it are all similarly oriented, it should be appreciated that the crystals will be randomly oriented with respect to each other.

Crystal growth takes place in three dimensions, the atoms attaching themselves in certain preferred directions, usually along the crystal axes. This growth gives rise to a characteristic treelike freezing pattern which yields a dendritic structure as shown in Fig. 3–10. A crystal nucleus forms as shown in (1) and then proceeds to send out shoots or axes of solidification as shown in (2), (3), and (4), forming the skeleton of a crystal in much the same way as frost patterns form. Atoms then attach themselves to the axes of the growing crystal in progressive layers, finally filling up these axes, and thus forming a completed solid, or crystal as shown in (5).

The external shape of metal crystals is mostly dependent on the in- fluence of other growing crystals which surround it. Thus, when the individual crystals have grown to the point where they come in contact with one another there will be a region of mismatch. The crystals found in all commercial metals are commonly called grains because of this variation in their external shape. A *grain*, therefore, is a crystal with almost any external shape but with an internal atomic structure based upon the space lattice with which it was born.

The nature of the grain boundaries is still more or less of an unknown, but it is generally assumed to be an interlocking border line where atoms of one crystal change orientation from the atoms of the adjacent crystals. In Fig. 3–11 is shown a schematic representation of a grain boundary between two crystals. The highly strained condition between

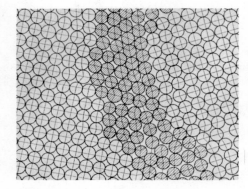

Fig. 3–11. Schematic of a grain boundary between two crystals. Cross-hatched atoms constitute the grain boundary. (By permission: *Engineering Metallurgy*, McGraw-Hill Book Co.)

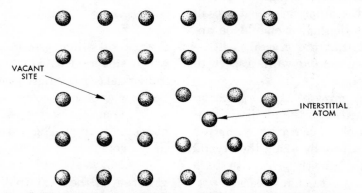

VACANT SITE

INTERSTITIAL ATOM

Fig. 3–12. Crystal defects due to vacancies and interstitial atoms.

grains, arising from mismatch between space lattices, may account for the greater strength observed for grain boundaries as compared to the strength of individual crystals.

It has been estimated that during very slow crystal growth, such as 1 mm per day, atoms must be laid down at the rate of one hundred per second on the surface, and in precise order. It is not surprising, therefore, that imperfections, on an atomic scale, exist in metal crystals. The most important crystal imperfections are *vacancies, interstitials,* and *dislocations.* Vacancies and interstitial defects are shown in Fig. 3–12.

Vacancies are simply empty atom sites produced during solidification whereas interstitials may be produced by distortion during severe plastic deformation. Both vacancies and interstitial atoms produce local distortion and interrupt regularity of the space lattice. Vacancies are also believed to explain the mechanism of *solid state of diffusion;* it is well known that two metals, for example, nickel and copper, will diffuse into each other when placed in intimate contact and heated to a high temperature. Diffusion in solid metals is explained by assuming that vacancies move through the lattice producing random shifts of atoms from one lattice position to another. Vacancy motion is nil at low temperatures, but at high temperatures is very rapid. For example, at a temperature just below the melting point of copper vacancies are about 10 atoms apart and move approximately 30 billion times per second, while at room temperature, vacancy movements occur only about once every eleven days.

One other crystal defect, and that which explains the wide discrepancy between theoretical and observed yield stresses, is the dislocation. Theoretical calculations show that a pure crystal of magnesium, for example, should have a yield strength of approximately one million pounds per square inch, whereas in reality, it may yield at 100 psi. Dislocations

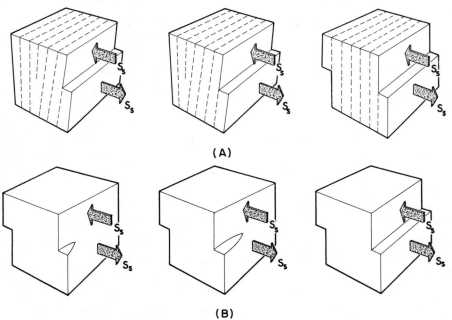

(A)

(B)

Fig. 3–13. A, edge dislocation. B, screw type dislocation. (By permission: *Physical Metallurgy Principles*, D. Van Nostrand Co.)

may be thought of as imperfections of a lattice which result from a mismatching of atomic planes within a crystal. Dislocations are of two types, the *edge dislocation* and the *screw dislocation*. Both types are shown schematically in Fig. 3–13.

The evidence for dislocations in metals is quite well established and helps to explain many phenomena in metals which could not otherwise be easily understood. Grain boundaries are thought, by some, to be complex arrays of dislocations. By examination of Fig. 3–13, it can be seen how dislocations cause movement of only a few atoms at a time instead of an entire plane of atoms. This, then, accounts for the large difference between theoretical yield strength and observed values.

Deformation of Metals

3.5 Single and Polycrystalline States. The size and number of grains formed during solidification is governed by the number of nuclei formed, which, in turn, is largely determined by the rate of cooling. If the melt is cooled slowly only a few nuclei will form and the grains will be large. Conversely, rapid cooling will produce a large number of small grains.

In commercial metals control of grain size is very important. Grain size affects toughness, strength, forming and heat treating character-

istics, as well as other properties. A fine grained metal is stronger and tougher than a coarse grained one, and in many applications, proves more satisfactory than a coarse grained metal. A polycrystalline or many-crystal aggregate has no smooth, continuous slip line as does a single crystal, but has changes in orientation of slip planes at each grain boundary. Thus, the effect of a fracture traveling through a polycrystalline aggregate is similar to a break in the masonry line of a brick wall. Each grain keys to its neighbor and builds up resistance to slip or failure. On the other hand, coarse grained metals exhibit greater plasticity than fine grained metals. The coarse grains have longer slip planes, and therefore, greater capacity to slip during plastic deformation than do fine grained metals with shorter slip planes. Coarse grained steel will harden deeper than fine grained steel and absorb carbon faster during carburizing operations. Thus, grain size control is very important in both forming and heat treating operations.

3.6 Elastic and Plastic Deformation. When a metal is subjected to a stress below its elastic limit, the resulting deformation is temporary and the object will return to its original shape after removal of the applied stress. Stressing a metal beyond its elastic limit results in permanent de-

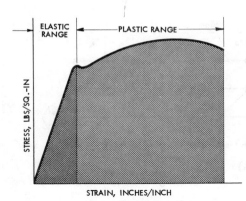

ELASTIC RANGE | PLASTIC RANGE

STRESS, LBS/SQ.-IN

STRAIN, INCHES/INCH

Fig. 3–14. Stress-strain diagram showing elastic and plastic ranges in metal working.

formation. Elastic and plastic ranges are shown in Fig. 3–14.

This ability of metals to deform plastically makes possible metal forming operations such as forging, rolling, extrusion, drawing, forming, and bending. Once a metal is subjected to sufficient stress, it can be readily worked into intricate shapes. Even brittle materials, such as tungsten and beryllium, can be plastically deformed when subjected to sufficiently high stresses. Much effort is being made in industry to replace expensive machining operations with new methods involving plastic working of metals. Recent developments involve high energy rate forming, using explosives, magnetic fields, and compressed gases.

Perhaps the mechanism of plastic deformation may best be understood by the study of a single metal

crystal. As already discussed, when metal crystals are subjected to sufficiently large outside forces plastic deformation takes place; that is movement of atoms take place. Atom movement does not take place haphazardly but occurs in definite directions within the space lattice. Plastic deformation may occur by *slip* or by *twinning* as shown schematically in Fig. 3–15.

Slip takes place along certain crystal planes, called slip planes, and in certain crystallographic directions. This combination is called a slip system. When many slip systems are available plastic deformation is relatively easy, but with fewer slip planes and directions, slip is more difficult and the metal is said to be stronger and less ductile. Slip lines in a metal crystal are shown in Fig. 3–15 A. In F.C.C. metals, for example, there are 12 possible slip systems, 48 in B.C.C. metals, and 3 in the H.C.P. structure.

Some metals deform easier by *twinning* than by slip, while others deform by both slip and twinning. Twinning takes place along special planes called twinning planes and results in a new lattice orientation in the twinning region. The shear movements that are necessary to produce twinning in a B.C.C. lattice are shown in Fig. 3–15 B. Twins always occur in pairs, hence the name. However, only one twin is shown in the deformed magnesium grain, Fig. 3–16.

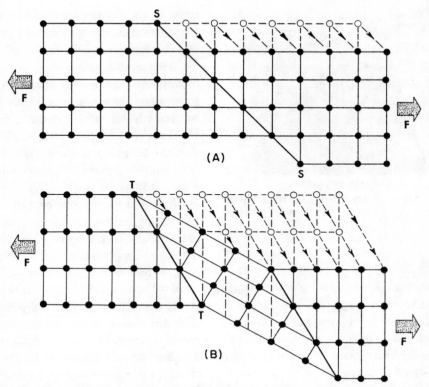

Fig. 3–15. Plastic deformation showing a crystal under tension elongating in the direction of applied stress by, A, slip and B, twinning. (By permission: *Engineering Metallurgy*, McGraw-Hill Book Co.)

Fig. 3–16. Photograph of slip lines and a twin band in magnesium, 250X. *(Trans. ASM, XLVIII, 1956, p. 994.)*

Deformation in a crystal takes place on planes which are oriented so that the shear component is maximum. This condition exists when the slip planes are at 45° to the direction of stress as shown in Fig. 3–17. The component of force required to initiate slip is called the *critical resolved shear stress.*

Now that we have examined the single crystal during plastic deformation let us look at polycrystalline materials. The interaction between adjacent crystals does not allow direct calculation of the shear stress on any slip plane. However, it does seem clear that slip begins on the most favorably oriented planes and continues to other planes as possibilities for deformation of the first become exhausted and as rotation concomitant with slip brings other less favorably oriented crystals into a position where they can be deformed. As the crystals rotate due to deformation, a *preferred orientation* starts to appear. Most of the crystals tend to line up with a certain crystal axis in the direction of deformation, which usually leads to different properties in different directions. These directional properties are referred to as *anisotrophy.*

This variation in properties, particularly resistance to further plastic deformation in sheet metals, causes irregular flow during forming. The most common example is the formation of "ears" in drawing cylindrical cups (Fig. 2-20). A common test for this "earing" tendency is the tear-length test shown in Fig. 3-18.

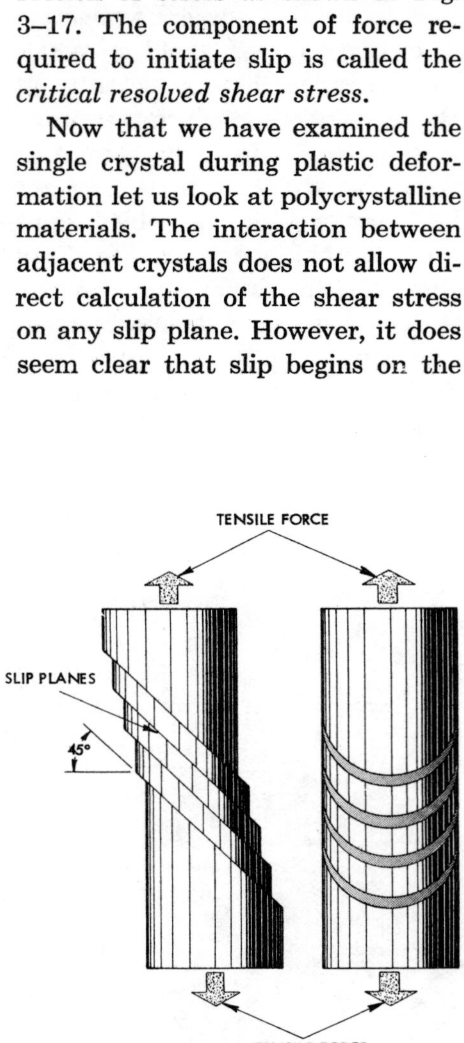

Fig. 3–17. Orientation of slip planes at 45° to direction of stress.

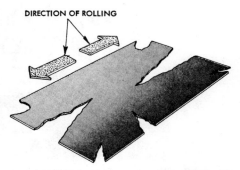

Fig. 3–18. Directional properties of a rolled steel sheet as shown by the tear-length test.

3.7 Fracture. The study of metal fracture has gained considerable impetus in recent years as metals are being subjected to higher stresses, higher temperatures, and highly corrosive atmospheres.

Metal fracture may occur rapidly and unexpectedly with little apparent reason or it may occur slowly and with appreciable deformation of the part under load. Metal fractures may be classed as:

1. Brittle cleavage fractures,
2. Ductile shear fractures,
3. Fatigue fractures,
4. Intergranular fractures.

Temperature, crystal orientation, critical stress, and lattice type affect the method by which fracture will occur. Under certain low temperature conditions, it is possible to split or cleave metal crystals such as zinc into two parts by a sharp blow with a knife-edged instrument. Low temperature brittleness of iron is attributed to metal *cleavage*. Propagation of cleavage is very rapid in metals with an upper limit apparently at the speed of sound which is approximately 15,000 feet per second in steel. Catastrophic brittle fractures have occurred in all-welded ships such as that shown in Fig. 3–19 where the ship actually broke in two, without warning, and almost instantaneously.

Ductile *shear fractures* occur in metals such as magnesium when

Fig. 3–19. All welded tanker that broke in two at the dock. (By permission: *National Bureau of Standards*)

crystals separate by continued shear along a slip band. In some cases the shearing process is preceded by twinning. Twinning brings slip planes into more favorable orientation so that shearing may occur. In polycrystalline metals both shear fractures and cleavage fractures may be observed. Cleavage surfaces show bright crystal facets whereas shear fractures leave dull, velvety fibrous surfaces. At room temperature mild steel tensile test specimens exhibit ductile shear fracture. However, at low lemperatuers they fail with a brittle type of fracture. The transition range between ductile and brittle fracture, as shown in Fig. 3–20, is an important characteristic of steels.

Typical Charpy V-notch fracture surfaces are shown in Fig. 3–21. These macro-photographs show a brittle fracture, a transition fracture which shows partly brittle and partly ductile behavior, and a completely ductile fracture.

There is no single temperature at which an average ferrous metal suddenly changes from a ductile to a brittle material. The transition occurs over a range of temperatures. The most important variables affecting low temperature toughness are microstructure and composition. Hardened and tempered steels with a spheroidized cementic structure can lower the transition temperature to as low as −300°F. whereas nor-

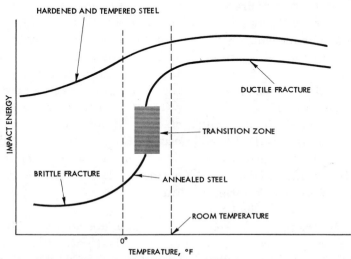

Fig. 3–20. Impact strength for annealed and hardened steels at varying temperatures.

Fig. 3–21. Typical Charpy fracture surfaces. A, brittle. B, transition. C, ductile.

malized and tempered steel may have a transition from ductile to brittle fracture at 0°F. Manganese, when added to steel, has been found to lower the transition temperature, whereas carbon seems to increase it.

A *fatigue fracture*, shown in Fig. 2–15, may occur if stresses are cyclically varied or reversed. During the process slip bands form followed by strain hardening. If stresses are sufficiently great microcracks may form at the ends of slip bands which have been most active or at other stress concentrations. The cracks may eventually expand and spread sufficiently to cause failure. In most cases the surfaces of a fatigue failure will have areas with two distinctly different appearances. One area will appear polished or burnished because of the rubbing of the crack surface during each stress cycle. When the specimen finally fractures the second area will have a granular or crystalline appearance. The crystalline

appearance has caused the erroneous conclusion that the metal crystallized in service and thereby became embrittled.

Intergranular fracture occurs when cracks spread along the grain boundaries. This type of fracture often occurs when metal are strained slowly at elevated temperatures. Intergranular cracks are not accompanied by necking down but are characterized by many cracks forming simultaneously at the metal surface and spreading inwardly. Because these cracks originate at the surface the surrounding atmosphere is important and may lower fracture strength if intergranular oxidation occurs. Coarse grained metals have been found to be stronger than fine grained ones at elevated temperatures.

3.8 Strain Hardening. After metals are plastically deformed and atom movement by slip or twinning has taken place, it would seem that the lattice is again in an unstrained condition. However, it has been observed that after plastic working, also referred to as cold working or strain hardening, metals are harder, stronger, less ductile and have reduced electrical conductivity and are more susceptible to corrosion as illustrated in Fig. 3–22.

X-ray diffraction studies substantiate microscopic evidence of residual stresses, lattice distortion, some grain fragmentation, and preferred orientation in deformed metal crystals. Distortion is greatest at slip planes, resulting in high internal energy at those locations. Thus, a deformed crystal has a distortion of peaks and valleys of internal energy, the peaks corresponding to points of most severe distortion.

Several theories have been proposed to help explain the nature of the lattice distortion resulting from cold working. The most recent and widely accepted explanation is the *dislocation theory*. This theory states that as energy is applied to deform a metal crystal dislocations are formed along the slip planes. As deformation progresses the number

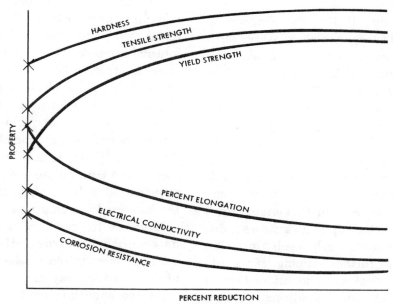

Fig. 3–22. Typical effect of cold working on metal properties.

91

of dislocations increase, and strong interactions arise as dislocations are forced to intersect dislocations on other slip planes. Current theory is that dislocations are only able to move by creation of vacant lattice sites, or interstitial atoms, and at the cost of further expenditure of energy. The net result is increasing resistance to further deformation. The old *Amorphous Cement Theory* explained that work hardening was caused by the formation of a hard, noncrystalline, amorphous (without form) cement, due to the rubbing together of the weak planes within the crystals. The theory was that the cement acted as a binder making the weak planes stronger and therefore increasing the strength of the metal as a whole.

Another theory, the *Fragmentation Theory*, explained work hardening as the action of local disorganization of the crystal structure into fragments. These crystal fragments, according to this theory, act as mechanical keys along the weak planes of the crystals keying them together and thereby making them stronger. All of these theories, while sounding quite different, are basically the same in that atom movement at the slip planes causes a distorted zone which results in resistance to further slipping. If we add to these effects the interaction of randomly oriented adjacent crystals we may have some reasonable explanation of the effects of strain hardening of metals.

3.9 Recovery, Recrystallization, and Grain Growth.

Metals which have been made harder and stronger by cold working can be restored to their original soft condition through an *annealing* process. This process consists of heating the metal to an elevated temperature followed by slow cooling. During annealing the deformed structure disappears and the properties return to normal. Three distinct stages have been observed during the annealing process, namely: (1) recovery, (2) recrystallization, and (3) grain growth. The effects of these three stages of annealing are shown in Fig. 3–23.

The first noticeable effect when reheating previously cold worked metal is the reduction of locked-in internal stresses. This reduction in internal stress occurs without noticeable change of physical properties and is not evident microscopically. However, submicroscopical elastic distortion of atomic planes is removed, and the atoms move to a more nearly equilibrium position due to increased mobility given them by the increased temperature.

Evidence for recovery is usually determined by x-ray diffraction methods as indicated by sharpening of lines on the x-ray film.

As the annealing temperature is raised beyond that required for re-

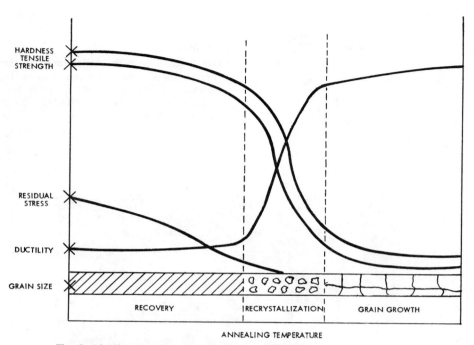

ANNEALING TEMPERATURE

Fig. 3–23. Changes in properties by annealing of cold worked metal.

covery there is a complete formation of new and unstrained metal grains. This new change, called *recrystallization,* is very evident because of its marked effect on the properties of the metal as shown in Fig. 3–23.

The changes that accompany recrystallization are in the direction of restoring properties of the metal prior to cold deformation.

The driving energy required to overcome the rigidity of a distorted lattice is sometimes compared to an "energy hill" in which the internal energy due to deformation is not quite sufficient to push the atoms up over the hill and into a distortion-free lattice. Heat from an external source supplies the additional energy necessary to move the atoms over the hump and results in the formation of nuclei for new stress-free grains as shown in Fig. 3–24. Nuclei appear to form at regions where atomic structure is most highly disarranged such as at grain boundaries, slip lines, and twin planes. The photomicrograph of Fig. 3–25 shows the beginning of recrystallization of deformed cartridge brass.

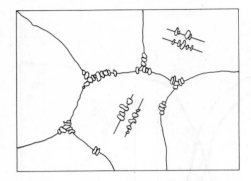

Fig. 3–24. Schematic of the recrystallization process. New, stress-free grains grow from distorted grains, beginning at grain boundaries and along severely distorted slip lines and twin planes.

Fig. 3–25. Beginning of crystallization of deformed 70-30 cartridge brass, 100X. (By permission: *Applied Metallurgy for Engineers*, McGraw-Hill Book Co.)

The approximate recrystallization temperatures are shown in Table 3–1. However, it should be noted that the recrystallization temperature is dependent upon several factors, including the degree of prior cold work, the annealing time, grain size, and purity of the metal.

An interesting aspect of recrystallization is the fact that a certain minimum amount of cold work (usually 2 to 8 percent) is necessary before recrystallization will occur. (Fig. 3–26.) The minimum strain which will allow recrystallization during subsequent heating is called the *crit-*

Table 3–1. Recrystallization temperatures.

MATERIAL	RECRYSTALLIZATION TEMPERATURE, DEGREES FAHRENHEIT	PERCENT PRIOR COLD WORK	INITIAL GRAIN SIZE
ALUMINUM, 99,999%	175		
ALUMINUM BRONZE, 5%	660	44	.075 mm
ALUMINUM BRONZE, TYPE B150*	1350	11	
BERYLLIUM COPPER	900	50	
BERYLLIUM COPPER	700	90	
CARTRIDGE BRASS	660	50	.045 mm
COPPER, ELECTROLYTIC*	435	69	
IRON, ELECTROLYTIC	750		
LEAD, 99.999%	BELOW 32		
MAGNESIUM, 99.85	350	30	
MAGNESIUM ALLOY, AZ8DX	650	10	
MONEL	1220	10	
NICKEL, 99.99%	600		
STEEL, LOW CARBON	1000		
TIN	25		
ZINC	50		

*RECRYSTALLIZATION TIME 1/2 HOUR AT TEMPERATURE

ical strain. Not all metals are equally susceptible to large grain growth. However, it is quite likely that formed parts will be strained a critical amount at certain locations and thus become subject to abnormally large grain growth during in-process annealing.

Below the critical strain, the grain size does not change during annealing. In metals which are strained considerably more than the critical amount the nucleation rate is high with a resultant large number of fine grains. Thus, in metal forming it is desirable to strain all portions of the metal uniformly, and considerably more than the critical amount.

When the temperature of cold-worked metal is held above the recrystallization temperature, the new-born grains grow rapidly in size by absorbing other grains, a process called *grain growth*. Because mechanical properties of metals are so closely related to grain size, control of grain size is an important study in the metallurgy of cold-working and annealing. Although much study has been given to the mechanism of grain growth, it is not yet completely understood. However, it is generally recognized that the driving force for grain growth in a recrystallized metal lies in the surface energy of the grain boundaries. Small grains have higher surface energy than large grains. Surface energy is reduced as large grains grow larger, and thus, they totally absorb smaller grains. In the process of grain growth, the total number of grains

95

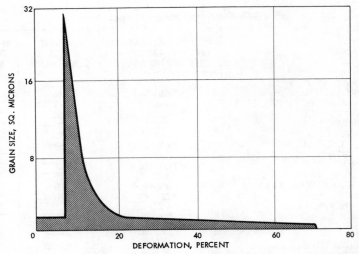

Fig. 3–26. Critical strain for growth of large grains in low carbon steel.

is reduced. When the boundary between two metals is plane, there is little tendency for change in grain size (grain boundary migration). In grains whose boundaries are curved, there is a tendency for migration toward the center of curvature of the boundary. However, because detailed knowledge of the grain boundary is lacking, the mechanism by which atoms on one side of a grain boundary cross the boundary and join the crystal on the other side is not known.

References

Avner, Sidney H. *An Introduction to Physical Metallurgy.* New York: McGraw-Hill, 1964.

Barrett, Charles S. *Structure of Metals,* 2nd ed. New York: McGraw-Hill, 1952.

Burton, Malcolm S. *Applied Metallurgy for Engineers.* New York: McGraw-Hill, 1956.

Mondolfo, L. E. and Zmeskal, Otto. *Engineering Metallurgy.* New York: McGraw-Hill. 1955.

Reed-Hill, Robert E. *Physical Metallurgy Principles.* Princeton, N. J.; D. Van Nostrand Co., 1964.

Metallurgical Examination

4.1 Introduction. A number of methods and tools have been developed in order to reveal the constitution and structure of metals and alloys. In the early 1800's, scientists first started looking at metals under the microscope. These early investigators discovered the relationship between grain size and hardness, the presence of deformation lines during work hardening, the hardening effect of alloying elements and precipitates, and finally, the relationship between microstructure and the properties of quenched and tempered steel. The results of these discoveries were later applied to process control —specifying, for example, the best structure for machining, or the best grain size for a deep drawing operation.

In recent years metallurgists have been able to capitalize on the information gained and develop new alloys with the desired microstructure and with custom-made properties.

A number of methods of metallurgical examination will be discussed including the development of new tools such as the electron microscope and x-ray diffraction apparatus which have greatly helped in changing metallurgy from an art to a science.

Microscopy

4.2 Specimen Selection. Careful consideration should be given to the selection of metallurgical specimens which are to be examined under the microscope. The specimen should be representative of the metal to be studied. If a structural member has failed in service and the cause of the failure is to be determined by metallographic examination, then the specimen should be taken from the region which will yield the maximum amount of information. In some instances, such as with rolled shapes, both longitudinal and transverse sections should be examined. Also, for comparative purposes, a specimen should be taken from a normal and sound section of the structure.

If the metal is soft the specimen to be examined may be removed by means of a hacksaw. Brittle metals may be fractured by a hammer blow and a suitable fragment selected as the specimen. Specimens may be secured from hardened metals by means of an abrasive cutoff machine shown in Fig. 4-1. The machine utilizes a rubber-bonded abrasive wheel approximately $\frac{1}{16}''$ thick operating at relatively high speed. The specimen is usually cooled during the cutting operation either by directing a

Fig. 4–2. Burn marks on a specimen overheated during cut-off.

stream of coolant to the point of cutting or by completely submersing the specimen in the cooling media. Care must be taken in selecting the proper abrasive wheel and providing ample cooling or the original structure of the metal may be drastically altered because of the heat generated during cutting. Wheels that are too hard tend to glaze and burn the specimen and wheels that are too soft wear rapidly. In general, the specimen should never be heated enough to feel warm to the touch. Overheating due to improper cutting is clearly evident in Fig. 4–2.

4.3 Specimen Mounting. Metallographic specimens that are too small or too awkwardly shaped to be held conveniently while grinding and polishing may be mounted in synthetic

Fig. 4–1. Abrasive cut-off machine used for metallurgical specimens. (Buehler Ltd.)

Fig. 4–3. Metallographic specimen mounted in Lucite. (Buehler Ltd.)

plastic materials such as bakelite, lucite, transoptic, or the recently developed diallyl phthalate. Fragile specimens may be mounted by placing them in small cups and pouring liquid casting resin around them. After the resin has solidified, the cups may be torn away, leaving the mounted specimen. Clear mounting materials as shown in Fig. 4–3 make it possible to see the orientation of the specimen during grinding and polishing.

Mounting of specimens in plastic may be easily accomplished by means of a specimen mounting press such as that shown in Fig. 4–4. The press, which may be used with both thermosetting and thermoplastic resins, consists of a heating unit for melting the plastic and a hydraulic ram for applying pressure during curing of the resin, and for ejecting the mounted specimen from the mold after it is cured. When thermosetting materials are used the mold can be

Fig. 4–4. Specimen mounting press. (Buehler Ltd.)

kept hot and mounting time is less than with thermoplastics. Thermoplastic materials must be cooled in order to solidify them whereas thermosetting plastics solidify under heat and pressure.

Improper molding pressure or temperature or improperly shaped specimens may result in defective mounts. Thermosetting plastics such as bakelite will have a soft, gravel-like appearance if they have not been adequately heated. Bulges may occur if pressure has not been maintained while the specimens are curing. Clear thermoplastic materials which are not heated through may have a cloudy appearance, due to unmelted particles, or if not cooled rapidly enough after melting, may result in loss of transparency. Specimens which are large in relationship to the size of the mount or which have sharp corners may cause cracking of the plastic during cooling due to differences in expansion rate. Manufacturers' recommendations for equipment and for plastic resins should be followed in order to achieve the best results in specimen mounting.

4.4 Grinding of Specimens. The surface of the specimen is first made flat by an abrasive belt machine similar to that shown in Fig. 4–5. These machines are frequently equipped with coolant attachments to insure cool cutting and to help wash abraded particles from the belt. Care should be taken to prevent overheating of the specimen during the rough grinding stage. Abrasive belts for wet grinding usually employ silicon carbide as the cutting media whereas aluminum oxide is used without a coolant. Abrasive grit size is usually in the 120-320 range. Rough grinding produces a flat, plane surface and removes deleterious effects due to the cutting-off operation. It is advisable during the rough grinding operation to bevel the sharp edges of the specimen or specimen mount to prevent tearing of grinding papers and polishing cloths during subsequent operations. Excessive pressure

Fig. 4–5. Abrasive belt surfacer for specimen grinding. (Buehler Ltd.)

during rough grinding will form deep scratches and will increase the depth of disturbed metal on the surface of the specimen. This disturbed layer on the surface is shown schematically in Fig. 4–6, and is often referred to as the *Beilby Layer*. The layer may extend from ten to fifty times the depth of the scratch produced by an abrasive grain.

The initially deformed layer is reduced in depth by the application of a finer abrasive. However, the finer

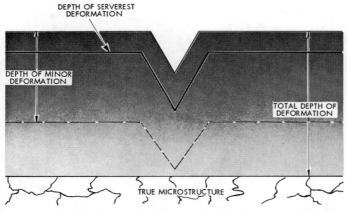

DEPTH OF SERVEREST DEFORMATION

DEPTH OF MINOR DEFORMATION

TOTAL DEPTH OF DEFORMATION

TRUE MICROSTRUCTURE

Fig. 4–6. Disturbed metal layer resulting from grinding and polishing of specimens. (L. E. Samuels)

abrasive also produces another deformed layer, this time of lesser depth. The process is continued with finer and finer grinding and polishing abrasives. The final layer is removed chemically by etching with an acid or some suitable agent.

During fine grinding the specimen is held so that the new finer scratches are introduced at approximately right angles to those resulting from its previous rough grinding operation. The purpose for this step is to make it easy to recognize the point at which the coarse scratches have been replaced by a series of new finer ones. Fine grinding is usually done on abrasive belts with grit size between 320 and 600. Following each grinding stage, the technician should wash his hands and the specimen

with soap and running water to prevent carrying coarse abrasive particles to the next specimen preparation. One large abrasive grain can ruin the surface of a finely ground and polished specimen. Cleanliness in metallurgical specimen preparation is essential and cannot be overemphasized. Fine grinding is often done using a water coolant and silicon carbide abrasive paper mounted on a rotating disc. Fig. 4–7. The three wheel unit shown may be equipped with a fine grinding wheel, a rough polishing wheel, and a final polishing wheel. Wheel speed for fine grinding is usually between 550-1150 rpm. A small stream of water directed at the center of the rotating wheel washes away abraded particles and insures cool grinding.

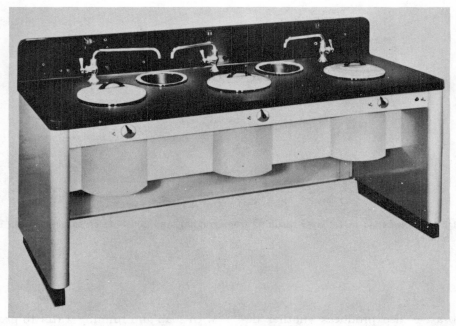

Fig. 4–7. Three wheel metallographic specimen preparation table. (Buehler Ltd.)

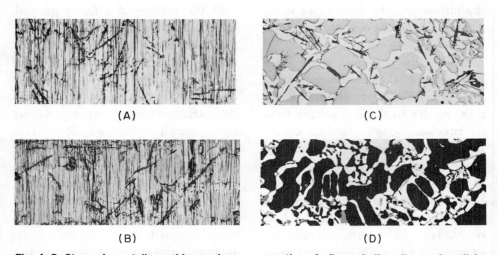

(A)

(B)

(C)

(D)

Fig. 4–8. Stages in metallographic specimen preparation. A, fine grinding. B, rough polishing. C, final polishing (unetched). D, polished and etched (etchant 10% $Fe(NO_3)_3$, 9 H_2O). 200X. (Buehler Ltd.)

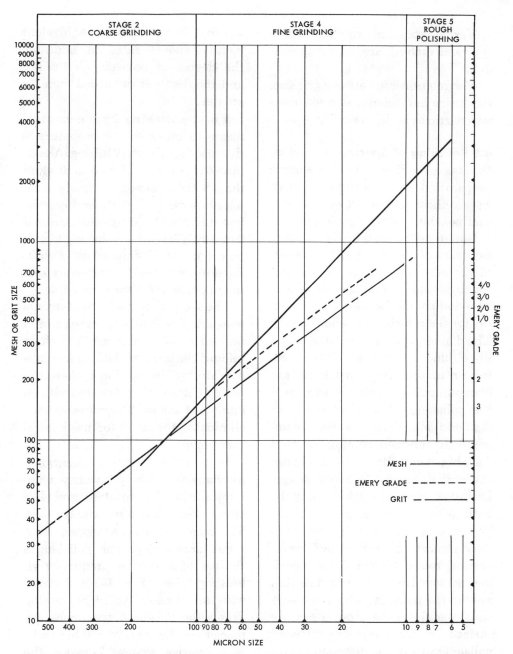

Fig. 4–9. Relationship between grit size, mesh size, and micron size for abrasive particles. For conversion to micron size locate grit size on ordinate, transpose this point to corrected graph line and locate this position on abscissa. (Buehler Ltd.)

103

Various stages of grinding, polishing, and etching are shown in Fig. 4–8.

The relationship between grit size, mesh size and micron size for abrasive particles is shown in Fig. 4–9.

4.5 Polishing of Specimens. Of all the stages required in quality sample preparation, perhaps the most critical are those concerned with the actual polishing of the specimen.

Rough polishing is best performed by means of diamond paste abrasives in the 4-10 micron size ranges. For rough polishing, a napless cloth such as nylon should be used as a covering for the rotating polishing wheels. The diamond paste is generally used in conjunction with an oil base extender to uniformly distribute the diamond particles over the surface of the polishing cloth. Polishing with diamond abrasives has been compared to micromachining in that the abrasive particles cut the metal rather than cause it to flow or smear. Diamond abrasives, which may be used with heavy hand or loading pressure, are cool cutting, and produce a minimum of disturbed metal. During rough polishing the specimen is moved in a clockwise direction around the polishing wheel to insure equal metal removal from the entire surface by not allowing prolonged polishing in any one direction. Abrasives other than diamond may be used but require a much larger grain

size in order to achieve an equivalent metal removal rate. In addition, the degree of polishing is inferior and the depth of disturbed metal is greater.

Final polishing by mechanical means follows a similar procedure to that used for rough polishing. Aluminum oxide, often referred to as alundum, is the most popular final polishing abrasive for ferrous and copper-based materials. Magnesium oxide is frequently used for polishing aluminum and magnesium alloys. Polishing abrasives used less often are diamond paste and chromium oxide.

The abrasive particles used in final polishing are generally carried on a napped or short pile cloth such as billiard cloth, microcloth, or kitten's-ear. Most polishing cloths can be obtained cut to size and coated with an adhesive backing. The adhesive back eliminates the need for mechanical clamping.

In addition to the conventional mechanical polishing, two new techniques, vibratory polishing and electropolishing, are also widely used. The most recently developed technique for metallographic polishing is the use of a variable frequency vibrating table (Fig. 4–10) covered with an abrasive polishing media. When the weighted specimens are placed on the vibrating table there is a relative motion between the abrasive surface and the specimen with a consequent polishing action.

Fig. 4–10. Vibrating metallographic polisher. (Buehler Ltd.)

Fig. 4–11. Electropolishing unit showing polishing cell and power supply. (Carl Zeiss Inc.)

Vibratory polishing is popular because of the extremely high quality of the scratch-free polished surface, the large number of specimens which can be prepared simultaneously, and the possibilities of releasing trained technicians from tedious polishing to devote their time to other matters.

Electropolishing is another process that has been gaining increasing acceptance as a means of metallographic specimen preparation. The increased use of electro-polishing may be traced to a number of distinct advantages that it has over mechanical polishing. Advantages of electro-polishing are (1) rapidity with which specimens may be polished, (2) elimination of cold-worked surface, (3) flatness of polished area, (4) application to a wide variety of materials and (5) polishing and etching can often be accomplished in one operation.

Essentially electropolishing is a reverse electroplating process in which minute projections and irregularities on the surface of a specimen are removed by electrolytic action. Electropolishing equipment shown in Fig. 4–11 consists of a polishing cell, which contains a circulating pump and the electrolyte solution, and a filtered DC power source. In operation the specimen is placed upside down on top of the polishing cell and is connected as the anode. Electrolyte is pumped up the center column of the cell.

Electric current density is higher at the projections on the specimen, and in a short time, the irregularities disappear and a smoothing and brightening action takes place. A specimen which has been ground by mechanical means is shown in Fig. 4–12A. The same specimen after electropolishing is shown in Fig. 4–12

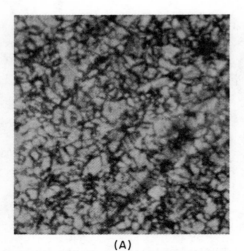

(A)

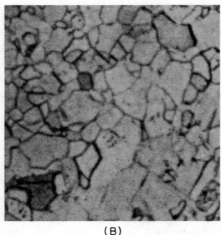

(B)

Fig. 4–12. A, deformed layer on zinc alloy produced by mechanical grinding. B, same specimen after electropolishing for 3 minutes shows no trace of deformed layer. (Carl Zeiss Inc.)

B. Evidence of deformation during mechanical polishing is shown by the recrystallized structure which was later removed by electropolishing.

There are a number of variables in electrolytic polishing including current density, voltage, time, temperature, and electrolyte which require actual laboratory tests to determine which performs best for the particular metal to be electropolished. It should be pointed out electropolishing is not a universal process suitable for all samples, but rather it is a superior alternative to mechanical polishing in many instances.

4.6 Etching of Specimens. Polished metal specimens usually show no structural characteristics. Etching of the metal surface is done to make visible the crystalline structure of the metal and to produce optical contrast between the various constituents. Etching is done by immersing the cleaned and polished specimen in a suitable etching solution such as those described in Table 4–1.

Etchants are composed of organic or inorganic acids, alkalies, or other complex substances in some solvent such as water, alcohol, glycerine, or glycol. These etching reagents are powerful and must be handled with care. Since each reagent has been developed for a specific purpose it must be chosen carefully if it is to reveal the structure that is desired. For example, picral is not a general purpose etchant but rather is designed to distinguish ferrite and iron carbide by preferential darkening of the carbide phase. Nital, on the other hand, is not suited for the above, but

Table 4–1. Selected etching reagents for microscopic examination of metals.

METALS	ETCHANT	COMPOSITION	REMARKS
IRON AND STEEL	NO. 1, NITAL	1 TO 5% NITRIC ACID 95 TO 99% METHYL ALCOHOL	CARBON STEELS – DARKENS PEARLITE, REVEALS FERRITE BOUNDARIES; GENERAL USE FOR HIGH SPEED STEELS; TIME: 5 TO 60 SECONDS.
	NO. 2, PICRAL	4 g. PICRIC ACID 100 ml. METHYL ALCOHOL	CARBON AND LOW-ALLOY STEELS HEAT TREATED OR NOT. TIME: 5 TO 120 SECONDS.
	NO. 3, FERRIC CHLORIDE AND HYDROCHLORIC ACID	5 g. $FeCl_3$ 50 ml, HCl 100 ml, H_2O	REVEALS STRUCTURES OF AUSTENITE NICKEL AND STAINLESS STEELS.
	NO. 4, HEAT TREATING	HEAT SPECIMEN ON HOTPLATE, FACE UP, 400 TO 700 DEG. F	PEARLITE FIRST TO PASS THROUGH A GIVEN COLOR FOLLOWED BY FERRITE; CEMENTITE LESS AFFECTED. ESP. USEFUL FOR CAST IRON. TIME: 10 TO 60 MIN.
COPPER AND ITS ALLOYS	NO. 5, AMMONIUM HYDROXIDE-HYDRO-GEN PEROXIDE. (MAKE FRESH DAILY)	5 PARTS NH, OH (sp.gr. 0.88) 5 PARTS H_2O 2-5 PARTS H_2O_2:(3%)	GENERAL ETCHANT FOR COPPER AND MANY OF ITS ALLOYS. TIME: 1 MIN.
	NO. 6, CHROMIC ACID	SATURATED AQUEOUS SOLUTION (CrO_3)	COPPER, BRASS, BRONZE, AND NICKEL SILVER.
	NO. 7, FERRIC CHLORIDE	5 g. $FeCl_3$ 96 ml. ETHYL ALCOHOL 2 ml. HCl	COPPER, ALUMINUM, MAGNESIUM NICKEL, AND ZINC ALLOYS. TIME: 1 SEC TO SEVERAL MIN.
ALUMINUM AND ITS ALLOYS	NO. 8, HYDRO-FLUORIC ACID	0.5 ml. HF (CONC.) 99.5 ml. H_2O	GENERAL ETCHANT; APPLY BY SWABBING. TIME: 15 SEC.
	NO. 9, SODIUM HYDROXIDE	10 g. NaOH 90 ml. H_2O	GENERAL ETCHANT; CAN BE USED FOR BOTH MICRO- AND MACRO-ETCHING. TIME: 5 SEC.
MAGNESIUM AND ITS ALLOYS	NO. 10, GLYCOL	75 ml. ETHYLENE GLYCOL 24 ml. H_2O 1 ml. HNO_3(CONC.)	FOR ALMOST ALL MAGNESIUM ALLOYS. TIME: 3-60 SEC.
NICKEL AND ITS ALLOYS	NO. 11, FLAT SOL'N. (MAKE FRESH DAILY)	50 ml. HNO_3(CONC.) 50 ml. GLACIAL ACETIC ACID	NICKEL, MONEL, AND OTHER NICKEL–COPPER ALLOYS. 5-20 SEC.
	NO. 12, AQUA REGIA	5 ml. HNO_3(CONC.) 25 ml. HCl (CONC.) 30 ml. H_2O	INCONEL
TIN, LEAD, AND ZINC ALLOYS	REFER TO "PRINCIPES OF METALLOGRAPHIC LABORATORY PRACTICE" BY GEORGE L. KEHL, McGRAW-HILL, 1949, PAGES 421-432, AND ASTM E3-58T ON "METHODS OF PREPARING METALLOGRAPHIC SPECIMENS."		

is used principally as a general etchant for steel and for delineating grain boundaries in ferrite.

The specimen is usually held by tongs and immersed with the polished face down into a small petri

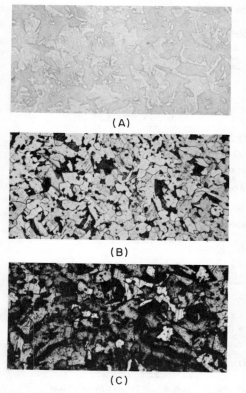

(A)

(B)

(C)

Fig. 4–13. Comparison of properly etched and improperly etched specimens. A, underetched. B, properly etched. C, overetched. 100X.

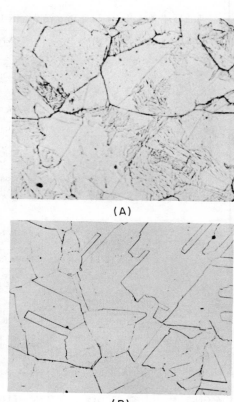

(A)

(B)

Fig. 4–14. A, pseudo-structure on stainless steel due to deformed surface metal. B, true structure revealed by alternate polishing and etching. 500X. (U.S. Steel Corp.)

dish partly filled with the reagent. The specimen may also be swabbed with cotton which has been saturated with the etchant. The progress of etching can be observed visually but should also be timed. The proper etching time must be found experimentally and may vary from a few seconds to a minute or more. A specimen which has been properly etched is shown in Fig. 4–13 and for comparison, under-etched and over-etched specimens.

If the specimen is not sufficiently etched after the first immersion the process may be repeated. If the specimen is overetched it must be repolished prior to re-etching. Immediately following the etching the specimen should be washed in warm water to stop the etching, then immersed in alcohol and finally dried in a blast of warm air. Rapid drying is important to prevent water spots.

One of the purposes of chemical etching is to remove deformed metal which may have developed during polishing. Fig. 4–14A shows a pseudo-structure in a stainless steel specimen due to the presence of disturbed metal. Fig. 4–14B shows the true structure of the same specimen after it has been alternately polished and etched several times.

Etching reveals structural characteristics by preferential etching.

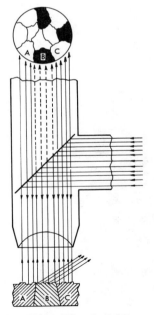

Fig. 4–15. Schematic of etching on grain boundaries and various crystal orientations. (By permission: *Principles of Metallographic Laboratory Practice*, McGraw-Hill Book Co.)

That means that some areas such as grain boundaries are more highly stressed than other areas and are more subject to chemical etching. Also, the etching rate is different for different crystallographic planes which results in varying degrees of reflected light with consequent shading of the various grains as illustrated by Fig. 4–15.

4.7 Microscopic Examination. The primary purpose of microscopic examination is to reveal details of metal structures which are too small to be seen with the unaided eye. Metallurgical microscopes such as those shown in Fig. 4–16A and B can be used to determine grain size, inclusions, previous heat treatments, possible causes for failures, deformation, and intergranular corrosion. In short, microscopic examination can reveal a great deal about the past history of a metal specimen and how it will act in service.

The metallurgical microscope consists essentially of an optical system and an illumination system. The optical system, Fig. 4–17, includes the eyepiece lens, relay system, and the objective lens.

The illuminating system, Fig. 4-17, consists of a high intensity light source, condenser lenses, an aperture diaphragm, a darkfield stop, and a plane glass reflector. The illuminating system may also include colored or polarizing filters. Green

(A)

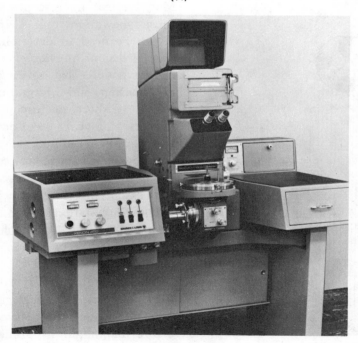

(B)

Fig. 4–16. A, table type metallurgical microscope. B, bench type research metallograph. (Bausch & Lomb)

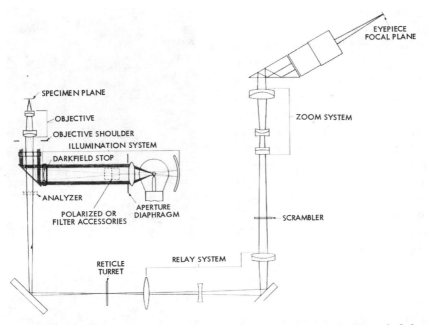

SPECIMEN PLANE

OBJECTIVE

OBJECTIVE SHOULDER

ILLUMINATION SYSTEM

DARKFIELD STOP

ANALYZER

POLARIZED OR
FILTER ACCESSORIES

APERTURE
DIAPHRAGM

RETICLE
TURRET

RELAY SYSTEM

SCRAMBLER

ZOOM SYSTEM

EYEPIECE
FOCAL PLANE

Fig. 4–17. Optical system for the metallograph shown in Fig. 4–16A. (Bausch & Lomb)

filters are often used to improve detail and polarizing filters to reduce surface glare and to improve grain boundary definition. The aperture diaphragm is used to control the amount of light entering the objective lens, and the field diaphragm (not shown) is used to minimize internal glare within the microscope and to enhance contrast of various constituents in the specimen. The quality of objective lenses used in the microscope is perhaps the most critical in influencing image quality. Objective lenses are usually classed as achromatic, fluorite, or apochro-

matic. Apochromatic objectives represent the finest lens and are corrected for optical errors. Objective lenses should be properly matched to the eyepiece lens to obtain optimum quality images.

Some metallurgical microscopes are equipped with revolving nosepieces which may hold up to four objective lenses. This setup makes it possible to change rapidly from one magnification to another without the inconvenience of unscrewing and changing lenses.

Magnifying power of both objective and eyepiece lenses is engraved on the lens mount. Total magnification of a microscope may be determined by finding the product of the eyepiece and objective lens magnification. For example: A 5X power eyepiece used in conjunction with a 40X objective lens gives a magnification of 200 diameters. Maximum magnification of optical microscopes is limited to about 2000 diameters. Higher magnification does not reveal greater detail and is called "empty magnification."

Contrary to popular opinion, high magnifying power does not necessarily reveal fine detail. It is the numerical aperture, (N.A.) or light-gathering ability of a lens which largely determines the ability to resolve fine detail. In Fig. 4–18 are shown two microphotographs of a steel specimen, one taken with an objective lens having a numerical aperture of 0.25 and one with a numerical aperture of 0.95. The detail is very much better when the lens with the higher numerical aperture is employed, especially at the particular magnification shown. At lower magnification the difference in resolution would be less dramatic.

Other factors which influence resolving power of an objective lens are (1) the index of refraction of the medium separating the specimen and objective lens, and (2) the wavelength of the light used to illuminate the specimen. The index of refraction between air and glass is 1.0, and

(A)

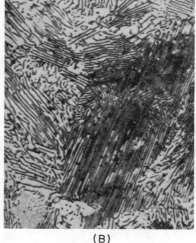

(B)

Fig. 4–18. Influence of numerical aperature on resolving power of an objective lens. A, N.A. 0.25. 600X. B, N.A. 0.95. 600X. (By permission: *Principles of Metallographic Laboratory Practice,* McGraw-Hill Book Co.)

between glass and cedar oil is 1.5. Thus, some high powered objective lenses (oil immersion type) have a drop of cedar oil placed between the lens and specimen to increase the resolving power of the lens. Illumination of the specimen with light of short wavelength (green or blue) also increases resolving power of the lens. It has been found by experience that the maximum total magnification necessary to observe details by a particular objective should not exceed 1000 times the numerical aperture of the lens. For example, a lens whose numerical aperture is 0.65 should not normally be used at greater than 650 magnification if the photomicrograph is to appear reasonably sharp and distinct when viewed from a distance of about 10 inches.

In all phases of microscopy cleanliness is essential, especially when it comes to optical parts. Lenses should be maintained free from fingerprints, dust, and oil films. High quality lenses are usually coated with an extremely thin film of magnesium fluoride to reduce light loss from reflection. Acid from a fingerprint can etch through the film and spoil the lens. Oil used with an oil immersion lens should be removed immediately after use and not allowed to dry on the lens. Oil and fingerprints are best removed by wiping the marred surface with lens tissue moistened in xylol (never alcohol or organic solvents). A little extra care with metallurgical equipment pays good dividends in terms of high quality work and long life for the equipment.

One of the most important contributions to microscopy has been the development of electron microscopes. The electron microscope. Fig. 4–19, represents the latest achievement in the search for systems capable of greater resolving power than is obtainable with optical microscopes. An electron microscope is capable of direct magnifications of 10,000 to 30,000 diameters, and with auxiliary optical equipment, may be extended to as high as 500,000 diameters.

The resolution attainable with the electron microscope of the order of .002 microns (0.000000078″) as compared to 0.2 microns (0.0000078″) for the optical microscope, is due to the very short wave length of the electron beam used to "illuminate" the specimen. The electron microscope is capable of great depth of focus, approximately 30 times that of oil immersion objectives. An electron photograph of Ferroxdure taken at 20,000 magnification is shown in Fig. 4–20.

The geometry for formation of an image by an electron microscope is similar to that of the optical microscope as shown in Fig. 4–21. In the electron microscope the electrons are focused onto the specimen to be examined by means of magnetic field coils which perform the same function as lenses in the optical microscope system.

After passing through the speci-

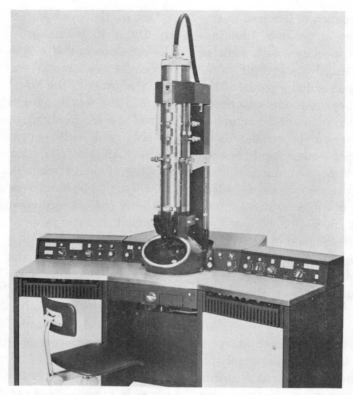

Fig. 4–19. Norelco EM300 Electron Microscope with resolution to 5 angstrom units and magnification to 500,000X. (Phillips Electronic Instruments)

Fig. 4–20. Structural appearance of Ferrox-dure (replica) observed under the electron microscope. 20,000X. (Phillips Electronic Instruments)

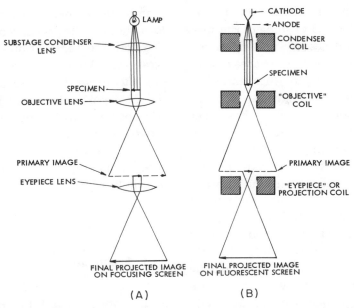

Fig. 4–21. Similarities of optical and electron microscope systems. A, light microscope. B, electron microscope. (By permission: *Principles of Metallographic Laboratory Practice*, McGraw-Hill Book Co.)

men or a thin, transparent replica of the specimen, the electron beam is focused by a second magnetic coil "objective" to form a primary image at a position directly above the projection "eyepiece" coil. The primary image is at a magnification of 50 to 100 diameters, depending upon the position of the specimen in the system and the field strength of the coil. That portion of the primary image which passes through the projection coil is re-magnified 20 to 300 diameters and projected onto a fluorescent screen. The image formed on the fluorescent screen can be visually ob-

served or optically magnified to a still greater degree.

Unfortunately, the standard electron microscope cannot be used directly with metallographic specimens. The usual procedure is to make a thin, transparent replica of the specimen which carefully defines the contour of the etched surface of a metallographic specimen. The newly developed scanning type electron microscope overcomes this problem as it can be used directly on the surface of the specimen. The magnification of the scanning type electron microscope is limited to about

50,000X but it has an exceptional depth of focus. Polishing of the specimen is unnecessary making it possible to examine the surfaces of fractured specimens, coatings, and other three-dimensional surfaces.

A number of techniques have been developed for preparation of a replica. One technique consists of softening a .005″ thick polystrene sheet with a suitable solvent and impressing it onto the specimen. After hardening the sheet is carefully stripped off and coated with a thin layer of carbon or silica by means of vapor deposition technique. The plastic sheet is then dissolved and the replica carefully placed on a copper wire grid for observation by means of the electron microscope. Initial specimen preparation may follow the same procedure as that for examination by optical means including mounting, grinding, polishing, and etching. The undisturbed, scratch-free surface from electrolytic polishing is preferred over conventional mechanically polishing surfaces when preparing specimens for the high magnification under the electron microscope.

4.8 Macroscopic Examination of Metals.

Macroscopic examination of metals involves visual or low-power (below 10X magnification) observation of metallurgical specimens. Macro examination gives an over-all view of the specimen such as the general size and distribution of non-metallic inclusions, uniformity of structure, the presence of fabricating defects, segregation, and large grain size. Microscopic examination, on the other hand, reveals detail about a particular small portion of the specimen which may or may not be representative of the entire piece.

Macro examination is relatively simple to carry out and consists essentially of:

1. Selection of the sample,
2. Surface preparation,
3. Etching.
4. Examination.

Selection of a specimen for macro examination should be made with proper consideration of the ultimate aim of the test.

Generally, thin slabs and discs of the specimen may be obtained by shearing, sawing, or flame cutting. It is important that machining marks be removed by filing or grinding to give a relatively smooth finish prior to etching. Also that the part to be examined be kept cool enough during cutting and grinding to prevent structural changes, which after etching may lead to misinterpretation of the macrostructure.

Etching reagents suitable for macroetching of common metals and alloys are given in Table 4-2.

Temperature affects the rate at which etching takes place. Higher temperatures accelerate etching rate. The precise influence of etching tem-

Table 4–2. Etching reagents for macroexamination of selected metals.

METALS	ETCHANT	COMPOSITION	REMARKS
IRON AND STEEL	NO. 1, HYDRO-CHLORIC ACID	50 ml. HCl (CONC.) 50 ml. H_2O	SHOWS SEGREGATION, POROSITY, CRACKS, DEPTH OF HARDENED ZONE, SOFT SPOTS, ETC. USED AT 160–180°F., 1–60 MIN.
	NO. 2, NITAL	5 ml. HNO_3 95 ml. METHLY ALCOHOL	SHOWS CLEANNESS, DEPTH OF HARDENING, CARBURIZED OR DECARBURIZED SURFACES AND STRUCTURE OF WELDS. ETCH 5 MIN. FOLLOWED BY 1 SEC IN HCl (10%).
COPPER AND ITS ALLOYS	NO. 3, NITRIC ACID	10 TO 35 PER CENT NITRIC ACID	SHOWS GRAIN SIZE, POROSITY, FLOW LINES AND INCLUSIONS. USED AT ROOM TEMPERATURE 5–20 MINUTES.
	NO. 4, KELLERS CONCENTRATED ETCH	10 ml. HF (CONC.) 15 ml. HCl (CONC.) 25 ml. HNO_3 (CONC.) 50 ml. H_2O	EXCELLENT ETCHANT FOR COPPER-BEARING ALLOYS. STORE CONCENTRATED SOLUTION IN PARAFFIN WAX BOTTLE.
ALUMINUM AND ITS ALLOYS	NO. 5, SODIUM HYDROXIDE	10 g. NaOH 90 ml. H_2O	CAN BE USED FOR BOTH MACRO- AND MICROETCHING. IMMERSE FOR 5 SEC. AT 160°F. AND RINSE IN COLD WATER.
	NO. 6, FLICK'S ETCH	10 ml. HF 15 ml. HCl 90 ml. H_2O	MICROSCOPIC EXAMINATION OF ALUMINUM ALLOYS. IMMERSE 10–20 SEC., WASH IN WARM WATER FOLLOWED BY DIP IN CONCENTRATED HNO_3.
MAGNESIUM AND ITS ALLOYS	NO. 7, ACETIC ACID	10 PER CENT AQUEOUS SOLUTION	MACROETCHING. SWAB WITH COTTON FOR 1/2 TO 2 MINUTES.
	NO. 8, TARTARIC ACID	10 PER CENT AQUEOUS SOLUTION	SHOWS FLOW LINES IN FORGINGS; REVEALS GRAIN SIZE IN CAST ALLOYS. IMMERSE SPECIMEN WITH POLISHED SURFACE UP.
NICKEL AND ITS ALLOYS	NO. 9, CONCENTRATED NITRIC ACID	USE CONCENTRATED, CLEAR WHITE ACID TO AVOID STAINING	SHOWS STRUCTURAL DETAIL, CRACKS, POROSITY, FLOW LINES, SULFUR EMBRITTLEMENT. IMMERSE SPECIMEN 3 TO 5 MINUTES FOR STRUCTURE; LONGER ETCHING REQUIRED FOR POROSITY OR FLOW LINES.
	NO. 10, AQUA REGIA	25 ml. HNO_3 (CONC.) 75 ml. HCl (CONC)	GENERAL USE FOR NICKEL-CHROMIUM AND NICKEL-IRON ALLOYS. IMMERSE 3 TO 5 MINUTES. PROLONGED ETCHING MAY PIT SPECIMEN.
TIN, LEAD, AND ZINC ALLOYS	NO. 11 AMMONIUM POLYSULFIDE	CONCENTRATED SOLUTION	TIN-RICH BABBITT METALS. IMMERSE 20 TO 30 MIN. AT ROOM TEMPERATURE.
	NO. 12 CONC. HYDROCHLORIC ACID	CONCENTRATED SOLUTION	PRODUCES GOOD GRAIN CONTRAST ON ZINC. IMMERSE AT ROOM TEMPERATURE
	NO. 13. NITRIC ACID	50 ml. HNO_3 50 ml. H_2O	FOR LEAD AND ITS ALLOYS. SHOWS WELDS, LAMINATIONS, ETC. ETCH IN BOILING SOLUTION 5 TO 10 MIN.

perature varies with alloy composition, heat treatment, and other factors and should be kept constant for each alloy if comparative results are wanted.

Etching time is an especially important variable in macro etching. Prolonged etching may darken the etched surface and obscure or obliterate certain parts of the structure.

Underetching may not completely develop the macrostructure and reveal sufficient details to permit accurate interpretation of the test.

Immediately following etching, the specimen should be removed from the reagent bath with tongs and thoroughly rinsed in warm running water. During rinsing the surface should be scrubbed with a brush having soft bristles to remove deposits formed during etching. The washed specimen may be dried by rinsing in alcohol followed by a blast of warm air from an electrically heated dryer.

Macroetched surfaces may be preserved from oxidation by spraying them with a thin coat of clear lacquer from an aerosol can.

Macroscopic examination can be most useful in revealing dendritic and columnar grains in castings, segregation of certain constituents, pipe in castings, internal rupture and flow lines of forgings, internal cracks, porosity, surface seams, grinding and quenching cracks, hardening depth, carburized and decarburized surfaces, and size of heat affected zone in weldments. When interpreting a macroetched specimen it is impor-

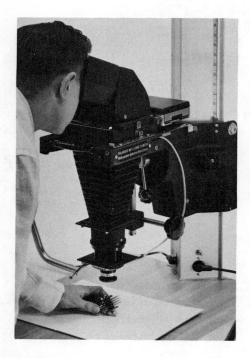

Fig. 4–22. Polaroid MP-3 multipurpose industrial camera used for macrophotography and with a microscope for photomicrography.

Fig. 4–23. Flow lines of an axle forging revealed by macroetching.

tant to learn as much about its past history as possible, including its method of manufacture, processing methods, thermal treatments, and service. Frequently it is desirable to photographically record macroetched specimens. A very easy-to-use multipurpose industrial camera is shown in Fig. 4–22. The camera can be fitted with accessory lenses to permit enlargements of up to 10X. Polaroid, roll film, or regular cut film backs are available which may be used with the camera.

Photographs of macroetched specimens in Fig. 4–23 through 4–27 show various types of structures and flaws revealed by macroexamination.

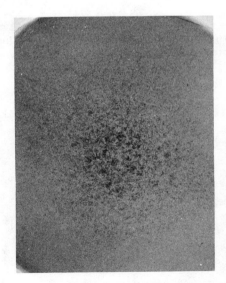

Fig. 4–24. Internal cracks or flakes in a 6″ square billet of tool steel. (Bethlehem Steel Co.)

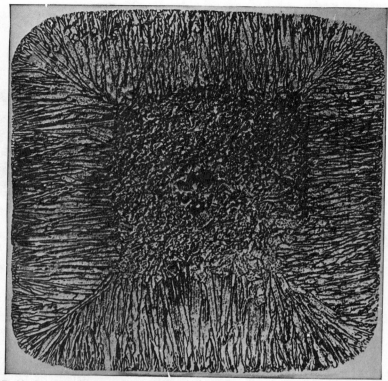

Fig. 4–25. Horizontal fracture of a steel ingot showing columnar grains. Cleavage planes are shown at the corners of the ingot while pipe (shrinkage) is shown in the center.

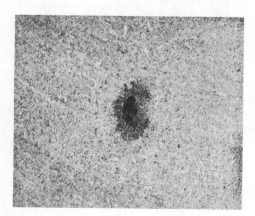

Fig. 4–26. Sectioned billet showing center segregation and pipe in AISI type 0 1 tool steel. (Universal Cyclops Specialty Steel Div., Cyclops Corp.)

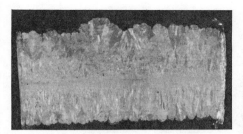

Fig. 4–27. Macrophoto showing columnar grains in electrolytically refined copper. Note starting sheet of pure copper. 3X. Etched 50% HNO_3.

Fig. 4–28. Large iron pyrite crystals found in nature. Crystalline structure is clearly shown. (Robert Rollins)

4.9 X-Ray Diffraction. Often large crystals are found in nature which clearly show hexagonal, cubic, and other crystalline structures, (Fig. 4–28). With metals, however, it is difficult to obtain large single crystals and other methods such as x-ray diffraction are often used for their examination. Because crystals are symmetrical arrays of atoms, they are able to act as three dimensional diffraction gratings. Low voltage x-rays have the proper wave length to be diffracted by crystals.

The resulting diffraction pattern, when properly interpreted, can reveal the inter-atomic spacing between atoms, the type of crystal structure, cubic, hexagonal, the degree of cold work to which a metal has been subjected, phase transformations, orientation of the crystals, directional properties, recrystallization temperatures, and other impor-

tant characteristics on an atomic scale.

The essential feature of x-ray diffraction is that when regularly spaced atoms in a metal crystal are irradiated by an x-ray beam, the electrons surrounding each atom are caused to vibrate with the same frequency as the x-ray beam. The vibrating electrons act as sources for a new set of electromagnetic waves. The reflected waves are broadcast or "scattered" in all directions. The scattered waves combine in certain directions and reinforce each other and may be treated as a single set of radiating waves originating from a point. In other directions waves may annul or cancel one another.

The geometry of the crystal and

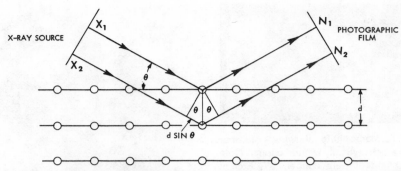

Fig. 4–29. Bragg's Law of x-ray diffraction. Reflected x-rays from parallel layers of atoms produce spots on photographic film.

initial x-ray beam which determines conditions for reinforcement or annulment of diffracted beams is shown in Fig. 4–29. The spacing between parallel crystallographic planes is represented by d. When the radiation X_1 and X_2 with a wavelength λ, strikes the planes at an angle θ, reflections result. When both diffracted rays are in phase, they reinforce one another and the film will be darkened: When out of phase, they cancel one another, and no darkening will occur. Reinforcing, and darkening of the film will occur when the following equation, known as Bragg's Law, is satisfied:

$$n\lambda = 2d \sin \theta$$

in which n = an integer number, 1, 2, 3, etc.

λ = wavelength of x-ray beam, angstroms

d = spacing between crystal planes, angstroms

θ = angle of incidence of the radiation, degrees

To ensure that Bragg's Law is satisfied when examining single metal crystals it is necessary to provide a range of values of either λ or θ. The various ways of doing this form the bases of the standard methods of diffraction used in crystal analysis.

In the *Laue Method* of crystal analysis the so called "white radiation" is composed of wavelengths from approximately 0.2 angstrom to 2.0 angstrom. The camera setup is shown in Fig. 4–30A. The spots on a film from a Laue back-reflection pattern are arranged on hyperbolas. The intersection of the cone of diffracted rays with the film forming the hyperbola is shown in Fig. 4–30B. Only one of the several cones of reflected rays is shown. The back reflection pattern of a face-centered cubic aluminum crystal is shown in Fig. 4–30

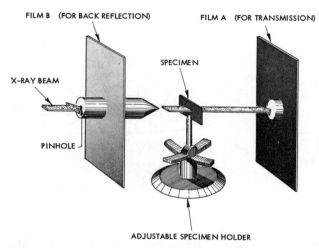

Fig. 4–30A. Laue method of crystal analysis camera setup.

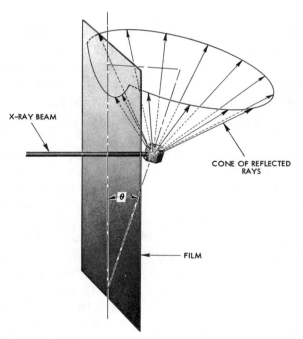

Fig. 4–30B. Laue method of crystal analysis, intersection of cone of diffracted rays with film back reflection.

123

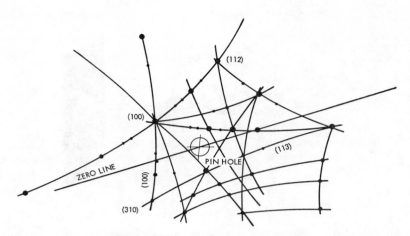

Fig. 4–30C. Laue method of crystal analysis, pattern of F.C.C. aluminum crystal with principal zones and spots identified.

C. The numbers shown in the photograph are indexes of the poles of the crystal planes.

Although the Laue pattern may appear to be difficult to interpret special stereographic projection charts have been developed to simplify the task. The Laue method is widely used for determining crystal orientation in the commercial production of piezoelectric crystals and solid-state components. Relative orientation of adjacent grains or imperfect regions within a crystal may also be studied.

The *powder method* of crystal analysis employs a powdered specimen of a polycrystalline metal contained in a 0.2 mm glass tube. The specimen is placed in a cylindrical camera with the film surrounding the specimen. The powdered metal specimen is subjected to a monochromatic radiation. The powdered metal provides crystals of random orientation so that many angles of θ are present to satisfy Bragg's Law. A powder diffraction pattern is shown in Fig. 4–31. The geometry of a powder camera is shown in Fig. 4–32.

The powder method of x-ray diffraction is frequently used to determine the size, shape, and orientation of the unit cell of the powdered specimen. It may also be used to determine positions of the atoms of the unit cell, the lattice parameter, and the number of atoms per unit cell. A spectrometer, Fig. 4–33, may be used for precisely measuring the intensities of radiation from single crystals or powdered specimens. The x-ray beam enters the chamber

124

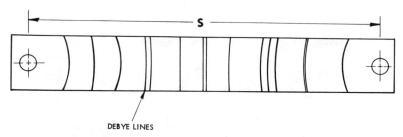

Fig. 4–31. Powder diffraction pattern showing distinctive Debye rings. The positions may be interpreted to give size, shape, and orientation of the cell. Atom arrangement within the cell may be determined by analysis of the intensity of the image.

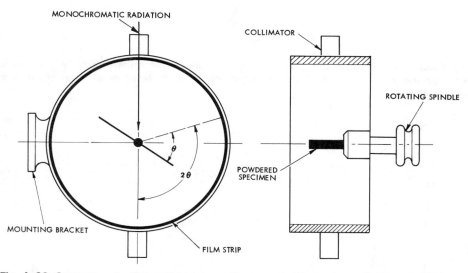

Fig. 4–32. Geometry of a Debye-Scherer powder camera. The camera diameter is 57.3 mm so 1 mm on the film corresponds to 1 degree. This helps in rapidly indexing the Debye lines.

through an adjustable slot. The reflected beam enters the ionization chamber where intensity is determined. Output may be plotted graphically as shown in Fig. 4–34.

The spectrometer gives more precise measurements of diffraction patterns than photographic films and has therefore become a very important technique for x-ray studies.

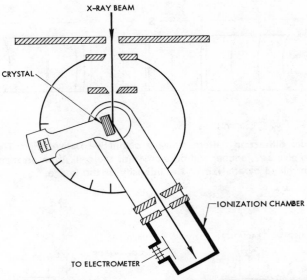

Fig. 4–33. X-ray spectrometer. This instrument accurately determines intensity of reflected x-ray beams when solving complex crystal structures or determining electron distribution within atoms.

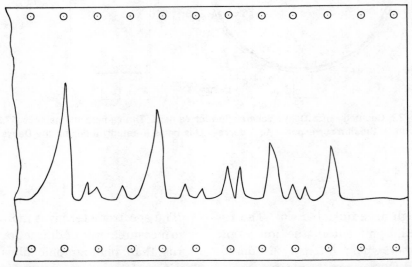

Fig. 4–34. Output from an x-ray spectrometer. Lines of high intensity show important crystallographic planes.

4.10 Thermal Analysis. As metals are cooled and undergo transformation from the liquid to solid state there is an evolution of heat which causes a temporary stop or arrest in temperature change. Conversely, as metals are heated an arrest in temperature may be observed as they transform from the solid to the liquid state. Changes in state, changes in lattice structure, and phase changes may often be detected by careful temperature measurements or by measurement of changes in volume. Changes during cooling and heating, referred to as thermal analysis, are often used in determining melting and freezing points of pure metals. Information gathered by this same method can be assembled to form a diagram as in Fig. 4-35.

The alloy equilibrium diagram, which will be discussed at length in Chapter 5, may be constructed from cooling curves of a series of metal alloys which have increasing percentages of alloying elements. Time-temperature cooling curves for a series of alloys between cadmium and bismuth are shown in Fig. 4–36A. The equilibrium diagram corresponding to the cadmium-bismuth cooling curves is shown in Fig. 4–36B.

Curves similar to those shown in Fig. 4–35 can be plotted by several methods other than thermal analysis including metallographic and x-ray diffraction methods.

The metallographic method is ideally suited for determining solid state transformation because of the visual picture of the alloy constituents. The method involves heating of small specimens to various tem-

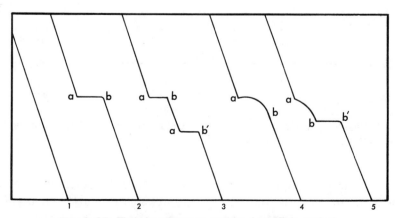

Fig. 4–35. Time-temperature curves (semi-log scales).

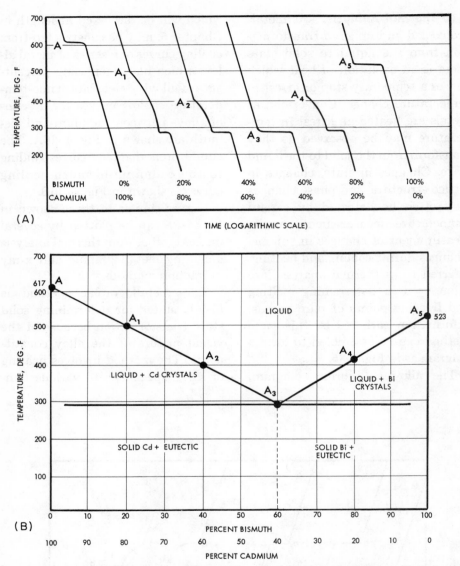

Fig. 4–36. A, time-temperature cooling curve for cadmium-bismuth alloys. B, alloy diagram constructed from the cooling curves.

peratures followed by rapid quenching. The structures of the quenched specimens are subsequently examined metallographically to determine approach to or attainment of equilibrium at the temperature of inter-

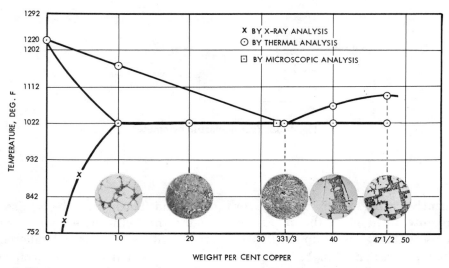

Fig. 4–37. A series of photomicrographs for a portion of the copper-aluminum system with the corresponding alloy diagram.

est. Photomicrographs for the copper-aluminum alloy system are shown in Fig. 4–37 together with its alloy diagram.

X-ray diffraction may also be used for determining various alloy phases but is involved and highly specialized. X-ray diffraction methods are discussed in paragraph 4-9.

4.11 Temperature and its Measurement. In metallurgy the measurement of temperature is of utmost importance. Temperature changes of only a few degrees may cause large changes in heat treatment response and the microstructure of metals. The ability to reproduce desired metallurgical treatment is based on temperature measurement and its control. In metallurgical work, temperature is often measured in degrees Fahrenheit (°F) although it is becoming more and more common to take readings in degrees Celsius (°C) (formerly Centigrade).

For low temperature work the Absolute temperature scale is used. In making a reading on the Absolute it is necessary to add 273° to the Celsius reading (°Kelvin) or 460° to the Fahrenheit reading (°Rankine).

The measurement of temperatures below 950°F has been referred to as *thermometry* while measurement of temperatures above 950° is frequently called *pyrometry*. Low temperature measurements are fre-

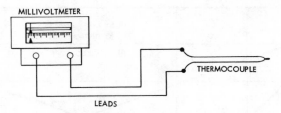

Fig. 4–38. Thermoelectric pyrometer.

quently made with mercury or creosote filled glass thermometers, creosote being used up to 950°F. For measurement of higher temperatures, instruments may be divided into two categories:

1. Thermoelectric and resistance pyrometers,

2. Optical and radiation pyrometers.

The thermoelectric pyrometer shown schematically in Fig. 4–38 consists essentially of two dissimilar wires welded together at one end, called the thermocouple, the opposite ends being connected to a millivoltmeter. As the thermocouple end of the wires is heated an electromotive force (Emf) or voltage is generated, and a current will flow, causing a deflection of the meter. The higher the temperature, the higher the voltage. The scale on the millivoltmeter usually indicates the temperature in degrees. The source of Emf generated by heating dissimilar wires is attributed to two distinct and independent sources, namely the Peltier, and Thompson effects. Essentially, the Peltier effect states that there will be a potential difference at the junction point between any two metallic wires of different composition. The magnitude of potential difference is governed by chemical composition of the wires and upon temperature. The Thompson effect states that a potential difference will exist between the two ends of a homogeneous metallic wire if a temperature gradient exists along the wire. Thus, the measured Emf is a resultant of the two combined effects. Common thermocouple pairs include iron-constantan ($-320°F$ to $1400°F$), chromel-alumel ($-320°F$ to $2200°F$ and platinum-platinum/rhodium ($32°F$ to $2700°F$). In order to prevent oxidation or corrosion of the thermocouple element at high temperatures, it is often contained in protective sheaths.

The electric resistance thermometer is suited for precise temperature measurements within the range of $-320°$ to $1100°F$. It consists essentially of a high-purity, uniform, platinum wire resistance coil enclosed within a suitable protective tube. The resistance thermometer operates on the well known fact that changes in temperature of a metal result in change of electrical resistance. The relationship between temperature and resistance can be determined and accurately measured by means of the Wheatstone Bridge Circuit. It is common to measure temperatures to an accuracy of 0.001° by means of the resistance thermometer.

For measuring temperatures higher than those for which thermocouples or resistance thermometers are adapted, such as steel melting furnace temperatures, pyrometers of the radiation or optical types may be used. The radiation pyrometers shown in Figs. 4–39, 4–40A and B measure the intensity of heat and light radiation from a radiating body. Energy of all wavelengths radiated from the source is focused upon the hot junction of a small thermocouple. The temperature reached by the thermocouple is proportional to the rate at which energy falls upon it; which, in turn, by the Stefan-Boltzmann Law, is proportional to the fourth power of the absolute temperature of the source.

This rise in temperature of the thermocouple generates a current which may be measured or converted to read temperatures directly. In using this instrument it must be remembered that all bodies do not emit energy at the same rate at a

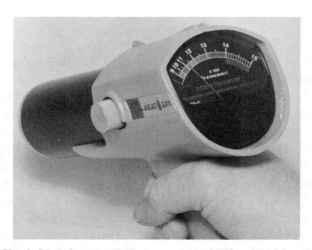

Fig. 4–39. Infrared radiation pyrometer. (William Wahl Corp.)

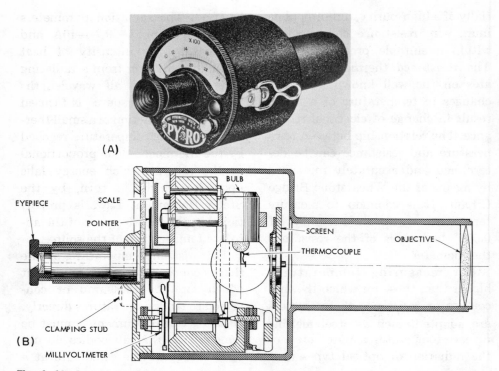

(A)

EYEPIECE

SCALE

POINTER

BULB

SCREEN

THERMOCOUPLE

OBJECTIVE

CLAMPING STUD

(B)

MILLIVOLTMETER

Fig. 4–40. A, pyro total radiation pyrometer. B, cross-section of the pyro total pyrometer showing thermocouple mounted on the optical axis of the instrument.

given temperature. As furnished by the makers, these instruments are calibrated for so-called "black body" conditions, such as exist inside an empty muffle furnace. For other conditions a correction factor called "total emissivity" must be applied. Values for emissivity may be found in various metallurgy or physics handbooks.

The optical pyrometer shown in Figs. 4–41A and B measures only visible (or spectral) radiation emitted by a glowing object. A lamp filament located within the instrument is heated by means of a variable current to match the brightness of the object whose temperature is being measured.

Temperature of the object is determined from the current required to heat the filament to the same brightness as the object. The indicated temperature must be corrected when measuring objects whose radiating properties differ from black

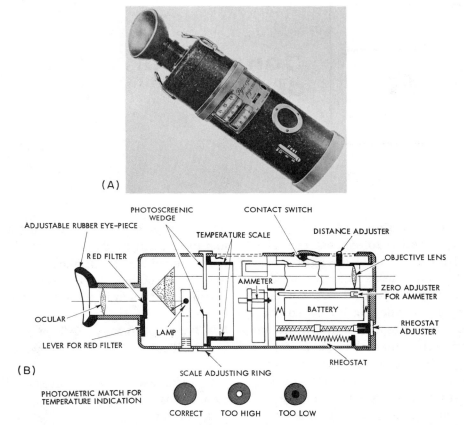

Fig. 4–41. A, pyro optical pyrometer. B, schematic showing optics and disappearing filament. (Pyrometer Instrument Co.)

body conditions. However, non-
black body specimens, when placed
within a glowing furnace and allowed

to reach temperature equilibrium
will closely approximate black body
conditions.

References

Eastman Kodak Co. *Photography Through
the Microscope.* Rochester, N. Y.: 1952.
—. *Photomacrography.* Rochester, N. Y.:
1962.
Hendrickson, R. C. and Iannone, B. N.

*Applications of Metallurgical Micro-
scopy.* Rochester, N.Y.: Bausch &
Lomb, 1960.
Kehl, George L. *The Principles of Metal-
lographic Laboratory Practice,* 3rd ed.
New York: McGraw-Hill, 1949.

Constitution of Alloys

5.1 Introduction. Metals are alloyed together to enhance corrosion resistance, tensile strength, magnetic properties, and a host of other physical, mechanical, or chemical properties. Alloying is now so common that pure metals are the exception rather than the rule as far as engineering materials are concerned. An alloy is a homogeneous solution, mixture, or compound composed of two or more metallic elements which do not separate naturally.

It is difficult to discuss the changes taking place during the cooling of an alloy from the liquid state to normal temperature without the aid of an alloy diagram. The alloy diagram, sometimes referred to as a constitution diagram or phase diagram, is to the metallurgist what the blueprint is to the toolmaker. As discussed in Chapter 4, the alloy diagram may be made up from a number of cooling curves, or from microscopic examination or by x-ray diffraction studies.

The alloy diagram is an aid to the metallurgist in predicting what the normal structure of an alloy will be upon slow cooling and the transformations that occur in the solid state.

5.2 Types of Alloys. Alloys may be classified as one of the following:
1. Solid solutions,
2. Intermediate alloy phases or compounds,
3. Mixtures.

When discussing solutions most students immediately think of liquids. However, in metallurgy solid solutions are frequently encountered. Solid solutions result when certain combinations of metals are melted and allowed to solidify. When the resulting metal is examined under the microscope it is impossible to distinguish crystals of either of the composite metals, and it is thus termed a "solid solution." In most alloy systems, solid solubility occurs by random *substitution* of solute atoms for

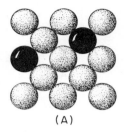

(A)

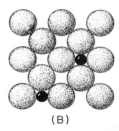

(B)

Fig. 5–1. A, substitutional solid solution alloy system. B, interstitial solid solution alloy system.

solvent atoms as shown in Fig. 5–1A or by solute atoms occupying *interstitial* positions as shown in Fig. 5–1B. The term "solvent" refers to the more abundant element and solute to the less abundant element.

5.3 Substitutional Solid Solutions. Factors, discovered by W. Hume-Rothery of Oxford University, which determine the extent of solid solubil-

ity of one metal in another include (1) crystal-structure factor, (2) relative size factor, (3) valence factor, and (4) chemical affinity factor. In order for two elements to have complete solubility they must have the same type of crystal lattice structure. In addition, a necessary condition for the formation of a substitutional solid solution between two alloyed metals is that their atomic radii must not differ by more than 15 percent of one another or their solubility will be very limited. Solubility also becomes more restricted as the difference in valences of the solute and solvent atoms increase. The affinity of metals for electrons also influences solid solubility. As this affinity between metals increases there is a tendency toward formation of an inter-metallic compound rather than a solid solution. Generally speaking, the farther apart the elements are in the periodic table the greater is their chemical affinity, and the greater the possibility of forming intermetallic compounds. The properties of substitutional solid solutions show gradual and continuous change as the concentration of the solute increases: strength increases, ductility usually decreases, and electrical resistance increases.

5.4 Interstitial Solid Solutions. The second type of solid solution is the interstitial solid solution. Examination of Fig. 5–1B shows that the solute atoms in interstitial alloys must be small in size to fit into the spaces or interstices between the larger solvent atoms in the lattice structure. Studies indicate that extensive interstitial solid solutions occur only if the solute atom has a diameter less than 0.59 (about half) that of the solvent. The most important interstitial atoms are hydrogen, boron, carbon, nitrogen, and oxygen, all of which have atomic radii less than one angstrom.

Atomic size is not the only factor influencing the formation of interstitial solid solutions. Interstitial atoms have been found to dissolve much more readily in transition metals of the periodic chart than in other metals. Commercially important transition metals include iron, nickel, chromium, vanadium, manganese, molybdenum, titanium, and tungsten. Of these, the most well known interstitial solid solution occurs between iron and carbon. Although the percentage of carbon that can be dissolved in B.C.C. iron at room temperature is small (less than .008%), it can be increased considerably (to 1.73%) by heating the iron above its lower critical temperature at which time it changes to the F.C.C. gamma iron. The increase in size of interstitial space between the B.C.C. and F.C.C. structure is shown in Fig. 5–2.

This change in solubility as iron undergoes an allotropic change from

(A)

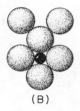

(B)

Fig. 5–2. Interstices of A, body centered cubic, and B, face centered cubic structures.

B.C.C. to F.C.C. structure is responsible for the ability to harden steel by thermal treatment.

In both the substitutional and interstitial solid solutions, distortion of the lattice structure exists in the region surrounding the solute atoms. This lattice distortion interferes with the movement of dislocations on the various slip planes and results in an increase in the strength of the alloy. This is the primary reason for strengthening of metals by alloying.

5.5 Intermediate Alloy Compounds. When compounds form the elements lose their individual identity and properties to a large extent, and the compound exhibits its own characteristic properties. For example, sodium (Na) is a very active metal that will burst into flame when exposed to the air. Chlorine (Cl) is a poisonous gas. Yet when one atom of each combine they form the harmless compound, NaCl, or common table salt, with entirely different properties than either component. Similarly, when an intermediate alloy compound forms in a binary alloy composed of two metals the intermediate alloy behaves much like a third metal, again with different properties. A compound, like a pure metal, has a definite melting point and a time-temperature cooling curve similar to a pure metal.

The three most common intermediate alloy phases or compounds are:

1. Intermetallic (valency) compounds,
2. Interstitial compounds,
3. Electron compounds.

5.5.1. Intermetallic or valency compounds generally form between chemically dissimilar metals and follow the rules of chemical valence. Their properties are essentially non-metallic in that they show poor ductility and poor electrical conductivity. They have very high melting points, a narrow range of compositions, and high hardness. Examples of intermetallic compounds are Mg_2Si, $CaAl_2$, $AlSb$, and Cu_2Se.

5.5.2. Interstitial compounds form

between transition metals and the small atoms of hydrogen, oxygen, carbon, boron, and nitrogen. The word interstitial means between the spaces and the small atoms mentioned fit into interstices of the lattice structure of the metal. These same elements form interstitial solid solutions which were described previously. Solid solutions differ from interstitial compounds in that solid solutions may form with a wide range of possible compositions whereas the interstitial compounds have a very narrow range of compositions that may be assigned a chemical formula. Examples of interstitial compounds are TiC, TiH, Fe_2N, Fe_3C, W_2C, and CrN. They have high hardness, very high melting points, and low ductility.

5.5.3. Electron compounds form near compositions of alloys in which there is a definite ratio of valence electrons to atoms and where size difference between alloying atoms does not exceed 15 percent. Electron compounds with electron-to-atom ratios of 3:2, 21:13, 7:4, 5:2 and 1.3:2.2 are shown in Table 5–1.

The electron-to-atom ratio for Cu_9Al_4 is calculated as follows: Each of the nine copper atoms has one valence electron and each of the four aluminum atoms has three electrons. The total number of valence electrons is 21, and the total number of atoms is 13, thus giving a 21:13 electron-to-atom ratio.

Table 5–1. Examples of electron compounds.

ELECTRON-ATOM-RATIO	3:2	21:13	7:1	5:2[1]	VARIABLE [2] 1.3:2.2
STRUCTURE	B.C.C.	COMPLEX CUBIC	C.P.H.	HEXAGONAL	CUBIC OR COMPLEX HEXAGONAL
EXAMPLES	AgCd	Ag_5Cd_8	$AgCd_3$	NiAs	MgCu
OF	AgZn	Cu_9Al_4	Ag_5A_3	NiS	$MgNi_2$
ELECTRON	Cu_3Al	Cu_3Sn_8	$AuZn_3$	NiSe	MgZn
COMPOUNDS	Cu_3Al	Cu_3Sn_8	$AuZn_3$	NiSe	$MgZn_2$
	AuMg	Au_5Zn_8	Cu_3Si	Ni_3Sb_2	$PbAu_2$
	FeAl	Fe_5Zn_{21}	$FeZn_7$	$NiSn_2$	$TiFe_2$
	Cu_5Sn	Ni_5Zn_{21}	Ag_3Sn	Ni_2Ge	$TiCo_2$

(1) THIS NICKEL ARSENIDE TYPE OF COMPOUND IS THE LINK BETWEEN ELECTRON COMPOUNDS AND CHEMICAL VALENCE COMPOUNDS. THESE COMPOUNDS HAVE RELATIVELY HIGH ELECTRICAL CONDUCTIVITY AND LOW PLASTICITY.

(2) SOMETIMES CALLED LAUE PHASES. THEY INCLUDE COMPOUNDS IN WHICH THE ELECTRON-ATOM-RATIO AND ATOMIC SIZE FACTOR ARE THE CONTROLLING FACTORS. THEY HAVE VERY HIGH HARDNESS AND BRITTLENESS AND RELATIVELY HIGH ELECTRICAL CONDUCTIVITY.

Electron compounds have properties similar to solid solutions including a wide range of compositions, high ductility, and low hardness.

5.6 Coring and Segregation. During the solidification of metal alloys dendritic skeletons similar to that shown schematically in Fig. 5–3 begin to form. The dendritic crystals grow by progressive solidification on primary, secondary, and tertiary axes. Since solute atoms tend to stay in the liquid solution the center of the crystal is richer in the higher melting point metal and the outside is richer in lower melting point metal. This composition gradient on a microscopic scale is called *coring*. Very rapid cooling of metal alloys results in widely differing compositions between the interior and exterior of the crystal. Since the rate of chemical attack varies with composition, polished and etched surfaces can reveal both coring and segregation.

Segregation is similar to coring but occurs on a macro scale. The center portion of an ingot or slowly cooled casting is richer in solute atoms and impurity atoms than the exterior which solidifies first. This type of segregation shown in Fig. 5–4 cannot be removed by annealing treatments alone, but must be accompanied by hot working. Inverse segregation sometimes occurs in copper-tin alloys in which the lower melting point tin is exuded out

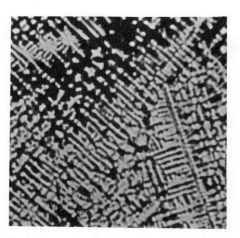

Fig. 5–4. Cast monel metal showing dendritic structure. 25X.

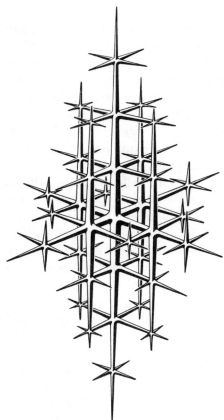

Fig. 5–3. Skeleton of three dimensional dendritic crystal. (By permission: *Foundry Engineering*, John Wiley & Sons)

through the solidified casting and forms beads of tin on the surface.

5.7 Solid State Diffusion. Metal atoms in the solid state oscillate continuously around fixed positions in the space lattice. Occasionally atoms move from one atom site to another.

This movement of atoms through a solid crystal is called *diffusion*. The rate at which solid state diffusion may occur is controlled by the driving force of the atoms to achieve a lower energy position and by the resistance to atom movement encountered with the particular metal.

If two dissimilar metals such as copper and zinc are placed in intimate contact diffusion of each metal into the other will occur, provided the resistance to atom movement is not too great. Experimental measurements by Kirkendall has demonstrated that the rates at which two metals diffuse into one another are not the same. The element with the lower melting point diffuses faster. Thus, zinc will diffuse into copper

139

faster than the copper can diffuse into the zinc.

The *vacant site theory* is generally agreed to be the mechanism by which atom movement occurs in substitutional solid solutions of body-centered, face-centered, and hexagonal metal. According to the theory an atom may move into an adjacent vacant site leaving behind another vacancy, and so on; thus allowing diffusion to take place by the shifting of vacant sites. The rate of movement of vacancies has been calculated for copper to be 30 billion times per second at 2150°F but only once every 11 days at room temperature. Vacancies in copper have been estimated to be approximately 10 atoms apart at 1970°F. For lead, the jump rate has been calculated to be approximately 22 per second at room temperature, with the mean spacing of 100 atoms between vacancies. Thus, appreciable solid state diffusion may take place in lead at room temperature, whereas little diffusion would take place in copper at room temperature.

A second method of solid state diffusion, that of *interstitial atom movement*, is simpler than diffusion of substitutional solid solutions in that interstitial diffusion does not require the presence of vacant atom sites in order for solute atom-movement. Interstitial atom movement occurs by atoms jumping from one interstitial site into a neighboring one.

Solid state diffusion is necessary for recrystallization, age hardening, and most solid state phase changes. The mechanism of solid state diffusion also helps to explain changes in powder metal parts upon heating, hardening of dental fillings, carburizing, and many other important metallurgical processes.

5.8 Type I, Solid Solution Alloys. This type of binary, or two component alloy, forms when there is complete solubility of the components in the solid and liquid states. The single cooling curve of this type of alloy is shown in Fig. 4–35, curve 4. Fig. 5–5 illustrates this alloy system by a diagram of the copper-nickel series. The following is a brief discussion of this diagram.

The upper line is known as the *liquidus*, for the alloy is completely liquid above this line. This liquidus line indicates the lowest temperature to which a given liquid composition can be lowered without freezing. It also indicates the composition of the liquid alloy on the verge of freezing at any known temperature. The lower line of the diagram is known as the *solidus*, for all compositions in the area of the diagram below this line are in the solid state. The solidus line indicates the composition of the alloy which freezes at the given temperatures within the limits represented by the solidus curve. The area between the liquidus

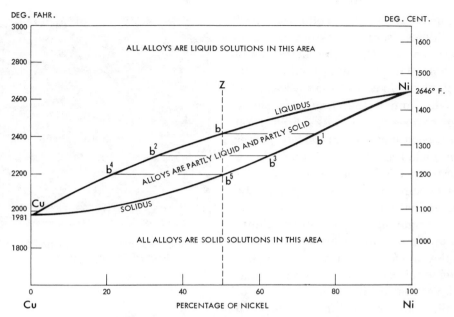

DEG. FAHR.

DEG. CENT.

ALL ALLOYS ARE LIQUID SOLUTIONS IN THIS AREA

LIQUIDUS

ALLOYS ARE PARTLY LIQUID AND PARTLY SOLID

SOLIDUS

ALL ALLOYS ARE SOLID SOLUTIONS IN THIS AREA

PERCENTAGE OF NICKEL

Fig. 5–5. Constitutional diagram of the copper-nickel alloys, Type I.

and solidus lines represents the mushy state of the alloy, partly liquid and partly solid.

Now, let us consider briefly the changes that take place during cooling, when an alloy of 50 percent copper and 50 percent nickel is at a temperature indicated by Z, Fig. 5–5, and the temperature is lowered.

The alloy remains a homogeneous liquid solution until the temperature drops to a value indicated by the intersection of the liquidus line at b. Here the alloy begins to solidify, forming not pure crystals, but solid solution crystals. It is to be remem-

bered that crystals form by progressive solidification, primary, secondary, and ternary axes forming a skeleton onto which the remaining liquid solidifies. The primary axis of the crystals which form from a 50-50 liquid are not pure, but consist of a solid solution, the composition of which is found on the solidus line at b^1, 78 percent nickel and 22 percent copper.

As the mass cools, the composition of the growing crystals changes along the solidus line from b^1 to b^5, while the remaining liquid alloy varies in composition along the liquidus line

141

from b to b⁴. The solid that first solidifies from the liquid contains less copper than the liquid as a whole. The remaining liquid is thus left richer in copper than it was originally, and it therefore possesses a lower freezing point. As solidification proceeds metal progressively poorer in nickel is deposited around the primary solid, that which solidifies last being richest in copper. The solidification, therefore, is not that of a single solution, but of an infinite number of solid solutions, and the solutions formed have a corresponding number of solidification temperatures, the result being a number of solid solutions of different chemical composition.

Hence, the structure of solid cast alloy differs from that of cast pure copper or nickel in that it consists of dendritic crystals, originally so-called from their branched tree-like appearance. These crystals vary in composition, being nickel-rich at the center and copper-rich on the outside. Their lack of homogeneity can be corrected by diffusion if the rate of cooling is sufficiently slow or if the alloy is rendered homogeneous by annealing at temperatures below the melting point. The rate of diffusion varies greatly for different metals, and the heterogeneous "cored" structure persists longest in alloys having the slowest rate of diffusion. It is persistent in nickel alloys. Fig. 5–6 illustrates this heterogeneity in

Fig. 5–6. Cast monel metal magnified 100X. Cored structure is very apparent.

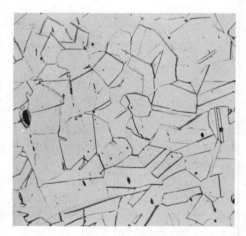

Fig. 5–7. Cast monel in Fig. 5–6 after rolling and annealing. Cored structure is eliminated by mechanical working. 100X.

a Monel metal, 67 percent nickel and 28 percent copper. Uniformity of composition of the crystal grains is assisted by mechanical working at elevated temperatures, such as by rolling, forging, and other operations, followed by annealing. All copper-nickel alloys in the rolled and annealed condition have a similar structure that of a homogeneous solid solution, as indicated in Fig. 5–7.

5.9 Type II, Eutectic Alloys. These alloys are made up of two metals in which there is complete solubility of the components in the liquid state and limited solubility in the solid state. In addition to eutectic alloys, other types of three-phase reactions in binary alloys include the peritectic and the monotectic alloys. A time-temperature curve for this type of alloy is shown in Fig. 4–35, curve 5. A simplified diagram for cadmium-bismuth, Fig. 5–8 can be used to describe the Type II alloy. Fig. 5–8

shows no solubility of metals in the solid state, which is an impossibility since it implies that pure component Cd has a melting range from temperature E to A. However, metals with extremely limited solubility would approach the shape shown which simplifies the task of presenting basic concepts concerning phase equilibrium diagrams.

The liquidus of the diagram in Fig. 5–8 is AEB, and the solidus is ACEDB. The line CED is often referred to as the eutectic line and point E as the *Eutectic* point. If a small amount of cadmium is added to molten bismuth the freezing point of the bismuth will be lowered. On the other hand, if a small amount of bismuth is added to molten cadmium the freezing point of the cadmium is lowered. It is apparent, since each metal lowers the freezing point of

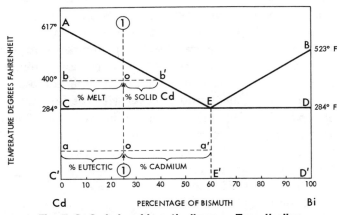

Fig. 5–8. Cadmium-bismuth diagram. Type II alloy.

143

the other, the lines connecting the freezing points must intersect at some point as shown by point E in Fig. 5–8.

This point of intersection, sometimes called the Eutectic point, is of the greatest importance. The composition that freezes at this point is called the eutectic alloy. The alloy of this eutectic composition melts and freezes at a constant temperature, in this respect behaving as a pure metal. The eutectic alloy has the lowest melting point of any composition in the series. The eutectic is not a homogeneous alloy, but consists of crystals of nearly pure cadmium and crystals of nearly pure bismuth, approximately 40 percent cadmium and 60 percent bismuth. Point E of Fig. 5–8 gives the chemical composition of the eutectic, and the extreme ends of the eutectic line, C and D, point out that the eutectic is made up of nearly pure cadmium and nearly pure bismuth crystals.

If the liquid solution contains less than 60 percent bismuth (the eutectic composition), when the temperature is lowered until the AE line is intersected, nearly pure cadmuim begins to separate from the liquid solution. The separation of cadmium and the lowering of the freezing point go along the curve AE, Fig. 5–8, until the remaining liquid solution contains 40 percent cadmium and 60 percent bismuth. If the solution contains more than 60 percent bismuth, then nearly pure bismuth separates from the liquid, and the separation of bismuth and the lowering of the freezing point go along the curve BE, Fig. 5–8, until the remaining liquid solution contains the eutectic composition.

No matter what the original analysis was, the liquid solution of cadmium and bismuth at temperature 284°F, (the eutectic temperature) always contains 40 percent cadmium and 60 percent bismuth. This so-called eutectic mixture freezes at a constant temperature, forming crystals of nearly pure cadmium and nearly pure bismuth.

From a study of the alloy diagram and through the use of the *"lever rule"*, we can determine approximately the relative amounts of each constituent present in any alloy at any selected temperature within the scope of the alloy diagram. To do this, select an alloy, such as 75 percent cadmium and 25 percent bismuth (See Fig. 5–8); and if we want to know its structural composition at any temperature below 284°F, draw a constant temperature line a-o-a'. This line a-o-a' is used as a lever with the fulcrum located at the alloy in question, 75Cd-25Bi, and therefore:

$$\frac{ao}{aa'} \times 100 = \text{Eutectic}$$

$$\frac{25}{60} \times 100 = 41.6\%$$

also;

$$\frac{oa'}{aa'} \times 100 = \text{Cadmium}$$

$$\frac{35}{60} \times 100 = 58.3\%$$

Both percentages are by weight.

In the area AEC we may find the relative amounts of melt and cadmium present by the same method. The lever is now drawn from b on the solidus line to b' on the liquidus line. Point o is the fulcrum of the lever. Therefore:

$$\text{b-o} = \text{percentage melt}$$
$$\text{o-b'} = \text{percentage of solid cadmium}$$

$$\frac{o\text{-}b'}{b\text{-}b'} \times 100 = \text{Cadmium}$$

$$\frac{15}{40} \times 100 = 37.5\%$$

and,

$$\frac{b\text{-}o}{b\text{-}b'} \times 100 = \text{melt}$$

$$\frac{25}{40} \times 100 = 62.5\%$$

The structural composition thus determined is one of weight percentages and may be considered fairly accurate for a volume check when the two constituents have about the same density; however, if the densities differ markedly from each other, a more general method of calculation must be used which takes into account the densities of the various constituents.

The photomicrographs in Figs. 5–9 to 5–12 show the structures developed in Type II, eutectic alloys,

as illustrated by the cadmium-bismuth alloys.

You will note that in this discus-

Fig. 5–9. Structure from slow cooling of molten alloy of 75% cadmium and 25% bismuth to room temperature. The alloy contains about 42% eutectic (wavy, flake-like structure) and 58% nearly pure cadmium (dark structure). (By permission: *Principles of Metallography,* McGraw-Hill Book Co.)

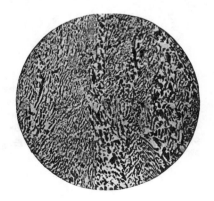

Fig. 5–10. Alloy containing the eutectic composition of 40% cadmium and 60% bismuth. 100% eutectic. (By permission: *Principles of Metallography,* McGraw-Hill Book Co.)

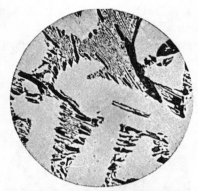

Fig. 5–11. Structure of an alloy containing 20% cadmium and 80% bismuth. The structure is about 50% eutectic and 50% nearly pure bismuth (light). (By permission: *Principles of Metallography*, McGraw-Hill Book Co.)

Fig. 5–12. Structure of an alloy containing 10% cadmium and 90% bismuth. Alloy is about 25% eutectic and 75% nearly pure bismuth. (By permission: *Principles of Metallurgy*, McGraw-Hill Book Co.)

sion we say that the crystals which separate are "nearly" pure cadmium or "nearly" pure bismuth. Careful analysis will show that the cadmium

crystals actually contain some bismuth, and the bismuth crystals, some cadmium. Only in a very few instances are the crystals which separate really pure metal. In most cases the separating crystals contain more or less of the other constituent. Such crystals are called solid solutions, indicating that the crystals contain more than one metal. The term "solid solution" does not refer to a mixture but a solution, in that the metals which are present in them cannot be separated by mechanical means.

As another example of this solid solution formation, let us consider the lead-antimony alloys rich in lead, which are represented in Fig. 5–13. If we consider the alloy represented by the point o, an alloy containing 3.2 percent antimony and held at a temperature of 660°F, it is entirely liquid. If we allow it to cool to the point C and skim the separated crystals, we will find on analyzing them that they are not pure lead but a solid solution of antimony in lead, containing the amount of antimony which is represented by the point G. Now as we cool and skim, we find that the amount of antimony in the skimmings increases up to a definite limit, which is called the limit of solubility, represented at M. The area Pb-M-N then represents a temperature and composition range in which the entire alloy will be solid solution.

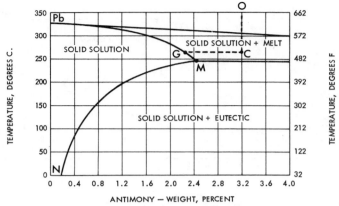

Fig. 5–13. High lead end of antimony-lead diagram. (American Society for Metals)

It will be noted that the limit of solubility of antimony in lead is reached at the point M, and that as we lower the temperature this solubility decreases along M-N as the temperature falls. This is analogous to the separation of salt from the "liquid" solution of it in water as a hot solution is cooled.

The solubility limit and the change of solubility with temperature vary with every pair of metals. In some cases the limit of solubility becomes infinite, that is, the two solid metals form solutions in all proportions, as water and alcohol form liquid solutions. Such a case is copper-nickel, Fig. 5–5.

5.10 Type III, Alloys. Very few alloys which form perfect molten solutions crystallize into two pure metals as described in Type II solidification.

Nevertheless, a large majority of the alloys which form molten solutions separate during freezing to form two kinds of crystals. These crystals are composed of the metals which are not pure, but which carry with them some part of the alternate metal in the state of solid solution. In other words, the two metals crystallize and separate, but the crystals are not pure; on the contrary, each contains some of the other metal as a solid solution.

The behavior of these alloys during freezing and the shape of their diagrams is of an intermediate nature when compared to the diagrams of the alloys of Type I and Type II. A diagram of this type is shown in Fig. 5–14 for the copper-silver system. Silver is capable of retaining 8.8 percent of copper in solid solution, and alloys in this series con-

147

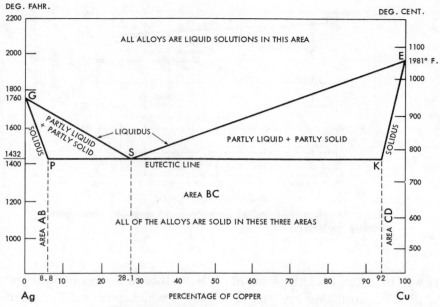

Fig. 5–14. Copper-silver diagram, Type III alloys.

taining less than 8.8 percent copper will solidify, if slowly cooled, in a manner typical of solid solutions, Type I, and will exhibit the characteristic cooling curves of the Type I alloys.

On the other hand, those alloys of this series having compositions which place them beyond the limit of solid solubility of the metals, at either end of the series, behave in most respects much like the alloys which are entirely insoluble in one another in the solid state. The only difference is that the solid which first crystallizes is not pure metal, but a

saturated solid solution of one metal in the other, similar to the saturated solid solution formed in the lead-antimony alloys. A complete diagram of such a system, therefore, is made up of three distinct parts: a central portion BC, resembling the diagram of Type II alloys, and portions AB and CD at each end, typical of the Type I alloys.

The areas AB and CD of Fig. 5–14, as well as the Type I alloys, Fig. 5–5, represent the same characteristic behavior of the alloys during solidification. Both form solid solutions upon solidification, which in the cast state are heterogeneous, with the crystals having a cored or dendritic structure. The solubility or miscibility of one metal in another in Types I, II, and III alloy systems, then, are as follows:

Type I. The metals are miscible in both the liquid and solid states; the crystals separating from the liquid contain both metals in solution.

Type II. The two metals are miscible in the liquid state but separate as nearly pure crystals of the two metals when the melt solidifies.

Type III. The two metals are miscible in the liquid state, but are only partially miscible in the solid.

5.11 Type IV, Monotectic Alloys.

Not all metals are miscible in the liquid state. Thus lead and zinc or lead and copper do not mix when melted together but form two layers, as do water and oil. This immiscibility is of commercial importance in making bearing bronze in which droplets of lead are mixed through molten copper like an emulsion of oil in water, and the whole is solidified. Two metals, however, are rarely completely insoluble. Each one dissolves some of the other, and as the temperature is increased, eventually a point is reached where any metal becomes miscible. Fig. 5–15 shows an equilibrium diagram for lead-zinc.

Alloys of the monotectic type are not widely used commercially because the tendency for the two liq-uids to separate causes problems.

5.12 Type V, Peritectic Alloys.

Under certain conditions of alloy behavior some metals undergo a reaction between the solid and liquid phases during the process of solidification, which results in the disappearance of the two phases, and in their place, a new solid phase is born. The new solid phase formed from a reaction between the liquid and solid phases may be a solid solution or an intermetallic compound. This reaction, resulting in the disappearance of two phases and the birth of a new one, has been called a *peritectic reaction*. Such an alloy behavior is illustrated by the diagram Fig. 5–16 where H corresponds to the formation of a compound *AmBn*. This compound *AmBn* forms along the CHS line of the diagram from the reaction of a nearly pure solid (metal A), shown by point C, and the liquid solution of composition S. If the original melt contained just 80 percent metal A and 20 percent metal B the result of this peritectic reaction gives birth to a pure compound designated *AmBn*. In the discussion of this diagram, we discover that nearly pure metal A separates from the liquid along the line GS, and upon cooling, from GS to CHS, we have a solid A and a liquid solution which may vary in composition from G to S. Upon cooling to the CHS line, the solid A reacts with the liq-

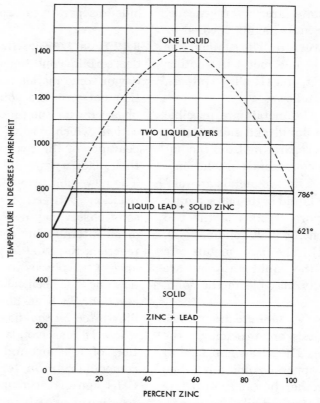

Fig. 5–15. Lead-zinc equilibrium diagram, Type IV alloys.

uid solution of composition S, forming the compound *AmBn*. In case the original liquid contained less than 20 percent metal B, the final structural composition would contain metal A plus the constituent *AmBn*. With an original composition between 20 and 40 percent metal B, the reaction along the CHS line would be evidenced by the formation of the compound *AmBn* and a liq-

uid. Cooling from the HS line to the PE line would result in more compound forming from the liquid solution and the remaining liquid becoming richer and richer in metal B until the temperature has dropped to the PE line, at which time the remaining melt would have a composition as shown by point E, or a eutectic composition. The remaining liquid at this time would solidify to

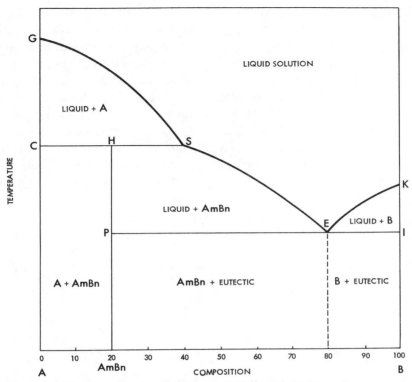

Fig. 5–16. Type V peritectic alloy diagram showing formation of compound *AmBn.*

form a eutectic structure. The final alloy would consist of crystals of *AmBn* and the eutectic.

The eutectic of this alloy solidifies along the PEI line, Fig. 5–16, with a eutectic composition which contains 20 percent A and 80 percent B, and consists of crystals of the compound *AmBn* and crystals of nearly pure metal B. The eutectic structure of this alloy would be similar in appear-

ance to that of any common binary alloys solidifying with a eutectic structure.

5.13 Type VI, Intermetallic Compounds. In general, intermetallic compounds are the principal hardeners in our industrial alloys, the compounds being hard and frequently brittle with little strength. Such compounds or hardeners are found

151

in alloys such as steel, aluminum alloys, and bearing metals, in which the compounds Fe_3C, $CuAl_2$, $NiCd_7$, CuSn, and SbSn, are found. Some of the hardest substances known to man are formed from compounds such as WC (tungsten carbide) and B_4C (boron carbide), both of which are harder than sapphire and nearly as hard as the hardest substance known, diamond.

The alloy diagram of the compounds forming metals may take on many different appearances. For discussion and introduction to this type of alloy behavior, Fig. 5-17 may be used, and the alloys of cadmium-nickel, iron-carbon, magnesium-tin, and many others may be used as illustrations of this type of behavior among the industrial alloys.

The compound formed in the al-

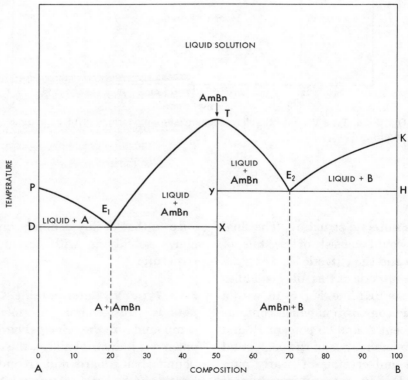

Fig. 5–17. Type VI alloys. These form two separate eutectics, E_1 and E_2 and a compound, *AmBn*.

loy diagram presented in Fig. 5–17 is designated as *AmBn*, and has a composition of 50 percent metal A, and 50 percent metal B. A melt of 50 percent A and 50 percent B will solidify at temperature T to form a homogeneous solid, an intermetallic compound. It will be seen from a study of this diagram that two Type II eutectic alloys are formed by this behavior. The components of the eutectic E_1, Fig. 5–17, are nearly pure metal A and the compound *AmBn;* whereas, the components of the eutectic formed at E_2, are nearly pure metal B, and the compound *AmBn.* We actually have two separate eutectic diagrams in one, and the discussion of such behavior is similar to that of a simple eutectic-forming alloy system with the intermetallic compound *AmBn* considered as a pure substance and as one of the components of the system. With this in mind, the significance of the various areas of the diagram becomes evident.

5.14 More Complex Alloy Systems. Many of the most important industrial alloys contain three or more metals in their composition. Certain important effects which a third element may have on a binary alloy are listed briefly.

1. The three elements may form a ternary eutectic with lower melting point than the binary eutectic. Thus, the addition of bismuth to the lead-tin eutectic produces an alloy which melts in boiling water.

2. The three elements may all go into solid solution, in which case the third element has no great effect on the binary alloy.

3. The third element may combine with one of the elements of the alloy or with the impurities to produce a profound effect. Thus manganese added to iron containing sulphur renders it forgeable by combining with the sulphur to produce manganese sulphide which separates from the alloy. Such alloys are very difficult to illustrate by the alloy diagram method. However, if only three components are present, an alloy diagram may be most conveniently plotted by means of triangular coordinates. In Fig. 5–18, each of the components of a ternary alloy is represented by one corner of an equilateral triangle. The sides of the triangle represent the alloy systems of three binary alloy diagrams, AB, AC, and BC. All of the possible combinations of the three components are then formed by the length of the perpendicular from any point Z, Fig. 5–18, to the opposite side of the triangle. For example, the composition of point Z will be seen to be 40 percent A, 40 percent B, and 20 percent C. To construct a diagram with the triangle as a base, temperatures are plotted at right angles to the plane of the triangle. The development of the diagram results in a

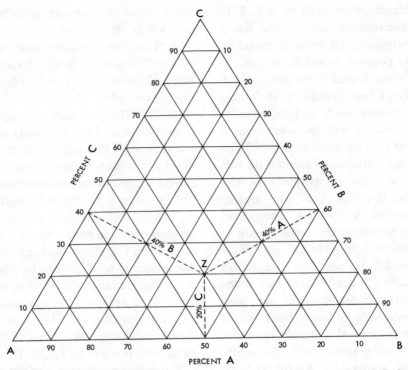

Fig. 5–18. Base triangle for a ternary alloy, illustrating a method of finding the composition of alloy Z. Composition is found by reading from the base representing 0% to the apex representing 100% for each of the three components.

solid model, each side of which represents one of the binary systems. The student should refer to a more complete text on alloy diagrams for a detailed discussion of this type of alloy diagram.

5.15 Changes in the Solid State. In dealing with the alloy diagram, we have discussed changes occurring during the process of solidification.

The study of the alloy behavior is complicated by the fact that solids, during heating and cooling, undergo certain changes in their structural makeup that may completely alter their properties. Such changes can be shown by the alloy-diagram principle. Some of the most common changes that occur include:

1. Changes in solid solubility upon cooling,

154

2. Changes from one allotropic form to another,

3. Formation of eutectoid structures,

4. Grain growth,

5. Recrystallization.

The first two changes, changes in solid solubility and changes from one allotropic form to another, are common changes that may be illustrated by the alloy-diagram method. The iron-carbon alloys, which will be discussed in more detail in Chapter 6, may be used as an illustration of

these two important changes taking place during the cooling of a solid. Fig. 5–19 is a simplified iron-carbon diagram showing the changes that take place within a solid solution, which exists in area AESG, when the solid solution is cooled to normal temperatures. The maximum solubility of Fe_3C in gamma (γ) iron is shown by point E. As cooling of the solid solution takes place, the solubility of gamma iron for Fe_3C follows along the ES line to point S. At point S, all iron becomes alpha

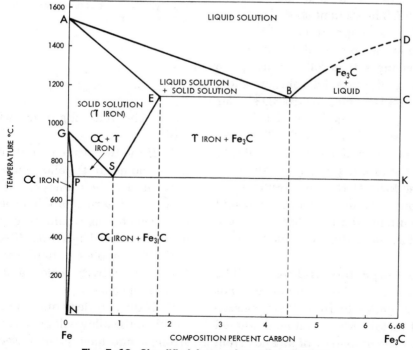

Fig. 5–19. Simplified iron-carbon alloy diagram.

(α) iron, changing from the gamma form, and the solubility of the new alpha iron phase becomes practically zero, represented by point P. Point G represents the change of pure iron from one allotropic form to another, the change involving a rearrangement of a face-centered cubic atomic lattice form of iron to a body-centered cubic form of iron. The structure born along the PSK line, from a solid solution, is similar to that formed along the EBC line from the liquid solution. However, one is born from a solid, and the other from a liquid. The structure formed at B is called the *eutectic*; whereas the structure born at S is called the *eutectoid*. The student should note the similarity in appearance of the diagram in the area of solidification and the changes that follow during the cooling of the solid solution alloy. Line AB is similar to GS, BD similar to SE, PSK similar to EBC, and AES similar to GPN. The two changes, one of changing solubility and the other, a change in the allotropic form of the iron, result in reactions that may be measured and account for the lines formed in the solid areas of this alloy diagram.

5.16 Properties of Alloys. The properties of alloys may or may not differ markedly from their constituent metals. Much can be predicted about the properties of an alloy from a knowledge of the equilibrium dia-

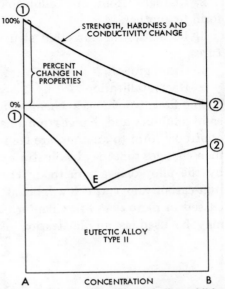

Fig. 5–20. Possible effect of alloy addition for Type II alloys upon strength, hardness, and conductivity.

gram. The following rules will be useful:

1. If the two elements form eutectic but do not form solid solutions or compounds, their alloys will have properties intermediate between the two elements as would be expected of a mechanical mixture. (See Fig. 5-20) This applies to hardness, electrical conductivity, color, and magnetic properties.

2. If the two elements form a solid solution, the alloy will be harder and stronger and have greater electrical resistance than would be obtained

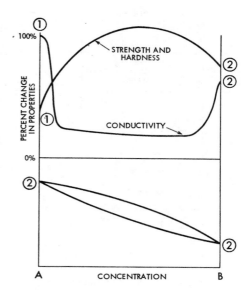

Fig. 5–21. Behavior that might be expected from a system of Type I alloys.

with a simple mixture of its constituent metals. (See Fig. 5–21). Its color and magnetic properties cannot be predicted. In general, it may be stated that the less the solubility of one metal in another, the greater will be the hardening and the more difficult it will be to dissolve a given percent of it in the second metal. Thus nickel which is soluble in all proportions in copper, does not affect the properties of copper to anywhere near the extent of phosphorus, which is only slightly soluble.

3. If the two metals form a compound nothing can be predicted of its properties from those of its constituents.

Finally, the properties of some alloys may be profoundly altered by heat treatment. These are the solid solution alloys which show a change of solid solubility with temperature, as described for lead-antimony. Such alloys may be greatly hardened by heating to the solid solution range and cooling suddenly, as by quenching in water, and subsequently aging at room or elevated temperature. The quenching preserves the solid solution, which, by the aging process, is broken up into very finely divided particles of one metal in the other. Such a finely divided mixture is much harder and stronger than the same mixture in coarser particles.

157

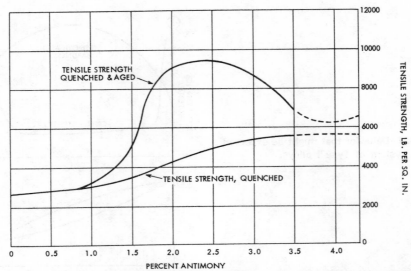

Fig. 5–22. Tensile strength of lead-antimony alloys after quenching from 450°F and aging at room temperature.

The tensile strengths of quenched and aged lead-antimony alloys are shown in Fig. 5–22.

This brief review should equip the reader to tell a great deal about the properties of alloys from published equilibrium diagrams.

References

Barrett, Charles S. *Structure of Metals, 2nd ed.* New York: McGraw-Hill, 1952.

Brick, Robert M., Gordon, Robert B. and Phillips, Arthur. *Structure and Properties of Alloys, 3rd ed.* New York: McGraw-Hill, 1965.

Clark, Donald S. and Varney, Wilbur R. *Physical Metallurgy for Engineers. 2nd ed.* Princeton, N.J.; D. Van Nostrand Co., 1966.

Cottrell, A. H. *Theoretical Structural Metallurgy, 2nd ed.* New York: St. Martins Press, 1957.

Reed-Hill, Robert E. *Physical Metallurgical Principles.* Princeton, N.J.: D. Van Nostrand Co., 1964.

Samans, C. H. *Engineering Metals and Their Alloys.* New York: Macmillan, 1964.

Shewmon, P. *Diffusion in solids.* New York: McGraw-Hill, 1963.

The Iron-Carbon Diagram

6.1 Introduction. Alloys of iron and carbon are the most widely used and least expensive of all metal alloys. The importance of iron-carbon alloys is due to the fact that their properties may be very decidedly changed by heating and cooling under controlled conditions. The processes for winning iron from its ore and for making cast iron and steel have been known for many centuries. In fact, metallurgical literature of 1540 A.D. interestingly, but crudely, described the procedure for making and hardening of steel like this:

Steel is nothing else but iron worked up with much art and much soaking in the fire until it is brought to a perfect mixture and given properties that it did not before possess.

Likewise, it may have taken up suitable material of a fatty tendency, also a certain moisture and thereby become whiter and denser. The long firing also opens up and softens its pores, which are drawn together again tightly by the power of the cold of quenching water. The iron is thus given hardness and the hardness makes it brittle. As iron can be made from any iron ore, likewise steel can be made from any pure iron.

Recent research has helped in replacing much of the art of metallurgy with scientific understanding of the basic principles underlying metal behavior.

6.2 Cooling Curve. One important discovery about iron was that it is *allotropic*, meaning that it can exist in more than one type of lattice structure depending upon temperature. The cooling curve for iron, Fig. 6–1, shows the temperatures at which important structural changes occur.

Iron solidifies at 2800°F into a B.C.C. form known as *delta* (δ) iron. Upon further cooling to 2554°F, the atoms rearrange themselves into a F.C.C. form known as *gamma* (γ) iron. Gamma iron is non-magnetic. As the temperature further drops to 1666°F, another phase change occurs

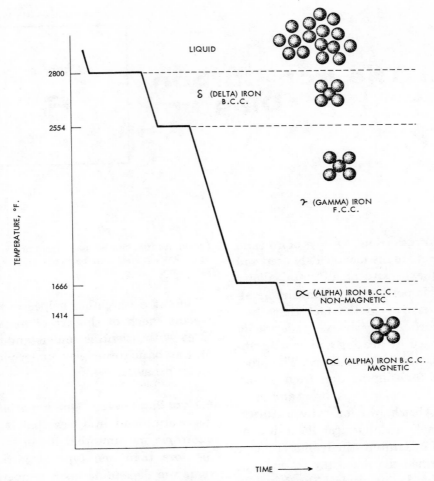

LIQUID

δ (DELTA) IRON
B.C.C.

γ (GAMMA) IRON
F.C.C.

∝ (ALPHA) IRON B.C.C.
NON-MAGNETIC

∝ (ALPHA) IRON B.C.C.
MAGNETIC

TEMPERATURE, °F.

2800

2554

1666

1414

TIME ——————▶

Fig. 6–1. Cooling curve for pure iron.

in that the F.C.C. structure changes to a B.C.C. non-magnetic structure known as *alpha* (α) iron. Below 1414°F alpha iron becomes magnetic accompanied by changes in internal energy and electrical conductivity. The temperature at which the above allotropic changes take place is in-

fluenced by the presence of alloying elements.

6.3 Iron-Carbon Diagram. It will be evident from the diagram Fig. 6–2, that iron-carbon alloys are Type III alloys with limited solubility (1.7 percent carbon) in the solid solution.

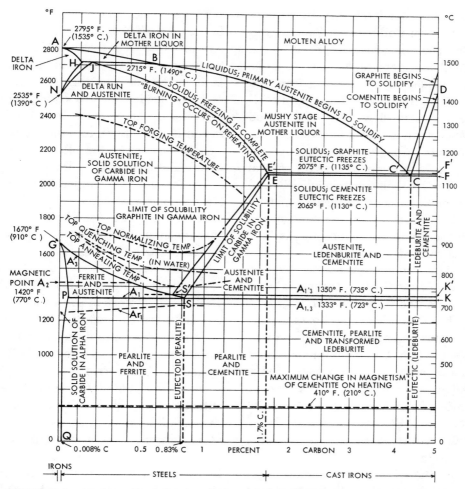

Fig. 6–2. Iron-carbon phase diagram. Carbon content is plotted on a logarithmic scale to magnify the lower end of the diagram. (By permission: *A Metal Progress Copyrighted Data Sheet***)**

The iron-carbon diagram not only shows the type of alloys formed under very slow cooling, but also indicates the proper heat-treating temperatures, and, in addition, how the properties of steels and cast irons may be so radically changed by heat treatments.

The iron-carbon diagram, sometimes referred to as an equilibrium

161

diagram, is not strictly in equilibrium even under conditions of very slow cooling. The iron carbide phase, Fe_3C, is a *metastable* phase. That is, it will decompose to form iron and free graphite flakes. This decomposition is very slow even at elevated temperatures and may take several years to form graphite. After operating at high temperatures for several years a steel pipe in an oil distillation tower was examined microscopically following removal from service and this decomposition of iron carbide was verified.

Pure iron, represented by the point A, and the iron-carbon eutectic represented by point C in Fig. 6–2, melt and solidify at a constant temperature. All other compositions represented in the diagram melt and freeze over a range of temperatures. Alloys containing 0.0 to 4.3 percent carbon begin to solidify on cooling to the line ABC by the separation of solid solution austenite crystals from the liquid. Alloys containing more than 4.3 percent carbon begin to solidify with the separation of Fe_3C (cementite) from the liquid, on cooling to the CD line. In the case of alloys containing 1.7 percent carbon or less, solid solution austenite begins to freeze out on cooling from ABC to AHJE. At AHJE, all of the alloy is solid, and consists of *dendritic* crystals. These dendritic crystals are formed of an infinite number of solid solutions; not of a single solid solution. The first to freeze is a solid solution relatively low in carbon content; the last to freeze is relatively rich in carbon. It is to be recalled that this lack of homogeneity due to coring and segregation can be corrected by diffusion if the rate of cooling is sufficiently slow or if the alloy is rendered homogeneous by hot working or by annealing at temperatures below the melting point.

For the present it can be assumed that the area NJESG of the diagram represents an alloy in one homogeneous phase, austenite.

In alloys containing from 1.7 to 4.3 percent of carbon solid solution austenite freezes out of the liquid, beginning at ABC and continuing as the temperature is lowered to ECF. At ECF, the eutectic line, there is some residual liquid, which is of the eutectic composition. This liquid then solidifies at a constant temperature, forming the eutectic mixture. The eutectic of the iron-carbon system consists of alternate layers of Fe_3C (cementite) and saturated solid solution austenite in a finely divided state known as *ledeburite*. The Fe_3C (cementite) and the solid solution austenite crystallize simultaneously at temperature ECF to form the eutectic structure. Thus, alloys containing from 1.7 to 4.3 percent carbon, just below the ECF line consist of mixed crystals of two constituents, austenite and eutectic.

The amounts of each vary from 100 percent austenite and 0.0 percent eutectic with 1.7 percent carbon alloy to 100 percent eutectic and 0.0 percent austenite with the 4.3 percent iron-carbon alloy.

In alloys containing more than 4.3 percent carbon, Fe_3C (cementite) freezes out on cooling from CD to CF. The residual liquid at CF is of eutectic composition and solidifies at a constant temperature, forming the eutectic constituent. These alloys just below CF consist of mixed crystals of cementite and eutectic. Very slow cooling will result in the formation of free graphite flakes due to the decomposition of iron carbide and rejection of excess carbon from the austenite.

In the discussion of Types I, II, and III alloys, it was considered that the structure of the solid alloy just below the solidus line and the eutectic line remained unchanged during the cooling to room temperature (except for diffusion taking place when dealing with a heterogeneous solid solution phase). The structure at room temperature was that of a cast alloy. However, with iron-carbon alloys, very marked changes occur during the cooling from the solidus AHJE and eutectic ECF to room temperature. These changes are brought about principally because iron has the ability to exist in several allotropic forms, and during cooling changes from one to another.

In the NJESG area and SECFK area, iron exists in the form known as gamma iron with its atoms arranged in the face-centered cubic lattice. Gamma iron has the ability to dissolve a maximum of 1.7 percent carbon at 2065°F which diminishes to a maximum of 0.83 percent at 1333°F. Gamma iron is more dense than alpha iron. Thus, the transformation from gamma iron to alpha iron is accompanied by an expansion of the lattice structure. Alpha iron exists in the areas below PS and SK.

Alpha iron has a body centered cubic lattice, is magnetic, and dissolves only a very limited amount of carbon (0.008 percent at room temperature.)

Two areas of special interest in the lower carbon range of the iron-carbon diagram, are those areas immediately below lines GS and SE. The line GS is sometimes called the ferrite solubility curve, *ferrite* being the solid solution in which alpha iron is the solvent. Upon cooling, gamma iron transforms to alpha iron at the temperatures indicated at the left of the curve GS. It can be seen that increasing the carbon content of the alloy lowers the transformation temperature considerably. When no carbon is present the transformation from gamma iron to alpha iron occurs at 1670°F, whereas the presence of 0.83 percent carbon lowers the transformation temperature to

163

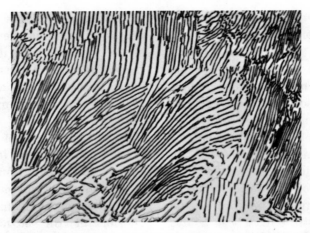

Fig. 6–3. Eutectoid 0.85% carbon steel. Pearlitic structure composed of alternate layers of cementite in a ferrite matrix. 1000X (U.S. Steel Corp.)

1333°F. Thus, during cooling, some of the gamma iron changes to alpha iron upon reaching the ferrite solubility curve. This leaves the solid solution richer in carbon, lowering its transformation temperatures. This process is continued until the composition reaches 0.83 percent carbon. This composition is known as the eutectoid composition from its similarity to the eutectic. The difference is that solid phases exist above the *eutectoid* whereas a liquid phase exists above the *eutectic* composition.

The other interesting area is that immediately below the line SE. This curve may be referred to as the cementite solubility curve, and represents the temperature at which Fe_3C (cementite) precipitates from the austenite upon cooling. At a temperature of 2065°F, austenite will hold up to 1.7 percent carbon in solid solution. This solubility diminishes to a maximum of 0.83 percent carbon at 1333°F. Thus it can be seen that, no matter what the initial carbon content of steel, some austenite reaches a temperature of 1333°F on slow cooling with a carbon content of 0.83 percent. This situation will exist only so long as the alloy remains at 1333°F. It cannot cool below this point without a complete separation of all the solid solution austenite into alternate layers of ferrite and cementite. The eutectoid mixture so formed is often referred to as *pearlite*, because it has the appearance of mother-of-pearl when viewed through the microscope. (See Fig. 6–3.)

6.4 Carbon Solubility. The reason for the formation of alternate layers of Fe_3C and ferrite below the eutectoid temperature is due to the fact that the F.C.C. gamma iron transforms to B.C.C. alpha iron. The F.C.C. gamma iron lattice shown schematically in Fig. 5–2B has interstitial holes with a radius of approximately 0.52 angstroms which can accommodate up to 1.7 percent interstitial carbon atoms. As the transformation to the B.C.C. alpha iron occurs, Fig. 5–2A, this number of holes increases but interstitial hole size is reduced to 0.38 angstroms. Since the radius of a carbon atom is greater than the interstitial space available, the carbon solubility is severely reduced in the B.C.C. structure. The carbon atoms forced out of the austenite upon transformation combine with iron to form lamellae of the intermetallic compound iron carbide, Fe_3C, also called cementite. The formation of pearlite is shown schematically in Fig. 6–4.

Growth rate for pearlite is dependent upon the temperature at which the transformation from austenite occurs. The lower the transformation temperature, the higher the growth rate, the closer the pearlite lamellae, and the harder the resulting structure. It has been reported in the literature that pearlite forming at 1250°F has a hardness of Rockwell C-15 with a pearlite spacing of 10^{-3}

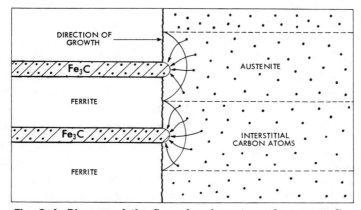

Fig. 6–4. Diagram of the flow of carbon atoms from austenite during formation of pearlite.

Fig. 6–5. Carbon steel with 0.25% carbon. Pearlite is the dark and ferrite is the light constituent. 150X. Nital etch.

mm, while pearlite formed at 1112°F has a hardness of Rockwell C-40 and a mean pearlite spacing of 10^{-4} mm.

The relative amounts of free ferrite and pearlite resulting from the transformation of austenite can be determined by use of the lever rule. For example, suppose we want to know the relative amounts of free ferrite and pearlite resulting from the slow cooling of a 0.25 percent carbon steel to room temperature as shown in Fig. 6–5.

0.0 percent carbon yields 100 percent free ferrite, F
0.83 percent carbon yields 100 percent pearlite, P

Therefore, for a 0.25 percent carbon steel,

$$P = \frac{.25}{.83} \times 100$$
$$= 30\% \text{ pearlite}$$
$$\text{and, } F = \frac{.83 - .25}{.83} \times 100$$
$$= 70\% \text{ free ferrite}$$

This method can be used at any temperature to estimate the relative amounts of each constituent present. Also, because the densities of the cementite (Fe_3C) and ferrite are about the same, this method is fairly accurate for estimating the volume percentages of each constituent.

When a steel containing greater than 0.83 percent carbon is cooled, for example, a 1.43 percent carbon tool steel, it is first composed of a solid solution of gamma iron and carbon. As it cools, however, it soon reaches a temperature where the gamma iron cannot hold this amount of carbon in solid solution, resulting in a separation of iron carbide, Fe_3C, from the solid solution. This temperature corresponds to the intersection of the SE line, Fig. 6–2, and is at approximately 1900°F. Precipitation of Fe_3C from the austenite continues during cooling until the eutectic temperature of 1333°F is reached, at which time the remaining austenite transforms into pearlite. The resulting structure then consists of a network of cementite surrounding eutectoid pearlite, as shown in Fig. 6–6. Steels containing less than 0.83 percent carbon are called *hypo-eutectoid* steels, while those containing

between 0.83 and 1.7 percent carbon are called *hyper-eutectoid* steels.

If an alloy containing more than 1.7 percent carbon is slowly cooled, the cementite, Fe_3C (containing 6.67 percent carbon) decomposes into iron and graphite. The resultant structure consists of fine, soft, graphite flakes in an iron matrix and is known as grey cast iron. More rapid cooling of the alloy may not allow time for the decomposition of the cementite and will result in an extremely hard, brittle structure known as white or chilled cast iron. It should be noted that most cast irons are not simple alloys of carbon and iron, but contain relatively large amounts of other elements such as silicon to promote carbide decompo-

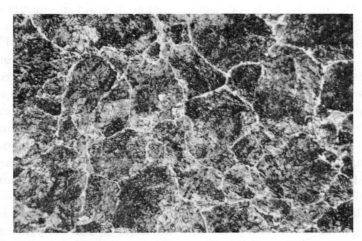

Fig. 6–6. Carbide network (light) surrounding pearlite (dark) resulting from slow cooling of a 1.25% carbon steel. 300X. Nital etch.

sition (grey cast iron) or manganese to prevent it (white cast iron).

6.5 Definition of Structures. Following is a definition of structures shown on the iron-carbon alloy diagram, Fig. 6–2, including photomicrographs, Table 6–1, page 169. Special names have been given to the various phases, along with their Greek-letter designations.

Austenite is a solid solution of carbon in F.C.C. Gamma (γ) iron. It is non-magnetic; can dissolve up to 1.7 percent carbon at 2065°F; unstable at room temperature; tensile strength, 150,000 psi; elongation, 10 percent in 2 inches; hardness, Rockwell C-40.

Delta Iron (δ) is the B.C.C. form of iron as it first solidifies at 2800°F. Exists only at high temperatures, can dissolve up to 0.10 percent carbon.

Ferrite is also known as alpha (α) iron with the B.C.C. structure. It exists below 1666°F and is non-magnetic above 1414°F. Can dissolve up to 0.025 percent carbon at 1333°F, but only 0.008 percent at room temperature; tensile strength 40,000 psi; elongation, 40 percent in 2 inches; hardness, less than Rockwell C-0.

Cementite is a very hard, brittle intermetallic compound of iron and carbon, Fe_3C containing 6.67 percent carbon. Low tensile strength but high compressive strength.

Ledeburite is the eutectic mixture of austenite and cementite and contains 4.3 percent carbon. It begins to form at 2065°F upon rapid cooling. The white mottle-like constituent is carbon with a ferrite background.

Pearlite is a distinctive two phase lamellar structure consisting of thin platelets of iron carbide in a ferrite matrix; the fully pearlite eutectoid structure contains 0.83 percent carbon and forms on slow cooling. Average properties are: 120,000 psi tensile strength; elongation, 20 percent in 2 inches; hardness, Rockwell C-20.

6.6 Grain Size Changes. At room temperature hypo-eutectic carbon steel with, for example, 0.25 percent carbon consists of pearlite and ferrite grains. During heating, no change in grain size occurs until the A_1 or *lower critical temperature* line is reached. See Fig. 6–2 and Fig. 6–7.

Upon reaching this lower critical temperature, pearlite recrystallizes to form a very fine-grained austenite as alpha iron changes to gamma iron. This change affects only the original pearlite, approximately 30 percent of the structure, the ferrite or 70 percent of the structure remaining unchanged.

As the temperature is raised from the lower critical temperature, A_1, to the *upper critical temperature*, A_3, the ferrite transforms from alpha to

Table 6–1. Photomicrographs of iron structures. (See page 168)

AUSTENITE, 250X

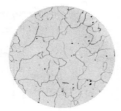

FERRITE (IRON), 50X

IRON CARBIDE (CEMENTITE)
NETWORK, 500X

PEARLITE, 500X

LEDEBURITE, 75X

gamma iron and is absorbed by the previously recrystallized austenite. During the change the gamma iron is physically disturbed and continues to recrystallize to form an extremely fine-grained structure. Any grain growth is prevented by diffusion and transformation of the surrounding ferrite. However after the temperature is raised beyond the upper critical temperature, A_3, and all iron has changed from alpha iron to gamma iron and gone into solution with austenite, grain growth begins and continues until the solidus line is reached.

Upon slow cooling, no marked change in grain size occurs until the temperature is lowered to the upper critical[1] temperature. Here the austenite precipitates fine ferrite grains at the boundaries of the austenite grains. As the temperature is lowered precipitation continues with gradual growth in volume and size of the newly formed ferrite grains. The grain size of the austenite is decreasing while the grain size of the ferrite is increasing. Upon reaching the lower critical temperature, any remaining austenite transforms into pearlite.

From this it may be seen that the

[1]It should be noted that both the upper and lower critical temperatures are suppressed during cooling. This is due to non-equilibrium conditions of cooling.

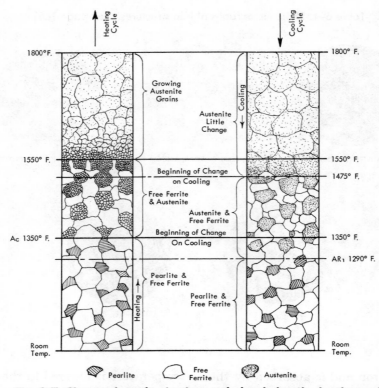

Fig. 6–7. Changes in grain structure and size during the heating and cooling of a hypoeutectic (0.35%) carbon steel.

size and shape of the original austenitic grains influence to a marked degree the resultant grain size and distribution of free ferrite and pearlite. A very fine grain size in the original austenite will result in a correspondingly fine aggregation of free ferrite and pearlite.

If a hyper-eutectoid steel containing pearlite and cementite is heated no change of grain size occurs below the lower critical temperature A_1. At

the lower critical temperature, pearlite changes to austenite (alpha iron changes to gamma iron and Fe_3C crystals dissolve), resulting in a very fine-grained austenite. The excess cementite is unaffected at this temperature but must be heated to the upper critical temperature, AC_m, before this excess cementite is dissolved by the fine-grained austenite. During this heating marked grain growth of the austenite is prevented by the absorption and diffusion of the excess cementite. However, some grain growth takes place which results in a coarser-grained austenite. When the upper critical temperature is reached and all excess cementite is dissolved, marked grain growth begins and continues up to the solidus temperature.

6.7 Critical Temperatures. The diagram shown in Fig. 6–8 is seen to be a portion of the iron-carbon alloy diagram and is often referred to as the critical temperature diagram. It is through the temperature range shown that steels undergo the described structural changes during cooling and heating. It is apparent from Fig. 6–8 that two lines are shown for both the lower and upper critical temperature lines. This splitting up of transformation temperature lines is due to the difference in transformation temperature during cooling as compared with transformation temperatures upon heating. That is, the transformation temperature of an 0.85 percent carbon steel during the change from austenite to

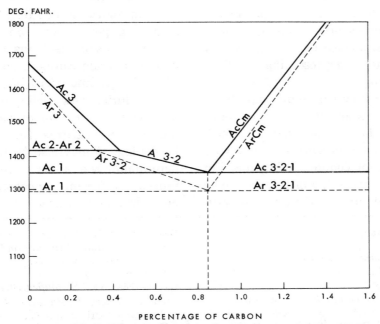

Fig. 6–8. Critical temperature diagram showing lag between heating and cooling transformation temperatures.

pearlite during cooling at a certain rate may be 1290°F, whereas the change from pearlite to austenite upon heating may occur at 1350°F. Thus, the difference between transformation temperatures may be 60°F even though very slow heating and cooling are employed. In reading the critical temperature diagram the A_c temperature lines represent the transformation temperatures upon heating and the A_r temperature lines represent cooling lines. The method of designating the various points is taken from the French. The halt or arrest during heating or cooling is designated by the letter "A" from the French word *arrêt*. An arrest on cooling is referred to by the letters "A_r", the "r" from the French word "*refroidissement*" meaning cooling. An arrest on heating is designated by "A_c", the "c" representing *chauffage* or heating.

Under practical conditions of heating and cooling there may be much more delay or lag than the 60°F cited. This difference increases with the rate of cooling or heating. For instance, the A_c temperatures can be raised about 300°F from their nor-mal occurrence by fast heating. The A_r temperatures can be lowered about 1000°F, in fact, can be forced below room temperature by rapid quenching of thin sections. The critical points, considered collectively, are known as the critical range. For instance, the lower critical range of an 0.45 percent carbon steel on heating extends from 1340°F to about 1410°F. (See Fig. 6–8.)

In summary, the iron-carbon diagram shows the transformation temperatures as iron undergoes allotropic changes. It also shows the solubility ranges of carbon in iron at various temperatures. Both the eutectic and eutectoid compositions are shown and can be used as the fulcrum for the lever rule when determining percentage compositions of the various phases. Thus, the iron-carbon diagram can be used to predict the composition of various phases during slow cooling or it can be used as a guide for proper hardening temperatures for quench hardening. It may also be used for determining the proper temperature ranges for forging, annealing, and normalizing operations.

References

Avner, Sidney H. *An Introduction to Physical Metallurgy*. New York: McGraw-Hill, 1964.

and Engineering Data. Bulletin No. 6.

Lyman, Taylor, ed. *Metals Handbook, 1948 ed.* Cleveland; American Society for Metals.

Brick, Robert M., Gordon, Robert B. and Phillips, Arthur. *Structure and Properties of Alloys, 3rd ed.* New York: McGraw-Hill, 1965.

LaSalle Steel Co., *Simplified Steel Terms*

Reed-Hill, Robert E. *Physical Metallurgy Principles*. Princeton, N.J.: D. Van Nostrand Co., 1964.

Heat Treatments for Steel

7.1 Introduction. The term "heat treatment" is a broad generic term and may have reference to any one of a number of heating and controlled cooling operations that will provide desired properties in a metal or alloy. It is therefore necessary to refer to a specific type of treatment such as martempering, process annealing, and spheroidizing, when discussing heat treating operations.

Metals and alloys are heat treated for a number of purposes, but primarily to (1) increase their hardness and strength, (2) to improve ductility, or (3) to soften them for subsequent metal forming or cutting operations. Various heat treatments may also be used to relieve stresses due to machining, forging, or casting, eliminate hydrogen picked up during chemical machining or electroplating, eliminate the effects of cold-work, and give a more uniform grain size.

Various heat treatments may be applied to both ferrous and non-ferrous metals and alloys. However, only certain alloys may be hardened by thermal treatment. Considerable attention will be given to the hardening of steel alloys because of their wide usage. Attention will also be given in following chapters to non-ferrous precipitation hardening alloys of aluminum, magnesium, stainless steels, copper, nickel, titanium, and beryllium.

7.2 Heat Treating Diagrams. A number of diagrams have been developed in order to help explain the changes that take place as various alloys are cooled. A simplified portion of the *iron-carbon diagram*, Fig. 7–1 shows the normalizing, annealing, and hardening temperatures for steels containing varying amounts of carbon. With reference to Fig. 7–1, it should be noted that the *aus-*

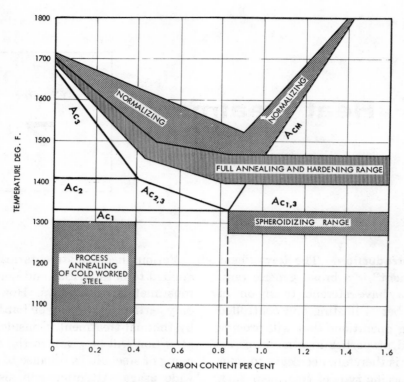

Fig. 7–1. Critical temperature diagram showing normalizing, annealing, spheroidizing, and hardening ranges for carbon steel.

*tenitizing** temperature for hypoeutectoid and eutectoid steels lies between the annealing and normalizing ranges and is usually about 50°F above the upper critical temperature. Hypoeutectoid steels with low carbon content require higher austenitizing temperatures than do eutectoid steels. For plain carbon hypereutectoid steels the recommended austenitizing temperature is usually between the Acm and Ac3 lines, Fig. 7–1. Higher temperatures will coarsen the austenite grains. Austenitizing is time-temperature dependent, the more important factor being temperature. Austenite forms by a process of nucleation and growth as the steel reaches the austenitizing temperature. A finely spheroidized (pin point) carbide

---

Austenitizing consists of heating a steel alloy to a high temperature and holding it there until the ferrite changes to austenite and the iron carbide, which is only slightly soluble in ferrite, dissolves completely in the austenitic solid solution. In the case of precipitation hardening alloys this is called *solution* treatment.

structure austenitizes most rapidly, followed next by fine pearlite, and then coarse pearlite or a spheroidized structure. Soaking time for carbon and low alloy steels may be as short as five minutes per inch of thickness. For high carbon steels containing alloys such as chromium, molybdenum, vanadium, or tungsten which have an affinity for carbon, the complex carbides are relatively slow to dissolve and soaking time should be increased over that of plain carbon or low alloy steels.

Austenitizing is the key to controlling the hardness in quenched steels, and it also affects eventual size change and toughness. If the temperature is too low, there may be incomplete solution of the carbides and full hardness will not be attained upon quenching. If too high, large grains may develop causing brittleness or cracking during heat treatment. Steel producers generally publish data sheets showing correct austenitizing temperatures for each alloy along with recommendations for annealing, normalizing, forging, hardening, and tempering.

Another important diagram which helps to explain the changes occurring during cooling is the *isothermal-transformation diagram*. This diagram, developed in Fig. 7-2 A, B, and C, shows transformation products resulting from cooling austenite, including time to beginning and end-ing of transformation. Thus the I-T diagram of a steel is a kind of map which charts the transformation of austenite as function of temperature and time and shows approximately how a particular steel will respond to any mode of slow or rapid cooling from the austenitic state.

At the austenitizing temperature steel is composed of a single constituent, austenite, which can exist indefinitely so long as the temperature is maintained above a certain minimum temperature. In eutectoid carbon steel, this minimum temperature, designated A_{e1}, is about 1335°F. When austenite is cooled below this temperature, it becomes unstable and changes into a mixture of ferrite and carbide. This change of austenite requires a certain incubation period before the transformation begins and a certain amount of time to complete it.

Thus, if a small steel sample is cooled quickly enough, it may be brought to any desired temperature below A_{e1}, virtually unchanged. In practice, a number of ½″ diameter samples approximately $\frac{1}{16}$″ thick are prepared from the same bar and are drilled with a small hole near one edge into which a small wire is threaded for ease of handling. The samples are heated then taken from the furnace and quenched into molten lead or salt baths whose temperature is maintained at the desired constant temperature.

175

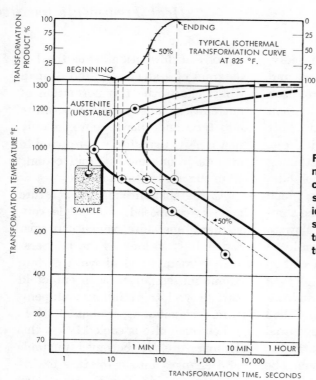

Fig. 7-2A. Step 1 in the development of the isothermal cooling curve by quenching a series of specimens into molten baths at various elevated temperatures. Curves show beginning and ending time for transformation at given temperatures.

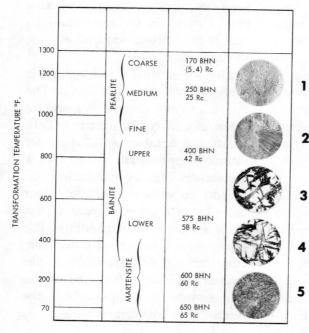

1 Fig. 7-2B. Step 2, identifying isothermal transformation products resulting from quenching specimen to temperatures shown and holding **2** until transformation is completed. All specimens are 2500X.

   (1) Coarse Pearlite
   (2) Fine Pearlite
3 (3) Upper Bainite, 50% transformed at 750°F.
   (4) Lower Bainite, 50% transformed at 550°F.
4 (5) Quenched Martensitic structure.

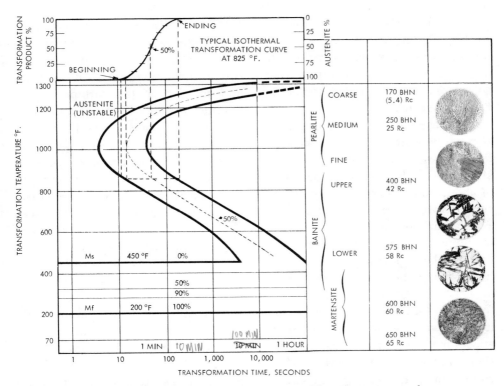

Fig. 7-2C. Complete isothermal transformation diagram with cooling curves and transformation products for eutectoid carbon steel. The transformation curve at the top left reveals that transformation begins rather slowly, accelerates rapidly, and then is reduced at the end. The horizontal portion of the curve near the bottom of the figure shows beginning and finishing temperatures for martensitic transformation.

After reaching the temperature of the molten bath, the samples are held for a period of time, and then at successive time intervals, individual samples are withdrawn and quickly quenched in cold water to stop further transformation to the ferrite-carbide structure. The samples are then examined for hardness and microstructure. As the temperature of the molten bath (and consequently the transformation temperature) is changed a series of data points are obtained which can be plotted to construct the so called I-T (isothermal transformation) diagram. This diagram is also called the Bain "S" curve or the TTT (time-temperature-transformation) diagram. One complete diagram may require heat treatment and metallographic examination on more than 100 samples.

The transformation products for

a eutectoid carbon steel are shown in Fig. 7–2A, the samples were quenched to the temperatures shown and held until transformation was completed. At the higher temperatures austenite transforms into coarse pearlite. As the transformation temperature is lowered the pearlitic flakes become finer and the structure harder. At a temperature of about 800°F a new type of structure is formed, called bainite after E. C. Bain, an outstanding pioneer in the field of metallurgy. Bainite is harder than pearlite and at the higher temperatures is called upper or feathery bainite. As the transformation temperatures are lowered, it takes on a needle-like or acicular appearance. With lower transformation temperatures, the carbides become smaller and more closely spaced and the individual bainite needles are thinner. Freshly formed bainite is dark etching and appears as an aggregate of ferrite and finely divided carbide particles.

As the temperature is lowered further austenite transforms more and more slowly until it reaches a temperature designated by M_s. At the M_s temperature (420°F for eutectoid carbon steel) portions of the austenite transform almost instantaneously into a product known as martensite. (Measurements in an iron-nickel alloy show martensitic transformation time to be of the order of 10^{-5} seconds). The transformation product formed is a super-saturated solid solution of carbon trapped in a body-centered tetragonal structure. Transformation is unlike that of formation of the carbide-ferrite mixture in that martensite transformation is diffusionless and involves no change in chemical composition from the original austenite. There is no nucleation followed by growth as with pearlite and bainite. See Fig. 7–3. Instead small discrete volumes of the austenite suddenly change from a F.C.C. structure into the tetragonal B.C.C. structure by a combination of two shearing actions. The unit cell is highly distorted because of the entrapped carbon and is the prime reason for the high hardness of martensite. The atoms of the tetragonal B.C.C. martensite are less densely packed than the F.C.C. austenite causing an expansion of approximately 4 percent during the transformation. This expansion during the formation of martensite results in high localized stresses resulting in plastic deformation of the austenite matrix. The start of martensitic transformation as shown by the M_s temperature in Fig. 7–2B progresses only while the specimen is cooling. Transformation ceases if cooling is interrupted before the M_s temperature is reached and retained austenite will be present in the structure.

Freshly formed martensite, unlike bainite, is slow etching and therefore

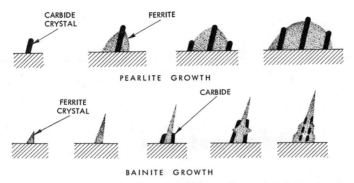

CARBIDE CRYSTAL   FERRITE

PEARLITE GROWTH

FERRITE CRYSTAL   CARBIDE

BAINITE GROWTH

Fig. 7–3. Nucleation and growth of pearlite and Bainite. Pearlite is nucleated by a carbide crystal. Bainite is nucleated by a ferrite crystal with carbide rejected as crystallites after reaching critical concentration.

appears in photomicrographs as a white needle-like or acicular structure sometimes described as a pile of straw. Upon tempering, at temperatures even as low as 212°F martensite changes to a rapid etching constituent and appears, accompanied by retained austenite. By suitable heat treatment the martensitic crystals formed first from quenching may be tempered into the dark etching state and the retained austenite may be subsequently transformed into white martensite, giving the contrast shown in Fig. 7–4.

The M_s temperature, or the temperature at which martensite begins to form, is dependent primarily upon chemical composition. The calculated M_s temperature for several alloys commonly used for carburized gears are shown in Table 7–1.

While the I-T diagram shows the time-temperature transformation at constant temperature, most heat treatments involve transformation as it occurs on continuous cooling. The I-T diagram is useful in planning heat treatments involving continuous cooling, but it cannot be used directly to accurately predict transformation products. However, the I-T diagram can be used to derive an approximate diagram as shown in Fig. 7–4 called the C-T or continuous cooling transformation diagram. Essentially, the C-T cooling curve is shifted slightly downward and to the right of the cooling curve for the I-T diagram. The reason for this shift is due to the slight delay in the beginning of transformation at higher cooling rates. Also, the I-T diagrams are usually inter-

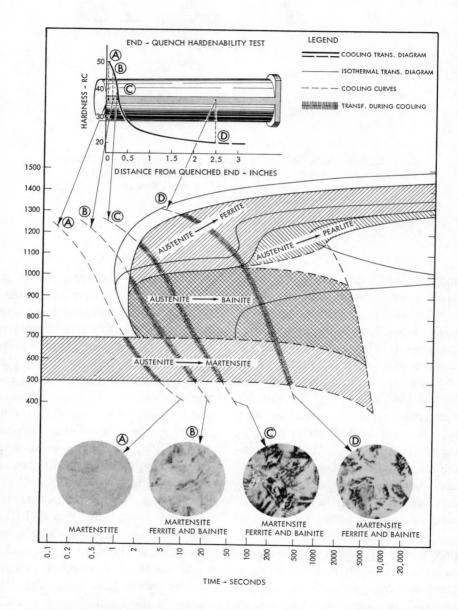

Fig. 7–4. Correlation of continuous cooling and isothermal transformation diagrams with end-quench test data for eutectoid carbon steel. (By permission: *U.S.S. Carilloy Steels, U.S. Steel Corp.*).

Table 7–1. Calculated M_s temperatures for carburized steel alloys used in the manufacture of gears.

STEEL ALLOY SERIES	CALCULATED* M_s TEMPERATURE, DEGREES FAHRENHEIT
10xx	233
41xx	203
86xx	185
46xx	181
31xx	168
T–1	164
43xx	163
23xx	138
Hy–80	133
25xx	107
93xx	94
49xx	93
33xx	74
28xx	3

*THE EMPIRICAL EQUATION USED FOR CALCULATING THE M_S TEMPERATURE INCLUDES THE EFFECT OF CARBON, CHROMIUM, NICKEL, SILICON AND MOLYBDENUM AND IS AS FOLLOWS: $M_s(°F) = 930-540$ (%C) $- 60$ (%Mn) $- 40$ (%Cr) $- 30$ (%Ni) $- 20$ (%Si) $- 20$ (%Mo).

FROM: "RETAINED AUSTENITE EFFECTS," LYNN E. ARNOLD, CINCINNATI SUB-ZERO PRODUCTS, INC., BY PERMISSION

preted by scanning from left to right along a particular temperature level, whereas the C-T diagram is interpreted by scanning downward along a cooling curve.

At the top of Fig. 7–4 a measured hardness curve has been superimposed on an end-quench hardenability bar. Four representative locations (A, B, C, D) along the bar corresponding to cooling rates of approximately 450°, 300°, 100°, and 5°F per sec and are shown by the corresponding cooling curves on the I-T and C-T diagrams.

Austenite at a particular location transforms when its cooling curve passes through a shaded zone of the C-T diagram. The type of microstructure resulting from transformation in each zone is given in the lower

portion of the chart. Thus, at point A on the end-quench bar, the hardness is high because the cooling rate was high enough to miss the "nose" of the curve on the transformation diagram and austenite transformed directly to the hard martensite structure. At point B, the hardness is lower because the cooling rate was slower and allowed the transformation of some austenite to fine pearlite as indicated by the intersection of cooling curve B with the shaded portion of the diagram. That cooling rate (somewhere between A and B) which just allows the cooling curve to miss the nose of the transformation curve is known as the *critical cooling* rate and permits a fully hardened martensitic structure. The cooling rates for points C and D are slow enough to allow complete transformation in the pearlite zone. The pearlite transformation product for point D will be coarser than at C. It should be here observed by reference to Fig. 7–4 that bainite which forms isothermally within the range of 850°F to 400°F is sheltered by an overhanging pearlite "nose" and thus bainite is not formed in any appreciable quantity during continuous cooling of this steel. That is, the rate of bainite formation is slower than that of pearlite formation and a steel cooled slowly enough to permit the formation of bainite will completely transform to pearlite before even cooling down to the bainite-forming temperatures.

The practical importance of Fig. 7–4 should be appreciated by the student of metallurgy. For example, if it were desired to fully harden a steel spring, quenching would have to be sufficiently rapid for the cooling curve to miss the nose of the transformation curve; otherwise, partial transformation would occur at the higher temperature and only a portion of the austenite would transform to martensite (beginning at the M_s temperature). Another example of the use of the transformation diagram would be when a fully annealed or soft pearlitic structure is desired for a eutectoid steel. Common practice is to heat a steel to the austenitizing temperature followed by cooling in the furnace. This may take as

much as 24 hours and is a time consuming and expensive practice. By examination of the transformation diagram Fig. 7–4, it can be seen that austenitizing a eutectoid steel at 1550°F following by holding in a second furnace at 1200°F for five minutes would permit complete transformation to pearlite. The steel could then be taken from the furnace and cooled in air, thus freeing both furnaces for additional work. Other variables which must be considered when applying data from a I-T diagram are noticeable effects arising due to (1) slight variation in the chemistry of an alloy from different "heats" of steel, (2) the effect of grain size, and (3) austenitizing temperature. Steels with larger grain size harden deeper than fine grained steels because large grain size tends to retard isothermal transformation.

Figures 7–5 through 7–8 are I-T diagrams with E-Q hardenability data for selected alloys. The effect of alloying elements in retarding transformation is pronounced in some alloys and allows slower quenching rates to achieve a fully hardened martensitic structure in large work pieces. The combined effect of nickel, chromium, and molybdenum are especially noticeable in affecting the E-Q hardenability curve of Fig. 7–7. It may be observed that the hardness of the test specimen is not appreciably diminished even 2½ in. from the quenched end which corre-

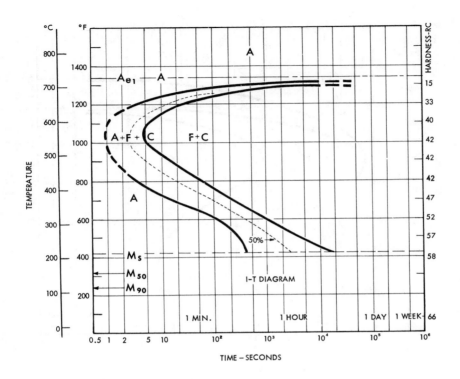

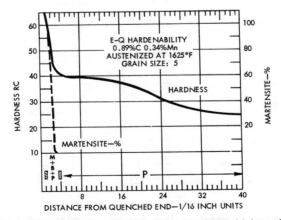

Fig. 7–5. Isothermal transformation diagram for 1095 high carbon steel austenitized at 1625°F., grain size 4-5. (By permission: *Atlas of Isothermal Transformation Diagrams*, U.S. Steel Corp.).

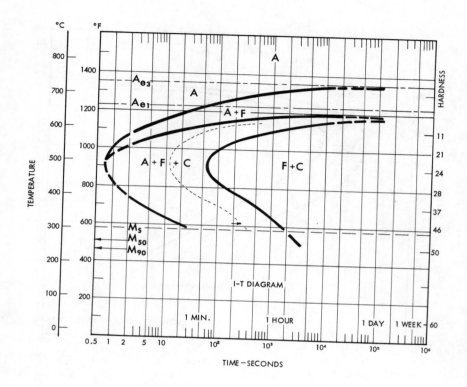

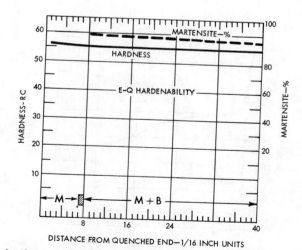

Fig. 7–6. Isothermal transformation diagram for 2340 nickel alloy steel austenitized at 1450°F, grain size 7-8. (By permission: *Atlas of Isothermal Transformation Diagrams*, U.S. Steel Corp.).

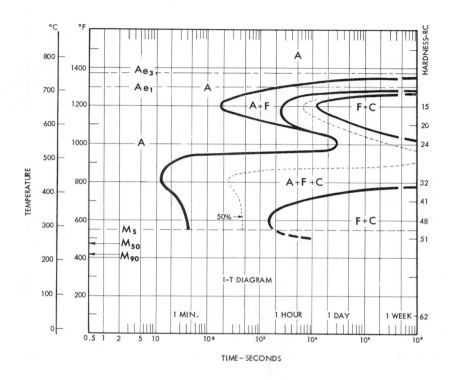

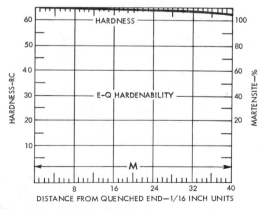

Fig. 7–7. Isothermal transformation diagram for 4340 nickel-chrome-molybdenum steel austenitized at 1550°F, grain size 7-8. (By permission: *Atlas of Isothermal Transformations*, U.S. Steel Corp.).

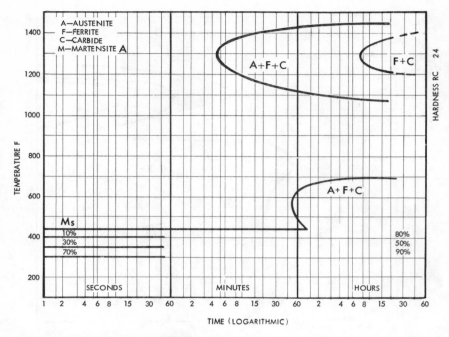

Fig. 7–8. Isothermal transformation diagram for air hardening die steel austeni-tized at 1850°F. (Crucible Steel Co.).

sponds to a cooling rate of only 3°F per second. Thus, this steel could be hardened throughout in relatively thick sections whereas low alloy steels such as those shown in Fig. 7–5 or 7–6 would have a high surface hardness with the hardness of the core diminishing rapidly. Fig. 7–8 shows the I-T diagram for a die steel. Note that the "nose" of the curve has been shifted far to the right by the addition of alloying elements giving sufficient time to cool the specimen in air and still obtain a fully hardened martensite structure.

Very little distortion is present because of the slow cooling rate required for this type of steel.

7.3 Annealing. The term annealing is a general one, usually referring to one of the heat treating processes designed to impart softness and ductility to a hardened or cold worked steel for easier machining, for further processing, or to relieve any internal stresses which may be present from casting, welding, forging, deforming, or machining.

Annealing consists of three stages, namely: (1) heating to the proper annealing temperature, (2) holding or "soaking" at the annealing temperature, and (3) controlled cooling from the annealing temperature.

Slow, uniform heating of a steel to be annealed is desirable for two reasons. First, there exists a temperature gradient between the outer surface and center of a piece of steel which is being heated. With very rapid heating, this temperature gradient will cause the outer surface of

the steel to reach the critical temperature range before the center reaches the same temperature. This means that the outer surface will change to austenite and contract while the center portion is still expanding. The result will be high tensile stresses at the surface and compressive stresses at the center of the steel. These internal stresses may be sufficiently great to produce warping or even cracking of the workpiece. The steels which are slowly heated will have a more uniform temperature through-

Table 7–2. Some common annealing processes.

PROCESS	HEAT TREATMENT	RESULTING STRUCTURE	APPLICATION
FULL ANNEALING (HYPO AND EUTECTIC STEELS)	HEAT TO 50–100°F ABOVE Ac_3 COOL 50 TO 100°F PER HOUR	LAMINATED PEARLITE	IMPROVEMENT OF LOW-CARBON STEELS FOR DEEP FORMING OPERATIONS
PROCESS ANNEAL	HEAT TO 1000°F TO 1300°F TEMPERATURE RANGE FOLLOWED BY AIR COOLING	RECRYSTALLIZATION OF FERRITE, DISTORTED PEARLITE REMAIN	USED AS AN INTERMEDIATE ANNEAL DURING COLD WORKING OF LOW CARBON STEELS INCLUDING SHEET STEEL AND WIRE
STRESS RELIEF	HEAT UNIFORMLY TO RECOMMENDED TEMPERATURE FOLLOWED BY AIR COOLING	NO APPRECIABLE CHANGE FROM ORIGINAL STRUCTURE	USED TO RELIEVE RESIDUAL STRESS RESULTING FROM COLD FORMING, WELDING, CASTING, HEAVY MACHINING, GRINDING, ETC.
NORMALIZING	HEAT ABOVE Ac_3 OR A_{cm} THEN AIR COOL	FINE PEARLITE	USED FOR PLAIN CARBON OR LOW ALLOY STEELS. USED EXTENSIVELY FOR GRAIN REFINEMENT OF CAST OR WELDED STEELS.
SPHEROIDIZING (HYPEREUTECTIC STEELS)	HEAT TO TEMPERATURE BETWEEN Ac AND $Ac 1,3$ AND Acm FOLLOWED BY SLOW COOLING THROUGH CRITICAL RANGE, HOLD JUST BELOW CRITICAL TEMPERATURE. AIR COOL	SPHEROIDIZED CARBIDES IN FERRITE MATRIX	IMPROVES MACHINABILITY OF HIGH CARBON STEELS, IMPARTS MAXIMUM DUCTILITY AND FORMABILITY

out. Thus, the surface and interior of the workpiece will both change to austenite at approximately the same time with minimum internal stress resulting from the transformation. Second, a large temperature gradient allows the outer surface and corners to reach high temperatures with consequent coarsening, while the grains in the center are being refined. Rapid heating of large workpieces will generally produce a structure with a non-uniform grain size.

The proper temperature for annealing is dependent upon the particular type of annealing process involved. The more common annealing processes are shown in Table 7–2.

7.3.1. Full annealing means heating hypoeutectoid and eutectoid steel to about 100°F above the upper

critical temperature, A_{c3}, and hypereutectoid steel 100°F above the $Ac_{1,3}$ temperature as shown in Fig. 7–1. These temperatures produce maximum grain refinement with uniform, equiaxed, reoriented, and recrystallized, stress-free grains. Heating above the recommended temperature for full annealing will result in large grains and the hypereutectoid steels will have a coarse, brittle, cementite network surrounding the pearlite grains. The cementite network or "eggshell" structure in hypereutectoid steel is shown in Fig. 7–9 and 7–10.

7.3.2. Process annealing is often used in the sheet and wire industries to soften the alloy for further cold forming and drawing. The process is normally carried out at tempera-

Fig. 7–9. Cementite network surrounding pearlite grains in hypereutectoid steel, 100X.

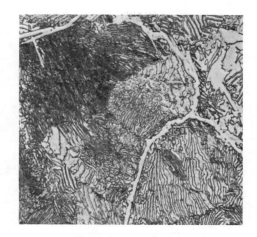

Fig. 7–10. The specimen in Fig. 7–9 magnified 500X.

tures between the recrystallization temperature and the lower critical temperature (1000°F to 1300°F). The intent of process annealing is to promote recrystallization, softening, and fine grains. For that reason, rapid heating and short holding time at temperature are used. Although the low temperatures used reduce scaling, some oxide is present unless a non-oxidizing atmosphere is used as with bright annealing. Fig. 7–11 and 7–12 show the microstructures of steel tubing that has been cold worked and then process annealed.

Stress relief is an annealing process for relieving internal stresses set up in a metal as a result of plastic deformation such as forming, machining, and grinding, or from residual stresses due to cooling of weldments, castings, and forgings. These internal stresses may be the cause of serious warping and even failure by rupture. Due to internal stresses castings that have been accurately machined and assembled into a complicated machine tool may warp (after going into service) and cause the finished machine to become inaccurate or prevent its proper operation. Stress relief consists of heating the workpiece to the recommended temperature, holding it for a sufficient length of time to become uniformly heated throughout, and cooling in air. Stress relieving treatments are often performed in such a way that the properties of material produced as a result of strain hardening, age hardening, and so on, are not substantially affected. Stress relieving operations conducted at lower temperatures (approximately

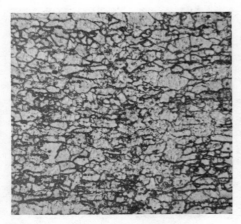

Fig. 7–11. Microstructure of very low carbon steel sheet cold reduced 69%. Grains are highly distorted and show directional properties, 250X.

Fig. 7–12. Material in Fig. 7–11 after process anneal at 1250°F. Ferrite grains have recrystallized during the process. Small pearlite areas remain unaffected.

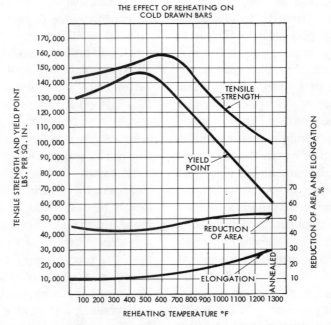

Fig. 7–13. The effect of stress relief annealing treatments at various temperatures on cold drawn steel bars. (LaSalle Steel Co.)

190

550°F) on cold worked steel is sometimes termed *strain tempering*, while those conducted at temperatures approaching the lower critical temperatures are sometimes called *strain annealing*. Stress relief treatments which have been recommended for ferrous metals are as follows:

Cast iron . . . 900 to 1100°F, ½ to 5 hours

Carbon steel (cold worked and heat treated) . . . 925 to 1200°F, 1 hour

Carbon steel weldment . . . 1200°F, 1 hour

The effect of stress relieving at various temperatures is shown in Fig. 7–13.

7.3.3. Normalizing is similar to annealing in that the workpiece is heated to a temperature above the upper critical range and held for a sufficient length of time to allow the structure to become completely austenitized. However, normalizing differs from annealing in that the rate of cooling is somewhat accelerated by allowing the steel to cool in air. The usual objectives of normalizing treatments are to secure a controlled, definite grain size and to produce a finer and stronger pearlite structure than is obtained by full annealing. Normalizing temperatures are approximately 100°F above the upper critical temperature.

The undesirable structure resulting from hot forging is shown in Fig.

7–14, and in Fig. 7–15 is shown the normalized structure for the same steel. Grain size resulting from various heat treatments such as normal-

Fig. 7–14. Undesirable microstructure of hot forged steel showing banded structure and mixed grain size, 75X.

Fig. 7–15. Material in Fig. 7–14 after normalizing. Undesirable structure has been corrected and grain size is uniform, 75X.

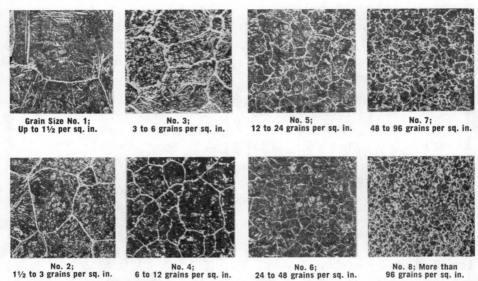

Grain Size No. 1;
Up to 1½ per sq. in.

No. 3;
3 to 6 grains per sq. in.

No. 5;
12 to 24 grains per sq. in.

No. 7;
48 to 96 grains per sq. in.

No. 2;
1½ to 3 grains per sq. in.

No. 4;
6 to 12 grains per sq. in.

No. 6;
24 to 48 grains per sq. in.

No. 8; More than
96 grains per sq. in.

Fig. 7–16. Standard grain size for steel, 100X. (By permission: *A Metal Progress Data Sheet***).**

izing or austenitizing may be determined by examination of the fractures of ¾ inch diameter specimens and comparing them with Shepherd Fracture Standards as described in paragraph 2.9. Another method of examination which is more reproducible is by microscopic examination at 100X magnification and comparison with printed ASTM grain size charts shown in Fig. 7–16, or more conveniently, by means of a microscope eyepiece reticle. Grain boundaries are caused to appear by suitable chemical or oxidation etching or by the McQuaid-Ehn test (See also ASTM E112-61) in which the

steel is carburized for 8 hours at 1700°F, cooled slowly, etched or microscopically examined. The carbide network resulting from the carburizing clearly outlines the grain boundaries.

7.3.4. Spheroidizing has been designed to convert hard lamellar or network carbides of high carbon steels into globular or spherical shapes for improvement of machinability or for obtaining maximum ductility. The spheroidized type of structure, shown in Fig. 7–17 may be obtained by one of the following three methods:

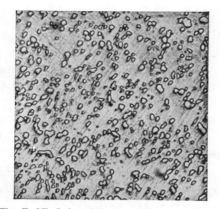

Fig. 7–17. Spheroidized carbide microstructure of S.A.E. 3250 steel, suitable for automatic screw machine work, 750X, 3% Nital etch.

and reheating operations. If the microstructure is initially coarse pearlite, the spheroidizing time may be 16 to 72 hours or longer at 1300°F. In order to facilitate spheroidizing, the initial structure should preferably be fine pearlite. Quenching the steel from the austenitic region often proves desirable prior to spheroidizing.

2. Small objects of high carbon steel may be spheroidized more rapidly by alternately heating them to temperatures slightly below and then above the lower critical temperature. The length of time the steel is held at each temperature and the number of cycles through which it is heated and cooled is dependent upon the carbon content and initial structure of the steel.

1. Prolonged heating at a temperature just below the lower critical temperature followed by relatively slow cooling. Although this method is the slowest, it is frequently used, since it does not involve quenching

3. Tool steels are generally spheroidized by heating to a temperature of 1380 to 1480°F for carbon steels and higher for many high alloy tool

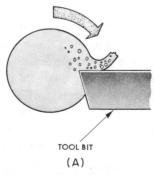

TOOL BIT

(A)

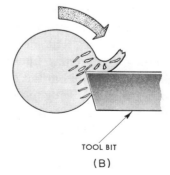

TOOL BIT

(B)

Fig. 7–18. Schematic showing ease of machining A, spheroidized structure compared with B, lamellar carbide structure.

steels, holding at temperature for 1 to 4 hours followed by slow cooling in a furnace.

A schematic illustration of the improvement of machining of spheroidized carbides over lamellar pearlite is shown in Fig. 7–18. Spheroidized tool steel is much softer and more easily machined than annealed tool steel with coarse pearlitic structure.

7.4 Hardening by Quenching. One of the important properties of carbon and alloy steels is that they can be hardened and strengthened by heating them to the austenitizing temperature followed by rapid cooling to room temperature. The resultant high hardness makes possible the use of steel for metal cutting tools which can maintain sharp cutting edges under severe operating conditions, for dies which can resist abrasion and wear, and for strengthening machining components and making them resistant to nicks and burrs. It should be mentioned that high hardness obtained from quenching of steel is also usually accompanied by excessive brittleness. The designer has to make a compromise in choosing between maximum hardness and maximum toughness. Toughness can be increased and hardness decreased by tempering treatments which will be discussed later.

There are three main requirements for satisfactorily hardening steel by the quench hardening method.

These requirements are:

1. Adequate carbon content in the steel alloy to form martensite.

2. The proper hardening temperature and time to allow the steel to become fully austenitic.

3. Sufficiently rapid cooling to prevent transformation until the M_s temperature is reached.

Each of the requirements for quench hardening will be discussed in some detail.

The first requirement for quench hardening is that there is adequate carbon present. Of all the elements alloyed with iron for altering mechanical properties carbon is the most powerful for controlling hardness, strength, and wear resistance. Most all carbon steels can be quench hardened but the hardness does not become appreciable until the carbon content of the steel reaches about 0.35 percent. As shown in Fig. 7–19, the hardening capability of carbon steel clearly increases with the increase of carbon. This increase in carbon retards transformation in making it easier to undercool austenite and harden the steel. Increasing the carbon content also increases the hardness of steel in both the pearlitic and martensitic conditions.

The hardness of martensite depends upon the carbon content and levels off at an *apparent* maximum of about Rockwell C66 at the composition of 0.60 percent carbon. However, the Vickers hardness test,

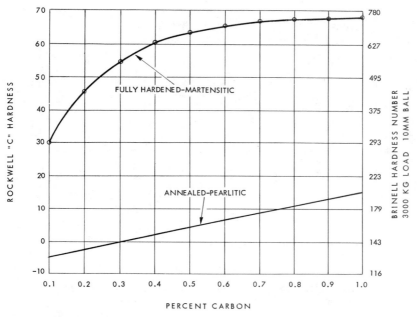

Fig. 7–19. Effect of carbon content on hardness of carbon steel in the martensitic and pearlitic conditions (approx.).

which is more sensitive in the higher hardness ranges because of its sharp pyramid rather than the cone shaped Rockwell indenter with its rounded point, shows that the hardness of martensite continues to increase even beyond a carbon content of 1.20 percent.

Since the Rockwell machine is widely used, much of the literature refers to a maximum hardness of steel of Rockwell C66 with a carbon content of 0.60 percent. The highest carbon content normally found in plain carbon steel is about 1.30 per-

cent and is used for razors, engraving tools, and similar application. The temperature of martensite transformation is very much affected by increased carbon content as shown in Fig. 7–20. The higher the carbon content of the austenite the lower is the temperature of martensite transformation. Particular attention should be paid to the M_f temperature; steels whose carbon content is greater than 0.60 percent will have some retained austenite present because the M_f temperature is below room temperature. The percentages

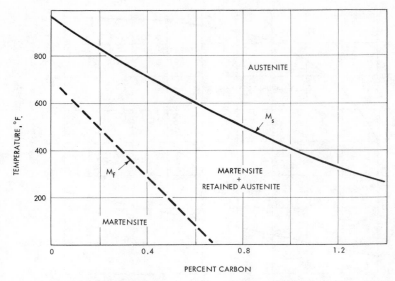

Fig. 7–20. Effect of carbon content on temperatures for the start and finish of martensitic transformation in an iron-carbon alloy.

of retained austenite for typical carbon steels which have been quenched to room temperature is shown in Table 7–3. It shows that for higher carbon steels the amount of retained austenite is quite appreciable and could be the source for serious distortion in precision tools and dies if it were to transform to martensite at a later time as a result of applied external stresses, heat, and other factors.

The retained austenite may be transformed to martensite by subzero treatment or by suitable tempering. Both treatments will be discussed later.

The second requirement for quench hardening is that the steel be heated to the recommended hardening temperature and held for a sufficient length of time to allow the steel to become fully austenitized. The recommended hardening temperatures are usually in the same range as for annealing, approximately 25-50°F above the upper critical temperature for hypoeutectic alloys and between $A_{1,3}$ and A_{cm} for hypereutectic alloys. Large workpieces are usually heated to the upper side of the range and small or thin sections to the lower side of the range of temperatures.

In order to reduce possibilities of cracking intricate or large sections

Table 7–3. Typical percentages of retained austenite for carbon steels quenched to room temperatures.

INITIAL CARBON CONTENT, PER CENT	RETAINED AUSTENITE, AT ROOM TEMPERATURE, PER CENT
.20	0–2
.40	0–4
.60	2–7
.80	5–14
1.00	15–23
1.20	30–38

of high alloy or tool steels as they are being heated, a *preheating* treatment is frequently employed. This preheating treatment involves heating a steel slowly and uniformly to a temperature below the hardening temperature then transferring it to the higher temperature hardening furnace and bringing it rapidly to the hardening temperature. There are two primary functions of preheating, namely: (1) to reduce thermal shock which is always present when a cold steel workpiece is placed directly into a hot furnace, and (2) to start the austenitizing reaction so that the carbides will be in a more favorable condition to dissolve at the higher temperature. This latter function is particularly important in hardening of high speed steels which require temperatures above 2000°F.

Because preheating does reduce thermal shock it is beneficial in minimizing the danger of excessive distortion, warping, and cracking. It also serves as a stress relief for internal stresses which may be present from machining of the workpiece. If preheating is done at a temperature below the lower critical temperature the internal stresses will be relieved prior to any transformation within the steel.

One or more preheating steps will also give a more uniform temperature throughout the entire cross section of the workpiece and allow it to pass through the transformation range as uniformly as possible. It should be remembered that heating a workpiece to some temperature above the lower critical temperature results in a decrease in volume. The lattice structure changes from the body-centered-cubic to the face-cen-

197

tered-cubic arrangement resulting in interior stresses within the workpiece. It is important for intricate shaped workpieces to go through this phase change slowly and uniformly to minimize distortion.

Preheating temperatures for carbon and low alloy tool steels are normally in the 1200-1300°F range, high carbon, high chromium and other highly alloyed steels between 1200-1450°F, and high speed steels between 1500-1600°F. When a protective atmosphere is not available

for the preheating furnace the temperatures should be adjusted to the lower side of the range to reduce excessive scaling and decarburization.

The final requirement for quench hardening of iron-carbon alloys is that they be quenched rapidly enough to undercool austenite to the martensitic transformation temperature, M_s, before normal pearlite has a chance to form. As discussed previously, the critical cooling rate required to successfully undercool austenite and secure full hardening is

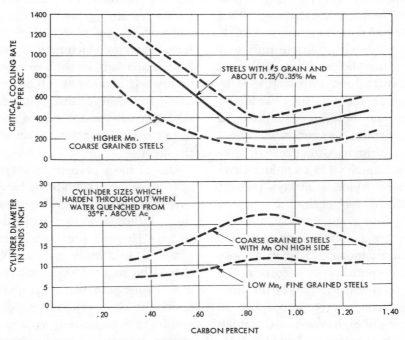

Fig. 7–21. Critical cooling rates for plain carbon steels as affected by carbon and manganese contents and grain size. (International Nickel Co.)

profoundly affected by the alloying elements present in the steel. Another important factor influencing critical cooling rate is the grain size of the steel.

The variations in critical cooling rates with carbon and manganese content and with grain size in plain carbon steels are shown in Fig. 7–21. Increase of carbon from low values of around 0.20 percent as in machinery steel to about 0.85 percent as in tool steels lowers the critical cooling rate from about 1300°F per second (as measured at 1290°F) to about 250°F per second. With an increase in carbon above 0.85 percent the critical cooling rate increases slightly. The effect of nickel in reducing the critical cooling rates of steels is shown in Table 7–4. This great reduction in critical cooling rates permits the hardening of small

parts with mild quenching media such as oil or even air, depending upon the type and amount of alloying elements present. This gives rise to such terms as *water hardening, oil hardening,* and *air hardening* steels.

The effect of grain size on critical cooling rate may be dramatically illustrated. Fig. 7–22. The plain 1.0 percent carbon steel specimens are one inch in diameter and are arranged with grain sizes ranging from coarse, No. 1, to fine, No. 8. The coarse grained specimen hardened all the way through because of its lower critical cooling rate. (That is, the knee of the I-T curve was further to the right.) The fine grained steel, on the other hand, is fairly shallow hardening because of its much higher critical cooling rate. In other words, the fine grained steel was partially transferred to pearlite

Table 7–4. Comparison of critical cooling rates for carbon and nickel steels of medium grain size.

CARBON CONTENT, PER CENT	CRITICAL COOLING RATE, DEG . F, PER SEC.	
	CARBON STEELS	3% NICKEL STEELS
0.20	1260	78
0.40	920	62
0.60	600	42
0.80	290	32
1.00	325	40
1.20	415	50

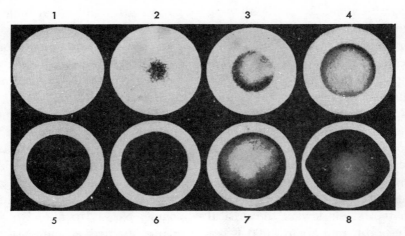

Fig. 7–22. Effect of austenitic grain size on depth of hardness of quenched 1″ rounds of 1% carbon steel. No. 1 coarse grained steel is fully hardened, No. 8, fine grained steel has a thin hardened case (white) with a soft core (dark). Specimens sectioned and etched, 1x.

before the M_s temperature was ever reached. Thus, only the outer surface of the fine grained steel was cooled rapidly enough to miss the knee of the I-T curve and avoid transformation at the higher temperature. The hardness penetration in any standard carbon steel may be affected to a large degree by specifying the hardening temperature. It will be recalled that the minimum grain size for austenite is just at the $Ac_{1,3}$ temperature of Fig. 7–1. Heating beyond that temperature results in grain growth of the newborn austenite. By heating only slightly above the $Ac_{1,3}$ critical temperature a very fine grain size results in relatively shallow hardening steel; whereas heating above the critical temperature and developing a coarser grained austenite will produce a deeper hardening steel. Large pieces, such as forging dies, which require a deep hardness penetration are usually heated to a higher hardening temperature to insure uniform surface hardness and rather deep hardness penetration; whereas small dies and thin sections may be hardened from a temperature close to the critical temperature.

7.5 Quenching Methods and Media. One of the important decisions that must be made in quench hardening of steels is selection of the process for cooling the steel. Cooling must be rapid enough to completely suppress the formation of pearlite or spheroidite in the 300 to 950°F temperature range. Too slow a quench will result in only partial transformation of austenite into martensite with resultant soft spots in the steel, while too drastic a quench may result in distortion, warping, or cracking. The critical cooling rate necessary to suppress transformation of austenite until the martensitic transformation temperature, M_s, is reached varies considerably from one type of steel to another. One type of steel may require an extremely rapid quench such as obtained with a water quench to match the critical cooling rate of the steel, while other steels may be cooled

rapidly enough in a slower quenching oil to achieve a fully martensitic transformation product. The composition of the steel is the most important item influencing the selection of the quenching media, although grain size is also important. Thus the effect of different alloying elements makes it possible to divide tool steels into:

1. Water hardening steels (shallow hardening),

2. Oil hardening steels (deep hardening),

3. Air hardening steels (minimum distortion).

Air hardening steels having high alloy content delay the start of transformation by several minutes, thus

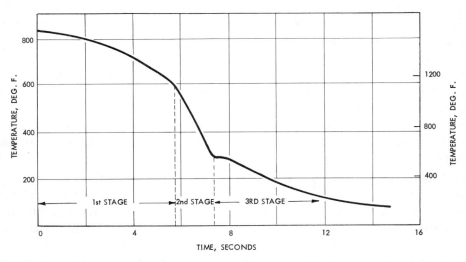

Fig. 7–23. Typical cooling curve for the center of a small diameter water quenched steel specimen showing three stages of cooling.

giving deep hardening characteristics and minimum distortion.

An interesting cooling curve for a small steel specimen, Fig. 7–23, illustrates how the cooling power of a water quenching medium changes as the surface temperature of the specimen decreases. Three distinctive stages of liquid quenching are shown and may be described as follows:

Vapor Film Cooling Stage. A thin, stable vapor film is formed as the hot steel is immersed in the liquid bath and surrounds the hot metal insulating it from the quenching media. Unless the film is broken cooling is by conduction through the vapor blanket and is relatively slow, Fig. 7–24 A. If the film is maintained, soft spots, warping, and cracking may result.

Vapor Transport Cooling Stage. This second stage occurs when the vapor film collapses and breaks away from the surfaces of the hot metal allowing the hot metal to come in contact with the quenching media. During this stage the cooling rate is most rapid. The quenching liquid wets the hot metal surfaces, bursts into vapor, and creates a boil of the quench bath. This stage is graphically portrayed in Fig. 7–24B as the action is violent enough to tear oxide scale from the surface of the specimen.

Liquid Cooling Stage. This stage shown by a reduction in the slope of the cooling curve, Fig. 7–23, occurs as the hot metal approaches the boiling point of the quenching liquid, Fig. 7–24C. Vapor no longer forms and cooling is by liquid conduction and convection resulting in a relatively slow cooling rate. However, during this stage of cooling a fast rate of quenching is not as essential as during the earlier stages, particularly during the second stage where the steel is passing the nose of the I-T curve with its maximum austenitic transformation rate.

The first stage of quenching is important as a great majority of the common hardening difficulties occur during this stage. A persistent vapor film may result in a soft steel, soft spots, warping, or cracking.

The temperature of the quenching bath is an important factor in determining how long the vapor film phase will last. The duration of the vapor film for various quenching liquids at progressively increasing temperatures is shown in Fig. 7–25.

The extremely rapid vapor transport stage of cooling is easily recognized. A specimen held in a pair of tongs and quenched in water or brine vibrates rapidly and emits a characteristic quenching sound. The old time heat treater said of this stage, "The water bites the steel."

Water is the most common quenching medium used in the hardening of the carbon and low alloy steels. Steels that have a very rapid transformation rate from austenite and are consequently shallow hard-

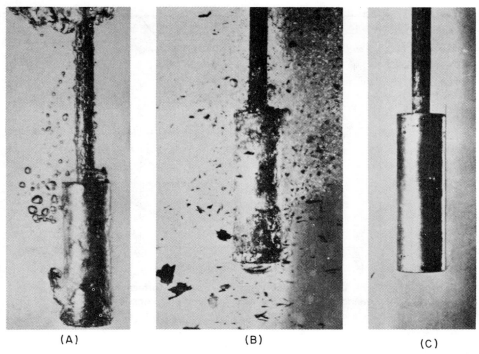

(A) (B) (C)

Fig. 7–24. Three stages of cooling during water quenching of steel cylinders. A, vapor blanket stage. Vapor blanket at the start of quenching from 1550°F moves wave-like toward the top of the specimen. B, vapor transport stage. As water "bites the steel" the action is violent enough to tear off oxide heat treat scale. Two seconds after start of quenching. C, liquid cooling stage. Vapor no longer forms and cooling is by conduction and convection, much slower than stage two. (*Metal Progress*).

ening require water or some aqueous solution as a quenching medium. In order that the quenching rate be great enough to consistently harden carbon steel the water should be kept at a temperature below 80°F and be continuously agitated during the quenching operation. Agitation of the cooling medium allows a more uniform and faster cooling action. A 5-10 percent sodium-chloride *brine* solution usually gives more rapid and uniform quenching than does straight water. The salt addition materially decreases the duration of the vapor film stage. It has been shown that crystals of salt are precipitated momentarily on the steel surface and then explosively thrown off during the initial stages of brine quench-

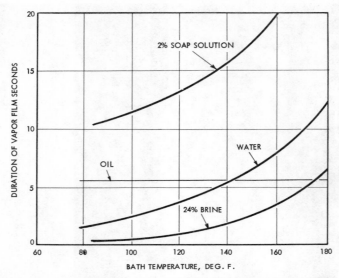

Fig. 7–25. Duration of the vapor film stage of surface cooling when quenching liquids are held at increasing temperatures.

ing. (Fig. 7–26A.) This causes a continuous disruption of the vapor film with improved wetting of the steel and increased heat removal. Brine quenching, because of its vigorous action, removes heat treat scale from specimens much more readily than when a water quench is used.

A 3-5 percent sodium hydroxide quenching bath has also been found to be a good medium for carbon steels. This bath cools even faster than the sodium chloride bath; however, more caution must be exercised in its use since sodium hydroxide (caustic soda) may cause skin burns and may cause blindness if splashed into the eyes.

Oil is frequently used as a quenching medium for hardening carbon steels of thin sections such as knives, razor blades, or spring wire. Its quenching rate results in the formation of martensite. Oil is also widely used for quenching heavy sections of alloy and tool steels. Oil is recommended as a quenching medium in preference to water whenever feasible because its use results in less danger of cracking, less distortion, and lower residual quenching stresses. The general characteristics of oil quenching which distinguish it from water quenching is the relatively stable vapor phase and a very slow rate of cooling during the final

(A)

(B)

Fig. 7–26. A, beginning of quench in 20% solution of sodium chloride. In brine or caustic solution quenching, action is distinctive; a cloud of salt crystals is thrown away from the hot metal with explosive violence causing extremely rapid cooling. *(Metal Progress)*. **B,** one-half second after beginning quenching in Russian oil. Bubbles of oil vapor form a thin, very stable insulating blanket around specimen resulting in a slow quench. *(Metal Progress)*.

stages of quenching. When compared with water mineral oils show a very stable vapor blanket, Fig. 7–26B, which is not destroyed even by vigorous agitation. Consequently the influence of agitation of the specimen in the oil quench bath is relatively small, other than minimizing the build-up of oil temperature in one part of the oil bath. It has fur-ther been found that the duration of the vapor film stage in oils is longer for oils having lower boiling points.

Various types of oils differ widely in quenching characteristics and should be carefully selected. Such properties as flash point, boiling point, density, and specific heat should be considered. Also, care should be used in designing the quenching system. The proper volume of oil, oil circulation, and bath temperature need to be considered. In general, it requires approximately one gallon of oil for every pound of steel quenched per hour. Therefore if 100 lbs of steel are to be hardened every hour, a tank of 100 gal capacity is required. Temperature of the tank should be maintained between 80° and 150°F. Keeping the oil warm actually increases its cooling power over cold oil, especially during the second stage of cooling, because of its lowered viscosity. Another and perhaps more important advantage of a warm quenching oil is in the reduction of residual quenching stresses.

Air cooling is employed with some high alloy steels such as high speed tool steel and air hardening die steel.

Steels to be hardened in air are usually heated to a rather high austenitizing temperature, held for a short period, then removed from the heating furnace and exposed to still air. The best practice is to place the tool on a screen so that air can circulate freely past it. Accelerated air cooling has been accomplished by using a common electric fan—but unless especially arranged the blast will cool one side of the tool more rapidly than the other with resulting distortion. Compressed air may also be used but must be guarded against having condensed water in the air line which may cause an air hardening steel to crack.

The relative cooling rates of various quenching media are shown in Table 7–5 using different degrees of agitation. It should be noted that

Table 7–5. Quenching severity for common media with various degrees of agitation. Still water is 1.0.

DEGREE OF AGITATION OF PIECE OR CIRCULATION OF LIQUID	SEVERITY OF QUENCH, "H"			
	AIR	OIL	WATER	BRINE
NONE	0.02	0.25–0.30	0.9 – 1.0	2
MILD	. . .	0.30–0.35	1.9 – 1.1	2 – 2.2
MODERATE	. . .	0.35–0.40	1.2 – 1.3	. . .
GOOD	. . .	0.4 –0.5	1.4 – 1.5	. . .
STRONG	. . .	0.5 –0.8	1.6 – 2	. . .
VIOLENT	0.25	0.8 –1.1	4	5

violent circulation of an oil bath may give as severe a quench as water with no circulation. Thus, the heat treater has a wide range of cooling rates from which to select. One serious problem has been in being able to repeat the same quench and obtain reproducible results. All too often, an excellent heat treating cycle has ended by hanging the workpiece in a tank of oil and relying on some sort of convection currents generated by the hot metal to cool the workpiece. However, during the past few years more attention has been paid to mechanization of heat treatment and development of mechanical agitation. These efforts are paying dividends and spray quenches, Gleason Presses, and high velocity jets have been the result.

The mechanism of any quench involves the extraction of heat from the surface of the piece. The moment the surface becomes cooler than the interior heat begins to flow from the inside to the cooler surface. Severity of quench "H" as discussed and as shown in Table 7–5, therefore indicates only how fast heat is abstracted from the surface of the specimen. Clearly, then, the rate of cooling of a particular piece being quenched is governed by (a) the type of quenching employed, (b) the surface condition of the workpiece, (c) the thermal conductivity of the workpiece, and (d) the size of the piece.

Since the various commonly used quenches have already been discussed a brief statement about the other factors related to quenching is in order. Steels that carry a heavy, insulating, oxide scale from the heating operation cannot be cooled as rapidly, under a given set of conditions, as those free from scale. This may result in rather erratic hardening. Soft spots may result from attempts to quench badly scaled steel. The remedy is to carry out the heating operation in a controlled atmosphere furnace. Interferences to the quenching action by tongs or holders used in handling the steel may also contribute to soft spots. Deep recesses in the steel surface where pockets of vapor may collect will prevent proper quenching action and likewise cause soft areas.

Thermal conductivity for medium carbon steels ranges from 0.121 cal/sq-cm/cm/C°/sec at 212°F to 0.072 at 1300°F. It should be observed that the thermal conductivity or ability to transmit heat decreases with increasing temperature with the exception of certain heat-resisting stainless steel and high alloy steels.

Thus the propensity to transmit heat from the interior to the surface of a quenched alloy varies considerably and must be taken into account when comparing the quenching rate of one alloy with another.

Size or mass of the workpiece to be quenched has an important bearing on cooling rate and has received considerable attention in the literature. As an example, let us consider the cooling rates of ½″ diameter and 2″ diameter steel bars as quenched from 1600°F into water at 70°F. Temperature measurements were made by means of thermocouple wires attached to the surface, inside at the half radius, and in the center of the specimens. Fig. 7–27 shows the cooling rate for the two specimens. The time values were plotted on a logarithmic scale to permit judging a quench in relation to an

I-T curve. That is to say, if a steel of known I-T curve characteristics is quenched in the form of a certain size bar with a known severity of quench, it then becomes possible to make an approximate judgement as to how much of the steel would harden and how much would transform at higher temperatures. The depth of hardening or hardenability and hardenability testing of steels was covered in paragraph 2.8 and should be reviewed at this point.

7.6 Tempering Treatments. The heat treating operation known as *tempering* nearly always follows quench hardening of steel. Tempering consists of heating quenched, hardened steel in the martensitic condition to some predetermined temperature between room temperature and the lower critical temperature, holding for an hour or so, followed by air or oil cooling. The rate of cooling from the tempering temperature is usually without effect.

Fully quench hardened martensitic steel is generally very brittle, has high internal stress, and possesses low toughness and ductility. In this condition, steel has very few industrial applications. By means of the tempering operation, brittleness of the hardened steel can be reduced while still maintaining a high degree of hardness and strength. (See Fig. 7–28.) In general, the reasons for tempering are to:

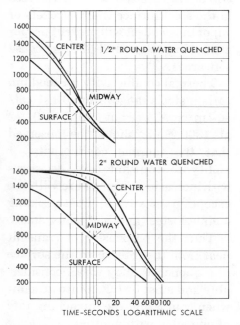

Fig. 7–27. Cooling rate for ½″ dia. and 2″ dia. steel test bars at the center, half radius, and surface positions.

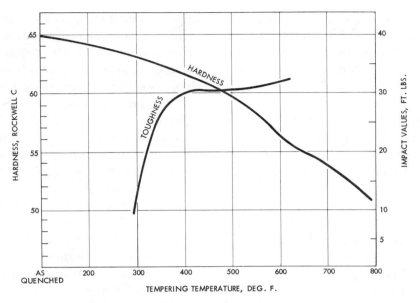

Fig. 7–28. Effect of tempering temperature on the hardness of type O-1 manganese non-deforming steel, oil quenched from 1475°F and tempered one hour.

1. Increase toughness,
2. Decrease hardness,
3. Relieve internal stresses,
4. Stabilize the structure,
5. Change the volume.

Tempering operations are usually carried out at relatively low temperatures with few problems of surface scale formation. However, higher tempering temperatures may require atmospheric controlled furnaces or liquid baths to prevent surface scale. Air furnaces, controlled atmosphere furnaces, and liquid tempering baths are all popular methods for heating parts during the tempering opera-

tion. Liquid baths containing oil, molten salts, or molten lead provide rapid heating, uniform temperature, protection against oxidation, and high production. Oils may be used in the 200-600°F range, salts in the 350-2400°F range, and lead in the 650-1600°F range. Although lead has a much higher heating rate than any of the salts, steel parts will float on top of molten lead and require special fixtures to keep the parts submerged. Molten salts on the other hand tend to decompose and have various chemical effects on the metals, especially at high tempera-

209

tures. Extreme caution must be observed to prevent water or wet parts from entering the tempering baths. Rapid steam formation may cause splattering of hot molten materials or even an explosion.

Temper colors of steel have been widely used in the past to judge the degree of hardness and toughness of small tools. The method now finds only very limited use in tool rooms where special pieces are tempered on a non-production basis. The temper colors appear as a result of slight oxidation of the steel surface. As the tempering temperature is increased, the oxide film becomes thicker and a series of colors from light straw through purple to light blue may be observed. The use of such a method of controlling the tempering operation is restricted somewhat since only surface temperature is obtained and the colors which appear are slightly different for different steels.

Quench hardened steel is considered to be in an unstable condition; that is, the white etching martensitic structure of the steel with its tetragonal atomic arrangement is eager to change to a more stable structure and actually undergoes this change when given an opportunity, such as when the temperature is raised during a tempering operation.

It is common practice to divide the reactions that occur during tempering into three stages. During stage one of tempering in which mar-

tensite, formed in plain carbon steels, is heated to a temperature of approximately 200°F, the tetragonal martensitic structure is said to begin decomposing into low carbon martensite, and a transition carbide called epsilon (ϵ) carbide with a hexagonal crystal structure is formed. The composition of this phase is believed to lie close to $Fe_{2.4}C$. It contains approximately 8.4 percent carbon instead of the 6.7 percent of cementite, Fe_3C, and is slightly harder. This first stage of tempering is accomplished by a slight reduction in the volume of the metal reaching an apparent maximum shrinkage of 0.0012 inches/inch. At a temperature of 400°F martensite has fully decomposed and is now known as tempered martensite and appears dark etching as viewed under the microscope.

During the second stage of tempering, as the tempering temperature is increased above 200°F and extending to about 500°F, the steel begins to soften as retained austenite transforms to bainite. The microstructure of the bainite that forms consists of ferrite and epsilon carbide. Bainite is somewhat similar to the decomposed martensite obtained in the first stage of tempering in that both contain epsilon carbide. The main difference, however, is in the matrix of the two constituents. In bainite the matrix is ferrite that is cubic in structure while decomposed martensite has a matrix of low-carbon tetragonal martensite. Stage two of tempering is accompanied by large positive dimensional changes of the order of .010 inches/inch as austenite transforms to bainite.

During the third stage of tempering, which occurs rapidly in the temperature range of 500-675°F, the low carbon martensite loses both its carbon and tetragonality as its carbon apparently diffuses to combine with epsilon carbide forming very fine cementite particles. These carbide particles cannot be resolved under the microscope and the structure is still called tempered martensite. It should be emphasized that tempered martensite does not take the form of plates or lamellae characteristic of pearlitic structures formed from austenite during slow cooling. Rather, a fine dispersion of more or less spheroidal particles form the size of which depends on time and temperature.

There is a marked decrease in hardness during this stage of tempering. Although tempering does occur in a series of distinct stages, the tempering curve shows a smooth decrease in hardness. The over-all volume or length change from tempering has been plotted for a particular oil hardening tool steel in Fig. 7–29. In general, total growth during tempering (stages one, two, and three) is of the order of 0.1 to 0.3 percent. One application of this

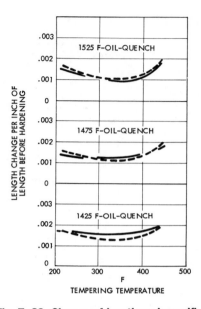

Fig. 7–29. Change of length and specific volume with tempering temperature for AISI O-1 oil hardening tool steel. Solid line indicates length, dashed line specific volume. Samples were 1″ dia. x 3½″ long and ⅞″ sq. x 3⅛″ long. (Crucible Steel Corp.).

growth phenomenon would be to reharden and temper worn or undersize reaming tools at a temperature designed to give maximum growth, then regrind them down to the proper diameter and extend their useful life.

There are a number of variables during tempering operations which play an important role in determining the final hardness, toughness, and corrosion resistance of the steel. Some of these variables are temperature, time, size of workpiece, and

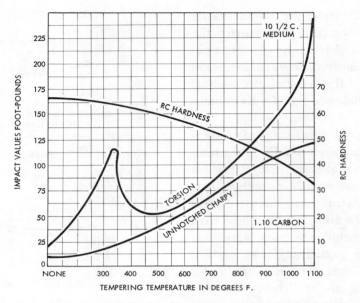

Fig. 7–30. Influence of tempering temperature on fully hardened 1.0% carbon steel as indicated by Rockwell hardness, torsional impact, and unnotched Charpy impact tests.

alloying elements. Fig. 7–30 shows the typical effect of tempering temperature on both hardness and toughness for carbon and low alloy steels. As temperature is increased hardness decreases and toughness is generally improved. It should be clearly understood that the purpose of tempering is not to lessen the hardness of quenched steel but to increase its toughness so as to avoid breakage and failure of heat treated steel. Tests for measurement of toughness of tempered steels have been made by impact testing and by torsional impact testing, Fig. 7–31, and some very interesting results have been obtained. If we assume that these methods of testing are measures of toughness then it becomes apparent that the gain of toughness upon tempering is not progressive over the entire tempering range. In fact, tempering to within a temperature range of 400°F to 600°F may show a loss of toughness. Because of this effect, tempering to within this range gives a condition known as *blue brittleness*. The cause for this condition has not been

Fig. 7–31. Torsion impact testing machine in which a ¼" dia. x 1" long specimen is broken by a torsional or twisting blow from a rotating flywheel. The test is extremely accurate and used for testing hardened tool steels. (The Warner and Swasey Co.).

fully explained although precipitation of certain phases at the grain boundaries have been blamed for it. Also, certain alloy steels, including chromium-nickel steels of relatively high carbon content, develop a seemingly brittle condition called *temper brittleness* when tempered in the 1000-1250°F range, followed by relatively slow cooling. Toughness in these steels may be maintained, however, if water quenched from the tempering temperature.

Tempering time plays an important role in determining the final hardness for tempered steel though it is not as important as temperature. The effect of tempering time at four temperatures on a manganese tool steel is shown in Fig. 7–32. It should be observed that the hardness is reduced quite appreciably during the first half hour but then decreases uniformly with time. Most steels are tempered for one hour after reaching the proper temperature. However, some tool steels of large sections are tempered for as long as 4-6 hours, and some which are tempered at high temperatures (1000-1200°F) are frequently given multiple tempers to permit transformation of all retained austenite.

The large quantity of retained austenite in certain high alloy and high speed tool steels transforms to martensite or to small carbide particles upon tempering to 1000-1100°

213

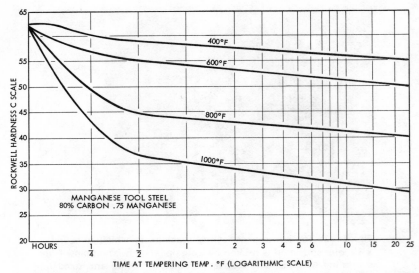

Fig. 7–32. Effect of tempering time on hardness for four tempering temperatures.

F and produces what is known as *secondary hardness*. (Fig. 7–33.) The resultant hardness is often even greater than that obtained at lower temperatures.

A general rule for tempering is that the workpieces should be tempered immediately after quenching and before they reach room temperature. This step may prevent cracking due to the high internal stresses caused by martensitic transformation.

As soon as the workpiece can be comfortably held in the hand (150°F), it should be placed in the tempering furnace. The properties of

several quenched and tempered steels are shown in Fig. 7–34. It is from these types of charts that designers may select the material and heat-treatment to achieve certain desired properties in a given part. The schematic Fig. 7–36A illustrates the customary quenching and tempering treatment as it relates to the I-T diagram.

7.7 Sub-Zero Treatments. It has been discovered that any austenite retained at room temperature may be transformed to martensite by cold treating, *i.e.* cooling the steel to sub-

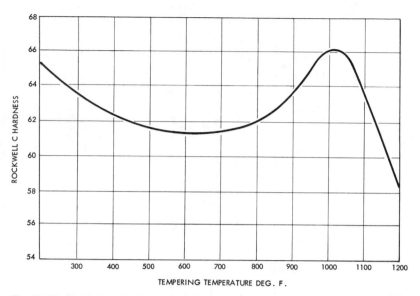

Fig. 7–33. Secondary hardness hump shown on a tempering curve for molybdenum-cobalt high speed steel oil quenched from 2250°F.

normal temperatures in the −70°F to −110°F range. Cold treating is carried out by cooling the workpieces in dry ice or in special refrigerating units such as the one shown in Fig. 7–35. Cold treating to transform retained austenite will not result in any loss of the original hardness of martensite and will result in maximum hardness being obtained. Conventional tempering operations always result in loss of some hardness. In order to lessen the danger of cracking the tool or workpiece should be given a low-temperature (300-350°F) tempering operation before the cold treatment and this should be followed by a second low temperature tempering treatment. Cold treatments are now being used for increasing the strength and hardness levels of carburized and hardened gear teeth, ball bearings, and other products having high surface carbon and which are subjected to high working stresses.

7.8 Martempering. Martempering is one of the three processes used to avoid or minimize problems of warping, distortion, and cracking, which accompany customary quench hardening. The other two processes are austempering and isothermal

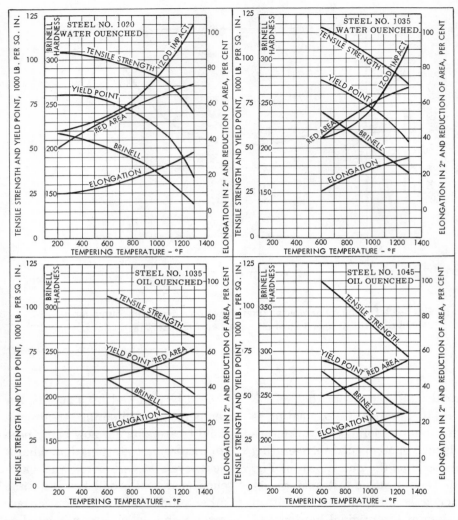

Fig. 7–34. Properties of quenched and tempered steels approximately 1″ dia. or thickness. (International Nickel Co.)

AISI 1020 steel water quenched from 1575/1625°F
AISI 1035 steel water quenched from 1525/1575°F
AISI 1035 steel oil quenched from 1525/1575°F
AISI 1045 steel oil quenched from 1475/1525°F

216

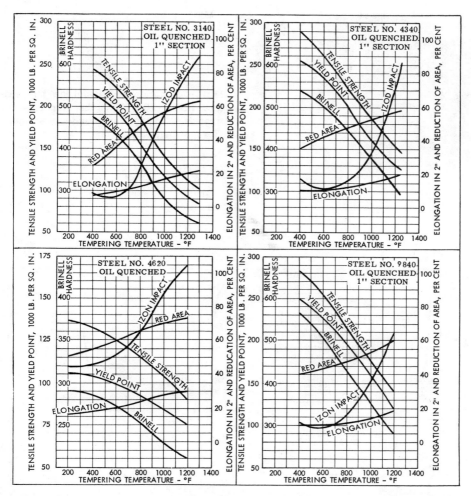

Fig. 7–34. (Cont.) Properties of quenched and tempered steels approximately 1″ dia. or thickness. (International Nickel Co.)

AISI 3140 steel oil quenched from 1475/1525°F
AISI 4340 steel oil quenched from 1500/1550°F
AISI 4620 steel oil quenched from 1550/1600°F
AISI 9840 steel oil quenched from 1525/1575°F

Fig. 7–35. Laboratory model low temperature (to −120°F) chamber for sub-zero treatment of ferrous and non-ferrous metals. (Tenney Eng. Co.).

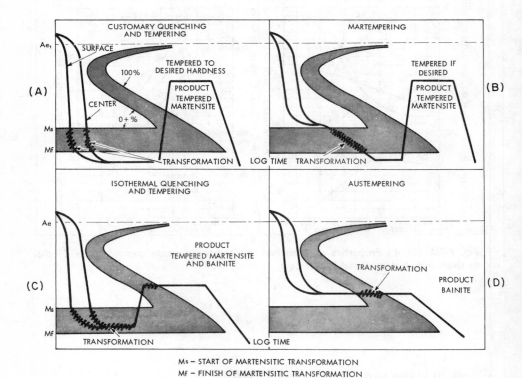

Ms – START OF MARTENSITIC TRANSFORMATION
Mf – FINISH OF MARTENSITIC TRANSFORMATION

Fig. 7–36. Schematic of the relationship of several tempering treatments to a typical I-T diagram. Customary quenching and tempering, martempering, austempering, and isothermal quenching. (U.S. Steel Corp.).

quenching and will be discussed in paragraphs 7.9 and 7.10.

The martempering treatment Fig. 7–36B is carried out by heating the steel to the proper austenitizing temperature and quenching rapidly in a molten salt bath held just above the M_s temperature. The steel is held long enough in the hot quenching bath to allow the surface and center to reach the same temperature. The steel is then removed from the quenching bath and cooled to room temperature. During the cooling from the M_s temperature to room temperature the austenite will transform to martensite. Since the temperature has been equalized throughout the steel the transformation will occur uniformly with minimum residual stresses and with greatly reduced danger of distortion and cracking. The heat treatment is completed by tempering the martensite to the hardness desired. The principal advantage of martempering, then, is that fewer internal stresses result from a lower temperature differential between the outside and center of steel during the transformation stage from austenite to martensite.

7.9 Austempering. The austempering cycle, shown schematically in Fig. 7–36D, is a hardening process based upon transformation of austenite to bainite. From a study of the I-T curve, Fig. 7–2, it becomes apparent that during conventional quenching there is strong likelihood of obtaining only two distinctive structures, that of fine pearlite (R C40-42) at the "nose" of the I-T curve or that of martensite (R C60-65) at room temperature. If one wishes to obtain a hardness of R C58, for example, the customary heat treatment is to quench the steel to make it fully hardened (R C65) then temper it to achieve the desired hardness, in this instance, R C58.

By the austempering treatment steel can be hardened to R C58, or other hardnesses, directly by the method of isothermal transformation. The austempering cycle is similar to that of martempering except that with austempering the steel is quenched to some selected temperature above the M_s and allowed to transform completely prior to cooling to room temperature. In the example cited above, that of obtaining a hardness of R C58, austempering would be carried out at approximately 500°F with one hour to complete the transformation. We then would have obtained a structure of 100 percent bainite with uniform hardness without the birth of martensite.

It is an established fact that steel with a hardness of R C58 obtained by austempering is much tougher than one of the same composition and hardness obtained by usual methods of hardening and tempering. Fig.

219

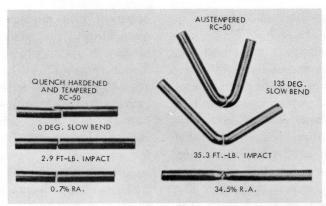

Fig. 7–37. Improved toughness and ductility of austempered rods (right) compared to rods (left) hardened by conventional quench and tempering treatment. (Edgar Bain Research Lab., U.S. Steel Corp.).

7–37 shows the relative properties of specimens hardened by conventional quenching and tempering and those hardened by austempering. The austempered specimens are much tougher and have much greater ductility than quenched and tempered specimens. Since martensite is never produced during austempering much of the danger of cracking is eliminated and the amount of distortion or warping is greatly reduced.

Successful austempering is dependent upon successful undercooling of austenite to the temperature of the hot quenching bath without the formation of any softer forms of pearlite in the 900-1200°F range. This limits austempering to relatively thin specimens generally un-der ½ in. in thickness which can be rapidly cooled. Austempering has been successfully used on springs, lock-washers, screws, pins, needles, cultivator shovels, and similar parts.

7.10 Isothermal Quenching. The isothermal quenching cycle is shown schematically in Fig. 7–36C. Transformation is allowed to proceed to completion at a temperature only slightly above the M_s temperature. The constituents formed from this treatment will be fairly hard. Increased toughness (and consequent reduction in hardness) can be achieved by raising the temperature, making it unnecessary to quench it before tempering. Slow cooling to room temperature completes the cycle. It should be apparent that the

I-T curve for any steel under consideration for hardening by isothermal quenching provides a guide to the temperature to be chosen for the isothermal quench and also gives an indication of the time requirements. It must be emphasized that section size of the workpiece is very important. If the section is large the cooling rate in the hot quenching bath will not be sufficient to prevent partial transformation at the elevated temperatures and only partial hardening will occur.

7.11 Common Heat Treating Problems. The most common problems related to heat treatment of steels are shown in Table 7–6. These problems may be manifest in several ways, including warping, dimensional change, cracking, failure to

Table 7–6. Common problems of heat treatment of steels.

PROBLEM	POSSIBLE CAUSES	REMEDY
A. WARPING	1. NON-UNIFORM QUENCHING PRACTICE	EMPLOY SPRAY OR AGITATED QUENCH
	2. IMPROPER SUPPORT DURING HEATING	SUPPORT WITH BRICK, CAST IRON CHIPS, OR SPENT COKE
	3. RELEASE OF MACHINING STRESSES	MACHINE EQUAL AMOUNTS FROM SURFACE OF PART OR ANNEAL PRIOR TO HEAT TREATMENT.
	4. UNBALANCED DESIGN.	CLAMP IN FIXTURE DESIGNED TO BALANCE MASS
	5. FAILURE TO STRAIN RELIEVE PRIOR TO HEAT TREATMENT	STRAIN RELIEVE
B. DIMENSIONAL CHANGES	1. RELEASE OF STRESSES FROM PREVIOUS COLD WORKING	STRAIN RELIEVE PRIOR TO HARDENING
	2. UNPREDICTED THERMAL STRESSES	BALANCE MASS WITH QUENCH FIXTURE
	3. SEVERE QUENCHING PRACTICE	CHANGE TO LESS SEVERE QUENCHING MEDIA OR WARM QUENCH BATH
	4. FAILURE TO TEMPER OR STABILIZE PROPERLY	EMPLOY STABILIZING OR SUB-ZERO TREATMENT
	5. DIMENSIONAL CHANGES FOR SOME ARE PREDICTABLE AND NORMAL	USE TABLES SUPPLIED WITH STEEL TO PREDICT SIZE CHANGE
	6. TRANSFORMATION OF RETAINED AUSTENITE	EMPLOY MULTIPLE TEMPERS OR SUB-ZERO TREATMENT
	7. OVERHEATING OR UNDERHEATING	CHECK FURNACE CONTROL AND RECOMMENDED TEMPERATURES
C. CRACKING	1. FAILURE TO TEMPER IMMEDIATELY AFTER QUENCHING	TEMPER BEFORE IT REACHES ROOM TEMPERATURE. APPROXIMATELY 150°F
	2. IMPROPER QUENCHING MEDIUM	USE LESS SEVERE QUENCH
	3. EXCESSIVE HARDENING TEMPERATURE	CHECK FURNACE TEMPERATURE AND RECOMMENDED TEMPERATURE
	4. LARGE GRAIN SIZE	NORMALIZE PRIOR TO HARDENING
	5. POOR DESIGN e.g. SHARP CORNERS, OR UNBALANCED MASS	DISCUSS WITH DESIGNER
	6. FAILURE TO PREHEAT PROPERLY	PREHEAT AS RECOMMENDED

Table 7—6. (cont.) Common problems of heat treatment of steels.

D. FAILURE TO HARDEN	1. QUENCH NOT DRASTIC ENOUGH	EMPLOY MORE DRASTIC QUENCH
	2. HARDENING TEMPERATURE TOO LOW OR NON-UNIFORMLY HEATING	CHECK RECOMMENDED TEMPERATURE
	3. MISLABLED STEEL	MAKE TEST RUN ON SAMPLE OR GET IT ANALYZED
	4. SEVERE DECARBURIZATION	USE CONTROLLED ATMOSPHERE OR BATH FOR HEATING
	5. TEMPERING TEMPERATURE TOO HIGH	USE RECOMMENDED TEMPERATURE
E. SOFT SPOTS	1. DECARBURIZED CASE	USE CONTROLLED ATMOSPHERE OR LIQUID HEATING BATH
	2. EXCESSIVE HEAT TREAT SCALE	SAME AS ABOVE
	3. QUENCH BATH TOO HOT	CHECK TEMPERATURE
	4. IMPROPER AGITATION	REVIEW RECOMMENDED PROCEDURES
	5. CONTAMINATED QUENCHING BATH	CLEAN, FILTER, OR CHANGE
F. EXCESSIVE BRITTLENESS	1. IMPROPER QUENCHING MEDIUM	USE RECOMMENDED QUENCH
	2. FAILURE TO TEMPER	TEMPER IMMEDIATELY AFTER HARDENING
	3. EXCESSIVE HARDENING TEMPERATURE	FOLLOW RECOMMENDED TEMPERATURE
	4. COARSE GRAIN SIZE	FOLLOW RECOMMENDED TEMPERATURE
	5. MECHANICAL STRESS RAISERS, e.g., SHARP CORNERS	DISCUSS WITH DESIGNER OR USE AIR HARDENING STEEL

harden, soft spots, and excessive brittleness. Possible remedies for these problems are given. Many heat treating problems result in unsatisfactory service or even in failure. Both are costly in terms of down time, replacement expense, inconvenience, and possibly serious accidents.

Much effort and money have gone into the development of mechanized heat treating equipment and automatic controls to make possible responsive, repeatable, and predictable heat treatments. Distortion and warpage, for example — heretofore major problems in heat treating— have been drastically reduced with automated equipment and appropriate instrumentation. Furnaces with controlled atmospheres improve dimensional accuracy, surface condition, and general cleanliness. Automatic time-proportioning pyrometer controllers give very accurate and uniform furnace temperatures. Improved quenching methods using impellers, jet pumps, and high velocity sprays, along with well designed fixtures also contribute much to a high level of quality. These advances have raised the quality of heat treating processes to such an extent that subsequent cleaning, grinding, or lapping, can often be eliminated.

Before discussing heat treating problems in particular it should be pointed out that satisfaction in production and service of almost any manufactured steel part involves three important aspects, namely:

Good part design,

Proper steel selection,

Suitable heat treatment.

Good part design, based on knowledge of working conditions and maximum operating loads, has made it possible to eliminate unneeded material and relieve local stress concentrations in many parts. Part failure occurs in a region of high localized tensile stresses. Good design practices avoid stress raisers such as notches, sharp corners, abrupt changes in section, inadequate fillers, threads, keyways, and thin fins. Even with the best possible design, stress concentration cannot be entirely eliminated. Threads and keyways, for example, introduce stress concentration factors (K = 2 or more). This means that tensile stress at the root of a screw thread would be double the stress in the body of the screw. In certain critical instances the deleterious effect of stress concentrations can often be combated by introducing compressive stresses into the surface of the part by nitriding, carburizing, or shot peening. However, the best procedure is to reduce stress concentrations to a minimum by good design practices before looking to other means of compensation. If a part design is poor then the best of steel and heat treatment will not give a satisfactory part.

Selection of steel of the proper grade for a particular application is a dynamic problem and must be reviewed periodically in light of price changes, fabricating costs, performance, and development of new materials.

An example of the diversity of types of steel used in spur gears for tractors produced by nine companies is illustrated in Table 7–7.

Thus, there is obviously no one best material for similar parts. Many factors relate to material selection including economics, performance, manufacturing and heat treating facilities, and the environment in which the part must operate.

Heat treatment problems for manufactured parts as described in Table 7–6 can generally be grouped into four areas:

1. Heating practices,
2. Quenching practices,
3. Steel quality and uniformity,
4. Residual internal stresses.

Heating practices should follow the recommendations of the manufacturer. Parts with large differences in cross section should be heated

223

Table 7–7. Materials used for spur gears by nine tractor makers.

(CONDENSED FROM WORTHINGTON AND RICH "3 KEYS TO SATISFACTION", CLIMAX MOLYBDENUM STEEL CO.)

NO. GEAR DESIGNS	MATERIAL	HEAT TREATMENT	HARDNESS
5	8620	CARBURIZE AND HARDEN	58–62, FILE HARD, 57–62 FILE HARD, 58 R_c MIN, 57 R_c MIN
1	8620H	GAS CARBURIZED AND DIRECT QUENCH	57 R_c MIN.
1	8640	CYANIDE HARDEN	50–55 R_c
1	4145	FULL HARDEN	47–53 R_c
1	4620	GAS CARBURIZE AND DIRECT QUENCH	75–90 SCLEROSCOPE
2	1020	GAS CARBURIZE AND DIRECT QUENCH	FILE HARD, 56–63 R_c
1	8615	CARBURIZE AND HARDEN	58–65 R_c
1	5135–40	CYANIDE HARDEN	FILE HARD
1	5145–50	CYANIDE HARDEN	53–56 R_c

slowly and uniformly. Preheating should be used for high speed tool steels which must be heated above 1900°F and tools with intricate sections. Furnace controllers should be adjusted and calibrated to hold workpiece temperatures to within ± 20°F. It has been assumed that steels heat much more rapidly on the surface than at the center and that the outside might be fully up to temperature while the center was still "black." This is not entirely true. Studies with thermocouples mounted on the surface and at the center of various sized specimens indicate that the surface and center reach the furnace temperature at nearly the same time. During the first stages of heating center temperature lags behind surface temperature by about 100°F. However, during the last 25-30°F of heating the temperatures are practically the same. Since the surface of a piece has no thickness, or no mass, it has no capacity to store heat. The surface merely transmits heat from the furnace to the interior of the workpiece. The only way for the surface to reach furnace temperature is for the center to stop absorbing heat from the surface. Thus the center and surface must arrive at the furnace temperature at very nearly the same time. (1″ diameter round or square specimens have a time lag of 6 seconds; 2″ diameter, 24 seconds; 4″ diameter, 96 seconds, and so on.)

These same studies have shown that, in general, carbon and alloy steels of uniform cross section can be put directly into the hot furnace without harm and heated at their own maximum natural rate. The only purpose for slow heating is to prevent an initial large temperature gradient between the interior and surface which might result in high thermal stresses. With normal heating practice, the temperature difference is probably never over 100°F which is not sufficient to harm most of the low carbon and low alloy steels. When the color of the furnace thermocouple and the specimen are alike to the normal eye, the workpiece is usually about 5° under the furnace temperature. Soaking the workpiece for about 5 minutes per inch of thickness will allow it to heat those last few degrees.

Quenching problems are greatest when the workpieces have abrupt changes in section, sharp curves, and re-entrant angles. When parts cannot be changed to have more gradual change in section, it is possible to design quenching fixtures which can be clamped around the parts to be heated and quenched with them. This allows more uniform cooling rates. Holes and key slots can also be plugged with fireclay or asbestos, thus blocking the coolant and giving a better heat distribution. A number of other solutions have been found by ingenious heat treaters to solve difficult quenching problems, examples of which are martempering, austempering, isothermal quenching, and interrupted quenching. Interrupted quenching is performed by cooling the steel in oil to 1000-1100°F (dull red) at which time it is removed from the oil and allowed to cool naturally in the air to below 150°F before tempering.

Steel quality and uniformity is another factor that is of concern to the heat treater and to the metallurgist. Tool steels are usually of high quality but occasionally contain flaws such as stringers, pipe, voids, or internal cracks. Testing equipment such as that described in Chapter 2 may be used to check the soundness of tool steel before it is purchased or used. The extra cost of purchasing certified clean, sound, uniform steel from a reputable steel supplier is good insurance for satisfactory heat treatment and performance. Even though high quality steel is used a soft skin or surface may develop during heat treatment. This soft surface on hardened steel is due to oxygen reacting with carbon in the steel in

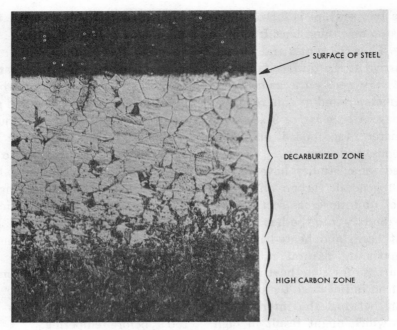

SURFACE OF STEEL

DECARBURIZED ZONE

HIGH CARBON ZONE

Fig. 7–38. Decarburized tool steel, 200X.

certain elevated temperature ranges. This results in the formation of carbon dioxide or carbon monoxide gas leaving the surface of the steel depleted of carbon or decarburized. A badly decarburized carbon tool steel is shown in Fig. 7–38.

Decarburization may be prevented by proper control of the atmosphere surrounding the steel during the heating operation. The use of lead baths or salt baths or the control of the gaseous atmosphere in a hardening furnace can prevent decarburization and oxidation from taking place. However, all steels as received from the steel mill retain decarburized skin which will cause trouble unless it is removed prior to the hardening operation. Steel should be purchased oversize so that the decarburized surface may be removed by machining or grinding before the hardening operation.

Residual internal stresses in hardened steel are the cause for many detrimental effects, including warping, distortion, and cracking. These same stresses can also be beneficial if directed properly by controlled quenching and have been commercially utilized to:

1. Increase mechanical properties of guns,

2. Improve service life of extrusion cylinders, forging dies, and drawing dies,

3. Improve fatigue life.

Residual internal stresses in hardened steel arise from (1) thermal stresses, and (2) transformation stresses. Such residual stresses can only be produced if sections of the object are plastically deformed during cooling. For example, when a steel cube is cooled by quenching in water there is a thermal gradient established between the outside and center of the piece. The outer portion in contact with the quenching water will cool rapidly and contract. The interior of the piece which is still hot will try to maintain its original dimensions, but since the two portions cannot separate there will be established internal tensile stresses in the outer portion and compressive stresses in the inner portion. These stresses will cause the hot center portion to plastically contract (upset) since it has the lower yield strength. This contraction will often lead to bulging or barbering of the surface of the object. (Fig. 7–39.) On further cooling the center portion will have a final length shorter than the surface portion because of the upsetting which occurred. This condition will then place the center portion in elastic tension and the surface in elastic compression. In a shallow hardening steel the expansion of the case during transformation to martensite will be superimposed on the effect of

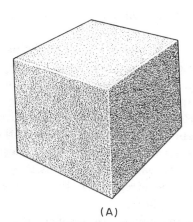

(A)

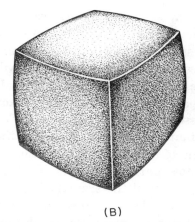

(B)

Fig. 7–39. Bulging effect of immersion quench on a steel cube. A, 1″ cube before quenching, B, after quenching a bulge of about .002″ on the thickness is present.

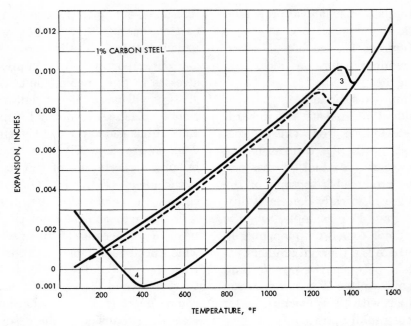

Fig. 7–40. Dilation curve for 1% carbon steel on heating and quenching (solid lines) and on slow cooling (dotted line). (American Society for Metals).

thermal stresses and increase the value of the surface compressive stresses and the center tensile stresses. A deep hardening steel will have a reverse effect and will result in partial alleviation of surface stresses or even their reversal. The dilation curve Fig. 7–40 shows the result of heating and quenching a shallow hardening steel. Factors which control the size and internal stresses and which correspond with numbered points on the graph are:

1. Thermal expansion upon heating,

2. Thermal contraction upon cooling,

3. Contraction upon changing to austenite during heating,

4. Expansion as martensite forms on quenching.

Residual stresses which arise from quenching medium and low carbon steels are shown in Fig. 7–41. Residual internal tensile stresses may exceed the strength of the steel and cause failure.

The purpose of tempering is to relieve the stresses which are set up during cooling and during martensi-

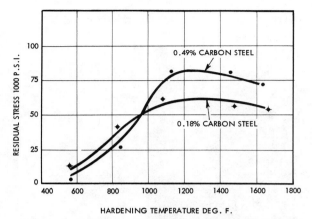

Fig. 7–41. Residual stress from quenching low and medium carbon steels from various temperatures.

tic transformation and make the piece tougher.

7.12 Heat Treating Equipment. In considering the various types of heat treating equipment on the market it is important to remember that all heat treating operations are a question of temperature, time, atmosphere, and cooling rate. If equal performance can be obtained from several types of equipment, then the choice can be made on the basis of capacity, cost, and convenience.

There are a number of different types of heat treating equipment on the market, the most popular of which may be grouped as follows:

Gas, oil or electric furnaces (semi-muffle, full-muffle, or radiant),

Induction heating (motor generator, spark gap, vacuum tube oscillator or solid state),

Molten bath (lead, salt),

Flame hardening.

Gas and oil fired furnaces may be either of the semi-muffle, full-muffle or radiant tube type. The semi-muffle furnace, Fig. 7–42, is so designed that the gases of combustion are deflected and do not impinge directly on the steel being heated, although they do circulate around it. In order to protect steel from oxidation during the heat treating operation, and to provide possibilities for full atmosphere control, the full-muffle type furnace is often used, Fig. 7–43. Muffle furnaces are so constructed that they provide a ceramic or heat

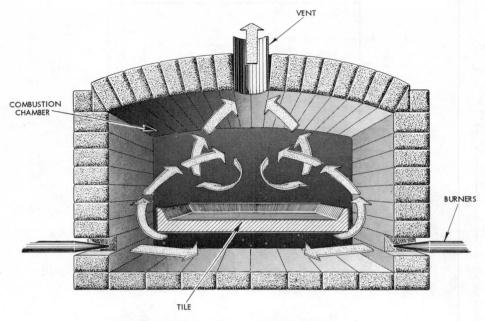

Fig. 7–42. The principle of the semi-muffle type furnace.

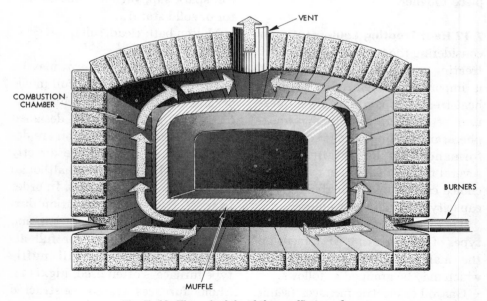

Fig. 7–43. The principle of the muffle type furnace.

Fig. 7–44. Electric muffle furnace with resistance and glo-bar hardening furnaces, air tempering furnace, endothermic atmosphere control, oil and water quenching tanks. (Waltz Furnace Co.).

resisting alloy chamber, known as the muffle, separate from the combustion chamber, into which the steel to be heated is placed.

The gases of combustion circulate around the chamber or muffle but cannot come in contact with the steel being heated. The electric muffle furnaces, Fig. 7–44, are of two types. The resistance type in which the heating element, or resistors, are usually placed around the outside of the muffle. These are adapted to the lower range of temperatures encountered in heat treating operations (up to 1800°F). The second type of electric muffle furnace is heated by "glo-bars," which are carbonaceous silicon carbide resistors in the form of straight bars inside the muffle. This type of furnace is adapted to the higher temperatures (up to 2400°F) needed for superheating high speed steels. Heating steel in a furnace in

231

the presence of air results in the oxidation of the steel surface with the formation of scale and loss of surface carbon (decarburization). The full-muffle furnace is commonly used as controlled atmosphere furnace when maximum protection from oxidation or decarburization is desired. In order to obtain a neutral atmosphere (non-oxidizing, non-carburizing), artificial gas atmospheres are added to the heating chamber. Bottled gases such as methane, propane, ammonia, or city gas are frequently burned in endothermic gas generators, Fig. 7–45, to provide the desired atmosphere. Endothermic gas generators are those in which heat is supplied in order to "crack"

the gases and provide atmospheres generally composed of hydrogen, nitrogen, water vapor, methane, carbon monoxide and carbon dioxide gases and occasionally small amounts of oxygen. With the exception of nitrogen, all of these gases react with steel causing oxidation, carburization, or decarburization.

The selection of the ideal gas atmosphere for the heating chamber depends upon many factors such as the type of steel being treated, the temperature of operation, and so on, and requires a careful study of the problem involved and an understanding of the reactions occurring at the heat treating temperature. In general, this problem is solved by

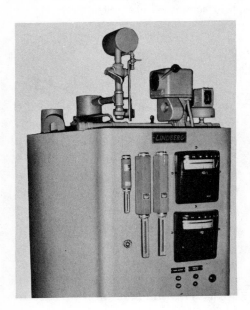

Fig. 7–45. Endothermic gas generator for preparation of protective furnace atmospheres (Lindberg Engineering Co.)

Table 7–8. Prepared gas atmospheres used in heat treating operations.

TYPICAL APPLICATION	GAS CONSTITUENTS—PER CENT BY VOLUME					
	CO_2	Co	H_2	CH_4	H_2O	N_2
NO. 1 CASE CARBURIZING LOW OR MEDIUM CARBON STEELS	----	20.7	38.7	0.8	----	39.8
NO. 2 BRIGHT ANNEALING COPPER	10.5	1.5	1.2	----	0.8	86.0
NO. 3 PROCESS ANNEALING COPPER, LOW CARBON AND STAINLESS STEELS	0.05	0.05	3–10	----	----	REMAINDER
NO. 4 P/M SINTERING—BRIGHT ANNEALING (CRACKED ANHYDROUS AMMONIA)	----	----	75.0	----	----	25.0

trying several gaseous mixtures and analyzing their effect upon the steel being treated. Some commonly employed gas atmospheres are shown in Table 7–8.

Another type of heating equipment commonly employed in heat treating operations is the pot type furnace used for liquid heating baths shown in Fig. 7–46. Baths of molten salts or lead transmit heat quickly and uniformly and afford atmospheric protection during the heating cycle.

The induction heating equipment shown in Fig. 7–47 is fast, maintains relatively scale-free surfaces, has good temperature control, and uniform heating. Hardening by high frequency induction makes possible localized heating and surface hard-

Fig. 7–46. Pot type electric furnace used for liquid heating, cyaniding, and martempering. (Pacific Scientific Co.).

Fig. 7–47. A set up for induction hardening of a gear. (Massachusetts Steel Treating Co.).

ening. Chief advantages are that it is fast, can accommodate many small or odd-shaped parts, can provide selective hardening, and lends itself to automatic control.

Furnace pyrometers are of three main types: indicating, recording, and controller types. Commercial types are shown in Fig. 7–48 through 7–50. The indicating type controllers show the temperature on a dial or by means of colored lights. The furnace must be manually adjusted to regulate temperature. The recording type pyrometer controllers record temperatures on a moving chart which is mechanically driven and gives a permanent record of temperatures. The wavy line indicates the time lag between the time and controller senses the temperature and can turn on the heating unit. Once

Fig. 7–48. Indicating furnace controlled with percentage timer. Input is regulated by turning load circuit on and off within the selected percentage range. Timers must be checked and re-set until input matches heat losses. (Thermolyne Corp.).

the furnace has achieved the desired temperature, the controller is turned off but because of the higher residual heat of the heating element, the fur-

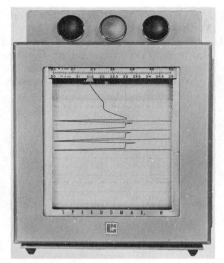

Fig. 7–49. Recording furnace controller. (Leeds and Northrup Co.).

Fig. 7–50. Indicating proportioning furnace controller. Maintains temperature within 1 degree. (Wheelco Instruments Div., Barber-Coleman Co.).

nace slightly overshoots the desired temperature.

Furnace pyrometer controllers not only indicate the temperature, but they automatically regulate it to within a predetermined zone. A recent modification of the on-off type controller is the time-proportioning temperature controller. This unit has an adjustable relay "on" time to "off" time proportional to the deviation from the set point. This controller is capable of straight line temperature control as close as ± 1°F.

References

Brick, Robert M., Gordon, Robert B. and Phillips, Arthur. *Structure and Properties of Alloys.* New York: McGraw-Hill, 1965.

Bethlehem Steel Co. *Tool Steel Treaters Guide, Booklist 148-A.* Bethlehem, Pa.

Carnegie-Illinois Steel Corp. *Suiting the Heat Treatment to the Job, #18226.* Pittsburgh, Pa.

Enos, George M. and Fontaine, William E. *Elements of Heat Treatment.* New York: John Wiley & Sons, 1955.

Gill, J. P. et al. *Tool Steels.* Cleveland, O.: American Society for Metals, 1946.

"Heat-Treating — Preheating, Austenitizing, Quenching, Tempering" *Tool Steel Trends.* Reprint American Iron and Steel Institute, New York.

LaSalle Steel Co. *Simplified Steel Terms and Engineering Data, Bulletin No. 6.* Chicago, Ill.

Lyman, Taylor, ed. *Metals Handbook, 1948 ed.* Cleveland, O.: American Society for Metals.

Palmer, Frank R. and Luerssen, *Tool Steel*

Simplified. Reading, Pa.; The Carpenter Steel Co., 1960.

Reed-Hill, Robert E. *Physical Metallurgy Principles*. Princeton, N.J.: D. Van Nostrand Co., 1964.

Seabright, Lawrence H. *The Selection and Hardening of Tool Steels*. New York: McGraw-Hill, 1950.

U.S. Steel Co. *The Process and Result of Austenite Transformation at Constant Temperature. Information Circular No. 1*. Kearney, N.J.: Research Laboratory.

Surface Treatments for Steel

Chapter

8

8.1 Introduction. Many metal objects made from ferrous and nonferrous metals may be subjected to some form of surface treatment which affects only a thin layer of the outer surface of the metal. Such treatment may be employed for developing greater corrosion resistance, increasing surface hardness and wear resistance, or for improvement of appearance and sales appeal. Surface treatments include hardening treatments, metallic, organic, oxide, and ceramic coatings, and mechanical finishing.

Surface Hardening

8.2 Carburizing. Probably the oldest heat treatment is that of carburizing by which the early metalworker made steel by adding carbon to wrought iron. Carbon was likely added by the carbonaceous gases from the charcoal in the ancient's forge fire, thus making the wrought iron hardenable by quenching.

Carburizing in today's world is used when a hard steel surface coupled with a tough core is wanted. Carburizing is usually done (1) by pack carburizing, (2) gas carburizing, or (3) liquid carburizing (cyaniding).

The *pack* carburizing process involves packing a low carbon steel such as AISI 1015, 1020, 2315, 3120, 4615, or 8620, into a heat resisting metal box and completely surrounding it with a carburizing compound. The container is then heated for several hours after which the box is allowed to cool. The specimens now containing high surface carbon are reheated and then quench hardened. During the pack carburizing process carbon rich gas is generated at high temperature from the carburizing compound. The gas, which is primarily carbon monoxide, is absorbed by the austenitic steel and diffuses slowly into the interior of the steel. The outer layer, high in carbon, is called the *case* while the balance of the unaffected low carbon steel interior is called the *core*. The solid carburizing compounds usually contain charcoal (80-85% by weight) as a base to which are added energizers such as barium carbonate (12-15%), calcium carbonate (1-3%), sodium carbonate (1%), and oil to bind the energizers to the charcoal.

The rate of carburizing at several temperatures is shown in Fig. 8–1. Carburizing is rarely done below 1600°F because of the longer time required or above 1700°F because of the possibilities of grain growth.

Microstructure of an unsatisfactory case of a pack carburized steel specimen is shown in Fig. 8–2, and a satisfactory case is shown in Fig. 8–3. Applications for the carburizing process include case hardening of gears, cams, crankshafts, firearm parts, roller bearings, ball bearings, conveyor chain, and some inexpensive tools. Pack carburizing is relatively slow and rather dirty and has been supplanted to a great extent by gas and liquid carburizing.

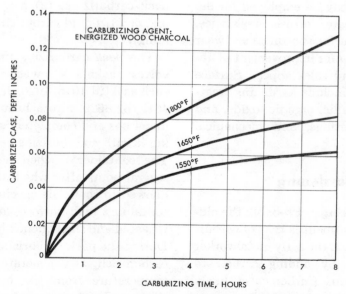

Fig. 8–1. Rate of carburization of a solid carburizing agent at several temperatures.

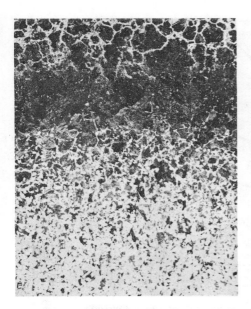

Fig. 8–2. Microstructure of pack carburized machine steel, 0.20% carbon. Carburized to produce an unsatisfactory case with a cementite network or eggshell hypereutectoid layer, 75X, Picral etchant.

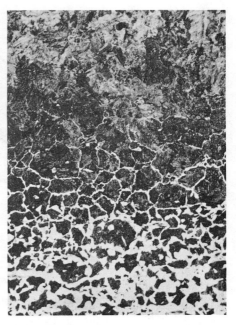

Fig. 8–3. AISI 1020 steel carburized to produce a satisfactory eutectoid case, 75X, Picral etchant.

Gas carburizing is done by placing the work in a heated retort or furnace to which the carburizing gas (methane, propane, or natural gas) is admitted. Exit gas is vented and burned. Carburizing time with natural gas is shown in Fig. 8–4. Continuous gas carburizing furnaces have been developed recently in which the carburizing, quenching, and tempering cycles are carried out sequentially in the same furnace as the work progresses on a conveyor from one operation to the next.

Liquid carburizing is performed by submerging the workpiece in a molten salt bath at a temperature range of 1550-1750°F. At the lower temperatures the process is sometimes called *cyaniding* as a large amount of nitrogen will enter the case forming beneficial nitrides. At higher temperatures more carbon will be absorbed and at a much faster rate from the carburizing bath than from a cyaniding solution. The bath for the cyanide treatment often contains as much as 96-98 percent

239

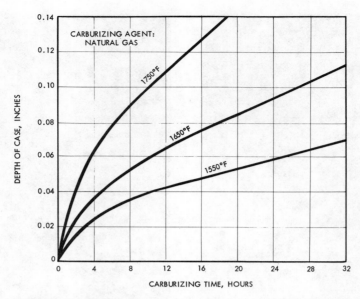

Fig. 8–4. Carburizing rates using natural gas as the carburizing agent.

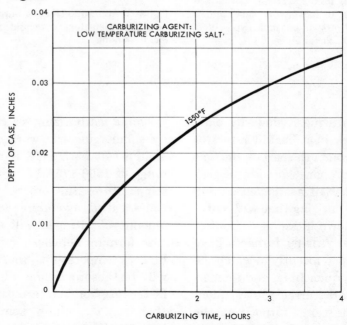

Fig. 8–5. Effect of carburizing time on case depth for low temperature carburizing. Salt on AISI 1020 steel.

sodium cyanide whereas the liquid carburizing bath may contain only 7-23 percent cyanide content. The effect of liquid carburizing time with low temperature carburizing salt on case depth is shown in Fig. 8–5. Light case baths operating at temperatures from 1550-1650°F and having up to 20 percent cyanide content produce case depth of .005"- .030". Baths for deep case carburizing operate at higher temperatures and contain less cyanide, often only about 7-10 percent. The steels which have been liquid carburized are usually direct quenched from the salt bath into oil or water and then tempered to the desired toughness and hardness.

8.3 Nitriding. Nitriding as a commercial process has been developed since 1925. It is similar to gas carburizing as the gas is fed into a heated gas tight retort or muffle furnace. The difference lies in the fact that nitrogen instead of carbon is added to the surface of the steel. Also, nitriding is done at a rather low temperature range of 930-1000°F, which is below the lower critical temperature, and as a consequence very little distortion or warpage occurs in the workpiece due to volume changes.

Anhydrous ammonia, which is purchased in tanks as a liquid and introduced into the furnace as a gas at slightly more than atmospheric pressure is broken down into monatomic hydrogen at the nitriding temperature according to the following reaction:

$$2\,NH_3 \rightleftharpoons 2N + 3H_2$$

The dissociation of the ammonia gas is usually held to approximately 30 percent but may be higher or lower. Uniform dissociation may be accomplished by means of a fan built into the muffle or nitriding chamber.

The gas mixture leaving the furnace consists of hydrogen, nitrogen, and undissociated ammonia and is either trapped in water or burned. The nitrogen enters the surface of the steel and unites with iron and other elements in the steel to form hard nitrides. Diffusion of the nitrogen inwardly is very slow and may take from 24 to 90 hours to produce the desired case depth.

The hardness of the case that may be expected from a nitriding operation as compared to a carburized and hardened steel is illustrated in Fig. 8–6. The maximum hardness obtained from a carburized and hardened case runs around Rockwell C67, whereas it is possible to obtain surface hardness values estimated to be in excess of Rockwell C74 by nitriding. The surface hardness of thin cases such as those obtained by carburizing or nitriding cannot be measured on the Rockwell C scale but are measured by machines such as the Rockwell Superficial Hardness

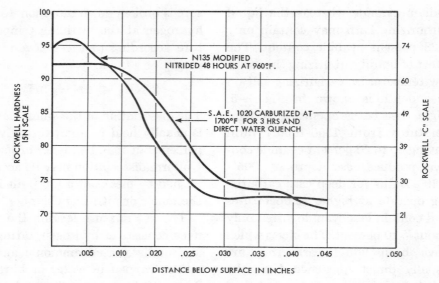

Fig. 8–6. Comparison of hardness at various depths of nitrided and carburized steels.

Fig. 8–7. Modified steel, nitrided at 960°F for 18 hours. Shows the white layer and case depth. 250X.

Tester using either the 15N or 30N scale. The hardness value on the Rockwell C scale given above was estimated from a hardness conversion table.

The nitrided case, Fig. 8–7, has a thin white layer on top which is extremely brittle. This white layer is usually ground off after the nitriding operation to avoid failure by chipping. However, the white layer has good corrosion resisting characteristics and from this standpoint it is desirable not to remove it.

In order to prevent nitriding of some surface areas, tin, nickel, bronze, and copper plating and pastes containing tin have been used successfully. Tin plating .0005″ thick is sufficient to prevent nitriding. Likewise, copper plating can be used to prevent carburization of steel parts in areas that are to be kept soft.

Occasionally it is necessary to obtain greater core strength and toughness to enable the thin nitrided case to withstand high crushing loads. This can be done by hardening, oil quenching, and tempering the workpieces before they are nitrided. Tempering is normally done at slightly below 1100°F. After tempering all oxide film and traces of a decarburized surface must be removed; otherwise the nitrided surface may peel or spall off. Following hardening and tempering the steel is nitrided and allowed to cool slowly to room temperature. Quenching is not required to produce a hard nitrided case.

It should be pointed out that special nickel steels called nitralloy steels containing aluminum, chromium, molybdenum, and vanadium have been developed for use in nitriding. These steels are age hardening at the customary nitriding temperature and thus provide automatic hardening of the case during the nitriding operations. Parts increase slightly in size during nitriding but can be lapped or carefully ground to size. An increase of .002″ for a specimen with a case depth of .030″ is representative of the growth which may be expected.

Advantages of nitriding as a surface hardening operation are:

1. Very high surface hardness,

2. Good resistance to wear and corrosion,

3. Retention of hardness at elevated temperatures,

4. Low distortion and warping of parts and

5. Improved endurance limit under bending stresses.

The principal disadvantages of nitriding are the long cycle of time required and the necessity of using special alloy steels. Nitriding is used extensively for aircraft engine parts, inspection gages, cylinder liners, cams, and similar applications.

8.4 Other Diffusion Processes.

There are several other processes in addition to carburizing and nitriding which are designed to improve wear resistance by diffusion of some element into the surface layers. They are:

Carbonitriding,
Chromizing,
Siliconizing, and
Aluminizing.

Carbonitriding is a process for case hardening a steel part in a carburizing gas atmosphere containing ammonia in controlled percentages. Both carbon and nitrogen are added to the steel; the nitrogen serving chiefly to reduce the critical cooling rate of the case and allow hardening with oil or forced air cooling. Carbonitriding is normally carried out above the A_{c1} temperature for steel. Penetration rate is apparently quite close to that obtained in carburizing.

The widest application for carbonitriding is as a low cost substitute for cyaniding. Operating costs may be as low as one fourth that of cyaniding.

Chromizing involves the introduction of chromium into the surface of the metal to improve corrosion and heat resistance. This process is not restricted to ferrous metals but may be applied to nickel, cobalt, molybdenum, and tungsten. When it is applied to iron or steel, the surface is converted into a stainless steel case.

If the steel contains above 0.60 percent carbon hard chromium carbides will be formed increasing wear resistance. However, low carbon steels (0.10-0.20%) are best suited for chromizing. The chromizing process most widely used employs the principle of transfer of chromium through a gaseous $Cr-Cl_2$ phase at temperatures ranging from 1650 to 2000°F. At the high temperatures employed in chromizing some grain growth and distortion may occur. Coatings about 0.0005″ thick may be obtained after 3 hours at 1650°F. Increased case thickness may be obtained at higher temperatures. Chromized steels have been used for turbine buckets, hydraulic rams, pistons, pump shafts, and drop forging dies.

Siliconizing, or Ihrigizing, was developed by H. K. Ihrig of Globe Steel Tubes Co., Milwaukee, Wis. This process involves the impregnation of iron or low carbon steel with silicon to form a case containing about 14 percent silicon. The process is carried out by heating the work in contact with a silicon bearing material such as silicon carbide or ferro-silicon in an ordinary carburizing pot at a temperature of 1700 to 1850°F. After the parts have reached the desired temperature chlorine gas is added as a catalyst to speed the liberation of silicon from the silicon carbide. The silicon then immediately diffuses into the metal under treatment. Case depth of 0.025 to

Fig. 8–8. Bar with silicon case cut in half. One surface (left) lightly etched and the other (right) boiled in dilute nitric acid leaving corrosion resistant shell only. (American Society for Metals).

0.030 in. is usually produced in low carbon steel in two hours. Siliconized cases are difficult to machine although the hardness is only Rockwell B80 to 85. The silicon case Fig. 8–8 is resistant to scaling up to 1800°F. Oil impregnated siliconized cases exhibit very good wear resistance especially under corrosive conditions. Siliconized cases have been successfully applied in applications involving corrosion, heat, or wear: pump shafts, cylinder liners, valve guides, forgings, thermocouple tubes, and fasteners.

Aluminizing or Calorizing is a patented process of alloying the surface of carbon or alloy steel with aluminum by diffusion. The process consists essentially of packing the articles in a powder compound containing powdered aluminum and ammonium chloride then sealing them in a gas tight revolving retort in which a neutral atmosphere is maintained. The retort is heated for 4 to 6 hours at 1550 to 1700°F. After removal from the retort the article is then heated for 12 to 48 hours at a temperature of 1500 to 1800°F to allow further aluminum penetration to depths of 0.025 to 0.045 inches. The usual case content of 25 percent aluminum provides good resistance to heat and corrosion. Calorized parts are reported to remain serviceable for years at temperatures not exceeding 1400°F and have been used at temperatures of up to 1700°F for heat treating pots. Typical calorized metal parts include pots for salt, cyanide, and lead, bolts for high temperature operation, tubes for superheated steam, furnace parts, and so on.

8.5 Flame Hardening. In recent years two new processes for surface hardening steel and cast iron parts have been developed. These processes are flame hardening and induction hardening. Neither of the processes add elements to the surface of the workpiece and must therefore be

245

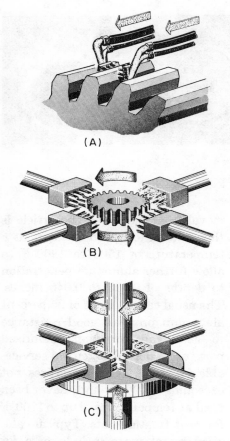

(A)

(B)

(C)

Fig. 8–9. Flame hardening by A, progressive method in which the flame moves across the area to be hardened, B, spinning method in which the work is revolved, and C, the combination method in which the work turns and the flame travels. (Linde Air Products).

used with materials capable of being hardened by heating and quenching. Recommended carbon content is 0.40-0.50 percent. Higher carbon content steels may be used but care

must be exercised to prevent surface cracking. Both processes are used when a hard surface with a tough core is desired. Flame hardening is especially useful on large parts such as lathe beds, large gears, or other parts which do not lend themselves to heating in a furnace. Flame hardening is also used on parts requiring selective hardening. In the flame hardening process shown in Fig. 8–9 heat is applied to the surface of the workpiece by means of an oxyacetylene flame. Heat is applied at such a high rate that only the surface of the metal is brought up to the hardening temperature. As the torch moves slowly along or around the workpiece a water spray follows, quenching and hardening the heated surface. The speed of the torch is adjusted to control the depth of heat penetration and resulting hardened case depth. Maximum hardness is approximately the same as can be obtained by conventional heating and quenching, with the possibility of much less distortion on many parts.

Flame hardened cases vary from about ⅛ in. depth to about ½ in. depth. Stress relieving at 400°F is recommended for all flame hardened articles except those of air hardening steel which are quenched in still air. Flame hardening is used to selective harden gear teeth, engine push rods, shafts, lathe ways, cams, and hand tools.

8.6 Induction Hardening. In the induction heating process as applied to the surface hardening of steel, heating is accomplished through the use of primary inductor coils placed around the surface to be hardened. As a high frequency alternating current passes through the inductor a high frequency magnetic field is set up around the piece to be hardened. The workpiece then, in effect, becomes the secondary of a high frequency induction apparatus. (Fig. 8–10A, B). The magnetic field induces high frequency eddy currents and hysteresis currents in the workpiece, causing it to heat rapidly. Of primary importance for induction surface heating is the fact that high frequency currents tend to travel at the surface of the conductor. This is known as the skin effect and it penetrates less deeply the higher the frequency. For example: At 1,000 Hz (cps) penetration is .059″ whereas at 1,000,000 Hz. (cps) it penetrates only .002″. The total depth of heat penetration is dependent upon the frequency employed, the input current involved, and the time allowed for conduction of the heat from the surface to the interior of the workpiece. Coil design to achieve maximum coupling is still pretty much of an art although much research has been done recently.

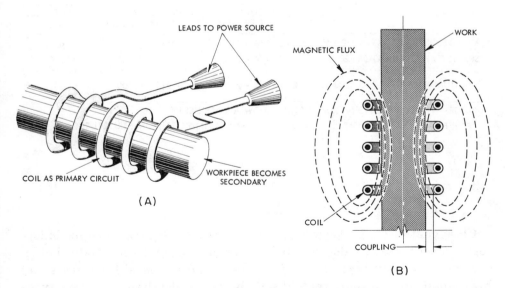

Fig. 8–10. Principles of high frequency induction heating. A, Coil and workpiece, B, magnetic coupling. (By permission: *Tool Engineers Handbook, 1959 ed.,* American Society of Tool and Manufacturing Engineers.).

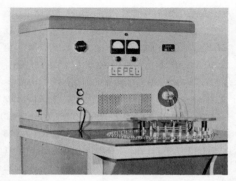

Fig. 8–11. High frequency induction heating unit equipped with rotating table, conveyor coil, and timer. Hardening or brazing can be done in 5-10 seconds. (Lepel High Frequency Labs.)

Fig. 8–12. Etched section of induction hardened gear. Show case depth and selective hardening capabilities of this process. (Tocco Div., Parks-Ohio Industries.)

Table 8–1. Frequency and hardening depth of induction heating units.

TYPE	FREQUENCY	USUAL SECTION SIZE, INCHES	HARDENING DEPTH, *INCHES	HEATING TIME
MOTOR GENERATORS	10,000 CYCLES	5/8 - 2	.050 – .150	2-4 sec.
SPARK GAP OSCILLATOR	20 TO 60 ck	1/4 - 2	0.015 – 0.100	APPROX. 2-10 SEC.
VACUUM TUBE	200 kc TO 2 mc	1/4 - 2	0.010 – 0.100	APPROX. 2-10 SEC.

NOTE: HIGHER FREQUENCIES GIVE SHALLOWER CASE DEPTH

*FOR HIGH POWER DENSITIES (15 kw/SQ. IN. MIN.) AND OPTIMUM STRUCTURE.

Close control of the heated zone together with greatly reduced distortion and very short heating time for high production rates are three distinct advantages of this method over general furnace hardening methods.

The electric oscillator-type induction heating unit is illustrated in Fig. 8–11 and the usual frequencies and hardening depths for the three types of induction heating units are shown in Table 8–1. In Fig. 8–12 is shown an etched section of an induction hardened spur gear illustrating the case depth and selective hardening effect.

Typical applications for induction hardening are gears, crankshaft bearing surfaces, cam shafts, tractor links, roller pins, ball and roller bearings, chain links, and firearm parts.

Coatings

8.7 Metallic Coatings. Metallic coatings are commonly applied to the base metal by electroplating, hot dipping, spraying, and impregnation. Metallic coatings are used to impart some particular characteristic to the surface of the metal such as improved corrosion resistance, wear resistance, hardness, or to enhance the appearance of the metal. The particular metallic coating selected and the process by which it is applied are usually matters of economics, availability of equipment, and knowledge and ability to successfully apply the coating. Several impregnation (diffusion) processes have already been discussed in paragraph 8.4.

8.8 Electroplating. This is a method of depositing a metal onto an electrically conductive surface in order to provide resistance to corrosion, special appearance such as color or lustre, or to increase the dimensions of the plated surface. Plating is applied commercially to steel, copper, brass, zinc, aluminum, and to some plastics. A laboratory or prototype electroplating unit is shown in Fig. 8–13A.

Plating materials commonly used are copper, nickel, chromium, tin, cadmium, and silver. Essentially, electroplating is done by placing the part to be plated into a solution, called the electrolyte, and sending a direct current through the solution. The electrolyte contains dissolved salts of the metal to be deposited and a weak acid or base to increase electrical conductivity of the solution. The part to be plated is made the cathode (−), Fig. 8–13B, and a suspended slab of the metal to be deposited is made the anode (+). When a low voltage direct current is applied metallic ions migrate to the cathode workpiece and upon losing their positive charge are deposited as a metal coating. The surfaces which are to be electroplated must be chemically clean since the metal coating is retained on the surface by virtue of atomic forces. If a smooth lustrous appearance (Fig. 8–14) is desired the surface to be plated must be cleaned and polished prior to the plating operation.

8.9 Hot Dipped Molten Metal. These coatings are applied to wire, sheet, and castings to provide corrosion protection. Aluminum, tin, lead, and zinc are the metals most fre-

Fig. 8–13A, Typical laboratory of prototype electroplating unit complete with controls-timer, recirculating and filtering system. (The Meaker Co.).

quently used for this process. In all cases, heat is required to cause some diffusion of the coating into the base metal. Surface preparation for hot dipping usually involves degreasing and acid or caustic cleaning. Hot dipped aluminum sheet steel has been used widely for automobile mufflers and tailpipes. The coating on hot dipped tin and tin alloy sheet is much thicker and uses more metal than its electroplated counterpart

250

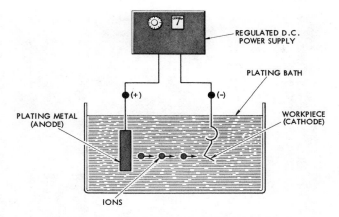

Fig. 8–13B, Schematic of an electroplating circuit.

Fig. 8–14. Highly reflective surface resulting from polishing prior to electroplating of silver plated service. (International Silver Co.).

Fig. 8–15. Dendritic structure (called spangle) on standard zinc galvanized sheet steel. Alloy coatings are free from spangle. (By permission; *Design News*, Sept. 30, 1964.).

used for food and beverage containers. Hot dipping is still used for wire, some roofing materials, and gasoline tanks. Lead coatings on steel sheet, wire, and fabricated articles such as bolts, washers, nuts, and hooks, are possible applications. Lead coated sheets are easily formed by spinning and drawing because the lead serves as an effective friction reducing lubricant.

Zinc hot dip coatings are widely known as galvanized coatings and are extensively used to protect ferrous structures and outdoor hardware against corrosion. In mild outdoor exposures zinc affords a high degree of protection. Zinc coatings deteriorate rapidly in sea water or under acidic conditions. However, positive protection against rusting is assured as long as zinc is present on the surface and has not been consumed during the rust preventing process. Fig. 8–15 shows the dendritic pattern, which forms much like frost on a window pane, as the zinc dip coating cools.

8.10 Metallizing. Metal spraying or metallizing, first invented in 1910 by a Swiss engineer, M. W. Schoop, is a process which has attained considerable importance within the last few years. Early sprayed metal coat- ings were brittle and did not bond well to the base metal, and only recently have most of the problems related to metallizing been solved. Metallizing is the process of spraying molten metal, or ceramic powders from a gun onto a surface to form a coating. (See Fig. 8–16A.) The sprayed metal coatings are used

(A)

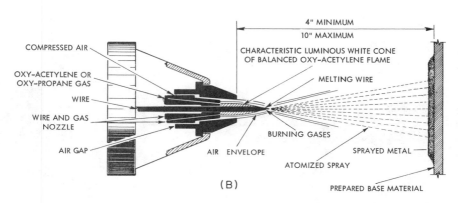

(B)

Fig. 8–16 A, Metallizing of a pump part using corrosion resisting powdered metals. B, cross section of wire nozzle and air gap of wire feed metallizing gun. (Metco, Inc.).

extensively to produce wear resistant or corrosion resistant surfaces or to build up parts which are undersize. Sprayed metal coatings have also been used for electrical shielding, conductive elements on glass for radiant heaters, soldering connections on ceramics, decorative uses, and printed circuit boards. Sprayed metal, refractory, and ceramic powders, have been used for hard facing, heat resistant coatings, electrical insulation, and other applications.

The various methods of preparing the surface to be metallized include threading, grooving, roughing, and grit blasting. Cylindrical pieces which have been grooved or threaded, Fig. 8–17, and then sprayed with a thin molybdenum coating as a bond for subsequent application of other sprayed metals, have developed very good shear strength of up to 16,000 psi and tensile strengths of around 3,000 psi.

Problems of bonding can be minimized by keeping the surface to be metallized scrupulously clean and by preheating. Preheating a surface (to a temperature of 175-200°F) before spraying will drive off all atmospheric moisture which may be on the metal surface and will also prevent condensation of moisture from the metallizing flame. Preheating of large shafts prior to metallizing will reduce danger of cracking of the coating and result in an improved bond.

8.11 Vapor Deposited Metallic Coatings. Ultra-thin coatings produced by condensation of metallic vapor under high vacuum are finding increasing usage, especially in the development of electronic solid state

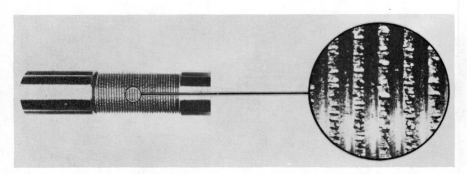

Fig. 8–17. Cylindrical piece rough threaded and then roughened with shaft preparation tool similar to a knurling tool. (Metco, Inc.).

devices. Articles to be plated may include paper, fabric, wax, and glass, in addition to metals. The metal to be deposited is vaporized by decomposition of metallic compounds, by sputtering techniques, or by evaporation of molten metals. The articles to be plated are carefully cleaned and then placed in a vacuum chamber. (Fig. 8–18A.) The chamber is

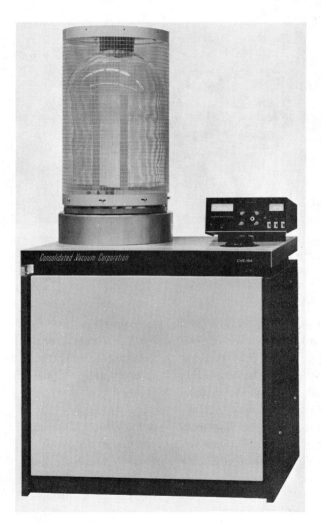

Fig. 8–18. A, high vacuum equipment for vapor deposition of thin films.

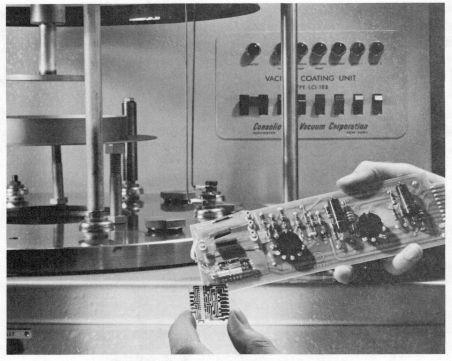

Fig. 8–18. B, articles that have been thin film coated. (Consolidated Vacuum Corp.).

usually flushed out with an inert gas such as argon and then is evacuated to a very low pressure. The coating is vaporized by one of the methods mentioned and is allowed to condense on the article to be plated. Masks may be used over areas where plating is not desired. Typical articles which have been metallized by vapor deposition include electrical capacitors, the 200-inch mirror at Palomar Observatory, alloy diaphragms for carbon broadcasting transmitters, high fidelity sound recording waxes, surgical gauzes, and integrated electronic circuits. Some coated articles are shown in Fig. 8–18B.

8.12 Oxide Coatings. Certain oxide films on metal surfaces are intended to build up resistance to corrosion or to change the appearance of the metal object. Aluminum and magnesium may be treated by the anodizing process to develop very thick (up to .005 in.) films which provide excellent corrosion and abrasion resistance and which can be colored with organic dyes.

Oxide coatings can be produced on iron and steel objects by heating them in alkaline oxidizing solutions in the 283-315°F temperature range, rinsing in water, and oiling. Fused nitrate baths consisting of conventional quenching and drawing salt mixtures will produce black oxide coatings in 30 to 120 seconds at 620 to 710°F. Heat tinting may also be done by heating steel with a bright, clean surface to temperatures of 540-600°F in a conventional furnace.

Copper may be given a black oxide coating by treatment with strongly alkaline solutions containing chlorite or persulphate. Many proprietary solutions are commercially available for development of decorative and corrosion resisting oxide coatings on metal parts such as spark plug parts, gun parts, small tools, and accessories.

8.13 Chemical Conversion Coatings. Phosphate and chromate treatments are given to steel, iron, aluminum, zinc, and magnesium alloys, among others, to convert the surface into a corrosion resistant non-reactive form. The conversion consists of a chemical modification of the metal surface so that the coating formed is an integral part of the parent metal. The process of producing phosphate coatings by immersion of metal parts in a hot solution of primary manganous phospate to which manganese dioxide has been added was developed early in the 1920's by the Parker Company and became known as "Parkerizing." Later the process was speeded up by the use of dihydrogen phosphate and copper salt and called "bonderizing," the name reflecting its use as a base for paint.

Zinc phosphate coatings have been used successfully on parts such as pistons, piston rings, cam shafts, and other parts to minimize wear and seizure during their initial wearing-in period.

Chromate conversion coatings are widely used for protection of zinc and cadmium plated war materiel used in tropical climates. Chromate coatings are also used on zinc base die castings and aluminum. Chromate films provide protection to the metals coated by (1) excluding water and (2) inhibiting any corrosion which may possibly start. Chromate films as initially formed are non-crystalline, non-porous and gel-like. While wet they can be dyed red, blue, green, or black. They should be aged 12-24 hours immediately following film formation and dyeing to allow the film to solidify. Clear, undyed chromate films are nearly always coated with transparent lacquer to enhance appearance and protection.

8.14 Organic Coatings. Paints and related organic coatings are applied for decorative as well as protective purposes. Paints protect metals from their environment by means of a continuous inert and adherent film. Although the use of paints made from natural pigments, with vehicles and binders of glue, egg white, linseed oil, gum, and resins, was practiced for many centuries in various parts of the world, the first manufacture of paint in America began in the early part of the nineteenth century. Since the beginning of manufacturing in this country there have been numerous steps forward in the coating industry including the development of synthetic resins, new plasticizers, new solvents, and new coating methods.

In the past it has been customary to classify organic finishing materials as paints, enamels, and lacquers. Although this classification is less distinctive than it used to be because of increased diversity and complexity of modern formulations, it nevertheless will be useful to give a brief description of each class as currently understood.

Paints consist essentially of pigments dispersed in a vehicle such as drying oil with the addition of a thinner or solvent to adjust to the proper viscosity for application. Film formation involves evaporation of the solvent and conversion of the drying oil to a gel state through an oxidation reaction. Oxidation or drying may be accelerated by the addition of catalysts known as driers. Ordinarily, drying is a relatively slow process at normal temperatures and the paint films formed are soft and readily deformable during early stages of drying. Paints are usually low in gloss and are principally used for protection of structures such as bridges, storage tanks, pipes, factories, and other structures.

Enamels differ from paints in the degree of dispersion of the pigment and in the type of vehicle. Enamels have a higher dispersion of pigment in the vehicle than do paints and require less mixing. The vehicle in enamels may be of the oil-resin mixture with synthetic or natural resins or it may be of the straight synthetic resin type. Vehicles may be oxygen convertible, heat convertible, or both in drying. Both catalysts and heat may be used to speed drying. Enamels generally dry more rapidly than paints and have harder, tougher coatings. It is in this class of coatings that the vast multitude of new resins are classified. Enamels are widely used in industrial finishing. Because of their fine pigment dispersion and good gloss, decorative as well as protective surfaces may be easily obtained. Enamels resist mechanical abuse and are thus used for metal products such as automotive, electrical, machine tool and office equipment finishes.

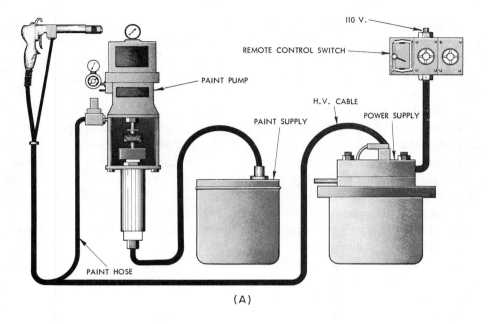

(A)

(B)

Fig. 8–19. A, electro-hydraulic electrostatic spraying system. Electrostatically charged paint particles are attracted to workpiece eliminating overspray. Hydraulic system atomizes spray. B, process hand gun with rotating atomizer increases paint mileage. (Ransburg Electro-Coating Corp.).

Lacquers in former times consisted of unpigmented cellulose esters or ethers and plasticizers. Modern usage is in the direction of classifying non-cellulosic solvent drying type resin finishes also as lacquers. For example, clear heat-setting phenolic resin finishes are also being classed as lacquers. Pigmented lacquers are called lacquer enamels and provide broadly useful protective coatings.

In contrast to paints and enamels which dry by oxidative polymerization, lacquers dry by solvent evaporation. Owing to their rapid drying characteristics, lacquers are employed widely for spray painting and dipping. The rapidity with which lacquers set make brush application exceedingly difficult. New lacquer type finishes including the vinyls, acrylates, allyls, silicones, chlorinated rubber, non-drying alkyds, and maleic-modified ester gums have been combined with cellulose derivaties to give coatings with improved qualities.

Application of organic finishes is by spraying, dipping, brushing and roller coating. New spraying methods shown in Fig. 8–19A and 8–19B cover both high pressure airless and electrostatic spraying. Both processes have minimum overspray and conserve paint.

Surface preparation is the most important factor in the success of good organic finishes. This generally means one which is free from oils and oxide films and which has been roughened by abrasive or chemical treatments such as sand blasting, anodizing, or parkerizing. The next factor is adequate coating thickness, and application of a primer coat is good insurance. Studies have shown that paint coatings of less than 5 mils in thickness are undependable when applied to exposed structural steel. A number of testing methods have been developed to test organic surfaces including the salt spray test, atmospheric exposure test, and some accelerated weathering tests.

8.15 Hardfacing. In the hardfacing process, air hardening steels such as high speed steels, natural hard alloys such as stellite, tungsten carbide, and boron carbide, and special stainless steels may be fused to metals requiring a harder or tougher wear resistant (abrasive), shock resistant and corrosion resistant surface. The hardfacing alloy may be fused to almost any metal part by either the oxyacetylene torch, the electric arc, or the powdered metal spray method. In applying hardfacing alloys it is important to prepare the surfaces to be hard faced by grinding, machining, or chipping, to clean and remove rust, scale, or other foreign substances. The work to be hard faced is often preheated before applying the hardfacing alloy. This preheating may be carried out in a furnace or by use of an oxyacetylene flame. If arc welding methods are used the atomic hydrogen process is recommended; however, good results can be obtained with straight arc welding equipment when a short arc and low current are used. If the oxyacetylene blowtorch is used it is recommended that the flame be adjusted to a reducing or excess of acetylene flame. The technique employed when using the oxyacetylene blowtorch is to heat the surface to be hardfaced to a sweating temperature, heating only a small section at a time. The hardfacing rod is then brought into the flame and allowed to melt and spread over the sweating area. If the operation is properly carried out, the hardfacing rod spreads and flows like solder. Additional hardfacing alloy may be added and the required built-up thickness accomplished. With ideal welding conditions, the bond between the hardfacing alloy and the parent metal is very strong, in most cases being stronger than the hardfacing alloy alone. Fig. 8–20 illustrates the type of structure found in a hardfaced steel surface using Haynes Stellite. Though there is practically no penetration between Haynes Stellite and steel the bond is actually stronger than the deposited metal. The deposited metal is pure Haynes Stellite,

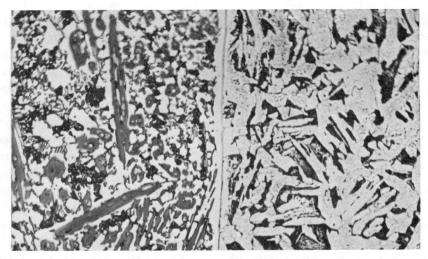

Fig. 8–20. Photomicrographs of bond between Haynes Stellite hardfacing and steel. Stellite (left) is applied with oxyacetylene torch.

261

undiluted with iron from the base metal, and hence possesses maximum wear resistance. This hardfacing process may be applied to nearly all steels and many alloys. Many of the uses include hardfacing tractor shoes, dies, cutting tools, plowshares, exhaust valve seats, pulverizer hammers, dipper bucket teeth, and parts subject to abrasion or impact.

8.16 Vitreous Coatings. Porcelain enamel and ceramic coated metals are finding increasing applications as corrosion resisting and high temperature engineering materials. Porcelain enamels present a hard, smooth, glossy surface of almost any color and provide excellent protection from atmospheric and chemical corrosion at temperatures ranging up to 1000°F. Porcelain enamels are essentially fused silicates or glasses containing color oxides and opacifying agents. They are available in compositions which will resist practically any chemical except those which dissolve silicates such as strong alkalies and hydrofluoric acid. The metal surface to be enameled must be of high quality with a minimum of surface defects. Steel, cast iron, copper, aluminum, and other metals have been successfully coated. The cast iron bath tub is said to have originated in Scotland and the porcelain enameled stove in Germany. Other applications for porcelain enameled coatings are tanks for the chemical, pharmaceutical, and food industries, glass lined railroad tank cars, stoves, refrigerators, table tops, road signs, stove fronts, cast iron sanitary ware, and linings of hot water tanks. The two general enameling processes in use are (1) the "wet" process in which the enamel is applied to the metal as a slurry by dipping, brushing or spraying, and (2) the "dry" process used on castings in which the enameling material is applied as a powder. Firing of the piece follows application of the enameling material, and involves heating the piece to a temperature of 1150-1600°F (varies greatly, dependent upon enamel and type of material) for 1 to 15 minutes followed by cooling to room temperature. As many as one to eight coats may be given depending on the number of colors involved and the coating thickness desired.

Ceramic coatings differ from porcelain enameled coatings in that they were formulated for continuous use at temperatures in excess of 1800°F. Ceramic coatings have been especially valuable for lengthening the life of expensive high temperature alloys or in permitting the substitution of less expensive, ceramic coated alloy steels. These coatings are particularly resistant to erosion by high velocity gases, chipping, and thermal shock and are free from problems of cracking and blistering.

One proprietary ceramic coating applied to Type 321 stainless steel is used to wrap insulating blankets in jet engine blast tubes. This coated sheeting has flexibility comparable with medium heavy tin foil. One ceramic metal combination of nickel-magnesia termed "cermet" is capable of withstanding temperatures of up to 3500°F for limited periods. One problem has been high temperature furnaces to fire the coatings and another has been distortion of the base metal. Developments in powdered metal spray techniques as discussed in paragraph 8.10 are making it possible to apply high temperature coatings without distorting the workpiece. Coatings which are often flame sprayed and which make dense, hard, wear resistant coatings include aluminum and zirconium. These materials have been coated on graphite crucibles as well as metal articles. Although the field of ceramic coating is a relatively new one it is expanding rapidly and filling a need in severe environment applications such as encountered in space flights and high temperature processing.

Surface Cleaning and Preparation

8.17 Degreasing. The most important requirement in successful coating application is cleanliness and preparation of the surface to which the coatings are to be applied. This generally involves degreasing, descaling, and deburring.

Degreasing operations are designed to remove grease or oil employed for temporary protection during storage or compounds used for processing operations such as drawing, threading, cutting, and surfacing. Degreasing is most frequently done with heated caustic, phosphates, and detergent solutions. The workpieces may be either dipped or sprayed. Alkali cleaners may also be operated electrolytically, the workpiece generally being made cathodic. Solvent degreasing has become popular in recent years, especially in small shops and where heavy accumulations of grease and oil must be removed.

Relatively new developments of solvent degreasing are vapor degreasing and ultrasonic degreasing. The equipment for vapor degreasing, shown schematically in Fig. 8–21, consists essentially of a heating chamber for the liquid, a vapor compartment, and vapor condenser cooling coils for controlling the vapor level. In operation, the workpiece is lowered into the boiling solvent where the bulk of the grease is removed; it is then lifted into the cooled vapor compartment where clean vapors condense onto the workpiece. The condensate runs off the workpiece, washing residual oils and greases with it, producing a very clean, oil free surface. Solvents commonly used in vapor degreasing systems are Freon, Trichloroethylene,

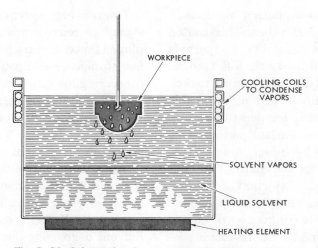

Fig. 8–21. Schematic of vapor degreasing system. Work is immersed to remove soil then held in vapors which condense and wash off residual dirt.

and various hydrocarbons such as naptha and kerosene.

Ultrasonic cleaning systems are new developments for part cleaning. The system, Fig. 8–22, and 8–23, employs high frequency mechanical vibrations transmitted into and through a suitable cleaning fluid by means of a transducer. As the parts to be cleaned are lowered into the solvent the high frequency movement of the transducer results in rapid generation and violent collapse (implosion) of minute bubbles around the parts. The impacts of the collapsing bubbles, called cavitation, generates pressures of up to 20,000 psi and erode or wear surface soil from the immersed parts. Two stages of ultrasonic cleaning are sometimes necessary in order to obtain the desired degree of cleanliness. The first stage is used to remove the major portion of the soil and the second stage, in a relatively clean solution, is to complete the cleaning.

Glass beakers containing samples of different cleaning solutions and parts can be partially submerged into the cleaning tank solution to provide a method for testing of different solvents without the necessity of preparing enough to fill the entire tank. Ultrasonic energy will be transmitted through the containers and into the liquid within them. This same method can be used when the cleaning application requires several

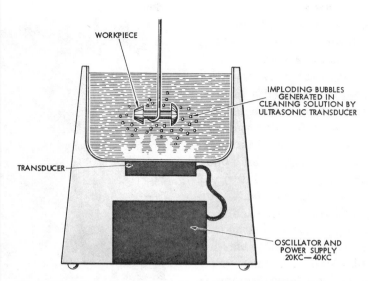

WORKPIECE

IMPLODING BUBBLES GENERATED IN CLEANING SOLUTION BY ULTRASONIC TRANSDUCER

TRANSDUCER

OSCILLATOR AND POWER SUPPLY 20KC — 40KC

Fig. 8–22. Schematic of ultrasonic cleaning system. High frequency generation and collapse of bubbles provides scrubbing action.

Fig. 8–23. Ultrasonic cleaning system employing high efficiency multi-frequency electrostrictive transducer and high pressure spray rinse tank. (Crest Ultrasonics Corp.).

stages of cleaning with different solutions.

8.18 Descaling. The process of removing thick oxide scale and corrosion products from metal surfaces by acid pickling or by mechanical abrasion is known as descaling. Chemical pickling is done by immersing the metal in dilute acid until the scale has been loosened or dissolved. Since acids usually generate hydrogen and tend to cause hydrogen embrittlement in steels, inhibitors are sometimes used in the bath.

Another practice is to heat the metal to about 200°F to drive off any hydrogen gas which may be present following the acid pickling operation. Descaled surfaces are exceedingly susceptible to oxidation and are frequently given a passivation or conversion treatment. For example: dipping descaled steel in a hot dilute solution of chromic-phosphoric acid (4-6 oz chromic-phosphoric acid per 100 gal of water) will produce a fast drying passivated surface that will resist oxidation.

Mechanical abrasion methods of descaling include abrasive blasting, shot peening, wire brushing, grinding, polishing, and flame cleaning. In general these methods remove more metal and produce rougher surfaces than do chemical pickling methods. Typical mechanical descaling equipment is shown in Fig. 8–24 and 8–25.

In the past abrasive methods of descaling have involved considerable hand labor. New automatic loading and unloading devices and work-piece manipulators are now making mechanical descaling very competitive with chemical methods. Shot peening, Fig. 8–25, has long been an important descaling treatment for forgings, castings, weldments, and heat treated steel parts. During the shot peening operation, surface layers are subjected to desirable compressive surface stresses which improve fatigue life, and reduce possibilities of failure due to residual tensile stresses in the surface.

An interesting aspect of mechani-

(A)

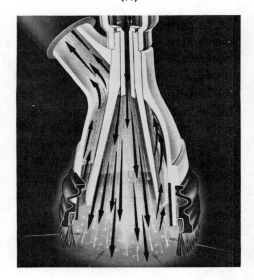

(B)

Fig. 8–24 A, abrasive blasting machine used in descaling by air-propelled glass beads. B, cutaway of gun showing how glass beads are propelled, then returned to supply by built-in vacuum system. (Vacu-Blast Corp.).

266

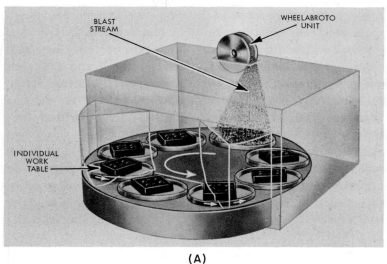

Fig. 8–25. A, phantom view of airless shot peening projector. B, operating principle of multi-table shot peening machine. (Wheelabrator Corp.)

cal abrasive descaling is the fact that surface area is increased due to the contour of peaks and valleys of the roughened surface. Several surfaces roughened by abrasive blasting with different grits ranging from 400 mesh silicon carbide to 60 mesh steel grit gave a minimum of a twenty-fold surface area increase. It has been estimated that anodized aluminum pieces have approximately a twelve-fold increase in surface area due to the fine pore structure of the anodic coating. It is well recognized that paints and other coatings adhere to roughened surfaces with greater tenacity than to polished surfaces, even though both are equally cleaned. One obvious reason for this difference is the fact that roughened surfaces provide more actual bonding area per square inch than do polished surfaces.

8.19 Deburring. Barrel tumbling, vibratory finishing, and electrode-burring are methods used for removing flash, burrs, sharp edges, and scratches from metal parts. Debur-

ring may precede descaling, or may be combined with it. For example: shot peening may be used for removal of scale and at the same time provide a deburring action. Likewise, vibratory or electrochemical finishing may perform the desired deburring action and at the same time give a clean, descaled surface.

Barrel finishing, shown schematically in Fig. 8–26A, is an operation in which parts are placed into a barrel or drum together with an abrasive medium and then rotated. Movement of the parts, as they tumble and roll over each other and the accompanying abrasive action of the medium produces a finishing effect similar to that obtained by abrasive

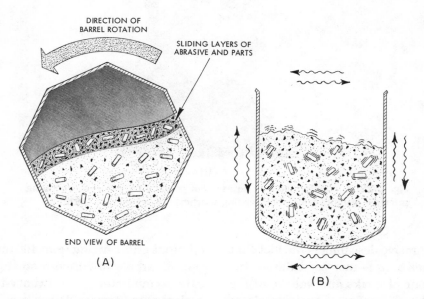

Fig. 8–26. A, schematic of barrel tumbling. B, schematic of vibratory finishing (Norton Co.).

Fig. 8–27. Three cu. ft. capacity vibratory finishing mill driven by off-balanced electric motor. (Southwest Engineering Co.).

blasting. This process may also be used to remove fins, flash, and scale from parts rugged enough to withstand the tumbling action. Tumbling is often done wet to promote the abrasive action of the medium.

Vibratory finishing is similar to tumbling except that the entire mass of abrasive, workpiece, detergent, and water is vibrated at frequencies ranging from 900 to 3600 cycles per second. The vibratory action is shown in Fig. 8–26B. One particular type of vibratory finishing mill which utilizes a spiral orbital motion is shown in Fig. 8–27. Vibratory finishing is especially useful on fragile parts which would be damaged by

269

tumbling. It also permits deburring inside of parts which is practically impossible with conventional tumbling.

Electrodeburring or electrochemical deburring is a specialized modification of electropolishing and is particularly suited to applications where burr location or quantity makes mechanical deburring economically impractical. Electrodeburring is the reverse of electroplating, with the workpiece being made the anode in an electrically conductive bath.

Metal removal rate from the burr is much faster than from the part because the sharp burr has higher current densities and is attacked from both sides. Some parts require special fixtures to concentrate the current on burr areas. Hardness of the workpiece has no effect on removal rate and hydrogen embrittlement is not a problem since the workpiece is anodic. Typical parts ideally suited for electrodeburring are small gears, intricately shaped stampings, and die castings.

References

American Foundry Equipment Co. *Shot Peening,* Mishawaka, Ind.

Burns, R. M. and Bradley, William. *Protective Coatings for Metals.* New York: Reinhold, 1962.

Graham, A. Kenneth. *Electroplating Engineering Handbook.* New York: Reinhold, 1955.

Ingram, H. S. and Shephard, A. P. *Metallizing Handbook, Vol I, Wire Process 7th ed.* New York: Metco, 1959.

Ingram, H. S. Shephard, A. P. *Flame Spray Handbook, Vol II, Powder Process.* New York: Metco, 1964.

Lyman, Taylor, ed. *Metals Handbook, 1948 ed.* Cleveland, O.: American Society for Metals.

Norton Co. *Barrel-Finishing (Rotational and Vibrational).* Worcester, Mass. 1961.

Mohler, J. B. and Sedusky, H. J. *Electroplating for the Metallurgist, Engineer and Chemist.* New York: Chemical Publishing Co., 1951.

Carbon and Alloy Steel

9.1 Introduction. Just over one hundred years ago the first commercial heat of steel was produced in the United States by means of the Bessemer Converter. The converter, a device to blow air through molten high carbon pig iron and oxidize impurities, was developed by Henry Bessemer and first operated in Wyandotte, Michigan, in 1864. In early colonial days small blast furnaces were used to produce iron. The small amount of steel that was produced probably was made by the ancient method of soaking iron in a coke fire until it absorbed sufficient carbon to change it to steel. The steel was then hammered into the desired shape and while still hot, quenched in water to harden it.

It is difficult to imagine what it would be like without the abundant supply of steel that we have learned to take for granted. A plentiful steel supply provided the material with which the railroads spanned the continents, skyscrapers erected, automobiles, ships, and tractors constructed, and thousands of other inventions developed for the use of man.

Even though the past century with steel has been one of great change, the century ahead promises to be one of significant progress and challenge. Steels will be required for extreme operating conditions of temperature, strength, and environment. A revolution in steel making has already begun with emphasis on higher quality, higher production, and lower cost.

9.2 Modern Steel Making. In the twentieth century more than 3 billion tons of steel have been produced with the United States as the undisputed leader in world steel production since 1890. The discovery of large bodies of ore in the Mesabi

Range of the Lake Superior Region along with creative enterprise has made this record possible.

At the present time the steelmaking industry is undergoing some rather important changes. Several new processes have been introduced which are improving the quality of steel, increasing productivity, and in many instances, decreasing unit costs.

These new steelmaking processes which are challenging the open-hearth method of steelmaking along with casting and rolling of ingots include use of the basic oxygen converter, spray steelmaking, electric arc slag refining, vacuum melting, continuous casting, and rolling of steel strip from powdered metals. It is beyond the scope of this book to study these developments in detail

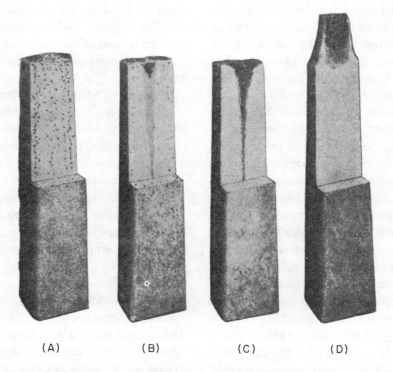

(A)   (B)   (C)   (D)

Fig. 9–1. Main types of steel ingots. A, rimmed steel. B, semi-killed steel. C, killed steel showing centerline pipe. D, killed steel with hot top to prevent centerline shrinkage.

since we are more concerned with the properties of the steel and how it behaves during heat treatment, forming, fabrication, and service. However, many of the properties of steel are related to the particular method used in its production.

One illustration of how the production process affects the properties of the steel may be seen by referring to Fig. 9–1. Steel, when it is taken in its molten state from the furnace, is poured into thick walled cast iron molds and allowed to solidify. The molds are tapered inside to facilitate removal of the solid ingot. Control of the amount of gas evolved during solidification determines the types of steel obtained. Gas evolution or "rimming" is dependent upon carbon content in the molten heat, use of deoxidizers such as aluminum, manganese, or silicon, and "capping" of ingots with a concave metal cover during solidification to increase gas pressure and inhibit gas bubble formation.

Killed steels are chemically deoxidized, have little gas entrapment, high center-line shrinkage of the ingot and are characterized by relatively uniform chemical composition and properties. This type of steel, because of its uniformity, makes sheets, strip, and plates with excellent properties for severe deep-drawing and difficult forming operations. A large portion of the top of the ingot contains shrinkage cavities and is

"cropped" off before rolling to prevent forming of long slag stringers and imperfections in the rolled steel.

Semi-killed steels are only partially deoxidized in an attempt to minimize pipe and control the shrinkage found in fully killed ingots. Semikilled steel ingots are characterized by primary and secondary blow holes but absence of piping and center-line shrinkage. Blow holes are not harmful if they do not reach the surface and become oxidized, and if the ingot is extensively worked so the holes will be flattened and welded shut.

Rimmed steels have considerable gas evolution during initial solidification of the outer rim of the ingot. Chemical composition of the ingot varies from outside to inside and top to bottom. The smooth, ductile rim or outer surface makes rimmed steels ideal for rolling sheet products well adapted to deep drawing operations.

Steels that are rolled from the ingot after it has been removed from the mold and reheated in "soaking pit" furnaces undergo severe plastic deformation. Homogenization of the steel is accomplished by rapid diffusion at the high temperature and by the mechanical kneading. Steels rolled into structural shapes or into sheet at elevated temperatures are known as *hot rolled* steels and have a tightly adhering black oxide surface coat. Hot rolled steels are accurate in smaller sizes to $\pm$ $\frac{1}{64}$

inch. Steels which have been hot rolled followed by pickling in an acid bath to remove oxide scale and subsequently rolled at normal room temperature are called *cold rolled* steels.

Cold rolled steels are work hardened from the rolling operation, have a smooth surface finish and are usually accurate in size to ± .002 inch. *Cold drawn* tubing, wire, and rod are similar to the cold rolled products but have slightly better geometry, roundness, angles, and so on, because they are drawn through hardened dies of the shape desired.

9.3 Classification of Carbon and Alloy Steels. There are a number of methods used to group or classify steels based on their chemical composition, mechanical properties, heat treatment response, ease of machining, ability to be fabricated into specified parts, and usage. The purpose of classification is to help both the producer and user. When only a few steels were available, such as carbon tool steels, or automotive steels, there was not much of a problem in selecting steels. Today, however, there are hundreds of different compositions of steel available, making the problem of selecting a steel for a particular application quite a difficult one. Classification systems for steels make it easier for the average user to select suitable steels for a particular application. They also help the producer in the develop-

ment of new steels. For example, the Society of Automotive Engineers has recently developed five new steels to fill a wide gap in the hardenability range between SAE 1026 and 1027 carbon steels. A few of the more common methods of classifying steels along with the effect of various alloying elements will be briefly discussed. It should be pointed out that steels have been broadly grouped as plain carbon steels, low alloy structural steels, stainless and heat resisting steels, tool steels, and specialty steels. Classification of tool steels and the effect of alloying elements will be discussed in Chapter 10.

Carbon Steels

9.4 Carbon in Steel. Although all steels contain carbon, the terms *carbon steel* or *plain carbon steel* are used to distinguish steel in which carbon, more than any other constituent, determines its properties and to which no special alloying elements, such as nickel, chromium, or molybdenum, have been added in any appreciable amounts. The carbon content of steel may vary from a few hundredths of one percent to nearly two percent carbon. The lower percentages are in the range of commercially pure iron, and the properties that very low carbon steel exhibits are those of nearly pure iron. The upper limit of carbon (1.40%) has been determined by practice, and this limit is controlled by two very obvious effects: (1) the decrease in plasticity and resultant brittleness in the higher percentages of carbon and (2) the danger of breakdown of cementite (Fe_3C) into graphitic carbon and ferrite. With the carbon content beyond 1.40 percent carbon, a weak and brittle steel results, developing properties similar to cast iron. The object of increasing the carbon is to increase the hardness of the steel. Under these circumstances if the carbon were to change from cementite to graphite the objective of increasing the carbon content would be lost. Graphitic carbon is a soft and friable form of carbon and acts as a softener, decreasing the hardness and strength of iron. The tendency of cementite to graphitize increases with the increase in amount of carbon or cementite; therefore, plain carbon steels containing the upper limits of carbon are more liable to graphitization than the lower carbon ones. However, under normal conditions of heating and cooling, steels containing less than 1.40 percent carbon may be safely handled without the formation of any graphitic carbon.

Carbon is the principal element controlling the structure and properties that might be expected from any carbon steel. The influence that carbon has in strengthening and hardening steel is dependent upon the amount of carbon present and its microstructure. Slowly cooled carbon steels have a relatively soft and brittle pearlitic microstructure, whereas rapidly quenched carbon steels have a strong, hard, brittle, martensitic microstructure.

9.5 Minor Constituents in Steel. We are familiar with the marked change that carbon makes on the structure of iron, changing it from a simple metal to a very complex alloy, producing an increase in the hardness and strength, and reducing the plasticity. Along with carbon all steels contain varying amounts of manganese, silicon, sulphur, phosphorus, and other impurities. These other constituents may influence the behavior of the steel adding to the variables that should be considered when steels are selected for some specific use. We will consider briefly some of the effects produced by these minor constituents.

Manganese is essential to make good steel. It promotes soundness of steel ingot castings through its deoxidizing effect and by preventing the formation of harmful iron sulphide promotes forgeability of the steel. Manganese content in regular steels varies from about 0.30 to 0.80 percent, but in special steels it may run as high as 25.0 percent. It combines readily with any sulfur in the steel, forming manganese sulfide (MnS), and preventing the sulfur from combining with the iron. If sul-

275

fur is allowed to combine with iron to form sulfide of iron, the effect would be to produce a steel that would be brittle when hot, known as *hot short* steel, resulting in difficulty and perhaps failure during forging operations. Any excess manganese (over the amount necessary to satisfy all the sulfur) combines with whatever carbon is present, forming a carbide of manganese (Mn_3C), this carbide associates with the Fe_3C in cementite. Mn_3C has properties similar to Fe_3C, increasing the hardness and strength, and lowering the plasticity of steel. Manganese is a very important alloying element in many of our special steels.

Silicon content of carbon steels varies from 0.05 percent to 0.30 percent. Special alloy steels may contain over 2.25 percent silicon. Silicon dissolves in iron in both the liquid and solid state forming a solid solution with the iron at room temperature. With a silicon content of 0.30 percent present in carbon steels the silicon has very little influence on either the structural characteristics or the mechanical properties of the steel. However, the presence of silicon in the steel during the steel melting operations increases the soundness of the steel ingots by opposing the formation of blowholes and acts as an aid during the deoxidizing process.

Sulfur content in steel may vary from a trace to 0.30 percent. In the usual types of steel the sulfur content is held below 0.06 percent. Sulfur should always be combined with manganese, for which it has a great affinity, to form manganese sulfide, MnS. This manganese sulfide occurs as pale elongated particles, bands, or strings in steel forgings or rolled stock. Manganese sulfide weakens steel by breaking up the continuity of the steel; and in the shape of elongated particles it develops directional properties. Increased sulfur reduces weldability. However, manganese sulfide has a greasy, lubricating effect, and breaks up the chips during the machine cutting of steel, thereby increasing the ease of machining. Steels that are very free cutting usually contain a higher content of sulfur than ordinary steels.

Phosphorus content of satisfactory steels varies from a trace to about 0.05 percent. As in the case of silicon, phosphorus remains dissolved in the solid iron and forms a solid solution at room temperature. It is believed that phosphorus increases the tendency toward a coarse-grained steel and therefore may weaken the steel. Also, phosphorus has a marked tendency to segregate in the last sections of a casting to solidify, producing a heterogeneous, dendritic structure. This effect of producing a segregated structure may result in marked banding of the steel during subsequent hot working operations, consequently developing marked directional properties during forging. Phosphorus in low carbon, free machining steels improves machinability.

Copper in small amounts is added to steel to improve atmospheric corrosion resistance. The presence of copper is detrimental to forge welding but not arc or acetylene welding.

Lead greatly improves machinability of steels. The lead is insoluble in steel and is randomly distributed throughout the steel in small globules which smear over the tool face when the steel is being machined to reduce cutting friction.

Oxides of iron (FeO) manganese (MnO), silicon (SiO_2), aluminum (Al_2O_3), and others may become trapped in the freezing steel. These impurities are often referred to as dirt in steel. In general these impurities are considered harmful because they are associated with the iron in a mechanical manner, being entangled throughout the structure of the steel. These oxide particles break up the continuity of the structure and impart directional properties to the forged steel. Also, as might be expected, they form points of weakness in the steel and may be the cause of the beginning of a fracture. Many failures of steel parts have been traced to particles of dirt in the steel.

Gaseous Impurities such as nitrogen, hydrogen, carbon monoxide, and oxygen are always present in varying amounts in steel. These gases may occur as bubbles in the steel or may be dissolved or combined. The effects of these gases have not been accurately determined. However, from the results of all investigations, it is thought that all these gases are harmful as they decrease the plasticity and embrittle the steel. No doubt, the more free the steel is from any of these gaseous impurities, the higher the quality and the more apt it is to be satisfactory.

9.6 Selection of Carbon Steels. Although the quality of steel is not determined by carbon content solely, the carbon content largely dictates the end use of the steel. The quality of any carbon steel is largely determined by the exacting control used by the steel maker; that is the metallurgical control used during the steel making operations and the testing and inspection given the semifinished or finished steel. Close control of the amount of carbon and control of the impurities, oxides, and slag, all constitute factors that contribute to good steel. In the selection of a carbon steel for any specific use both the carbon content and the amount of minor constituents will influence the choice. In order to have some suggestions to follow in the selection of a steel, Table 9–1 should be studied. It will serve as an indicator of the influence of carbon percentage in the choice of a steel.

Terms related to the production

Table 9–1. Carbon content of steels for different uses.

Carbon Range Per Cent	Uses of Carbon Steel
0.05–0.12	Chain, stampings, rivets, nails, wire, pipe, welding stock, where very soft, plastic steel is needed
0.10–0.20	Very soft, tough steel. Structural steels, machine parts. For case-hardened machine parts, screws
0.20–0.30	Better grade of machine and structural steel. Gears, shafting, bars, bases, levers, etc.
0.30–0.40	Responds to heat treatment. Connecting rods, shafting, crane hooks, machine parts, axles
0.40–0.50	Crankshafts, gears, axles, shafts, and heat-treated machine parts
0.60–0.70	Low carbon tool steel, used where a keen edge is not necessary, but where shock strength is wanted. Drop hammer dies, set screws, locomotive tires, screw drivers
0.70–0.80	Tough and hard steel. Anvil faces, band saws, hammers, wrenches, cable wire, etc.
0.80–0.90	Punches for metal, rock drills, shear blades, cold chisels, rivet sets, and many hand tools
0.90–1.00	Used for hardness and high tensile strength, springs, high tensile wire, knives, axes, dies for all purposes
1.00–1.10	Drills, taps, milling cutters, knives, etc.
1.10–1.20	Used for all tools where hardness is a prime consideration; for example, ball bearings, cold-cutting dies, drills, wood-working tools, lathe tools, etc.
1.20–1.30	Files, reamers, knives, tools for cutting brass and wood
1.25–1.40	Used where a keen cutting edge is necessary; razors, saws, instruments and machine parts where maximum resistance to wear is needed. Boring and finishing tools

and selection of carbon steel plates, sheet, strip, bars, and structural shapes, include: *Edge conditions* referring to the method of producing the edge, such as rolling, shearing, or flame cutting, and to the accuracy or squareness of the edge; *Quality* reflecting internal soundness, uniformity of properties, chemical composition, freedom from surface imperfections, which are important in determining the ultimate use of the steel; *Surface finish* or surface treatments referring to the degree of smoothness (or lustre) of steel and to special surface treatments such as bonderizing; and *Hardness* or temper of cold rolled sheets which have been subjected to varying degrees of cold reduction.

9.7 AISI and SAE Classification. A four or five digit numerical index system has been developed to classify steels by chemical composition. The system was first developed by the Society of Automotive Engineers for use in the automotive industry and later adopted and expanded by the American Iron and Steel Institute to include structural applications for machines. The classification also includes a limited number of high carbon and alloy steels that can be used for tools.

The first digit indicates the type of steel; thus 1 indicates a carbon steel, 2 a nickel steel, 3 a nickel-chromium steel, and so on. In the case of simple alloy steels the second digit indicates the approximate percentage of the predominant alloying element. The last two digits (or sometimes three digits in certain corrosion and heat resisting steels) indicate the average carbon content in points, or hundredths of one percent. AISI steels also have a prefix which indicates the steel-making process used. The prefix **A** denotes basic open-hearth alloy steel; **B**, acid-Bessemer carbon steel; **C**, basic open-hearth carbon steel; **D**, acid open-hearth carbon steel; and **E**, electric furnace steel. Thus, an AISI A2340 steel indicates a nickel alloy steel made in the basic open-hearth

and containing approximately 3 percent nickel (3.25 to 3.75) and 0.40 percent carbon (0.38 to 0.43).

The basic index system for the various AISI and SAE steels is:

Type of Steel	Series Designation
Carbon steels	1XXX
Plain carbon	10XX
Free machining, resulfurized (screw stock)	11XX
Free machining, resulfurized, rephosphorized	12XX
Manganese Steels	13XX
High Manganese Carburizing Steels	15XX
Nickel Steels	2XXX
3.50 percent nickel	23XX
5.00 percent nickel	25XX
Nickel-Chromium Steels	3XXX
1.25 percent nickel, 0.60 percent chromium	31XX
1.75 percent nickel, 1.00 percent chromium	32XX
3.50 percent nickel, 1.50 percent chromium	33XX
Corrosion and heat resisting steels	30XXX
Molybdenum Steels	4XXX
Carbon-molybdenum	40XX
Chromium-molybdenum	41XX
Chromium-nickel-molybdenum	43XX
Nickel-Molybdenum	46XX and 48XX
Chromium Steels	5XXX
Low chromium	51XX
Medium chromium	52XXX
Corrosion and heat resisting	51XXX
Chromium-Vanadium Steels	6XXX
Chromium 1.0 percent	61XX
Nickel-Chromium-Molybdenum	86XX and 87XX
Manganese-Silicon	92XX
Nickel-Chromium-Molybdenum	93XX
Manganese-Nickel-Chromium-Molybdenum	94XX
Nickel-Chromium-Molybdenum	97XX
Nickel-Chromium-Molybdenum	98XX
Boron (0.0005% boron minimum)	XXBXX

The standard composition for AISI and SAE carbon and low alloy steels are given in Tables 9–2 to 9–7.

Table 9-2. Standard carbon steels.
Bar, rod and semifinished products
Nonresulfurized steels.

AISI or SAE No.	Composition, % C	Mn	P Max	S Max	AISI or SAE No.	Composition, % C	Mn	P Max	S Max
1005	0.06 max	0.35 max	.040	.050	1042	0.40-0.47	0.60-0.90	.040	.050
1006	0.08 max	0.25-0.40	.040	.050	1043	0.40-0.47	0.70-1.00	.040	.050
1008	0.10 max	0.30-0.50	.040	.050	1044	0.43-0.50	0.30-0 60	.040	.050
1010	0.08-0.13	0.30-0.60	.040	.050					
1011	0.08-0.13	0.60-0.90	.040	.050	1045	0.43-0.50	0.60-0.90	.040	.050
					1046	0.43-0.50	0.70-1.00	.040	.050
1012	0.10-0.15	0.30-0.60	.040	.050	1048	0.44-0.52	1.10-1.40	.040	.050
1013	0.11-0.16	0.50-0.80	.040	.050	1049	0.46-0.53	0.60-0.90	.040	.050
1015	0.13-0.18	0.30-0.60	.040	.050	1050	0.48-0.55	0.60-0.90	.040	.050
1016	0.13-0.18	0.60-0.90	.040	.050					
1017	0.15-0.20	0.30-0.60	.040	.050	1051	0.45-0.56	0.85-1.15	.040	.050
					1052	0.47-0.55	1.20-1.50	.040	.050
1018	0.15-0.20	0.60-0.90	.040	.050	1053	0.48-0.55	0.70-1.00	.040	.050
1019	0.15-0.20	0.70-1.00	.040	.050	1055	0.50-0.60	0.60-0.90	.040	.050
1020	0.18-0.23	0.30-0.60	.040	.050	1059	0.55-0.65	0.50-0.80	.040	.050
1021	0.18-0.23	0.60-0.90	.040	.050					
1022	0.18-0.23	0.70-1.00	.040	.050	1060	0.55-0.65	0.60-0.90	.040	.050
					1061	0.55-0.65	0.75-1.05	.040	.050
1023	0.20-0.25	0.30-0.60	.040	.050	1064	0.60-0.70	0.50-0.80	.040	.050
1024	0.19-0.25	1.35-1.65	.040	.050	1065	0.60-0.70	0.60-0.90	.040	.050
1025	0.22-0.28	0.30-0.60	.040	.050	1066	0.60-0.71	0.85-1.15	.040	.050
1026	0.22-0.28	0.60-0.90	.040	.050					
1027	0.22-0.29	1.20-1.50	.040	.050	1069	0.65-0.75	0.40-0.70	.040	.050
					1070	0.65-0.75	0.60-0.90	.040	.050
1029	0.25-0.31	0.60-0.90	.040	.050	1072	0.65-0.76	1.00-1.30	.040	.050
1030	0.28-0.34	0.60-0.90	.040	.050	1074	0.70-0.80	0.50-0.80	.040	.050
1034	0.32-0.38	0.50-0.80	.040	.050	1075	0.70-0.80	0.40-0.70	.040	.050
1035	0.32-0.38	0.60-0.90	.040	.050					
1036	0.30-0.37	1.20-1.50	.040	.050	1078	0.72-0.85	0.30-0.60	.040	.050
1037	0.32-0.38	0.70-1.00	.040	.050	1080	0.75-0.88	0.60-0.90	.040	.050
1038	0.35-0.42	0.60-0.90	.040	.050	1084	0.80-0.93	0.60-0.90	.040	.050
					1086	0.80-0.93	0.30-0.50	.040	.050
1039	0.37-0.44	0.70-1.00	.040	.050	1090	0.85-0.98	0.60-0.90	.040	.050
1040	0.37-0.44	0.60-0.90	.040	.050	1095	0.90-1.03	0.30-0.50	.040	.050
1041	0.36-0.44	1.35-1.65	.040	.050					

INTERNATIONAL NICKEL CO.

Low Alloy Steels

9.8 Purposes of Alloying. Low alloy steels may be thought of as carbon steels to which have been added elements to enhance the characteristics and properties of the steel. Total percentage of alloying ele-

ments does not usually exceed about 5 percent for low alloy steels. Carbon steel does contain small amounts of silicon, manganese, sulfur, and phosphorous which are difficult to remove during the manufacture of steel; however, unless these elements are increased beyond the usual amounts

Table 9-3. Resulfurized and rephosphorized steels.

Standard carbon resulfurized steels.

AISI or SAE No.	Composition,* %				AISI or SAE No.	Composition,* %			
	C	Mn	P Max	S		C	Mn	P Max	S
1108	.08-0.13	0.50-0.80	.040	.08-0.13	1137	0.32-0.39	1.35-1.65	.040	.08-0.13
1109	.08-0.13	0.60-0.90	.040	.08-0.13	1139	0.35-0.43	1.35-1.65	.040	0.13-0.20
1110	.08-0.13	0.30-0.60	.040	.08-0.13	1140	0.37-0.44	0.70-1.00	.040	.08-0.13
1116	0.14-0.20	1.10-1.40	.040	0.16-0.23	1141	0.37-0.45	1.35-1.65	.040	.08-0.13
1117	0.14-0.20	1.00-1.30	.040	.08-0.13	1144	0.40-0.48	1.35-1.65	.040	0.24-0.33
1118	0.14-0.20	1.30-1.60	.040	.08-0.13	1145	0.42-0.49	0.70-1.00	.040	.04- .07
1119	0.14-0.20	1.00-1.30	.040	0.24-0.33	1146	0.42-0.49	0.70-1.00	.040	.08-0.13
1132	0.27-0.34	1.35-1.65	.040	.08-0.13	1151	0.48-0.55	0.70-1.00	.040	.08-0.13

* When SILICON is required, the following ranges and limits are commonly used:

Steel Designation	Silicon, %
Up to 1110 inclusive...............0.10 max	
1116 and over.....................0.10 max 0.10-0.20 or 0.15-0.30	

Lead: when required, lead is specified as an added element.

Standard carbon resulfurized and rephosphorized steels.

AISI or SAE No.	Composition,* %				
	C	Mn	P	S	Pb
1211	0.13 max	0.60-0.90	.07-0.12	0.10-0.15	—
1212	0.13 max	0.70-1.00	.07-0.12	0.16-0.23	—
1213	0.13 max	0.70-1.00	.07-0.12	0.24-0.33	—
B-1111**	0.13 max	0.60-0.90	.07-0.12	0.10-0.15	—
B-1112**	0.13 max	0.70-1.00	.07-0.12	0.16-0.23	—
B-1113**	0.13 max	0.70-1.00	.07-0.12	0.24-0.33	—
12L13	0.13 max	0.70-1.00	.07-0.12	0.24-0.33	0.15-0.35
12L14	0.15 max	0.85-1.15	.04-0.09	0.26-0.35	0.15-0.35

* These steels are not furnished to specified silicon content.
** Acid Bessemer Steel.

found in carbon steel, the steel is not considered an alloy steel.

We may assume that carbon is the most important element in either carbon or alloy steels. The carbon content determines the hardness of any steel as measured by the standard hardness testing machines. In other words, the hardness of any steel will depend chiefly upon its carbon content and heat treatments, regardless of the amount of special

Table 9-4. Alloy steels.
Bars, billets, blooms and slabs.

AISI or SAE No.	Composition, %					
	C	Mn	Ni	Cr	Mo	Others and Remarks
1330	0.28-0.33	1.60-1.90	—	—	—	—
1335	0.33-0.38	1.60-1.90	—	—	—	—
1340	0.38-0.43	1.60-1.90	—	—	—	—
1345	0.43-0.48	1.60-1.90	—	—	—	—
2317	0.15-0.20	0.40-0.60	3.25-3.75	—	—	—
2330	0.28-0.33	0.60-0.80	3.25-3.75	—	—	AISI only
2335	0.33-0.38	0.60-0.80	3.25-3.75	—	—	—
2340	0.38-0.43	0.70-0.90	3.25-3.75	—	—	—
2345	0.43-0.48	0.70-0.90	3.25-3.75	—	—	—
E2512	0.09-0.14	0.45-0.60	4.75-5.25	—	—	SAE No. 2512
2515	0.12-0.17	0.40-0.60	4.75-5.25	—	—	—
E2517	0.15-0.20	0.45-0.60	4.75-5.25	—	—	SAE No. 2517
3115	0.13-0.18	0.40-0.60	1.10-1.40	0.55-0.75	—	—
3120	0.17-0.22	0.60-0.80	1.10-1.40	0.55-0.75	—	—
3130	0.28-0.33	0.60-0.80	1.10-1.40	0.55-0.75	—	—
3135	0.33-0.38	0.60-0.80	1.10-1.40	0.55-0.75	—	—
3140	0.38-0.43	0.70-0.90	1.10-1.40	0.55-0.75	—	—
3141	0.38-0.43	0.70-0.90	1.10-1.40	0.70-0.90	—	—
3145	0.43-0.48	0.70-0.90	1.10-1.40	0.70-0.90	—	—
3150	0.48-0.53	0.70-0.90	1.10-1.40	0.70-0.90	—	—
E3310	0.08-0.13	0.45-0.60	3.25-3.75	1.40-1.75	—	SAE No. 3310
E3316	0.14-0.19	0.45-0.60	3.25-3.75	1.40-1.75	—	SAE No. 3316
4012	0.09-0.14	0.75-1.00	—	—	0.15-0.25	—
4023	0.20-0.25	0.70-0.90	—	—	0.20-0.30	—
4024	0.20-0.25	0.70-0.90	—	—	0.20-0.30	S 0.035-0.050
4027	0.25-0.30	0.70-0.90	—	—	0.20-0.30	—
4028	0.25-0.30	0.70-0.90	—	—	0.20-0.30	S 0.035-0.050
4032	0.30-0.35	0.70-0.90	—	—	0.20-0.30	—
4037	0.35-0.40	0.70-0.90	—	—	0.20-0.30	—
4042	0.40-0.45	0.70-0.90	—	—	0.20-0.30	—
4047	0.45-0.50	0.70-0.90	—	—	0.20-0.30	—

AISI or SAE No.	Composition, %					
	C	Mn	Ni	Cr	Mo	Others and Remarks
4118	0.18-0.23	0.70-0.90	—	0.40-0.60	0.08-0.15	—
4130	0.28-0.33	0.40-0.60	—	0.80-1.10	0.15-0.25	—
4135*	0.33-0.38	0.70-0.90	—	0.80-1.10	0.15-0.25	—
4137	0.35-0.40	0.70-0.90	—	0.80-1.10	0.15-0.25	—
4140	0.38-0.43	0.75-1.00	—	0.80-1.10	0.15-0.25	—
4142	0.40-0.45	0.75-1.00	—	0.80-1.10	0.15-0.25	—
4145	0.43-0.48	0.75-1.00	—	0.80-1.10	0.15-0.25	—
4147	0.45-0.50	0.75-1.00	—	0.80-1.10	0.15-0.25	—
4150	0.48-0.53	0.75-1.00	—	0.80-1.10	0.15-0.25	—
4161	0.56-0.64	0.75-1.00	—	0.70-0.90	0.25-0.35	—
4317	0.15-0.20	0.45-0.65	1.65-2.00	0.40-0.60	0.20-0.30	—
4320	0.17-0.22	0.45-0.65	1.65-2.00	0.40-0.60	0.20-0.30	—
4337	0.35-0.40	0.60-0.80	1.65-2.00	0.70-0.90	0.20-0.30	—
E4337	0.35-0.40	0.65-0.85	1.65-2.00	0.70-0.90	0.20-0.30	—
4340	0.38-0.43	0.60-0.80	1.65-2.00	0.70-0.90	0.20-0.30	—
E4340	0.38-0.43	0.65-0.85	1.65-2.00	0.70-0.90	0.20-0.30	—
4419	0.18-0.23	0.45-0.65	—	—	0.45-0.60	—
4422	0.20-0.25	0.70-0.90	—	—	0.35-0.45	—
4427	0.24-0.29	0.70-0.90	—	—	0.35-0.45	—
4608	0.06-0.11	0.25-0.45	1.40-1.75	—	0.15-0.25	Si 0.25 max
4615	0.13-0.18	0.45-0.65	1.65-2.00	—	0.20-0.30	—
4617	0.15-0.20	0.45-0.65	1.65-2.00	—	0.20-0.30	—
4620	0.17-0.22	0.45-0.65	1.65-2.00	—	0.20-0.30	—
X4620	0.18-0.23	0.50-0.70	1.65-2.00	—	0.20-0.30	—
4621	0.18-0.23	0.70-0.90	1.65-2.00	—	0.20-0.30	—
4640	0.38-0.43	0.60-0.80	1.65-2.00	—	0.20-0.30	—
4626	0.24-0.29	0.45-0.65	0.70-1.00	—	0.15-0.25	—
4718	0.16-0.21	0.70-0.90	0.90-1.20	0.35-0.55	0.30-0.40	—
4720	0.17-0.22	0.50-0.70	0.90-1.20	0.35-0.55	0.15-0.25	—
4812	0.10-0.15	0.40-0.60	3.25-3.75	—	0.20-0.30	—
4815	0.13-0.18	0.40-0.60	3.25-3.75	—	0.20-0.30	—

Table 9-5. Boron H-steels.

AISI or SAE No.	Composition, %				
	C	Mn	Ni	Cr	Mo
50B40H	0.37-0.44	0.65-1.10	—	0.30-0.70	—
50B44H	0.42-0.49	0.65-1.10	—	0.30-0.70	—
50B46H	0.43-0.50	0.65-1.10	—	0.13-0.43	—
50B50H	0.47-0.54	0.65-1.10	—	0.30-0.70	—
50B60H	0.55-0.65	0.65-1.10	—	0.30-0.70	—
51B60H	0.55-0.65	0.65-1.10	—	0.60-1.00	—
81B45H	0.42-0.49	0.70-1.05	0.15-C.45	0.30-0.60	.08-0.15
86B45H	0.42-0.49	0.70-1.05	0.35-0.75	0.35-0.65	0.15-0.25
94B15H	0.12-0.18	0.70-1.05	0.25-0.65	0.25-0.55	.08-0.15
94B17H	0.14-0.20	0.70-1.05	0.25-0.65	0.25-0.55	.08-0.15
94B30H	0.27-0.33	0.70-1.05	0.25-0.65	0.25-0.55	.08-0.15
94B40H	0.37-0.44	0.70-1.05	0.25-0.65	0.25-0.55	.08-0.15

INTERNATIONAL NICKEL CO.

alloying element present in the steel. A Rockwell hardness of C67 is about the maximum hardness that any plain carbon or alloy steel develops. Although the maximum hardness of steel may not be materially increased by the addition of any alloying element the properties such as ductility, magnetic properties, machinability, hardenability, and so on, may be very materially changed. In general, it is easier to harden alloy steels than plain carbon steels; and, because of this ease of hardening and because of the greater uniformity from any heat treatment operation, alloy steels are often selected for a job in preference to the plain carbon steel. Some of the reasons for the selection of alloy steels may include the following:

1. Increased hardenability,
2. Reduced danger of warpage,

Table 9–6. Behavior of alloying elements in annealed steel.

ELEMENT	DISSOLVED IN FERRITE	COMBINED IN CARBIDES	IN NONMETALLIC INCLUSIONS AND IN COMPOUNDS	UNCOMBINED
NICKEL	Ni	. . . .	$Ni\text{-}Si$, Ni_3, A_2	. . .
SILICON	Si	. . . .	SiO_2	. . .
ALUMINUM	Al	. . . .	Al_2O_3, etc.	. . .
COPPER	Cu	. . . .	. . . .	Cu when $> 90.8\%$
MANGANESE	Mn ◄——►		Mns	. . .
CHROMIUM	Cr ◄——►		Cr_xO_y	. . .
TUNGSTEN	W ◄——►		. . .	. . .
MOLYBDENUM	Mo ◄——►		. . .	. . .
VANADIUM	V ◄——►		V_xO_y	. . .
TITANIUM	Ti ◄——►		Ti_xO_y	. . .
NIOBIUM	Nb ◄——►		. . .	. . .
LEAD	. . .	. . .		Pb
TELLURIUM	. . .	. . .		. . .
SELENIUM	. . .	. . .	. . .	. . .
SULFUR	S?	. . .	(MnFe)S	. . .
BORON	. . .	. . .	OXIDES; NITRIDES	. . .

ADAPTED FROM E. C. BAIN AND W. H. PAXTON, ALLOYING ELEMENTS IN STEEL, 2ND ED., AMERICAN SOCIETY FOR METALS, METALS PARK, OHIO, 1961

Table 9–7. Specific effects of alloying elements in steel.

ELEMENT	SOLID SOLUBILITY		INFLUENCE ON FERRITE	INFLUENCE ON AUSTENITE (HARDENABILITY)	INFLUENCE EXERTED THROUGH CARBIDE		PRINCIPAL FUNCTIONS
	IN GAMMA IRON	IN ALPHA IRON			CARBIDE-FORMING TENDENCY	ACTION DURING TEMPERING	
ALUMINUM Al	1.1% (INCREASED BY C)	36%	HARDENS CONSIDERABLY BY SOLID SOLUTION	INCREASES HARDENABILITY MILDLY, IF DISSOLVED IN AUSTENITE	NEGATIVE (GRAPHITIZES)	......	1. DEOXIDES EFFICIENTLY 2. RESTRICTS GRAIN GROWTH (BY FORMING DISPERSED OXIDES OR NITRIDES) 3. ALLOYING ELEMENT IN NITRIDING STEEL
BORON B	0.0033%	......	RETARDS RATE OF NUCLEATION OF FERRITE (OR PEARLITE)	NOTABLY IMPROVES HARDENABILITY BY DELAYING BEGINNING OF TRANSFORMATION	MAY FORM OXIDES OR NITRIDES	ACCELERATES ANNEALING	1. SMALL BORON ADDITIONS INCREASE HARDENABILITY EQUIVALENT TO MUCH LARGER ADDITIONS OF ANY OTHER ELEMENT.
CHROMIUM Cr	12.8% (20% WITH 0.5% C)	UNLIMITED	HARDENS SLIGHTLY; INCREASES CORROSION RESISTANCES	INCREASES HARDENABILITY MODERATELY	GREATER THAN Mn; LESS THAN W	MILDLY RESISTS SOFTENING	1. INCREASE RESISTANCE TO CORROSION AND OXIDATION 2. INCREASE HARDENABILITY 3. ADDS SOME STRENGTH AT HIGH TEMPERATURES 4. RESISTS ABRASION AND WEAR (WITH HIGH CARBON)
COPPER Cu	8–10%	1.4%	SHIFTS EUTECTOID COMPOSITION TO THE RIGHT	TENDS TO STABILIZE AUSTENITE	NIL	......	1. IMPROVES ATMOSPHERIC CORROSION RESISTANCE.
COBALT Co	UNLIMITED	75%	HARDENS CONSIDERABLY BY SOLID SOLUTION	DECREASES HARDENABILITY AS DISSOLVED	SIMILAR TO Fe	SUSTAINS HARDNESS BY SOLID SOLUTION	1. CONTRIBUTES TO RED HARDNESS BY HARDENING FERRITE
MANGANESE Mn	UNLIMITED	3%	HARDENS MARKEDLY; REDUCES PLASTICITY SOMEWHAT	INCREASES HARDENABILITY MODERATELY	GREATER THAN Fe; LESS THAN Cr	VERY LITTLE, IN USUAL PERCENTAGES	1. COUNTERACTS BRITTLENESS FROM THE SULFUR 2. INCREASES HARDENABILITY INEXPENSIVELY
MOLYBDENUM Mo	3% + (8% with 0.3% C)	37.5% (LESS WITH LOWERED TEMPERATURE)	PROVIDES AGE-HARDENING SYSTEM IN HIGH Mo-Fe ALLOYS	INCREASES HARDENABILITY STRONGLY (Mo > Cr)	STRONG; GREATER THAN Cr	OPPOSES SOFTENING, BY SECONDARY HARDENING	1. RAISES GRAIN-COARSENING TEMPERATURE OF AUSTENITE 2. DEEPENS HARDENING 3. COUNTERACTS TENDENCY TOWARD TEMPER BRITTLENESS 4. RAISES HOT AND CREEP STRENGTH, RED HARDNESS 5. ENHANCES CORROSION RESISTANCE IN STAINLESS STEEL 6. FORMS ABRASION-RESISTING PARTICLES
NICKEL Ni	UNLIMITED	10% (IRRESPECTIVE OF CARBON CONTENT)	STRENGTHENS AND TOUGHENS BY SOLID SOLUTION	INCREASES HARDENABILITY MILDLY, BUT TENDS TO RETAIN AUSTENITE WITH HIGHER CARBON	NEGATIVE (GRAPHITIZES)	VERY LITTLE IN SMALL PERCENTAGES	1. STRENGTHENS UNQUENCHED OR ANNEALED STEELS 2. TOUGHENS PEARLITE-FERRITIC STEELS (ESPECIALLY AT LOW TEMPERATURE) 3. RENDERS HIGH-CHROMIUM IRON ALLOYS AUSTENITIC

Table 9-7. (cont.) Specific effects of alloying elements in steel.

Element						Specific effects
PHOSPHORUS P	0.5%	2.8% (IRRESPECTIVE OF CARBON CONTENT)	HARDENS STRONGLY BY SOLID SOLUTION	INCREASES HARDENABILITY	Nil	......	1. STRENGTHENS LOW-CARBON STEEL 2. INCREASES RESISTANCE TO CORROSION 3. IMPROVES MACHINABILITY IN FREE-CUTTING STEELS
SILICON..... Si	2% + (9% WITH 0.35% C)	18.5% (NOT MUCH CHANGED BY CARBON)	HARDENS WITH LOSS IN PLASTICITY (Mn Si P)	INCREASES HARDENABILITY MODERATELY	NEGATIVE (GRAPHITIZES)	SUSTAINS HARDNESS BY SOLID SOLUTION	1. USED AS GENERAL-PURPOSES DEOXIDIZER 2. ALLOYING ELEMENT FOR ELECTRICAL AND MAGNETIC SHEET 3. IMPROVES OXIDATION RESISTANCE 4. INCREASES HARDENABILITY OF STEELS CARRYING NONGRAPHITIZING ELEMENTS 5. STRENGTHENS LOW-ALLOY STEELS

TITANIUM Ti	0.75% (1% + WITH 0.20% C)	6% + (LESS WITH LOWERED TEMPERATURE)	PROVIDES AGE-HARDENING SYSTEM IN Ti-Fe ALLOYS	PROBABLY INCREASES HARDENABILITY VERY STRONGLY AS DISSOLVED. THE CARBIDE EFFECTS REDUCE HARDENABILITY	GREATEST KNOWN (2% Ti RENDERS 0.50% CARBON STEEL UNHARDENABLE)	PERSISTENT CARBIDES PROBABLY UNAFFECTED. SOME SECONDARY HARDENING	1. FIXES CARBON IN INERT PARTICLES
							  (a) REDUCES MARTENSITIC HARDNESS AND HARDENABILITY IN MEDIUM-CHROMIUM STEELS
							  (b) PREVENTS FORMATION OF AUSTENITE IN HIGH-CHROMIUM STEELS
							  (c) PREVENTS LOCALIZED DEPLETION OF CHROMIUM IN STAINLESS STEEL DURING LONG HEATING

TUNGSTEN W	6% (11% WITH 0.25% C)	33% (LESS WITH LOWERED TEMPERATURE)	PROVIDES AGE-HARDENING SYSTEM IN HIGH W-Fe ALLOYS	INCREASES HARDENABILITY STRONGLY IN SMALL AMOUNTS	STRONG	OPPOSES SOFTENING BY SECONDARY HARDENING	1. FORMS HARD, ABRASION-RESISTANT PARTICLES IN TOOL STEELS 2. PROMOTES HARDNESS AND STRENGTH AT ELEVATED TEMPERATURE
VANADIUM V	1% (4% WITH 0.20% C)	UNLIMITED	HARDENS MODERATELY BY SOLID SOLUTION	INCREASES HARDENABILITY VERY STRONGLY, AS DISSOVED	VERY STRONG (V Ti OR Cb)	MAXIMUM FOR SECONDARY HARDENING	1. ELEVATES COARSENING TEMPERATURE OF AUSTENITE (PROMOTES FINE GRAIN) 2. INCREASES HARDENABILITY (WHEN DISSOLVED) 3. RESISTS TEMPERING AND CAUSES MARKED SECONDARY HARDENING

3. Improved strength and toughness at high and low temperatures,

4. Resistance to grain growth at elevated temperatures,

5. Improved wear, corrosion, and heat resistance,

6. Improved magnetic properties,

7. Improved fatigue resistance,

8. Improved machinability.

Alloy steels are somewhat more difficult to produce than carbon steels because of the greater number of elements which must be kept within specified ranges and are therefore more expensive. The full value of alloy steels, as represented by higher cost, is usually obtained by using them in the heat treated condition.

9.9 Effect of Alloying Elements. Alloying elements, when added to steel, may combine in one of the following ways:

1. Form a solid solution with ferrite,

2. Combine with carbon to form carbides,

3. Deoxidize the steel by forming oxides,

4. Remain uncombined in the steel.

Alloying elements such as nickel, copper, and silicon combine with iron at all temperatures forming solid solutions. Solid solubility was discussed in Chapter 5 and is generally greatest with elements whose atomic size does not differ greatly from iron atoms and which are of the face-cen-

tered cubic structure. Solid solution forming elements slightly distort the lattice structure of the iron thus strengthening the ferrite, increasing both tensile strength and hardness of the steel. The probable hardening effect of ferrite strengthening and hardening elements is shown in Fig. 9–2. It has been found that the addition of several alloying elements in small percentages is more effective than larger amounts of one or two alloying elements. The effectiveness of alloying elements in strengthening iron, based upon equal additions by weight, and in ascending order is chromium, tungsten, vanadium, molybdenum, nickel, manganese, and silicon.

Alloying elements such as man-

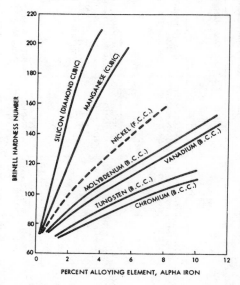

Fig. 9–2. Probable hardening effects of alloying elements dissolved in body centered cubic (B.C.C.) pure iron.

ganese, chromium, tungsten, vanadium, titanium, and molybdenum combine with the cementite of the steel forming both simple and complex carbides with the iron and carbon. Because of their hardness and relative brittleness, the carbide-forming elements, which separate out from the solid solution during slow cooling of steel as free carbides, seldom develop the properties desired. The value of carbide-forming elements in steels is that the carbides may be dissolved during heat treatment and retained in solution upon quenching.

The general effect of elements dissolved in austenite is to increase hardenability by decreasing the transformation rate; which means that pieces can be cooled more slowly, or in larger sections, and still satisfactorily transform to martensite or bainite. The relationship between Brinell Hardness and tensile strength of three alloy steels containing chromium, nickel, and molybdenum is shown in Fig. 9–3. It is interesting to note that the tensile strength is approximately 500 times the Brinell hardness number for quenched and tempered low alloy

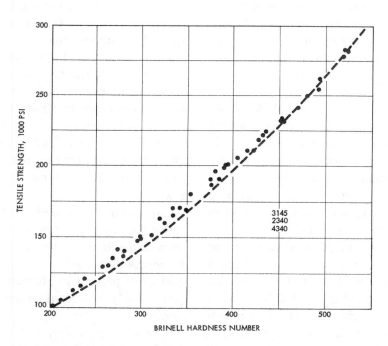

Fig. 9–3. Relationships between Brinell hardness and strength of quenched and tempered low alloy steels. (American Society for Metals)

steels. Several alloying elements which dissolve in austenite have the effect of increasing hardenability. Their value is approximately in the following descending order: boron, chromium, vanadium, molybdenum, nickel, manganese, and silicon. The effect of aluminum on hardenability seems to be relatively small.

Technically there is probably some solubility of all elements in ferrite. In the absence of carbon the elements which normally form carbides will be found dissolved in ferrite. Therefore, the carbide forming tendency is appreciable only when a significant amount of carbon is present.

The behavior of various alloying elements is shown in Table 9–6. The relative tendency for certain elements to dissolve in ferrite or to combine as carbides is shown by the size of the arrowhead.

It should be stressed that no element is wholly located in a single constituent[1]; and also, that the ultimate distribution of elements depends upon the concentration of other elements. Thus, a carbide forming element such as tungsten may be forced into ferrite solid solution by the addition of titanium or niobium unless sufficient carbon is present to combine with most of the titanium-niobium and still provide sufficient carbon for the less active carbide-forming element, tungsten.

Alloy elements in quenched steel have two functions: they permit the use of lower carbon content for a given application because of increased strength and hardenability of the alloy steel and at the same time they reduce susceptibility to quench cracking because of the greater plasticity of the low carbon martensite. The second function of alloying elements in quenching is to permit slower rates of cooling for a given section thus reducing the thermal gradient, and, in turn, reducing residual internal stresses.

The general effect of carbide forming elements such as chromium, molybdenum, and vanadium in tempered steels is to retard the softening rate. This means that the alloyed steels will require a higher tempering temperature to obtain a given hardness. The higher tempering temperatures for a given hardness result in increased toughness of the alloy steel. The general effect of alloying elements on physical properties and heat treatment response is as follows:

Aluminum because of its great affinity for oxygen is frequently used as a deoxidizer in steel making. It promotes fine austenitic grain size, and when present in amounts of approximately 1 percent promotes nitriding.

Boron a plentiful, non-metallic, hard element, with a melting point in excess of 4000°F, is used in steel

---

[1]A microscopically distinguishable part of an alloy or mixture.

288

for one purpose—to increase hardenability. When present in only a few thousandths of 1 percent boron intensifies the hardenability characteristics of other elements present in the steel. Effectiveness of boron is reduced as the carbon content of the steel is increased. When the carbon content is above 0.60 percent the use of boron is not recommended.

Chromium fosters hardenability, improves abrasion and wear resistance, and promotes carburization. Chromium is used extensively to improve hardenability, especially in air hardening steels. The presence of chromium is a vital factor in the development of both heat resisting and corrosion resisting steels.

Cobalt additions to steel increase hardness of the ferrite but reduces hardenability. Cobalt contributes to "red hardness," or the ability to resist softening at elevated temperatures.

Copper (0.20-0.50%) is used as an alloying element in steel to increase resistance to atmospheric corrosion and to act as a strengthening agent. It has also been found that paint frequently lasts longer on such steels than on non-copper-bearing types.

Lead additive in the 0.15 to 0.35 percent range improves machinability of steel but does not alter the mechanical properties of the steel. Lead has a self-lubricating action which reduces friction between the tool and

chip during metal cutting operations. Leaded alloy steels may be formed, torch cut, welded, or heat treated. Care must be exercised, however, during heat treating. Heating should be done in a well ventilated area to reduce concentration of lead vapor, and avoid creating a health hazard.

Manganese is one of the most basic elements in alloy steels. It is an energetic deoxidizer, is beneficial to surface quality, and minimizes susceptibility to tearing and cracking during high temperature mechanical working. Manganese enhances hardenability, toughness, and strength. However, high manganese content with increasing carbon tends to lower both ductility and weldability.

Molybdenum is often used in conjunction with chromium, manganese, nickel, cobalt, tungsten, and vanadium alloying elements in steel. Molybdenum promotes hardenability and widens the range of effective heat treating temperatures. Moreover, it has a strong carbide forming tendency which hampers grain growth at hardening temperatures. Molybdenum steels are especially noted for high tensile and creep strength at elevated temperatures. Quench hardened molybdenum steel is fine grained and unusually tough at all hardness levels.

Nickel increases toughness and impact properties of steel particu-

289

larly at low temperatures. It also lowers the critical temperature of steel and widens the temperature for successful heat treatment. Nickel steels are particularly suitable for case hardened parts requiring strong, tough, wear resistant cases and ductile core properties. High strength levels are obtained with lower carbon content with a resultant improvement in toughness, plasticity, and fatigue-resistance.

Niobium (Columbium) is used extensively as small additions (1-2%) in stainless steels to stabilize them against intergranular corrosion. Higher additions (up to 5%) also increase strength at elevated temperatures.

Selenium is used as a free-machining addition to stainless steels. Selenium often replaces sulfur, giving a cleaner steel with greater ductility, formability, and corrosion resistance. Selenium may also be used to prevent graphitization.

Silicon is used as a deoxidizer and as a hardener in both carbon and alloy steels. Three important effects of silicon in steel should be noted carefully: (1) It raises the critical temperature, (2) it increases the tendency for graphitization and decarburization, and (3) it promotes resistance to high temperature oxidation when combined with nickel, chromium, and tungsten.

Tellurium is used to improve machinability of carbon and alloy steels.

It is also used to prevent graphitization in cast iron castings.

Titanium has the greatest observed carbide forming tendency except possibly niobium (Columbium). Even small proportions of titanium reduce the carbon content of austenite appreciably and at 1.5 to 2.0 percent titanium in a 0.50 percent carbon steel will render it unhardenable by quench hardening. Its uses as an alloying element in steel include deoxidation, grain refinement, prevention of strain-aging in deep-drawing steel, stabilization of carbon, age hardenability in stainless steels, and increased strength for heat resisting steels.

Tungsten forms hard, abrasion resisting carbides and develops high temperature ("red") hardness in tool steels. It contributes to creep strength in some high temperature alloys.

Vanadium is a highly valuable alloying agent. It is an extremely powerful deoxidizer, though seldom used primarily for that purpose. Vanadium inhibits grain growth at elevated temperatures and promotes fine grained steels, thus imparting strength and toughness. Vanadium also tends to form stable carbides which do not agglomerate during tempering.

Specific effects for various elements are given in Table 9–7.

The steels produced by the addition of alloying elements develop

Table 9–8. Corrosion resistance of basic types of stainless steel to various media.

MEDIUM	302	316	430	410
ORGANIC SUBSTANCE:				
ACETONE	a	a	a	b
BENZOL	a	a	a	a
CARBON TETRA-CHLORIDE	m	m	m	m
ETHYL ALCOHOL	a	a	a	a
FRUIT JUICE	a	a	a	a
TOMATO JUICE	m	m	m	m
TRICHLORETHYLENE	m	m	m	m
ACIDS:				
ACETIC	m	m	m	m
BORIC	m	a	m	...
CARBOLIC	m	m	c	c
CHROMIC	c	c	c	c
CITRIC	a	a	..	b
HYDROCHLORIC	c	c	c	c
NITRIC (conc.)	a	a	a	a
PHOSPHORIC	a	a	c	..
SULFURIC (conc)	a	a	c	c
SALTS:				
ALUMINUM CHLORIDE	c	..	..	c
ALUMINUM SULFATE	a	a	..	b
AMMONIUM HYDROXIDE	a	a	a	a
BARIUM CARBONATE	a	a	..	..
CALCIUM CARBONATE	a	a	a	a
CALCIUM CHLORIDE	m	m	c	c
COPPER CYANIDE	a	a	a	a

MEDIUM	302	316	430	410
COPPER SULFATE FERRIC CHLORIDE (10%)	a	a	a	..
FERRIC SULFATE	c	m	c	c
MAGNESIUM CHLORIDE	a	a	a	..
POTASSIUM DICHROMATE	m	m	c	c
SILVER NITRATE	a	a	a	a
SODIUM CHLORIDE	a	a	a	a
(2% AERATED)	a	a	b	..
SODIUM HYDROXIDE	a	a	a	a
MISCELLANEOUS				
ALUMINUM (MOLTEN)	c	c	c	..
AMMONIA	a	a	a	..
BROMINE	c	c	c	c
CHLORINE (WET AND DRY)	c	c	c	c
CIDER	a	a	a	a
COPPER SULFATE ELECTRO-PLATING SOLUTION	a	a	..	..
LEAD (molten)	c	c	..	..
MEATS	a	a	..	..
MERCURY	a	a	..	..
NICKEL SULFATE ELECTRO-PLATING SOLUTION	a	a	..	..
SAUERKRAUT BRINE	m	m	..	..
SEA WATER	m	m	..	c
VEGETABLE JUICE	a	a	a	a
WATER	a	a	a	a

DATE FROM "STAINLESS STEEL HANDBOOK," ALLEGHENY STEEL CORP., PITTSBURG, PA. 1959.

CODE a = unaffected
b = slightly attached
c = attacked
m = must be evaluated in conditions of service

structures and properties that may be divided into four groups. These groups which include (1) pearlitic, (2) martensitic, (3) austenitic, and (4) cementitic alloy steels are represented graphically in Figs. 9–4 and 9–5.

9.10 Group I—Pearlitic Alloy Steels. The pearlitic type of alloy steel contains a relatively small percentage of the alloying element, although with a low percentage of carbon the amount of the special element may be as high as 6 percent. Fig. 9–4. The structure and characteristics of pearlitic alloy steels are similar to carbon steels and the microscopic analysis may reveal all pearlite, or a mixture of pearlite and free ferrite, or a mixture of pearlite and free cementite, all depending upon the amounts of carbon and the special element. Generally, the alloying ele-

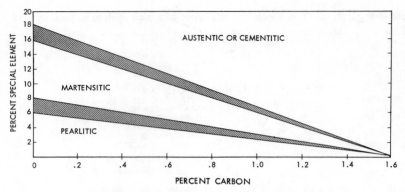

Fig. 9–4. Constitutional diagram of alloy steels. (After Guillet)

ment lowers the eutectoid ratio of carbon to iron, and it therefore requires less than 0.83 percent carbon in a special steel to produce 100 percent pearlite or eutectoid steel. The lower carbon alloy steels of this group usually are used for structural purposes; however, they may be hardened by casehardening methods and used for tools.

The medium-to-high carbon steels of this group are subjected to heat treatments and used in highly stressed parts of structures and machines, or many applications in tools. Because a great many of these alloy steels have a slower rate of transformation from austenite to pearlite during cooling they are more easily made martensitic and may often be quenched in oil in place of water or brine. They may be hardened fully by oil quenching with less danger of warping and cracking. However, due

to their slower rate of transformation these steels are apt to be hardened through their section leaving no soft core, which, in a tool, may increase the danger of breakage in service.

9.11 Group II – Martensitic Alloy Steels. When the carbon content and the amount of the special alloying element exceed that of the pearlitic type of alloy, a series of compositions are found that form a natural martensitic structure upon air cooling as shown in Fig. 9–4. The addition of a greater amount of the special alloying element than found in the pearlitic alloy steels results in the slowing up of the transformation rate that allows undercooling of the austenitic structure to around 200° F, even when cooled in still air. The alloying element acts as a fast quench in the lowering of the transforma-

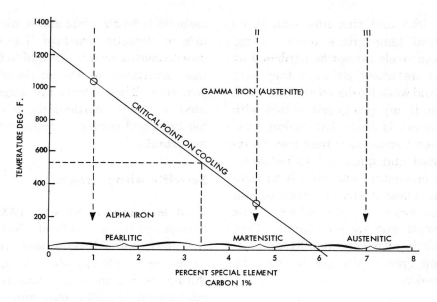

Fig. 9–5. Influence of special elements on the critical temperature of alloy steels. (After Sauveur)

tion from austenite to pearlite. This is indicated in Fig. 9–5, which illustrates the influence of the special element upon the transformation temperature.

If the transformation from austenite takes place above 600°F, a pearlitic structure is formed. If the transformation upon cooling takes place below 400°F, but above room temperature, a martensitic type of structure is formed. The properties of martensitic alloy steel are like those of fully hardened and martensitic carbon steel, although they are much more stable above room tem-

perature and resist tempering effects and are therefore used for tools and dies subjected to heating. These steels are usually hardened by air cooling from above their critical temperature and annealed by unusually slow cooling in a furnace.

9.12 Group III — Austenitic Alloy Steels. Austenitic alloy steels, as indicated in Figs. 9–4 and 9–5, are steels that remain austenitic in structure (gamma iron) upon slow cooling from the temperature of solidification. These steels do not undergo any change in the condition of the iron and therefore exhibit no critical temperature upon cooling. These steels cannot be hardened by heat treatment although they may be cold work hardened and annealed. Also, if any precipitate occurs with these steels upon slow cooling from a high temperature they may be reheated and quenched to redissolve the precipitate and keep it in solution. These austenitic steels exhibit great shock strength and low elastic strength and are very ductile. They work harden very rapidly and develop great resistance to wear by abrasion.

9.13 Group IV — Cementitic Alloy Steels. As indicated in Fig. 9–4, some alloying elements on being added to steel in increasing amounts, fail to convert the steel into an austenitic type. A steel containing 18.0 percent of a special element and 0.6 percent of carbon, upon slow cooling from above its critical temperature, would have a structure of ferrite or martensite with numerous particles of cementite embedded in the ferritic or martensitic matrix. Such a steel has been referred to as a cementitic or Group IV type of steel. The cementitic type of alloy steel is usually a difficult type of steel to use. It requires special care to properly anneal the steel to make it machinable. Also, these steels are subjected to such hardening heat treatments as to cause most of the cementite or

carbides to be absorbed and retained in a martensitic structure. The excess cementite or carbides add to the wear resistance of the martensitic structure. These steels are largely used in tools, particularly where hardness and resistance to wear are important.

Specific Alloy Steels

9.14 Manganese Steels, 13XX. Manganese has a beneficial effect in the steel both directly and indirectly, and is always added to steel during its manufacture. Manganese combines with sulfur, iron, and carbon. With no carbon present, it forms a solid solution with both gamma and alpha iron. Manganese has only a mild tendency to combine with the carbon and therefore to form a carbide (Mn_3C), which accounts for one of the principal advantages obtained from alloying manganese with steels: the manganese dissolves in the ferrite matrix adding materially to its strength. All steels contain manganese from a few hundreths of one percent in many of the low carbon structural and machine steels to about 2 percent in steels that remain pearlitic in structure.

As the manganese content of the steel is increased, the steel changes from a pearlitic to a martensitic, to an austenitic steel. The pearlitic steels are the most important. A very important steel is one containing up to 1.5 percent manganese. This type is oil hardening and retains its shape; therefore, it is known as non-deforming steel. It is used in many types of tools and dies that are to be hardened fully.

Manganese structural steel contains from 1 percent to 2 percent manganese, and from 0.08 percent to 0.55 percent carbon.

In tool steels increasing the manganese content from 0.30 percent to 1.0 percent changes the steel from water hardening to oil hardening, due to the greater hardenability of the steel with the higher manganese content.

Martensitic manganese steels are not used because of great brittleness and hardness.

Austenitic manganese steels, sometimes called Hadfield steels, contain up to 15 percent manganese and 1 percent to 1.5 percent carbon. In the cast condition this steel is very weak and brittle because of the presence of free carbides, and it is necessary to subject it to a heat treatment. The cast steel is heated to about 1900°F and quenched in water which increases the strength to 160,000 lbs per sq in., with an elongation of 60 percent to 70 percent in 2 inches. The properties are those of austenite, with a low elastic limit, but it possesses great wearing power combined with much ductility. The wearing power is due to work hardening under cold deformation.

Tests have shown that the initial wear of steel crusher jaws made from this type of steel is greater than the rate of wear after a few tons of stone have been crushed.

The reason for this increase in hardness and wear resistance is due to the ability of manganese steels to work harden rapidly and also due to conversion of retained austenite to martensite. Manganese steels find application in power shovel buckets and teeth, grinding and crushing machinery, and railway track.

9.15 Nickel Steels, 2XXX. Nickel when added to steel dissolves to form a solid solution with iron, lowering the critical range to a marked degree, and thus forms a steel that may be made pearlitic, martensitic, or austenitic by simply varying the percentage of nickel. The lower carbon pearlitic-nickel steels contain from 0.5 percent to 6.0 percent nickel, and are particularly suited for structural applications because of greater toughness, strength, and resistance to corrosion.

Martensitic-nickel steels, because of brittleness and hardness, are little used. They contain from 10 percent to 22 percent nickel. This range of nickel compositions was selected as the most probable to form martensitic structures upon slow cooling. It is only an approximate range.

The diagrams in Figs. 9–4 and 9–5 are only working diagrams which serve to illustrate the theory of the metallic behavior of special elements in iron-carbon alloys. These diagrams should not be used to determine the structural composition of any specific composition of alloy steel if the analysis is to be very accurate. If accuracy is wanted, a student should consult a constitutional diagram of the particular composition of alloy steel being studied. Constitutional diagrams of many of the common industrial alloy steels are now available. The diagrams in Figs. 9–4 and 9–5 may be considered as introductory diagrams to the study of special alloy steels.

The tonnage nickel steels have a microstructure similar to the plain carbon steels; that is the structure consists of ferrite, pearlite, and cementite in various amounts, depending upon the ratio of nickel to carbon in the steel. Low carbon steel may contain up to about 10 percent nickel and remain pearlitic, whereas with higher carbon contents the nickel is decreased to about 1 percent and the steel remains pearlitic. In pearlitic nickel steels (nickel from 0.5% to 5.0%,) nickel has a marked effect in slowing down the rate of transformation from austenite to pearlite during the cooling cycle from above the critical range. This retarding of the transformation rate allows the use of a slower cooling rate during the hardening operation. These steels can be fully hardened

by oil quenching or use of hot salt bath quenches which reduce the warping and danger from cracking during the hardening operation. Also, the slower transformation rate of these nickel steels makes it possible to heat treat heavy sections and obtain uniform hardening results. The addition of other elements, such as molybdenum and chromium, to the nickel steel increases the effectiveness of the nickel.

Nickel also retards the rate of grain growth at elevated temperatures. This effect is valuable when low carbon nickel steels are subjected to carburizing treatments for casehardening. Very little grain growth occurs in nickel steels during the carburizing cycle, preventing the formation of a coarse grained core in the finished casehardened part. By maintaining a fine grain in the core during the carburizing cycle, the necessity of a regenerative heat treatment can be avoided, and a single quench and draw can be employed to finish the carburized part.

Nickel in steel also improves the resistance to fatigue failure, increases the resistance to corrosion, and improves the toughness and impact properties. Nickel steels are harder than carbon steels, a characteristic which is of value in parts that are used for wear resistance.

Nickel steels, with nickel from 1.5 percent to 3 percent, are used for structural applications. These steels, as forgings or castings, develop excellent mechanical properties after a simple annealing or normalizing operation. Steels containing approximately 5 percent nickel are famous for their superior behavior in parts that are subjected to severe impact loads. These steels, low in carbon, may be casehardened and make excellent gears. In the higher carbon contents these steels are hardened and tempered before using.

Although many straight nickel alloy steels are used the value of nickel as an alloying element may be increased by the addition of other alloying elements. Both chromium and molybdenum are used with nickel in some of these special steels. The nickel-chromium, and nickel-chromium-molybdenum alloy steels are capable of developing superior mechanical properties and respond very well to heat treatments. In the higher alloy compositions, 2 percent nickel, 0.8 percent chromium, 0.25 percent molybdenum, these steels have excellent hardening characteristics and develop a uniform hardness in heavy sections.

Many corrosion resistant steels will be found within the range of 10 percent to 22 percent nickel but the nickel is usually combined with chromium producing an austenitic steel.

The austenitic nickel steels present a most fascinating study. Research engineers are constantly finding new alloys of this group. A few of the most common are as follows:

1. Nickel, 25% to 30%, is used for corrosion resistance.

2. Nickel-iron alloys, 20% to 30%, are non-magnetic by normal cooling and can be made magnetic by cooling to liquid air temperature.

3. Nickel, 30% to 40%, as a very low coefficient of expansion and is called Invar.

4. Nickel, 36%, and chromium, 12%, form an alloy called Elinvar that has a non-variant elastic modulus with temperature change.

5. Over 50% nickel, steel develops high magnetic properties.

9.16 Nickel-chromium Steels, 3XXX. These steels generally have a ratio of approximately 2½ parts nickel to 1 part chromium. The combination of nickel and chromium imparts some of the characteristics of each to the alloy. The effect of nickel is to increase toughness and ductility while chromium imparts wear resistance and improved hardenability. It is important to remember that the combined effect of alloying elements is usually greater than the sum of the effects of the elements when considered independently.

Low carbon nickel-chromium alloy steels are frequently carburized. The chromium carbides provide a hard wear resistant case, while both nickel and chromium improve toughness of the core. Alloys with 1.5 percent

297

nickel and 0.60 percent chromium are used for worm gears, piston pins, and machine tool components. Heavy duty applications such as aircraft gears, shafts, and cams require higher alloy content in which nickel content is increased to 3.5 percent and chromium is increased to 1.5 percent.

Nickel-chromium steels containing 0.40-0.50 percent carbon are used in the manufacture of automotive connecting rods and drill shafts.

The very high nickel-chromium alloy steels have excellent corrosion resisting properties and will be discussed in paragraph 9.23.

9.17 Molybdenum Steels, 4XXX. Molybdenum dissolves in both gamma and alpha iron, but in the presence of carbon it combines to form a carbide. The behavior of molybdenum in steel is about the same as tungsten but twice as effective. Since large deposits of molybdenum were discovered in this country the commercial development has been rapid.

Molybdenum, like tungsten, improves the hot strength and hardness of steels. The complex carbide formed by the addition of molybdenum stabilizes martensite and resists softening upon heating, similar to the action of tungsten. Molybdenum reduces grain growth at elevated temperatures and slows down the transformation rate from austen-

ite to pearlite. This characteristic of molybdenum steels, increased hardenability, allows the use of air quenching for many of the alloy tool steels instead of oil quenching in the hardening operation. The sluggishness of the molybdenum steels to any structural change during heat treating operations leads to longer soaking periods for annealing or hardening and slower cooling rates in order to obtain a well annealed structure. Molybdenum is often combined with other alloying elements in commercial alloy steels. The addition of other special elements seems to enhance the value of molybdenum in steels. Molybdenum-chromium, molybdenum-nickel, molybdenum-tungsten, and molybdenum-nickel-chromium are common combinations used. The use of molybdenum as a supplementary element in steel is more extensive than its use alone.

Molybdenum structural steels of the SAE 4000 series find many applications intended for heat treated parts. The molybdenum in these steels strengthens the steels both statically and dynamically, eliminates temper brittleness almost completely, refines the structure, and widens the critical range.

Molybdenum is alloyed with both structural steel and tool steel. The low carbon molybdenum steels are often subjected to casehardening heat treatments. Medium carbon molybdenum steels are heat treated for parts such as gears, roller bearings, and aircraft and automobile parts that are subjected to high stresses. Tool steels for general use and for hot working are often alloyed with molybdenum. Greater hardenability and improvements in physical properties are outstanding reasons for the selection of a tool steel containing molybdenum. As a stabilizer for martensite molybdenum can be used as a substitute for tungsten in hot working and high speed steels.

9.18 Chromium Steels, 5XXX. Of all the special elements in steel, none is used for such a wide range of applications as chromium. Chromium dissolves in both gamma iron and alpha iron, but in the presence of carbon it combines with the carbon to form a very hard carbide. Chromium in steel imparts hardness, wear resistance, and useful magnetic properties, permits a deeper penetration of hardness, and increases the resistance to corrosion.

When chromium is used in amounts up to 2 percent in medium and high carbon steels it is usually for the hardening and toughening effects and for increased wear resistance and increased resistance to fatigue failure. Chromium raises the critical temperature and slows down the rate of transformation from austenite to pearlite during the hardening process. This slower transformation rate allows the use of a slow quench in hardening and produces deep hardening effects. With additions of less than 0.50 percent chromium to the steel there is a tendency to bring about a refinement of grain and to impart slightly greater hardenability and toughness.

Chromium is added to the low carbon steels for the purpose of effecting greater toughness and impact resistance at subzero temperatures, rather than for reasons of increased hardenability. The lower carbon chromium steels can be casehardened, resulting in greater surface hardness and wear resistance. The SAE 5120 type of steel is used in the casehardened condition for general purposes. Engineering steels of the SAE 52100 class find many applications where deep and uniform hardening from oil quenching is needed. These steels develop good wear and fatigue resistance with moderately high strength. Steels of this type, the SAE 52100 series, are used for gears, pistons, springs, pins, and bearings.

Chromium is used quite frequently in tool steels to obtain extreme hardness. Applications include files, drills, chisels, roll thread dies, and ball bearings, which contain approximately 1 percent carbon and 1.5 percent chromium. The extreme hardness and resistance to wear associated with the chromium steels seems to be a specific property of the chromium carbides such as Cr_7C_3, Cr_4C, and $(Fe, Cr)_3C$ found in these steels.

Increasing the chromium content of steels beyond 2 percent and up to 14 percent increases the hardenability, the wear resistance, the heat resistance in steels used for hot work, and the corrosion resistance. The high carbon high chromium steels, with chromium from 12 percent to 14 percent and carbon from 1.50 percent to 2.50 percent, are used when wear resistance is of prime importance, as in knives, shear blades, drawing dies, and lathe centers.

Although high percentages of chromium, such as those in the high carbon high chromium steels, lead to the property of corrosion resistance, this property requires that the chromium carbides be dissolved at a high temperature and retained in solution by quenching. In order to obtain corrosion resistance, it is necessary to keep the chromium carbides in solution or to lower the carbon content to a value that will reduce the formation of chromium carbides. So-called stainless steel is a chromium alloy steel with a relatively low carbon content.

9.19 Chromium-Vanadium Steels, 6XXX.

Chromium-vanadium steels are characterized by fine grain size, sound, uniform structure, good hardenability, high strength, and toughness. Vanadium raises the endurance limit significantly in very low carbon steels as well as in medium and moderately high carbon steels.

Low carbon chromium-vanadium steels are used in the casehardened condition for pins and crankshafts. Medium carbon chromium-vanadium steels are noted for high toughness and strength and find application in axles and springs. Fig. 9–6

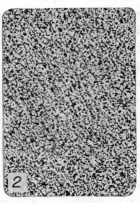

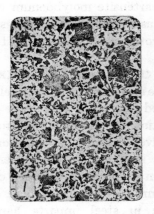

Fig. 9–6. Effect of vanadium addition on grain size of steel slowly cooled from 1550°F. 1, without vanadium, 2, with 0.26% vanadium. (American Society for Metals)

shows the effect of vanadium on grain size refinement of a slowly cooled steel. The high carbon chromium-vanadium steels have high hardness and wear resistance and are used for bearings, wrenches, and tools.

9.20 Nickel-Chromium-Molybdenum Steels, 8XXX. Nickel-chromium-molybdenum steels with medium carbon content have very high hardenability and are used extensively in aircraft structural parts. Low carbon grades used in heavy duty gears for tractors and heavy equipment are carburized to give a hard wear resistant case coupled with a strong, tough core. The cross

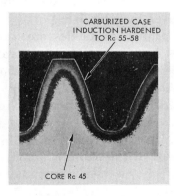

Fig. 9–7. Sectioned gear which has been carburized, the core quench hardened to R C45 and the case induction hardened to R C55-58. (By permission: *An Introduction to Physical Metallurgy*, McGraw-Hill Book Co.)

section of a gear tooth which has been carburized, then core quench hardened to R C45, followed by induction hardening of the case to R C55-58 is shown in Fig. 9–7.

9.21 Silicon-Manganese Steels, 9XXX. Silicon and manganese are present in all steels and are inexpensive deoxidizers. Steels containing more than 0.60 percent silicon are classed as silicon steels. Silicon dissolves in ferrite increasing strength and toughness. Manganese contributes markedly to strength and hardness. Manganese, like nickel, lowers the critical range and decreases the carbon content of the eutectoid composition.

A properly balanced combination of manganese and silicon produces a steel with unusually high strength, good ductility, and toughness. Silicon-manganese steels are widely used for coil and leaf springs and also for chisels and punches.

9.22 Boron Steels, XXBXX. The need for conservation of strategic elements such as manganese, nickel, chromium, and molybdenum has prompted the use of boron as an intensifying agent in the production of hardenable alloy steels. A few thousandths of one percent of boron causes a marked increase in hardening of certain steels. A very small amount of boron may replace several hundred times its own weight of critical alloying elements. Hardenability of alloy steels increases linearly as a function of boron content up to about 0.001 percent. The hardenability response of boron steels is greater for those where carbon content is in the range of 0.15 to 0.45 percent carbon. However, eutectoid and hypereutectoid steels show no improvement in hardenability from boron additions. Boron additions to steel cause a delay in the beginning of transformation but have negligible effect on the time required for completion or upon transformation temperature. This provides the advantage of shorter annealing times. Boron steels, in summary, provide substitute steels during periods of alloy shortage, and at the same time possess good hot and cold working properties, more economical annealing properties, and increased machinability.

9.23 Stainless Steels. Stainless steel, originally intended for high-grade cutlery and tools, has since developed into a steel of inestimable value for engineering purposes because of its corrosion resisting and heat resisting properties. Metal waste due to corrosion is an extremely important engineering problem and one which has been partially solved by the use of chromium and nickel alloying elements in steel. In addition to corrosion resistance and

stainless characteristics certain stainless steels are noted for their resistance to oxidation and scaling and for their maintenance of strength at elevated temperatures. As an example, type 310 stainless heat resisting alloys will resist scaling at temperatures of up to 2100°F under continuous service conditions.

The corrosion resistance of chromium alloys is due to the formation of a tightly adhering oxide film on the surface of the metal. This film which is stable, and tough, is believed to be self-healing, in that a new film will immediately form should the film be removed or damaged. This oxide film is very effective against most common corrosive media such as acids, bases, organic substances, and various salts. The corrosion resistance of four basic types of stainless steels to various media are given in Table 9–8.

The family of stainless steels comprises four basic classes: austenitic, martensitic, ferritic, and precipitation hardening types.

9.23.1 Austenitic stainless (type 2XX and 3XX) Tables 9–9 and 9–10, have the highest corrosion resistance of the stainless family because of the high chromium and nickel content (at least 23% total.) The austenitic stainless steels are so highly alloyed that they remain in the austenitic condition even at low temperatures and may not be hardened by heat treatment. However, these steels may be cold worked to develop a wide range of mechanical properties ranging from 90,000 to 350,000 psi tensile strength. These types are essentially non-magnetic in the annealed condition but may become slightly magnetic after cold working. Austenitic stainless steels are extremely shock resistant, can be readily hot worked, and are difficult to machine unless they contain sulfur, selenium, or lead.

Austenitic stainless steels are susceptible to *intergranular corrosion* when held sufficiently long at temperatures between 800 and 1600°F because of chromium carbide precipitation at the grain boundaries. Fig. 9–8. These chromium carbides are less resistant to corrosion than the balance of the austenitic structure. The loss of corrosion resistance due to carbide precipitation may be avoided by using stainless steels with lower carbon content or those containing titanium or niobium additions which form stable carbides. Austenitic stainless steels which have been subjected to a heating cycle as in welding or forging may be "stabilized" by a solution heat treatment to remove danger of intergranular corrosion. The stabilization, or solution heat treatment, usually consists of heating the stock to a temperature range of 1850°F to 2100°F for approximately a half hour to get the carbides back into solution, followed by rapid cooling such as water

quenching. A microscopic examination of austenitic steel after a stabilizing treatment will reveal a homogeneous structure free from any carbide precipitate. Fig. 9–9.

9.23.2 Martensitic stainless, Type 4XX, are primarily the straight chromium steels containing sufficient carbon to permit formation of a martensitic structure by quench hardening. These steels contain between 11.5 and 18 percent chromium and 0.15 to 0.75 percent carbon. The martensitic stainless steels are magnetic, can be satisfactorily hot or cold worked, especially in low carbon grades, are not difficult to machine, are resistant to abrasion, and have good toughness. Martensitic stainless steels show good resistance to atmospheric corrosion and to some chemicals; their best corrosion resistance is attained when quench hardened from the recommended temperature. The high chromium content of martensitic stainless steels retards transformation so steels of this type can be hardened by oil quenching and some by air cooling. Following quench hardening martensitic stainless steels are usually tempered at temperatures between 350° and 1200°F, depending upon the toughness desired. Annealed stainless steel of the martensitic types consist of complex carbides in an alpha-iron structure. Corrosion resistance of steel in this condition is poor and it should be reheated to the

303

Table 9–9. The stainless steel family, types and uses.

STRUCTURE	HARDENABILITY	GROUP	TYPE NO.	MODIFICATION FROM BASIC TYPE	GENERAL PROPERTIES AND USES
AUSTENITIC	HARDENABLE ONLY BY COLD-WORKING	CHROMIUM-NICKEL-MANGANESE	201	Cr, Ni AND Mn LOWER	SUBSTITUTE FOR TYPE 301 BUT WITHOUT MACHINABILITY AND DRAWING CHARACTERISTICS
			202	BASIC TYPES C LOWER TO PREVENT CARBIDE PRECIPITATION	SUBSTITUTE FOR TYPE 301 USE AT ELEVATED TEMPERATURES
			204L	C LOWER	FOR WELDING APPLICATIONS
			301	Cr, Ni LOWER FOR MORE WORK HARDENING	A GENERAL-UTILITY STEEL, EASILY WORKED; FOR TRIM, HOUSEHOLD UTENSILS, STRUCTURAL PURPOSES
		CHROMIUM-NICKEL	302	BASIC TYPE	EASILY FABRICATED, DECORATIVE OR CORROSION RESISTANT USE
			302B	Si HIGHER	INCREASED SCALING RESISTANCE AT E TEMPERATURES
			303	P, S ADDED FOR MACHINABILITY	HEAVY FEEDS OR DEEP CUTS
			303Se	Se ADDED FOR MACHINABILITY	MACHINABILITY AND HIGH SURFACE FINISH
			304	C LOWER TO AVOID CARBIDE PRECIPITATION	FREE SPINNING, COLD DRAWING
			304L	C LOWER FOR WELDING APPLICATIONS	WELDING WITHOUT HEAT TREATMENT
			304	Ni HIGHER FOR LESS WORK HARDENING	PLASTIC DEFORMATION, WELDING
			308	Cr, Ni HIGHER, C LOWER FOR CORROSION AND SCALING RESISTANCE	WELD WITHOUT STRESS RELIEF
			309	Cr, Ni HIGHER FOR CORROSION AND SCALING RESISTANCE	HIGH TENSILE AND CREEP STRENGTH GOOD DUCTILITY AND MALLEABILITY
			309S	C LOWER TO AVOID CARBIDE PRECIPITATION	WELDING WITHOUT STRESS RELIEF
			309C	Nb ADDED TO AVOID CARBIDE PRECIPITATION	WELDING, CORROSIVE SERVICE
			310	Cr, Ni HIGHEST TO INCREASE SCALING RESISTANCE	HIGH TEMPERATURE SERVICE
			314	Si TO INCREASE SCALING RESISTANCE	HIGH TEMPERATURE, CARBURIZING ATMOSPHERE
			316	Mo FOR CORROSION RESISTANCE	HIGHEST TENSILE, CREEP AND RUPTURE AT ELEVATED TEMPERATURES
			316L	C LOWER FOR WELDING	MINIMUM CONTAMINATION IN PHARMACEUTICAL MANUFACTURING
			317	Mo FOR CORROSION RESISTANCE AND STRENGTH AT HIGH TEMPERATURE	ELIMINATE INTERGRANULAR CORROSION IN HEAVY WELDS
			318	Nb TO AVOID CARBIDE PRECIPITATION	NOT SENSITIVE TO INTERGRANULAR CORROSION WITHIN CARBIDE PRECIPITATION RANGE
			321	Ti TO AVOID CARBIDE PRECIPITATION	HEAVY WELDMENTS WITHIN 800-1500°F RANGE
			347	Nb TO AVOID CARBIDE PRECIPITATION	STABILIZED, FREE MACHINING
			347Se	Se TO IMPROVE MACHINABILITY	
			348	SIMILAR TO 347 BUT LOW TANT	

Table 9-9. (cont.) The stainless steel family, types and uses.

Family	Hardenability	Base	Type	Characteristic	Uses
MARTENSITIC	HARDENABLE BY HEAT-TREATMENT	CHROMIUM	403	Cr ADJUSTED FOR SPECIAL MECHANICAL PROPERTIES	FORGED TURBINE BLADES
			410	BASIC TYPE	CORROSION AND HEAT RESISTANT HARDENABLE BY AIR QUENCH
			414	Ni TO INCREASE CORROSION RESIST, MECHANICAL PROPERTIES	SPRINGS, KNIFE BLADES, RULES
			416	P, S ADDED FOR MACHINABILITY	AUTO. SCREW MACHINING, NON-GALLING
			416Se	Se FOR MACHINABILITY	AUTO. SCREW MACHINING GRADE
			418	Sp W FOR HIGH TEMPERATURE PROPERTIES	
			420	C HIGHER FOR CUTTING PURPOSES	ABRASION RESISTANT, QUENCH-HARDENABLE
			420F	P, S FOR MACHINABILITY	FREE MACHINING VERSION OF 420
			431	Cr, Ni FOR BETTER RESISTANCE AND PROPERTIES	BEST CORROSION RESISTANCE OF HARDENABLE GRADES
			440A	C HIGHER FOR CUTTING APPLICATIONS, DUCTILITY	INSTRUMENTS, CUTLERY, VALVES
			440B	C HIGHER	HIGHER HARDNESS THAN 440A
			440C	C STILL HIGHER FOR WEAR RESISTANCE	HARDENED STEEL BALLS, OIL INDUSTRY APPLICATIONS
			501	CORROSION RESISTANT RATHER THAN STAINLESS	GOOD STRENGTH AT ELEVATED TEMPERATURES
			502	CORROSION RESISTANT RATHER THAN STAINLESS	GOOD STRENGTH AT ELEVATED TEMPERATURES
FERRITIC	NON-HARDENABLE	CHROMIUM	405	Al TO PREVENT HARDENING	NON-HARDENING WHEN AIR COOLED
			406	Al INCREASED	ELECTRICAL RESISTANCES, NON-HARDENING
			430	BASIC TYPE	EASILY FORMED, AUTOMOBILE TRIM, CHEMICAL EQUIPMENT
			430F	P, S OR Se FOR EASIER MACHINING	FREE MACHINING VARIETY OF 430
			442	Cr TO INCREASE SCALING RESISTANCE	HIGH TEMPERATURE SERVICE, DIFFICULT TO FABRICATE
			443	Cu, FOR EASIER FABRICATION	CHEMICAL EQUIPMENT AT HIGH TEMPERATURES
		NICKEL	446	Cr MUCH HIGHER FOR SCALING RESISTANCE	HIGH RESISTANCE TO CORROSION AND SCALING TO 2150°F.
PRECIPITATION HARDENING	PRECIPITATION AGE HARDENABLE		17-4PH	CHROMIUM-NICKEL COPPER PRECIPITATION HARDENING	ULTRA-STRENGTH, EXCELLENT CORROSION RESISTANCE
			17-7PH	ALUMINUM PRECIPITATION HARDENING	EXCEPTIONALLY HIGH STRENGTH SIMPLE HEAT TREATMENT
			18-Ni-Co-Mo	NICKEL-COBALT-MOLYBDENUM PRECIPITATION HARDENING	MARAGING STEELS WITH U.S. OF UP TO 300,000 PSI DEPENDENT UPON COMPOSITION AND TREATMENT

FROM DATA CONTAINED IN "STAINLESS STEEL HANDBOOK," ALLEGHENY LUDLUM STEEL CO., METALS HANDBOOK, ASM, "ARM60 STAINLESS STEELS," AND "THE MARVEL OF MARAGING," G. P. CONTRACTOR, JOURNAL OF METALS, AUG. 1966

Table 9–10. Typical mechanical properties of basic stainless steel types.

GROUP	AUSTENITIC		MARTENSITIC	FERRITIC	PRECIPITATION HARDENING	
TYPE	202	302	410	430	17.4	18 Ni–Co–Mo (250)
ANALYSIS, 9%						
CHROMIUM	17.0–19.0	17.0–19.0	11.5–13.5	14.0–18.0	15.5–17.5	-----
NICKEL	4.0–6.0	8.0–10.0	0.50 MAX.	0.50 MAX.	3.0–5.0	18.0–19.0
OTHER ELEMENTS	$N_2$0.25 MAX	-----	-----	-----	Cu 3.5–5.0 Nb+ Ta 0.15–0.45	Ti 0.3–0.5 Al0.05
CARBON	0.15 MAX.	0.15 MAX.	0.15 MAX.	0.12 MAX.	0.07 MAX.	0.03 MAX.
MANGANESE	7.5–10.0	2.0 MAX.	1.0 MAX.	1.0 MAX.	1.0 MAX.	0.12
SILICON	1.0 MAX.	1.0 MAX.	1.0 MAX.	1.0 MAX.	1.0 MAX.	-----
TEMPERATURE, °F						
FORGING–START	2300	2200	2100	2100	2000	2000
ANNEALING OR SOLUTION TREATMENT	1850–2000	1850–2050	1500–1650	1400–1500	1900	1500
HARDENING	COLD WORK	COLD WORK	1700–1850	NON–HARD-ENABLE	900–1100	900
QUENCHING	-----	-----	OIL OR AIR	-----	AIR	AIR
TEMPERING	-----	-----	OVER 1100°F	-----	-----	-----
STRESS RELIEF	-----	-----	UNDER 700°F	-----	300–800	-----
MECHANICAL PROPERTIES ANNEALED:						
STRUCTURE	AUST.	AUST.	F–C	F–C	COMPLEX	COMPLEX
YIELD STRENGTH, PSI.	40,000	30,000	32,000	35,000	111,000	105,000
UTS, PSI	100,000	80,000	60,000	60,000	150,000	140,000
ELONGATION %, 2 IN.	40	50	20	20	6–15	16
R.A., %	—	60	50	40		68
ELASTIC MODULUS X 106	29.0	29.0	29.0	29.0		27
BHN	210 MAX.	200 MAX.	200 MAX.	200 MAX.	355 MAX.	270–300
IMPACT, IZOD FT.-LB.	85 MIN.	85 MIN.	85 MIN.	3–85	----	
MECHANICAL PROPERTIES HEAT TREATED:						
YIELD STRENGTH PSI.	----	----	35,000–180,000	----	200,000	250,000
UTS, PSI	----	----	60,000–200,000	----	185,000	257,000
EL. % IN 2 IN.	----	----	25–2	----	14	12
BHN	----	----	120–240	----	420	490
MAXIMUM OPERATING TEMPERATURE °F CONTINOUS SERVICE	1550°	1700°	1300°	1500°	900°	900°

austenitic range to dissolve the chromium carbide followed by quenching.

9.23.3 Ferrite stainless containing from 11.5 to 27 percent chromium and a maximum of 0.25 percent car-

Table 9–11. Composition and properties for typical ultrastrength steels.

BRAND NAME AND PRODUCER	T.S. X1000 PSI	Y.S. X1000 PSI	EL. %	R.A. %	IZOD IMPACT FT. LB.	AVERAGE COMPOSITION, PERCENT						
						C	Mn	Si	Cr	Mo	Ni	V
AISI E4340 (ALLEGHENY LUDLUM)	220	190	12	35	12	0.40	0.75	0.27	0.80	0.25	1.82	....
AMS 6427B (ALLEGHENY LUDLUM)	220	190	12	35	12	0.30	0.87	0.27	0.87	0.42	1.82	0.07
H 5 220 (TIMKEN)	249	201	11	36	24	0.30	0.70	0.55	1.17	0.45	2.05	....
HY-TUF (CRUCIBLE)	239	183	14	46	33	0.25	1.30	1.50	....	0.40	1.80	....
AISI E4340 (ALLEGHENY LUDLUM)	260	217	8	30	10	0.40	0.75	0.27	0.80	0.25	1.82	....
HS 260 (TIMKEN)	271	221	7	24	16	0.40	0.70	0.60	1.17	0.45	2.05	....
NO. 5-317 (CARPENTER)	306	237	6	13	11	0.50	0.50	0.20	1.00	....	1.75	....
UHS 260 (CRUCIBLE)	270	230	11	40	18[2]	0.35	1.25	1.50	1.25	0.35	....	0.20
TRICENT (BETHLEHEM)	289	245	9	34	23	0.43	0.77	1.62	0.82	0.37	1.82	0.07
USS "STRUX"[1] (U.S° STEEL)	301	251	9	29	13[2]	0.43	0.87	0.65	0.92	0.52	0.75	0.06
LABELLE H+ (CRUCIBLE)	314	255	9	29	12[2]	0.43	1.35	2.25	1.35	0.40	....	0.30

"ULTRA-STRENGTH STEELS," CLIMAX MOLYBDENUM CO., N.Y., 1960.

(1) BORON ADDED

(2) V-NOTCH CHARPY, ROOM TEMPERATURE

bon are magnetic and consist structurally of a solid solution of iron and chromium. Ferritic stainless steels are used where resistance to corrosion and heat is needed, but where service conditions are not too severe and where slight discoloration of the surface during service can be tolerated. Steels of this type cannot be hardened by heat treatment. They

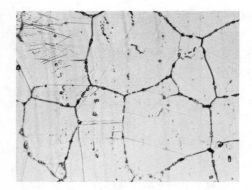

Fig. 9–8. Photomicrograph showing carbide precipitation at grain boundaries in austenitic stainless steel held at 1220°F. 500X. (Armco Steel Co.)

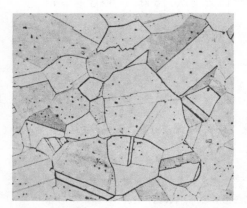

Fig. 9–9. Austenitic stainless steel with a homogeneous structure free from carbide precipitation.

are susceptible to grain growth at elevated temperatures, and with larger grain size, suffer a loss in toughness. The ferritic stainless steels may be cast and forged hot

or cold, but they do not machine easily unless they contain free-machining additives such as molybdenum, phosphorus, sulfur, selenium, or tellurium. Ferritic stainless steels possess high resistance to corrosion in an ordinary atmosphere, providing they have been highly polished and are free from foreign particles.

9.23.4 Precipitation hardening alloys were announced late in 1959 by the International Nickel Company. These high nickel steels were originally developed with 20-25 percent nickel and titanium plus aluminum as the hardening agents. These alloys have a number of desirable properties including high strength and ductility, good machinability, and good weldability. The low temperature heat treatment is simple and gives low distortion, freedom from transformation cracking, and has no decarburization problem. In general, the hardening mechanism for these steels involves a combination of martensitic transformation and precipitation hardening. The martensite, because of its low carbon content, is soft and tough.

The major increase in hardness and strength for this alloy is produced by aging at 900°F for 4 hours followed by air cooling. During aging there is a precipitation of nickel-aluminum-titanium compounds which distort the lattice structure, strengthening and hardening the alloy, Fig. 9–10. Other precipitation hardening combinations now being used besides titanium-aluminum include cobalt, molybdenum-titanium, copper-molybdenum, and chromium-nickel-copper. The effect of aging temperature on mechanical properties of a titanium-aluminum precipitation hardening alloy is shown in Fig. 9–11.

9.24 Ultra-Strength Steels. Ultra-strength steels are those steels which are able to retain ductility and toughness at tensile strengths above 200,000 psi. Steadily increasing engineering requirements are demanding higher strengths and strength-weight ratios. Ten years ago, 180,000 psi tensile strength was the maxi-

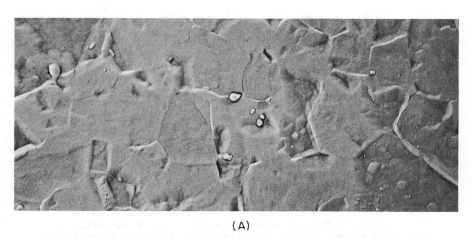

(A)

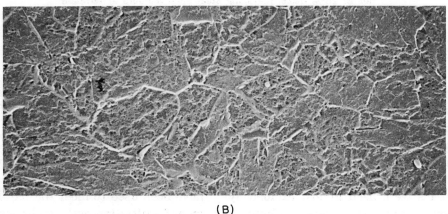

(B)

Fig. 9–10. Electron micrographs of 18% maraging stainless steel. A, solution annealed 1 hour at 1500°F. B, aged 3 hours at 900°F after annealing. 8000X. (International Nickel Co.)

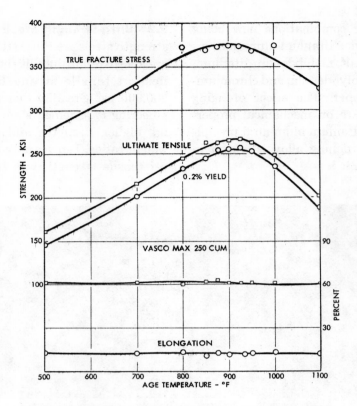

Fig. 9–11. Effect of aging temperature on the tensile properties of 18% nickel maraging steel aged 3 hours at temperatures shown after 30 min. solution treatment at 1500°F. (Vanadium Alloys Steel Co.)

mum that could be used. Today, advances in design, processing, and metallurgy have paved the way for alloy steels having minimum tensile strength of 200,000 to 300,000 psi. Until recently, these ultra-strength steels were used almost exclusively for aircraft parts such as landing gear and structural members. These materials are now available for other applications where their advantage in weight and space saving offset their higher cost.

Composition and properties for typical ultra-strength steels are given in Table 9–11. Two different means have been used to meet the requirement for ultra-strength steels: modification of the heat treatment of a standard alloy (AISI E4340); or modification of composition. Because of the low tempering temperatures required, ultra-strength steels are suitable for service at temperatures usually below 400°F. Typical applications include portable tools, materials handling equipment, highly stressed machine components, and aerospace vehicle components.

References

American Iron and Steel Institute. *The Making of Steel*. New York, 1962.

Bethlehem Steel Corp. *Quick Facts About Alloy Steel, 7th ed.* Bethlehem, Pa., 1964.

Climax Molybdenum Co. *Ultra-strength Steels*. New York, 1960.

Rogers Publishing Co. "Carbon Steel Mill Products" *Design News Supplement*, Sept. 30, 1964.

Taylor, H. F., Flemings, M. C. and Wulff, J. *Foundry Engineering*. New York: John Wiley & Sons, 1959.

Tool Steels

10.1 Introduction. Man has been using metal tools for many centuries to aid in the production of food and clothing and in the construction of buildings and labor saving devices. Although steel had been used in implements of war for a long time it was not until the 1700's that it was used for tools. Steel tools were far superior to the copper or bronze tools that had been used previously because of the ability of steel to maintain sharp cutting edges and to resist wear. The first steel drills and lathe tools were made from plain carbon steel. In the late 1800's the need arose for cutting tools which could maintain their hardness at higher temperatures resulting from faster machining speeds. The result was that high speed steels were developed. The first high speed steel was known as Mushet High Speed and required no heat treatment. It was merely forged to shape and allowed to cool. Mushet steel could not be annealed, nor could it be machined into drills, taps, and reamers. It soon passed out of existence.

The next high speed steels contained high percentages of tungsten and were able to maintain their hardness even at a dull red heat. These steels could be annealed and made readily machinable. Later special tool steels were developed with increased wear resistance, high strength properties, great toughness, anti-galling qualities, and high hardness. Today, there are a great variety of tool steels available. These tool steels represent the highest quality of steel available. Special precautions are taken in tool steel manufacturing to keep porosity, segregation, non-metallic inclusions and impurities at a minimum. The tool steel represents only a small part of the investment in a complex mold or die, the major investment being for machining costs. It would be poor management, indeed, to purchase low quality, inexpensive steel for a die then spend several hundred hours machining it, only to find internal flaws required the piece to be scrapped. Many tool steels are sold under trade names such as *ketos*, *solar*, and *Regal*, rather than on analysis.

The large number of tool steels now available from various steel producers makes their classification difficult. Although several systems of tool steel classification have been devised, only the AISI classification system will be discussed.

10.2 AISI Tool Steel Classification. The most commonly accepted classification system is the one established by the American Iron and Steel Institute. The AISI system of tool steel identification is based on quenching method commonly employed, application, special characteristics, and steels for particular industries. The commonly used tool steels are grouped, under this system, into seven major classifications:

1. High speed:
   Tungsten base . . . . T
   Molybdenum base . . . . M
2. Hot work:
   Chromium base H10-H19
   Tungsten base H20-H39
   Molybdenum base H40-H59
3. Cold work:
   High carbon, high chromium . . . . . D
   Medium alloy air hardening . . . . . . A
   Oil hardening . . . . O
4. Shock resisting . . . . S
5. Mold steels . . . . P
6. Special purpose:
   Low alloy L
   Carbon tungsten . . . . F
7. Water hardening . . . . W

Characteristic properties of the various types of tool steels are shown in Table 10–1.

10.3 High Speed Tool Steels. The high speed steels maintain their hardness at temperatures of up to 1100°F. Due to this property, tools made from high speed steels can operate satisfactorily at speeds which cause the cutting edges to reach a temperature that would soften and ruin ordinary tools. The principal alloying elements are tungsten, molybdenum, chromium, and vanadium. One of the most commonly used high speed steels is the 18-4-1 steel which contains 18 percent tungsten, 4 percent chromium and 1 percent vanadium. The carbon content in high speed steels is usually around 0.50—0.90 percent but runs as high as 1.50 percent in the T-15 and M-15 types. Because of the expense and difficulty of obtaining tungsten, which came primarily from China at the time, experiments were conducted to find a substitute. It was found that molybdenum could be successfully substituted for tungsten in these steels, with up to 8 percent molybdenum and no tungsten. Today, the more popular types of high speed steels contain approximately 6 percent tungsten and 5 percent molybdenum. Thus, high speed steels are divided into two types: Tungsten type T and molybdenum type M steels. Composition of the various

Table 10–1. Comparative properties of tool steels.

TYPE	GROUP	QUENCH MEDIUM*	WEAR RESISTANCE	TOUGHNESS	WARPAGE RESISTANCE	HARDENING DEPTH	RESISTANCE TO SOFTENING AT HIGH TEMP	REL COST	MACHINING RATING	RESISTANCE TO DECARBURIZATION
M	HIGH SPEED	O, A, A	VERY HIGH	LOW	A, S: LOW / O: MEDIUM	DEEP	HIGHEST	HIGH	45–60	LOW–MEDIUM
T	HIGH SPEED	O, A, S	VERY HIGH	LOW	A, S: LOW / O: MEDIUM	DEEP	HIGHEST	HIGHEST	40–55	LOW–HIGH
H	HOT WORK: CR BASE	A, O	FAIR	GOOD	O: FAIR / A: GOOD	DEEP	GOOD	HIGH	75	MEDIUM
	W BASE	A, O	FAIR TO GOOD	GOOD	O: FAIR	DEEP	VERY GOOD	HIGH	50–60	MEDIUM
	Mo BASE	O, A, S	HIGH	MEDIUM	A, S: LOW / O: MEDIUM	DEEP		HIGH	50–60	LOW–MEDIUM
D	COLD WORK	A, O	BEST	POOR	A: BEST / O: LOWEST BEST	DEEP	GOOD	MED. HIGH	40–50	MEDIUM
A	COLD WORK	A	GOOD	FAIR		DEEP	FAIR	MEDIUM	85	MEDIUM–HIGH
O	COLD WORK	O	GOOD	FAIR	VERY GOOD	MEDIUM	POOR	LOW	90	HIGH
S	SHOCK RESISTING	O, W	FAIR	BEST	O: FAIR / W: POOR	MEDIUM	FAIR	MED. HIGH	85	LOW–MEDIUM
P	MOLD STEEL	A, O, W	LOW TO HIGH	HIGH	VERY LOW	SHALLOW	LOW	MEDIUM	75–100	HIGH
L	SPECIAL PURPOSE LOW ALLOY	O, W	MEDIUM	MED. TO VERY HIGH	LOW	MEDIUM	LOW	LOW	90	HIGH
F	SPECIAL PURPOSE CARBON-TUNGSTEN	W, B	LOW TO VERY HIGH	LOW TO HIGH	HIGH	SHALLOW	LOW	LOW	75	HIGH
W	WATER-HARDENING	W, B	FAIR-GOOD	GOOD	POOR	SHALLOW	POOR	LOWEST	100	HIGHEST

*W-WATER, B-BRINE, O-OIL, A-AIR, S-MOLTEN SALT

Table 10–2. Composition of molybdenum and tungsten high speed steels.

AISI TYPE	COMPOSITION, PER CENT								TYPICAL APPLICATIONS
	C	Mn	Si	W	Mo	Cr	V	Co	
	HIGH SPEED, MOLYBENUM, TYPES								
M1	.80	---	---	1.50	8.00	4.00	1.00	---	GENERAL PURPOSE, LOW COST DRILLS TAPS, LATHE TOOLS; POPULAR GRADE
M2	.85	---	---	6.00	5.00	4.00	2.00	---	PUNCHES, DIES; MOST POPULAR GRADE
M3-1	1.05	---	---	6.00	5.00	4.00	2.40	---	UNIVERAL CUTTING TOOL MATERIAL
M3-2	1.20	---	---	6.00	5.00	4.00	3.00	---	INCREASED WEAR RESISTANCE FOR CUTTING TOOLS
M4	1.30	---	---	5.50	4.50	4.00	4.00	---	HEAVY DUTY CUTTING TOOLS
M6	.80	---	---	4.00	5.00	4.00	1.50	12.00	
M7	1.00	---	---	1.75	8.75	4.00	2.00	---	SMALL CUTTING TOOLS, WOODWORKING TOOLS
M10	.90	---	---	---	8.00	4.00	2.00	---	SIMILAR TO M7, POPULAR GRADE
M15	1.50	---	---	6.50	3.5	4.00	5.00	5.00	
M30	.80	---	---	2.00	8.00	4.00	1.25	5.00	INCREASED RED HARDNESS IN CUTTING TOOLS
M33	.90	---	---	1.50	9.50	4.00	1.15	8.00	HEAVY DUTY HIGH SPEED STEEL FOR MACHINING MATERIALS OF HIGH HARDNESS
M34	.90	---	---	2.00	8.00	4.00	2.00	8.00	SIMILAR TO M33
M35	.80	---	---	6.00	5.00	4.00	2.00	5.00	
M36	.80	---	---	6.00	5.00	4.00	2.00	8.00	
M45	1.25	---	---	8.00	5.00	4.25	1.60	5.50	
M46	1.25	---	---	2.00	8.25	4.00	3.20	8.25	
	HIGH SPEED, TUNGSTEN, TYPES								
T1	.70	-	-	18.00	-	4.00	1.00	-	GENERAL PURPOSE TUNGSTEN HIGH SPEED STEEL FOR DRILLS, TAPS, LATHE TOOLS
T2	.80	-	-	18.00	-	4.00	2.00	-	LIGHT CUTS AT HIGH SPEEDS
T4	.75	-	-	18.00	-	4.00	1.00	5.00	FOR CUTTING HARD, GRITTY, OR TOUGH MATERIALS
T5	.80	-	-	18.00	-	4.00	2.00	8.00	HIGH RED HARDNESS, HEAVY DUTY CUTTING
T6	.80	-	-	20.00	-	4.50	1.50	12.00	VERY HIGH RED HARDNESS
T7	.75	-	-	14.00	-	4.00	2.00	-	ROUGHING CUTS, SOMEWHAT ERRATIC IN HARDENING
T8	.75	-	-	14.00	-	4.00	2.00	5.00	
T9	1.20	-	-	18.00	-	4.00	4.00	-	
T15	1.50	-	-	12.00	-	4.00	5.00	5.00	HEAVY DUTY CUTTING IN ABRASIVE MATERIALS

*BY PERMISSION FROM AISI "STEEL PRODUCTS MANUAL, TOOL STEELS," APRIL 1963.

types is given in Table 10–2. The replacement of tungsten with molybdenum has resulted in the production of high speed steels at much lower cost. The addition of cobalt to high speed steels permits cutting hard, abrasive materials and gives the ability to maintain hardness at higher temperatures.

The tungsten high speed steels are

Metallurgy Theory and Practice

hardened by heating to a temperature between 2250° and 2370°F, depending upon the particular type, followed by cooling in oil, air, or molten salt. Preheating the steel at 1500-1600°F before placing it in the high temperature furnace is recommended to minimize oxidation, grain growth, and thermal cracking of intricate sections. High speed steels should be held at the hardening temperature for only 2-5 minutes before quenching. The greatest hardness (R C61-66) is obtained by tempering at about 1050°F where secondary hardening occurs. Tempering time is from one to two hours and may be repeated, after cooling to room temperature, two or three times to insure transformation of all retained austenite.

Molybdenum type high speed steels perform as well as the conventional 18-4-1 type tungsten steels but are somewhat more difficult to heat treat. They are more susceptible to oxidation and grain growth at the hardening temperature. There is also a tendency for decarburization in the molybdenum high speed steels. Hardening temperatures for molybdenum high speed steels range from 2150° to 2275°F depending on alloy content. These temperatures are somewhat lower than required for the tungsten base high speed steels.

High speed steels containing cobalt additions are somewhat more susceptible to cracking during quenching than the tungsten type steels. The hardening temperatures for these steels is about 50°F higher than those of the tungsten type.

The microstructure for as cast, annealed, as hardened, oil quenched and tempered type M2 steel is shown in Fig. 10–1. Grain size in the as hardened condition may be clearly shown by a nital etch. The development of austenitic grain boundaries in tempered materials is difficult to achieve since ordinary techniques reveal a predominantly martensitic structure which usually obscures the original austenite boundaries. An etching technique involving the recently developed *Snyder-Graff Grain Size Reagent* has made it possible to reveal grain outlines in tempered materials. This reagent is composed of 9 percent hydrochloric acid and 3 percent nitric acid in methyl or ethyl alcohol. Etching time usually is 15 seconds to 5 minutes but may be longer, depending upon the specimen. With some samples the etching treatment is followed by a light polishing operation to remove the effect of the etchant on the matrix and more clearly reveal grain outlines.

The tempering diagram, Fig. 10–2, for AISI type M2 steel is representative of the high speed steels with its characteristic secondary hardening hump.

Typical uses for high speed steels include drills, taps, reamers, end mills, milling cutters, hobs, farm

316

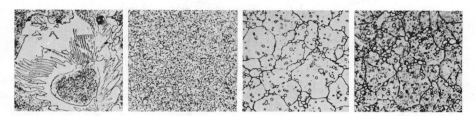

Fig. 10–1. High speed tool steel, Type M2. 1000X. Left to right, as cast showing skeleton-like eutectic carbide; annealed; as hardened; grain size #15; Nital etch; and hardened and tempered, Snyder-Graf-grain size etchant. (Cyclops Steel Corp., Specialty Steel Div.).

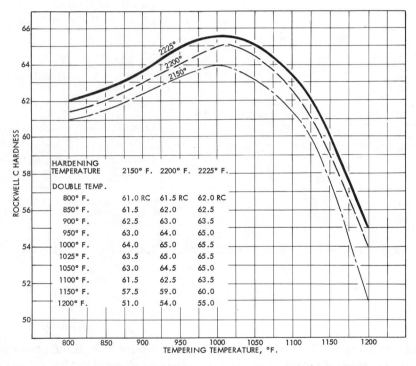

HARDENING TEMPERATURE	2150° F.	2200° F.	2225° F.
DOUBLE TEMP.			
800° F.	61.0 RC	61.5 RC	62.0 RC
850° F.	61.5	62.0	62.5
900° F.	62.5	63.0	63.5
950° F.	63.0	64.0	65.0
1000° F.	64.0	65.0	65.5
1025° F.	63.5	65.0	65.5
1050° F.	63.0	64.5	65.0
1100° F.	61.5	62.5	63.5
1150° F.	57.5	59.0	60.0
1200° F.	51.0	54.0	55.0

Fig. 10–2. Tempering curve for AISI type M2 high speed steel. (Cyclops Steel Corp.).

tools, saws, lathe tools, thread chasers, broaches, punches, drawing dies, and wood working tools.

10.4 Hot Work Tool Steels. The hot work steels are those used for cutting, gripping, forming, or mold-

317

Table 10–3. Composition of hot work tool steels.

AISI TYPE	C	COMPOSITION, PER CENT							TYPICAL APPLICATIONS
		Mn	Si	W	Mo	Cr	V	Co	
CHROMIUM, TYPES								-	
H10	.40	-	-	-	2.50	3.25	.40	-	EXCELLENT RESISTANCE TO SOFTENING; GOOD TOUGHNESS
H11	.35	-	-	-	1.50	5.00	.40	-	USED FOR MANDRELS, DIES, PUNCHES; EXCELLENT TOUGHNESS; DIE CASTINGS, DIES, PUNCHES
H12	.35	-	-	1.50	1.50	5.00	.40	-	EXTRUSION TOOLING, FORGING DIES, HIGH HARDENABILITY; SIMILAR TO H11
H13	.35	-	-	-	1.50	5.00	1.00	-	DIE CASTING NON-FERROUS METALS
H14	.40	-	-	5.00	-	5.00	-	-	AIR HARDENING
H16	.55	-	-	7.00	-	7.00	-	-	AIR OR OIL HARDENING
H19	.40	-	-	4.25	-	4.25	2.00	4.25	EXTRUSION AND FORGING DIE INSERTS
TUNGSTEN TYPES									
H20	.35	-	-	9.00	-	2.00	-	-	SERVICEABLE UP TO 1000°F
H21	.35	-	-	9.00	-	3.50	-	-	WIDELY USED FOR HOT BLANKING, PUNCHING, SHEARING AND EXTRUSION
H22	.35	-	-	11.00	-	2.00	-	-	SLIGHTLY BETTER RESISTANCE TO SOFTENING THAN H21
H23	.30	-	-	12.00	-	12.00	-	-	EXTRUSION AND DIE CASTING DIES FOR NON-FERROUS ALLOYS
H24	.45	-	-	15.00	-	3.00	-	-	SHEAR BLADES, HOT DRAWING DIES FORMING DIES, DIE BLOCK INSERTS
H25	.25	-	-	15.00	-	4.00	-	-	SERVICEABILITY UP TO 1100°F
H26	.50	-	-	18.00	-	4.00	-	-	MAX. STRENGTH, SHOCK RESISTANCE FAIR
MOLYBDENUM TYPES									
H41	.65	-	-	1.50	8.00	4.00	1.00	-	LOW CARBON HIGH SPEED STEELS WITH TOUGHNESS AND RESISTANCE FIRE-CHECKING
H42	.60	-	-	6.00	5.00	4.00	2.00	-	
H43	.55	-	-	-	8.00	4.00	2.00	-	

BY PERMISSION FROM AISI "STEEL PRODUCTS MANUAL, TOOL STEELS," APRIL 1963.

ing of hot metals. They must be resistant to erosion, wear, cracking, and heat-checking under conditions of severe thermal shock. Hot work steels shown in Table 10–3 are composed of medium to high alloy steels with relatively low carbon content. They may be divided into three groups based on their predominant alloying elements. These three types include: chromium hot work steels, H10-H19, tungsten hot work steels, H20-H39, and molybdenum hot work steels, H40-H59.

Chromium hot work tool steels are favored for die casting dies, extrusion dies, and forging dies because of their high ductility, toughness, and resistance to splitting. On the other hand, tungsten hot work tool steels are superior in their wear resistance and resistance to softening in service. For this reason tungsten types are used as dummy blocks, brass extrusion dies, hot punches, and other applications involving long tool contact with the workpiece and where resistance to softening is of prime importance. Molybdenum types are generally intermediate in properties and are used where a compromise between toughness and resistance to softening is desirable.

Hardening of hot work steels is usually accomplished by preheating at 1500°F, transferring to the high temperature furnace (1825-2200°F), allowing it to come to the temperature of the furnace, holding for 2-5 minutes at temperature, followed by an air or oil or molten salt quench. If molten salt is used, the recommended bath temperature is about 1000°F. As with high speed steels, the use of a high temperature salt bath is suggested for rapid heating to the hardening temperature and for protection against decarburization. Long soaking times at the hardening temperature are to be avoided because of the danger of grain growth. Generally speaking, as hardening temperatures increase, soaking time should be reduced. Air quenched dies may have an objec-

tional rough surface due to scaling during cooling; in such instances, oil cooling may be used, or preferably, an interrupted oil quench. This type of quench will allow the tool to cool under the protection of the oil to about 1000°F, or until all color disappears, after which it is removed from the oil and air cooled. Regardless of the quenching method employed, tools should be transferred to the tempering furnace or bath while at about 150°F.

All tools should be tempered immediately after hardening to avoid cracking and warping. Double tempering is recommended for all hot work tools with complete cooling to room temperature between tempering operations. Tempering time is usually 2 to 4 hours depending upon the mass of the tool being treated. The hardness after tempering at 1000°F is approximately Rockwell C59.

Typical applications for hot work tool steels include aluminum die casting, coining, extrusion, forging and heading dies, die inserts, mandrels, dummy blocks, hot shears, punches, and ejector pins.

10.5 Cold Work Tool Steels. The family of tool steels which are used for cold working, die work, or which require non-deforming properties varies widely in chemical composition and in physical properties. Cold work steels are of three principal

Table 10–4. Composition of cold work tool steels.

AISI TYPE	COMPOSITION, PERCENT									TYPICAL APPLICATIONS
	C.	Mn	Si	W	Mo	Cr	V	Co	Ni	
COLD WORK HIGH CARBON HIGH CHROMIUM TYPES										
D1	1.00	–	–	–	1.00	12.00	–	–	–	BLANKNG, FORGING, DRAWING DIES PUNCHES, GAGES, KNIVES
D2	1.50	–	–	–	1.00	12.00	–	–	–	
D3	2.25	–	–	–	–	12.00	–	–	–	EXTREME WEAR RESISTANCE, DIES, SHEARS, ROLLS, PUNCHES
D4	2.25	–	–	–	1.00	12.00	–	–	–	AIR HARDENING WEAR RESISTANT STEEL
D5	1.50	–	–	–	1.00	12.00	–	3.00	–	AIR HARDENING
D7	2.35	–	–	–	1.00	12.00	4.00	–	–	AIR HARDENING, HIGHEST WEAR RESISTANCE
COLD WORK, MEDIUM ALLOY, AIR HARDENING TYPES										
A2	1.00	–	–	–	1.00	5.00	–	–	–	BLANKING, STAMPING, COLD FORMING DIES
A3	1.25	–	–	–	1.00	5.00	1.00	–	–	
A4	1.00	2.00	–	–	1.00	1.00	–	–	–	LESS DISTORTION WITH ADDITION OF Mn
A5	1.00	3.00	–	–	1.00	1.00	–	–	–	
A6	.70	2.00	–	–	1.00	1.00	–	–	–	GOOD SHOCK RESISTANCE
A7	2.25	–	–	1.00*	1.00	5.25	4.75	–	–	
A8	.55	–	–	1.25	1.25	5.00	–	–	–	
A9	.50	–	–	–	1.40	5.00	1.00	–	1.50	WIDELY USED FOR DIES EXTRUSION, COINING
A10	1.35	1.80	1.25	–	1.50	–	–	–	1.80	
COLD WORK, OIL HARDENING TYPES										
01	.90	1.00	–	.50	–	.50	–	–	–	COLD FORMING ROLLS, DIES, PUNCHES, GAGES, BUSHINGS, GENERAL PURPOSE
02	.90	1.60	–	–	–	–	–	–	–	SUBJECT TO GRAIN GROWTH
06	1.45	–	1.00	–	.25	–	–	–	–	
07	1.20	–	–	1.75	–	.75	–	–	–	BEST WEAR RESISTANCE

BY PERMISSION FROM AISI "STEEL PRODUCTS MANUAL, TOOL STEELS," APRIL 1963

*OPTIONAL

types: high-carbon high chromium, type D, medium alloy air hardening, type A, and oil hardening, type O.

Composition and typical applications for these steels are given in Table 10–4. The high alloy steels have high hardenability, low distortion in heat treating, and high wear resistance, but are difficult to ma- chine. When hardening these steels are preheated to 1500°F and then transferred to the hardening furnace. Hardening temperatures range from 1700° to 1950°F depending on the alloy content. Grain coarsening is not usually serious at temperatures below 1900°F. Soaking time at the hardening temperature is from 15 to

45 minutes followed by air cooling with the exception of Type D3 steel which is usually oil quenched.

Quenching should continue to at least 150°F at which time tempering should begin. There is a pronounced secondary hardness at a tempering temperature of about 950°F. This temperature gives optimum toughness and abrasion resistance. These high alloy steels have a considerable amount of retained austenite after quenching, which is only partially transformed to martensite during tempering treatments. Gages or close tolerance tools may change size to an unbearable extent over a period of time as transformation continues. Subzero treatments accelerate austenite-martensite transformation and produce more stable tools and gages; this treatment is also given when extremely high hardness is required. The cold treatment is made part of the heat treating cycle as follows: After quenching, the tools should be tempered at about 250°F (boiling water is sometimes used) and then allowed to cool in air. They should next be chilled to −70 to −100°F and held for approximately two hours. Following this, they should be allowed to come to room temperature and then be tempered to the required hardness. Occasionally the boiling water-freezing cycle may be repeated two or three times prior to tempering to achieve maximum stability.

The medium alloy cold work steels are of the air hardening type when air cooled from the appropriate hardening temperature. Hardening temperatures vary widely for the different types with A4 being hardened from 1500-1600°F and A7 from 1750-1800°F. These steels have very low distortion during heat treatment.

Distortion can be held to a minimum by relatively slow heating to the hardening temperature, heating to the low side of the hardening range, and uniform air cooling. It should be recognized that poor design, failure to stress relieve, nonuniform heating or overheating, and poor mechanical handling can render the best tool steel unsatisfactory in service.

Oil hardening types of cold work tool steel are relatively inexpensive and are widely used for blanking dies, plastic molding dies, shear blades, taps, reamers, master tools, and gages. All of the oil hardening cold work steels show minimum size change when properly quenched from a relatively low temperature, and are frequently known as "non-deforming" steels although this is not strictly true. For example, Type O1 cold work die steel, when properly hardened will expand slightly, but then return very close to its original size after tempering. If overheated, this steel will exhibit shrinkage after tempering.

Metallurgy Theory and Practice

Table 10–5. Recommended heat treating practices for cold work tool steels.

HEAT TREATMENT	TYPE 01	TYPE 02	TYPE 06	TYPE 07
NORMALIZING, °F	1600	1550	1600	1650
ANNEALING TEMPERATURE, °F	1400-1450	1375-1425	1410-1450	1450-1500
MAX RATE OF COOLING, °F/HR.	40	40	20	40
TYPICAL ANNEALED HARDNESS, Bhn	183-212	183-212	183-217	192-217
HARDENING PREHEAT TEMPERATURE °F	1200	1200	----	1200
HARDENING TEMPERATURE, °F	1450-1500	1400-1450	1450-1500	1550-1625
TIME AT TEMP. MIN.	10-30	5-20	3-5	10-30
QUENCHING MEDIUM	OIL	OIL	OIL	OIL
TEMPERING TEMPERATURE, °F	350-500	350-500	350-600	350-350
APPROX. TEMPERED HARDNESS.	62-57	62-57	63-52	64-58

ADAPTED FROM "TOOL STEELS" PRODUCT MANUAL, AMERICAN IRON AND STEEL INSTITUTE, 1963

The recommended heat treating practice for each of the oil hardening cold work steels is shown in Table 10–5, and a tempering curve for Type O1 steel is shown in Fig. 10–3.

10.6 Shock Resisting Tool Steels. The steels listed in Table 10–6 are able to withstand extreme shock or impact loading for applications such as punches, chisels, concrete breakers, rivet sets, forming dies, and bold header dies. The shock resisting family of tool steels have reduced carbon content to enhance toughness. The reduction of carbon, however, brings about a lower degree of hardness. To offset this loss of hardness, elements such as manganese, silicon, chromium, and molybdenum are added to promote hardenability.

Elements such as chromium and tungsten also improve abrasion resistance through the formation of

322

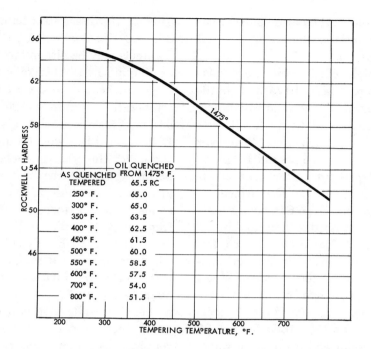

Within the figure:

Tempered	
AS QUENCHED	65.5 RC
250° F.	65.0
300° F.	65.0
350° F.	63.5
400° F.	62.5
450° F.	61.5
500° F.	60.0
550° F.	58.5
600° F.	57.5
700° F.	54.0
800° F.	51.5

OIL QUENCHED FROM 1475° F.

1475°

Fig. 10–3. Tempering curve for AISI type O1 oil hardening cold work die steel. (Cyclops Steel Corp., Specialty Steel Div.).

Table 10–6. Composition of shock resisting tool steels.

AISI TYPE	C	Mn	Si	W	Mo	Cr	TYPICAL APPLICATIONS
SHOCK RESISTING TYPES							
S 1	.50	–	–	2.50	–	1.50	CHISEL AND PUNCH STEEL
S 2	.50	–	1.00	–	.50	–	LOW Si FOR INCREASED TOUGHNESS
S 4	.55	.80	2.00	–	–	–	
S 5	.55	.80	2.00	–	.40	–	SHEAR BLADES, PNEUMATIC TOOLS, CARBIDE TOOL SHANKS
S 6	.45	1.40	2.25	–	.40	1.50	
S 7	.50	–	–	–	1.40	3.25	

BY PERMISSION, AISI, "STEEL PRODUCTS MANUAL, TOOL STEELS," APRIL 1963

hard carbides; they also have the capacity to resist softening at elevated temperatures.

The high silicon grades are relatively inexpensive, have high hardness, toughness, and wear resistance, but possesses relatively low resistance to decarburization. Wherever possible, allowance should be made for final grinding after heat treatment because of possible surface decarburization. Type S1 shock resisting steels are hardened by soaking at 1650-1750°F for 15-45 minutes, followed by oil quenching. After tempering at 400-1200°F, these steels can be expected to have a hardness of Rockwell C58-40. Type S2 steels are soaked at 1550-1650°F for 5-20 minutes followed by a brine or water

quench. Hardness after tempering at 350-800°F is Rockwell C60-50. Type S4 steels may be either water, brine, or oil quenched. When oil quenched, they are heated to near the upper temperature range (1600-1700°F) and held for a short time (5-10 min). Distortion is less with oil quenching than with the other quenches. Tempering temperatures and hardness are similar to those of the S2 type steels. Type S5 steel is oil quenched from 1600-1700°F, and Type S6 from 1675-1750°F. Type S7 steel which has deep hardening characteristics may be either air or oil quenched from 1700-1750°F. These grades of steel develop hardnesses of 60-45 when tempered at 350-1150°F, coupled with very high toughness.

Table 10–7. Composition of mold steels.

AISI TYPE	COMPOSITION, PER CENT								TYPICAL APPLICATIONS
	C	Mn	Si	W	Mo	Cr	Ni	AL	
P 1	.10	-	-	-	-	-	-	-	
P 2	.07	-	-	-	.20	2.00	.50	-	
P 3	.10	-	-	-	-	.60	1.25	-	HUBBED MOLDS FOR CARBURIZING AND QUENCH HARDENING
P 4	.07	-	-	-	.75	5.00	-	-	
P 5	.10	-	-	-	-	2.25	-	-	
P 6	.10	-	-	-	-	1.50	3.50	-	
P 20	.35	-	-	-	.40	1.25	-	-	CUT MOLDS FOR CARBURIZING
P 21	.20	-	-	-	-	-	4.00	1.20	

*BY PERMISSION, AISI "STEEL PRODUCTS MANUAL, TOOL STEELS," APRIL 1963.

The lower tempering range (300-500°F) is generally used for cold work applications and from 750-1150°F for hot work applications.

10.7 Mold Steels. The compositions of mold steels are given in Table 10–7. These steels or irons are used for molds to be used for plastics or for holder blocks for aluminum and zinc die casting dies. In the case of the molds themselves, the mold impression is either forced (hubbed or hobbed) into the soft mold blank with a hardened master pattern as in the case of type P1 to P6, or is machined in the higher carbon type P20 steels. After forming the cavity the mold is carburized and hardened to develop a surface hardness of 52-64 on the Rockwell C scale. Fig. 10–4 shows a die being hubbed in a hydraulic press. Since one master (hob) may be used for reproducing a number of identical cavities, this process is frequently more economical than conventional die sinking by machining.

Fig. 10–4. Hubbing (hobbing) a die block in a hydraulic press. Inset shows close-up of the hardened hob and the impression in the die block. The die block is contained in a reinforcing ring which is removed after the mold has been formed. (By permission: *Materials and Processes of Manufacturing*, Mac-Millan Book Co.).

Metallurgy Theory and Practice

10.8 Special Purpose Tool Steels.
The special purpose tool steels as
classified in the AISI system include
(1) low alloy type L steels, and (2)
carbon-tungsten type F steels. These
steels are intended for specialized
applications requiring specific metallurgical characteristics or as a substitute for expensive tool steels in
low production applications. Type L
steels are chromium cutting tool
steels used interchangeably with
type O, oil hardening cold work
steels, when increased wear resistance is needed.

Type F, or tungsten finishing
tools, are particularly useful for high
speed finishing. They provide relatively high wear resistance in water
hardening steels and have also found
application in dental burrs, rifling

tools, and drawing dies for wire, bar,
or tube. Because of their very good
abrasion resistance, type F steels are
somewhat difficult to grind after
hardening.

10.9 Water Hardening Tool Steels.
The water hardening tool steels are
essentially plain carbon tool steels
some of which have small percentages of chromium or vanadium
added to improve hardenability and
wear resistance. (See Table 10–8.)
The carbon content varies between
0.60 and 1.40 percent for water hardening steels depending upon the ultimate use of the steel. These steels
may be roughly placed in three subdivisions according to carbon content.

Carbon 0.60 to 0.75 percent—for

Table 10–8. Composition of water hardening tool steels.

AISI TYPE	COMPOSITION, PER CENT								TYPICAL APPLICATIONS
	C	MN	SI	W	MO	CR	V	CO	
W 1	.60/1.40*	–	–	–	–	–	–	–	HAND CHISELS, SHEAR BLADES PRESS DIES, CHUCK JAWS, HAND TAPS AND DIES
W 2	.60/1.40*	–	–	–	–	–	.25	–	SAME AS W 1 WITH V TO CONTROL GRAIN GROWTH
W 4	.60/1.40*	–	–	–	–	.25	–	–	CHROMIUM ADDITION TO CORRECT TENDENCY FOR SOFT SPOTS
W 5	1.10	–	–	–	–	.25	–	–	

BY PERMISSION, AISI "STEEL PRODUCTS MANUAL, TOOL STEELS," APRIL 1963
*VARYING CARBON CONTENTS MAY BE AVAILABLE.

326

applications where toughness is the primary consideration such as for shear blades, rock drills, hammers, and heading dies for short runs.

Carbon 0.75 to 0.95 percent—for applications where toughness and hardness are of equal importance, such as trimming dies, punches, chisels, and shear blades.

Carbon 0.95 to 1.40 percent—for applications requiring increased wear resistance and retention of the cutting edge. These steels are used for trimming dies, cutters, drills, large taps, and small cold chisels. Wood cutting tools, small taps, taps, turning tools, and razors may contain between 1.1 and 1.40 percent carbon.

In Chapter 7 it was shown that increasing the carbon content of carbon steel above about 0.60 percent does not result in an increase of

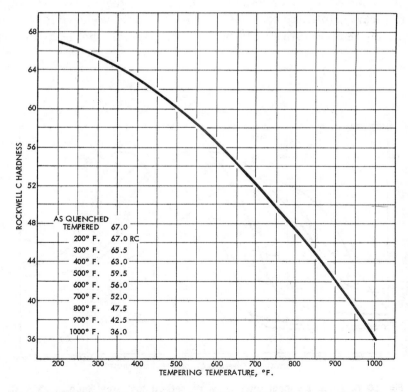

Fig. 10–5. Tempering curve for type W1 and W2 water hardening tool steels. (Cyclops Steel Corp., Specialty Steel Div.).

quenched hardness. The advantage of higher carbon content lies in the greater abrasion resistance because of the hard iron carbide particles formed from the excess carbon.

The water hardening steels are austenitized at 1400-1500°F for 10-20 minutes followed by water or brine quenching. A 10 percent brine quenching solution at 70°F is preferred for giving uniform hardness and clean surfaces with freedom from soft spots. Thin edged tools made from high carbon grades are sometimes oil quenched to reduce possibilities of cracking. In these cases, the austenitizing temperature should be

50 to 75°F higher than normally used.

All tools should be tempered immediately after quenching to prevent cracking, relieve hardening stresses, and to obtain the necessary degree of toughness. Tempering is frequently done in air circulating furnaces or in liquid baths to obtain uniform results. Fig. 10–5 may be used as a guide for tempering of Type W1 and W2 tool steels to achieve a specific hardness. The torsion impact toughness of water hardening tool steel at various tempering temperatures is shown in Fig. 10–6.

In general the straight carbon

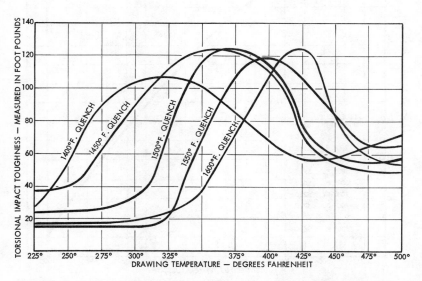

Fig. 10–6. Torsion impact toughness for carbon tool steel (app. 1.05% C) hardened at five different temperatures. (Courtesy: *Tool Steel Simplified*, Carpenter Steel Co.).

steels are less expensive than the alloy tool steels, are easier to machine, and are more resistant to decarburization. These steels must be water-quenched for high hardness and are therefore subject to considerable distortion. Their resistance to heat is poor and consequently cannot be used as cutting tools under conditions where high heat is generated at the cutting edge. Typical applications for water hardening tool steels include punches, beading tools, cold heading dies, axes, drills, reamers, countersinks, jewelers dies, files, and woodworking tools.

10.10 Selection of Tool Steels. The selection of a tool steel for a particular application is generally not limited to a single grade and may not be limited to a particular family of tool steels. Several different steels may offer a workable solution and the choice may be dictated by material cost, ease of machining, or availability.

Since carbon tool steels are the least expensive of all tool steels and are easily machined, they should be given first consideration when selecting a tool steel for a new application. These steels require no special techniques in hot working or heat treatment. However, they may lack some of the characteristics required in tooling applications. The main improvement in tool steels wrought by alloying may be grouped under four

headings: greater wear resistance, greater toughness or strength, increased hardenability, or resistance to softening at high temperatures.

If a carbon tool steel is not suitable for the intended application, the alloy steels should be examined to find the ones which have suitable characteristics. Comparative properties of the various types of tool steels are given in Table 10–1. This table rates the tool steels on the basis of wear resistance, toughness, hardenability, warpage resistance, and resistance to softening at high temperatures. In addition, relative cost, machinability rating, and resistance to decarburization are given. Although this table is not complete it can serve as a guide in selecting the steels having the desired qualities.

Suggested tool steels for a given application are listed in Table 10–9. These steels are only a suggestion as to types which should perform satisfactorily and is not intended as a recommended and alternate choice.

10.11 Tool Failures. Whenever a tool fails prematurely there is a reason for this failure. Premature failure may be caused from (1) faulty tool design, (2) faulty heat treatment, (3) improper grinding practice, (4) mechanical overloading, (5) faulty or (6) improper steel. Perhaps the first impulse after tool failure is to blame the failure onto the steel. However, past experience in-

Metallurgy Theory and Practice

TABLE 10-9   LIST OF SUGGESTED TOOL STEELS FOR COMMON TOOLING APPLICATIONS

APPLICATION	SUGGESTED TYPE	APPLICATION	SUGGESTED TYPE
ARBORS	L6, L2	DEEP DRAWING	A7, D2
AUTOMATIC SCREW MACHINE TOOLS	M2, M3-1	DIE CASTING	H13, H22
BEADING TOOLS	S1, S5	EMBOSSING	A2, D2
BITS, ROUTER	M2, M3	EXTRUSION, COLD	M3, D2
BORING BARS	L2, M2	FORGING, HOT	H12, H21
BURNISHING TOOLS	W1, D3	POWDER METALLURGY	A7, D2
BUSHINGS, GROUND	W1	PLASTIC MOLD, MACHINED	P20, H11
BUTTONS, LOCATING	W1	PRESS BRAKE	L1, L2
CAMS	D2, A2	STAMPING	A2, D2
CENTERS, LATHE, DEAD	M2	WIRE DRAWING	D3, M3
CENTERS, LATHE, LIVE	W1	DRILLS	M2, W1
CHISELS, HAND	W1, S5	DUMMY BLOCKS	M21, H10
COLLETS, SPRING	W1, O2, L6	FINISHING TOOLS	F2, M2
CUTTING TOOLS:		FINGERS, FEED, CUTOFF	L6, S2
		FIXTURES	W1, D2
BORING	M2, M3	GAGES, PLUG, RING, SNAP, THREAD	W1, A2, L1
BROACHES	M2, M3	HAMMERS, BALL PEEN	S2
BURRS, METAL CUTTING	M2, M3	HIBS (HOBS)	A2, D2
COUNTERBORES	M2, M3	JAWS, CHUCK	W2, S5
CUTOFF	M3, M33	KNURLING TOOLS	D2, A2
END MILLS	M2, M3	KNIVES, WOOD WORKING	W1, M2
FORM TOOLS	M2, M3	LINERS, MOLD	A7, D2
LATHE TOOLS	M2, M3	PILOTS, COUNTERBORE	F2, W1
MILLING	M2, M3	PNEUMATIC TOOLS	S1, S2
REAMERS	M2, M3	PUNCHES	S5, S2
DIES:		SCRAPERS	F2, W1
BENDING	A2, D2	SHEAR BLADES, COLD	S1, S5
BLANKING, COLD	A2, D2	SPINDLES	L2, L6
COINING	A2, D2	SPINNING	M2, M3
COLD HEADING	W1, W2	WEAR PLATES	A7, D2

dicates that changing to a different grade of steel should be the last resort. Each of six factors which may contribute to tool failure will be discussed.

Faulty tool design may lead to

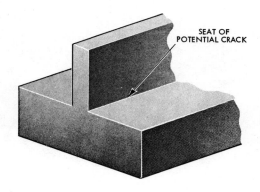

SEAT OF
POTENTIAL CRACK

Fig. 10–7. Avoid large changes in cross section to minimize heat treatment problems such as cracking. Use tapered sections and generous radii where possible.

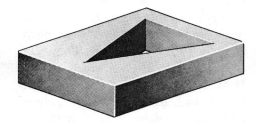

Fig. 10–8. Sharp corners and re-entrant angles cause high stress concentration during hardening. Avoid them if possible. Mold holes and pockets may be packed with clay to give more uniform cooling.

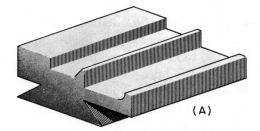

(A)

Fig. 10–9. The mass is balanced in a form tool by use of holes and radii. A, Shows incorrect design, B, shows correct design.

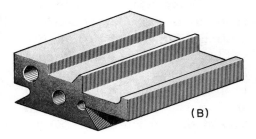

(B)

Fig. 10–10. Deep scratches and tool marks should be removed as they may be a source of cracking.

failure during heat treatment or while the tool is in service. Design features to be avoided are abrupt changes in cross-section, sharp corners, re-entrant angles, and deep scratches or tool marks. They are shown in Fig. 10–7 through 10–10. If it is essential to use adjacent heavy and light sections, designs with sharp corners, or parts with unevenly spaced holes, then the use of air hardening steel is recommended.

Faulty heat treatment causes the large majority of tool failures. Tools should be preheated at the recommended temperature prior to placing them in the hardening furnace. Excessively high hardening temperatures will cause grain coarsening with subsequent brittleness. Parts should be protected from decarburization by means of an atmosphere controlled furnace or by a protective salt bath. Quenching should be done in the recommended medium, and tempering should be started as soon as the quenched tool is cool enough to handle in your bare hands (150°F).

Tempering may be repeated if necessary to achieve minimum distortion in service and maximum toughness. Examination of the cracks in a tool which has failed may give some indication of the cause of failure. A crack which does not parallel the length of the original bar (or if it shatters the piece) may be due to one of two causes: too high a hardening temperature, or placing a tool which had been previously hardened into a hot furnace. Since these cracks were present while still in the hardening furnace, the surface of the fracture will be oxidized. Hardening cracks described as "spalling" or shelling off of corners and edges are generally caused by too low a hardening temperature, non-uniform heating, or tempering before the tool has cooled sufficiently in the quench. Cracks of very shallow depth on unground surface may be caused by a decarburized surface. Soft spots may arise from inadequate circulation of the quenching bath or from a contaminated bath.

Fig. 10–11. Grinding cracks in improperly ground hardened tool steel. Revealed by acid etching. (Carpenter Steel Co.)

concentration, improper alignment, or insufficient clearance. A hardened piece of tool steel has a certain amount of strength depending upon its composition, quality, and heat treatment. When the total of the residual internal stresses and the applied external stresses exceed the strength of the tool, it fails. If, for example, a hardened tool steel has a tensile strength of 400,000 psi but has high internal residual stresses due to the heat treating operation of 350,000 psi, then any load on the tool exceeding 50,000 psi will cause failure. Good tool design with ample fillets and rounds, coupled with good heat treating practices, and proper mechanical loading, should guarantee use of the part at its intended design strength.

Faulty steel may develop despite the careful control used in production and inspection. Porous areas resulting from shrinkage of the ingot during solidification are known as *voids* or *pipe*. Segregation or non-metallic inclusions running longitudinally with respect to the original bar stock are known as *streaks* or *laps*. *Tears* may be present in the tool steel because of plastic working such as forging or rolling at too low a temperature. A brittle carbide network may be present in large bar stock due to insufficient hot working. Flaws, such as those described, may be observed by microscopic examination of the parts which have

Improper grinding practice may set up surface stresses high enough to cause cracking in hardened tools. Light grinding cracks tend to run at right angles to the direction of grinding while heavy grinding cracks present a characteristic network pattern shown in Fig. 10–11. Grinding cracks may be alleviated by use of a softer grinding wheel, reducing the feed rate, and increasing the flow of coolant onto the workpiece.

Mechanical overloading of parts causing tool failures may be accidental or may be due to excessive stress

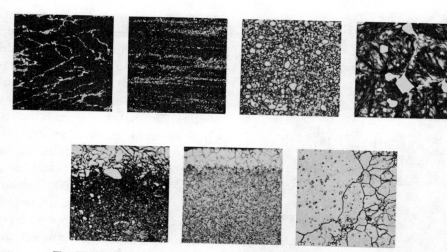

Fig. 10–12. Typical flaws found in tool steels. Top row: carbide network, 100X, Banding of carbides, 100X, Overheated high speed steel, 1000X, Underheated high speed steel, 1000X. Bottom row: carburized surface, Austenite case (white) over carburized zone (dark), 500X, Decarburized surface, ferrite (white) over part decarburized zone (light), 500X, and Mixed grain size, 500X. (Latrobe Steel Co.).

failed. Typical metallurgical flaws are shown in Fig. 10–12.

Improper steel selection is the last factor to examine. If none of the factors discussed seem to be at fault, then a change in the tool steel grade should be considered. The change may require a steel which has greater toughness, strength, or improved hardenability, and greater care in hardening.

References

American Iron and Steel Institute. *Tool Steels,* Products Manual Rev. 1966.

Clark, Donald S. and Varney, Wilbur R. *Physical Metallurgy for Engineers.* Princeton, N.J.; D. Van Nostrand Co., 1966.

Gill, J. P. et al. *Tool Steels.* Cleveland, O.: American Society for Metals, 1946.

Palmer, Frank R. and Luerssen, George V. *Tool Steel Simplified, 3rd ed.* Reading, Pa.: The Carpenter Steel Co., 1960.

Seabright, Lawrence H. *The Selection and Hardening of Tool Steels.* New York: McGraw-Hill, 1950.

**Cast
Irons**

Chapter

11

11.1 Introduction. The earliest successful iron founding has been credited to the ancient civilizations which occupied the plains of Mesopotamia. It was here that the Babylonians, Assyrians, and Chaldeans evolved the metallurgical knowledge necessary to make castings from high carbon iron. There is evidence that the Chinese capitalized on early discoveries which were undoubtedly passed along by migrating Mesopotamian craftsmen. The Chinese became the first people to produce iron castings successfully and regularly. Ancient records indicate that the Chinese were making iron castings as early as the 6th century before Christ. Many of the old castings were of a utilitarian nature such as cooking stoves, kitchen utensils, bells, agricultural implements, and tools, and ornamental objects. Various Chinese temples with cast iron roofs a thousand years old appear to be in first class condition today.

In the 12th century A.D., iron founding progress tapered off in China. The scene then shifted to European countries, principally Germany, Belgium, France, Italy, and England.

Even as late as the 1600's iron was considered a luxury item, and during the reign of Edward III, iron pots and skillets were numbered among the royal jewels. By the 1700's European production included bells, cannon, gunshot, pipe, stoves, and cast iron cooking utensils. In Fig. 11–1 are shown cast iron water pipes laid in 1644 to supply water to the Fountains of Versailles and still in use today. Fig. 11–2 shows examples of the beautiful cast iron art work produced in Berlin in 1775.

The first casting made in America was produced at the Saugus Iron Works near Lynn, Mass., shortly after its establishment in 1642. It was not long afterward that foundries appeared in Connecticut, Rhode

Fig. 11–1. Cast iron water pipe laid in 1644 to supply Fountains of Versailles. Still in use after 300 years. Initials are for Louis XIV of France. (Gray and Ductile Iron Founders' Society).

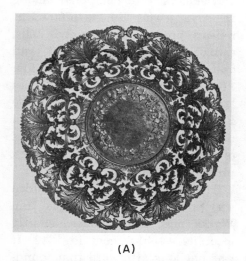

(A)

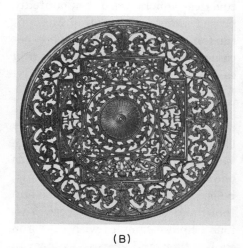

(B)

Fig. 11–2. Famous cast iron art objects produced in Berlin after 1775. A, fruit bowl. B, sandwich tray. From the Lamprecht collection. (American Cast Iron Pipe Co.).

336

Island, New Jersey, Virginia, the Carolinas, Pennsylvania, and Maryland. The foundries supplied neighboring farmers and industries with cast iron cooking utensils, stoves, tools, and agricultural implements. As the foundry industry grew in America, it became a real threat to England's profitable colonial export business. In 1750, in order to protect her markets, the English government passed laws requiring that American-made pig iron be shipped to England to be cast into finished products. This burdensome restriction is believed to be one of the important provocations which finally led to the Revolutionary War. When war struck, foundry men were both generous financially and active politically. Many served as military leaders in trying to gain independence from England. Six prominent foundrymen, including Colonel George Taylor, signed the Declaration of Independence in 1776.

Today designers, engineers, and users recognize the unique properties of cast iron as it fills a very important role in our modern world. There are cast irons available with excellent ductility and heat resistance, cast irons which can resist 30 percent boiling hydrochloric acid solutions and cast irons which can damp out vibrations in machinery.

Cast iron is essentially an alloy of carbon, silicon, and iron in which more carbon is present than can be retained in solid solution in austenite at the eutectic temperature. Thus cast irons usually contain some decomposition products such as free graphite or cementite. The carbon content in cast irons is usually more than 1.7 percent and less than 4.5 percent. The high percentage of carbon renders the cast iron brittle and unworkable except by casting. Thus the name, cast iron. Silicon, which acts as a graphitizing agent is usually present in amounts of one half to two percent, although it may be much higher in certain silicon irons.

Cast iron always contains a certain amount of manganese, phosphorus, and sulfur because of the refining methods used in its production. Special alloying elements such as copper, molybdenum, nickel, and chromium are added for the purpose of altering the chemical and mechanical properties of the iron. There are many variations in structure and properties available in cast iron. However, they may be classified into one of the following main groups:

1. Gray cast iron,
2. White cast iron,
3. Malleable iron,
4. Ductile (nodular) iron,
5. Alloy cast irons.

The terms *gray* and *white*, when applied to cast iron, refer to the appearance of the fracture of a casting. The gray iron fractures with a dark gray fracture, whereas the white cast iron shows a light gray, almost white

Fig. 11–3. Fracture appearance of white cast iron, left. Gray cast iron, right.

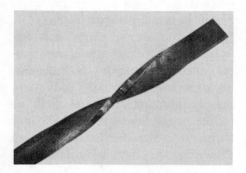

Fig. 11–4. Ductile iron test specimen showing extreme ductility.

fracture, illustrated in Fig. 11–3. Malleable iron gets its name from its ability to bend or undergo permanent deformation before fracturing. Ductile or nodular iron, Fig. 11–4 is similar to malleable iron in that it is much more ductile and tougher than either white or gray cast iron but differs in the method of production, and composition. Ductile iron can be produced with higher yield tensile strength than malleable iron. Alloy cast irons are those to which chromium, nickel, molybdenum, copper, or other elements have been deliberately added to modify the properties attainable by the normal elements—iron, carbon, silicon, manganese, phosphorus, and sulfur. For example: nickel may be added to gray cast iron in sufficient amount as to lower the eutectoid transformation temperature to such a degree that austenite will be retained at room temperature. This cast iron has unusual magnetic properties and corrosion resistance and is known as austenitic cast iron. Each of the five classifications of cast iron will be considered.

11.2 Gray Cast Iron. Gray iron is the most widely used of all cast metals. It is employed in greater tonnages in industry than all other cast metals combined. Typical applications for gray iron castings are for automotive engine blocks, machine tool castings, agricultural implements, cast iron pipe and fittings, bathtubs, household appliances, electric motor frames and the like. The good strength, wear resistance, castability, low notch sensitivity, and high damping capacity, together with manufacturing economy, explain the extensive use of cast iron. In the production of a casting, molten metal is taken from the furnace (usually a cupola or electric furnace) and poured into a prepared sand mold. After the metal has solidified the sand mold is broken up to allow removal of the casting. The gates and risers are cut from the casting after which it is then cleaned and inspected. Heat treating, when required, is performed after cleaning.

11.3 Classification of Gray Cast Irons. In general, gray cast irons are classified on the basis of tensile strength. Thus a class 20 cast iron

has a tensile strength of 20,000, a class 30 iron 30,000 psi, and so on. See Table 11–1. Tensile strength is a property that is easily understood and can be related to many other mechanical properties. A simple test for cast iron without elaborate equipment and with a test bar in the unmachined condition is known as the transverse bend test. A test bar of standard dimensions is cast and then placed on supports located 12, 18, or 24 inches apart. The load required to rupture the bar is measured and the value obtained can be reasonably correlated with the tensile strength of the iron. The maximum deflection at the center of the bar indicates ductility of the iron. Transverse rupture loads for the various classifications are also shown in Table 11–1.

Chemical analysis alone is not sufficient to designate a cast iron because of the wide range of mechanical properties that can be obtained with a given analysis.

11.4 Microstructure of Gray Cast Iron. The microstructure is as important as the chemical analysis in determining final properties of a casting. Properties such as machineability and wear resistance are almost entirely dependent upon the microstructure. The microstructure consists of two principal parts: the graphite flakes, and the matrix metal surrounding the flakes. The matrix structure of gray iron can be readily

Table 11–1. Minimum tensile and transverse strength for the various classes of gray iron.

ASTM CLASS	MIN. TENSILE STRENGTH, PSI	MIN. TRANSVERSE STRENGTH BREAKING LOAD, LBS.		
		TEST BAR A 7/8 DIA 18" SPAN	TEST BAR B 1.2 DIA 18" SPAN	TEST BAR C 2.0 DIA 24" SPAN
20	20,000	900	1800	6000
25	25,000	1025	2000	6800
30	30,000	1150	2200	7600
35	35,000	1275	2400	8300
40	40,000	1400	2600	9100
45	45,000	1540	2800	9700
50	50,000	1675	3000	10300
60	60,000	1925	3400	12500

TEST BAR A (7/8" DIA X 15" LONG) USED WHERE CONTROLLING SECTION IS 0.5" AND UNDER
TEST BAR B (1.2" DIA X 21" LONG) USED WHERE CONTROLLING SECTION IS 0.5" TO 1.0"
TEST BAR C (2.0" DIA X 27" LONG) USED WHERE CONTROLLING SECTION IS 1.0" TO 2.0"
ASTM A 48–56

changed by heat treatment but the graphite, once formed, is little influenced by heat treatment. Examination of the microstructure of cast iron is done in two steps: First, a representative specimen is polished with a fine abrasive, followed by viewing under the microscope at 25X and later at 100X to establish the size and distribution of graphite that is present. This is done by comparing the microstructure with standard charts (Fig. 11–5 and 11–6). The charts are made for comparison with unetched specimens at 100X. Sec-ondly, the specimen is etched with a recommended reagent to make the matrix structure visible. Microscopic examination of the etched specimen and identification of the matrix structure completes the examination. Three typical matrix structures for gray iron are shown in Fig. 11–7.

11.5 Metallurgy of Gray Iron. The metallurgy of cast irons is very complex and cannot be represented on the binary iron-carbon diagram which is used in discussing steels. However, the basic composition of

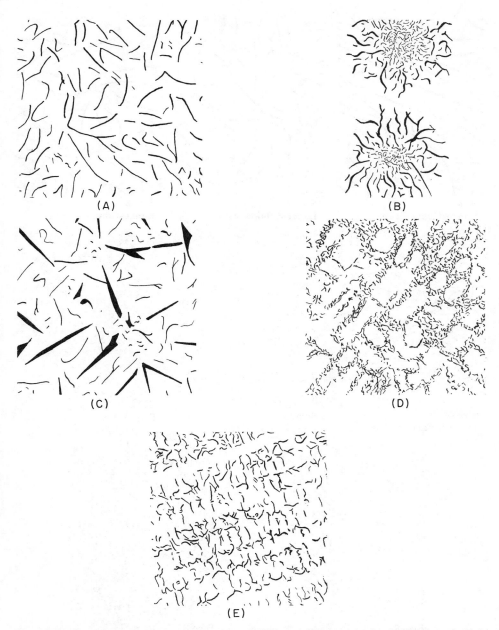

Fig. 11–5. Graphite flake types in gray irons. ASTM and AFS standards. 100X. Type A, Uniform distribution, random orientation. Type B, Rosette groupings, random orientation. Type C, Superimposed flake sizes, random orientation. Type D, Inter-dendritic segregation, random orientation. Type E, Inter-dendritic segregation, preferred orientation.

341

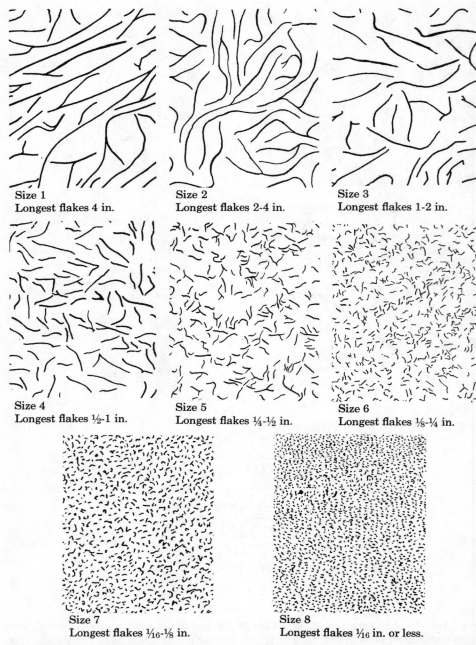

Size 1
Longest flakes 4 in.

Size 2
Longest flakes 2-4 in.

Size 3
Longest flakes 1-2 in.

Size 4
Longest flakes ½-1 in.

Size 5
Longest flakes ¼-½ in.

Size 6
Longest flakes ⅛-¼ in.

Size 7
Longest flakes 1/16-⅛ in.

Size 8
Longest flakes 1/16 in. or less.

Fig. 11–6. Graphite flake size in gray irons. 100X. (American Society for Testing and Materials and American Foundrymen's Society.)

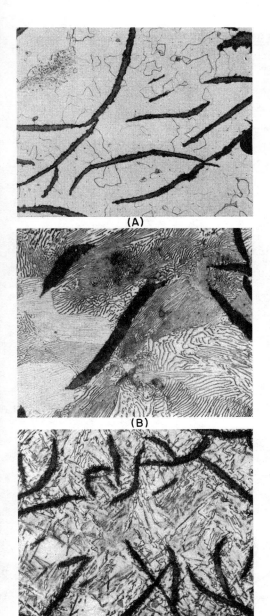

(A)

(B)

(C)

Fig. 11–7. Microstructures of typical gray irons. A, with a ferritic matrix, etched. 250X. B, matrix of pearlite, etched. 500X. C, acicular matrix, etched. 500X. (Gray and Ductile Iron Founders' Society)

gray irons is often described in terms of a carbon equivalent factor (C.E.) This factor gives the relationship of the percentage of carbon and silicon in gray iron to its capacity to produce graphite. It is calculated as follows:

$$C.E. = T.C. + \tfrac{1}{3} \,(\%\,Si)$$

Where:

$$T.C. = Total\ Carbon$$

(Phosphorus should be included in the formula but is here omitted because of the low phosphorus content of domestic pig irons.) Thus if an iron contained 3.4 percent carbon and 2.4 percent silicon, its carbon equivalent would be:

$$C.E. = 3.4\% + \tfrac{1}{3}\,(2.4\%) = 4.2\%$$

The carbon equivalent has metallurgical significance by indicating the mode of solidification of the iron and thus its basic microstructure. See Table 11–2. Irons with a carbon equivalent of over 4.3 are called hypereutectic. Those with less than 4.3 carbon equivalent are hypoeutectic cast irons. The average effect of carbon equivalent on tensile strength is shown in Fig. 11–8. Another method of representing the type of structure which will be formed for various carbon-silicon combinations is shown in Fig. 11–9. Silicon is by far the strongest graphitizer of all the elements that may be added to cast iron. Consequently, it is the predominant element in determining the relative properties of combined

343

Table 11–2. Cast iron types and their carbon equivalents.

TYPE	DESCRIPTION	RANGE OF SECTION THICKNESS, IN.	TOTAL CARBON RANGE, PERCENT	SILICON RANGE, PERCENT	CARBON EQUI-VALENT (C.E.)
S	SOFT	0.10 – 0.40	3.4 – 3.8	2.4 – 2.8	4.2 – 4.5
A	AUTOMOTIVE	0.15 – 1.0	3.25 – 3.35	2.2 – 2.4	4.0 – 4.2
P	PRESSURE	0.50 – 2.5	3.0 – 3.2	1.8 – 2.2	3.6 – 4.0
H	HEAVY SECTION	1.5 MINIMUM	3.9 – 3.1	1.0 – 1.8	3.3 – 3.6

BY PERMISSION FROM INTERNATIONAL NICKEL CO., INC.

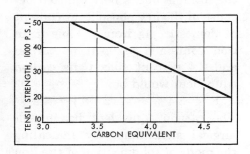

Fig. 11–8. Relationship between carbon equivalent and tensile strength of gray iron test specimens. (Gray and Ductile Iron Founders' Society)

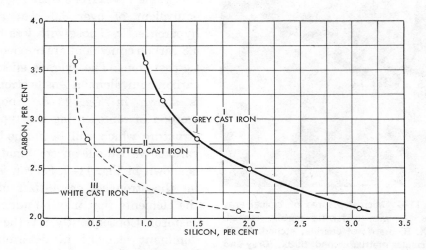

Fig. 11–9. Composition limits for gray, mottled, and white cast irons.

and graphitic carbon that will be present in the final casting. Thus, the foundry metallurgist may control the properties of the cast iron largely by adjusting silicon content relative to the amount of carbon and section size of the casting.

Under commercial casting conditions gray irons solidify under non-equilibrium conditions. The cooling rate (section size) and chemical analysis determine the microstructure of gray iron and consequently its properties.

The properties of hypoeutectic gray cast iron cooled at different rates is shown in Fig. 11–10. Very rapid cooling (also low silicon and low carbon contents) results in the retention of the metastable structures consisting of free carbide and

pearlite. Martensite or even some retained austenite may also be present. Very slow cooling promotes the stable system, which consists of ferrite and graphite. The rate of cooling also influences grain size and size of the graphite flakes. Slow cooling produces soft coarse structures, and rapid cooling, fine grained, hard structures.

11.6 Selection of Gray Iron Alloys. One practical application of the above information is in custom selection of alloy composition for a casting of given section thickness. For example: a gray iron with carbon and silicon adjusted to produce an ideal structure in a cast 2-in. section will have chilled edges and hard spots in a ½-in. section because of

COOLING RATE	MICROSTRUCTURE *	REMARKS	BHN
FAST	P + C (WHITE IRON)	EXTREMELY HARD AND BRITTLE	325–500
MODERATELY FAST	P + C + G (MOTTLED IRON)	GREATEST STRENGTH, HARD TO MACHINE	250
MODERATE	P + G	BEST HIGH TEST IRONS; CLOSE GRAINED	200
MODERATELY SLOW	P + G + F	FAIR STRENGTH, EASY MACHINING, FAIR FINISH	150
SLOW	F + G	LOW STRENGTH, OPEN GRAINED MACHINES SOFT	100

* P = PEARLITE   C = CEMENITE   G = GRAPHITE   F = FERRITE

Fig. 11–10. Effect of cooling rate on microstructure and properties of gray cast iron.

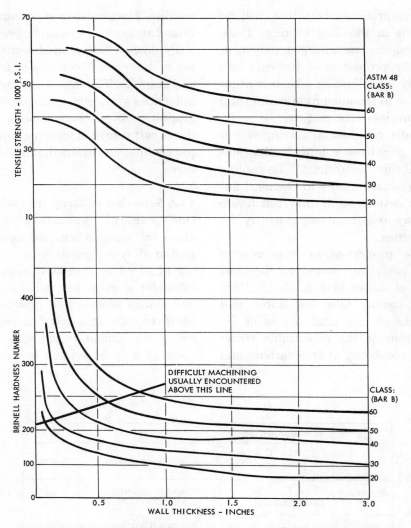

Fig. 11–11. Effect of wall thickness on tensile strength and hardness of typical gray irons. (Gray and Ductile Iron Founders' Society)

the non-uniform cooling rate. The thinner section cools faster, inhibiting formation of graphite flakes and permitting carbide formation. In a casting of uniform thickness the problem of obtaining a uniform structure is much simplified as both carbon and silicon content as well as cooling rate can be controlled. Whenever the walls of a casting vary in thickness, the use of alloy gray irons should be considered. In Fig. 11–11 is shown the effect of section thickness on hardness and tensile strength of typical gray irons. Here the proper combination of alloys can give greater uniformity of properties throughout the casting, and improve any or all of the properties of the base cast iron. Gray irons may be grouped into the four types shown in Table 11–3 according to the dominant section thickness of the casting. It should be noted that the castings

Table 11–3. Basic gray cast irons grouped according to dominant section thickness.

DOMINANT SECTION THICKNESS INCHES	TYPE	CLASS	TENSILE STRENGTH	TYPICAL COMPOSITION PERCENTAGE					APPLICATION
				T.C.	Si	Ni	Mo	Cr	
LESS THAN 0.5 IN.	S	BASE: IRON	25,000	3.5	2.5	---	---	---	LIGHT CASTING OF ABOUT 3/8 IN. MAXIMUM THICKNESS. TOTAL CARBON RANGE FROM 3.4 TO 3.8% AND SILICON FROM 2.4 TO 2.8%
		NCI-35-S	35,000	3.5	2.5	0.75	1.50	---	
		NCI-40-S	40,000	3.5	2.5	0.30	0.60	---	
0.15 TO 1.00	A	BASE IRON	30,000	3.3	2.3	---	---	---	AUTOMOTIVE CASTINGS, ENGINE HEADS, COMPRESSOR BLOCKS, TOTAL CARBON RANGES FROM 3.25 TO 3.40% AND SILICON FROM 2.2 TO 2.4%
		NCI-35-A	35,000	3.3	2.3	0.75	0.30	---	
		NCI-40-A	40,000	3.3	2.3	1.25	0.25	0.35	
		NCI-45-A	45,000	3.3	2.3	1.50	0.30	0.40	
		NCI-50-A	50,000	3.3	2.3	2.00	0.35	0.60	
0.5 TO 2.5	P	BASE IRON	35,000	3.1	2.0	---	---	---	PRESSURE CASTINGS FOR MEDIUM SIZE VALVE AND PUMPS (3" to 12") AND FOR ENGINE AND COMPRESSOR CASTINGS. TOTAL CARBON RANGES FROM 3.0 TO 3.25% AND SILICON FROM 1.5 TO 2.2%
		NCI-40-P	40,000	3.1	2.0	0.75	0.30	---	
		NCI-45-P	45,000	3.1	2.1	1.00	---	0.35	
		NCI-50-P	50,000	3.1	2.1	1.25	0.30	0.40	
		NCI-55-P	55,000	3.1	2.1	1.50	0.30	0.60	
1.5 MINIMUM	H	BASE IRON	35,000	3.0	1.5	---	---	---	FOR HEAVY SECTIONS HAVING THICKNESSES OF FROM 1/2 INCHES THICK AND UP. TOTAL CARBON RANGES FROM 2.8 TO 3.1% AND SILICON FROM 1.0 TO 1.8%.
		NCI-40-H	40,000	3.0	1.5	1.00	---	---	
		NCI-45-H	45,000	3.0	1.5	1.25	0.50	---	
		NCI-50-H	50,000	3.0	1.5	2.00	0.35	0.50	
		NCI-55-H	55,000	3.0	1.5	2.5	---	1.00	

(INTERNATIONAL NICKEL CO.)

Table 11-4. General properties of cast irons.

ENGINEERING PROPERTY	UNALLOYED BASE IRON	Ni ALLOY	Ni-Cr ALLOY	Ni-Mo ALLOY	Ni-Cr-Mo ALLOY
TENSILE STRENGTH	P	F	G	G-E	E
WEAR RESISTANCE	P-F	F-G	G	G	E
PRESSURE TIGHTNESS AND DENSITY OF MACHINED SURFACES	P	G	G	G	E
SUPPRESSION OF CHILLING IN THIN SECTIONS	P	G	F	G	E
REFINEMENT OF METAL MATRIX IN HEAVY SECTIONS	P	F	G	G	E
UNIFORMITY IN BOTH THIN AND HEAVY SECTIONS	P	F	G	G	E
MACHINABILITY	F	G	G	G	F
UNIFORMITY IN HEAT-TREATMENT	P	F	G	G	E
CORROSION RESISTANCE	F	G	G	G	G
RELATIVE RESISTANCE	LOW	MED.	MED.-HIGH	MED-HIGH	HIGH

E = EXCELLENT,   G = GOOD;   F = FAIR;   P = POOR

with the thinnest sections require the highest percentages of carbon and silicon.

The selection of a cast iron for a given application and section thickness may be made by referring to Table 11-3 and 11-4. One of the four basic types of iron should be chosen that has properties suitable for the intended application. Next, one of the specific alloy combinations should be selected by reference to the appropriate curves in Fig. 11-12—11-15. The minimum cost is generally associated with the less highly alloyed irons (lower curves). Costs may be kept down by picking the lowest and simplest alloy content that will do the job.

11.7 Effect of Major Alloying Elements in Gray Iron. The elements commonly present in or added to gray iron may be divided into two groups according to their effect on the iron during solidification and

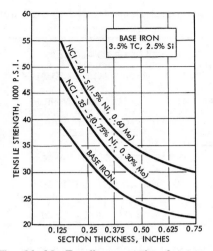

Fig. 11–12. Tensile properties for type S cast irons with various section thicknesses.

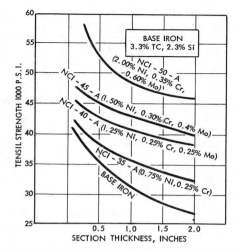

Fig. 11–13. Tensile properties for type A cast irons with various section thicknesses.

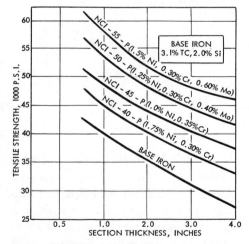

Fig. 11–14. Tensile properties for type P cast irons with various section thicknesses.

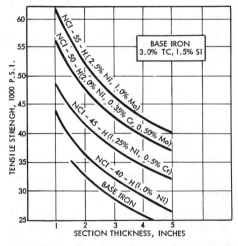

Fig. 11–15. Tensile properties for type H cast irons with various section thicknesses.

(International Nickel Co.)

cooling. Elements which promote graphite formation are called "graphitizers" and those which tend to retain the carbon in the form of iron carbides are termed "carbide stabilizers." The graphitizing effect of several elements as compared with silicon is shown in Table 11–5.

349

Table 11–5. Graphitizing and carbiding effects of several elements used in gray iron compared with silicon.

EFFECT	ELEMENT		
GRAPHITIZERS	SILICON	+ 1.00	(a) THIS FIGURE HOLDS UP TO ABOUT 2% ALUMINUM FROM 2 TO 4% THE THE GRAPHITIZATION VALUE GRADUALLY DECREASES TO ZERO
	ALUMINUM	+ 0.50 (a)	(b) MEAN VALUE. 0.1 TO 0.2% TITANIUM HAS A STRONGER ACTION THAN SILICON
	TITANIUM	+ 0.40 (b)	ON THE GRAPHITE FORMATION, WHILE LARGER QUANTITIES HAVE A LESSER
	NICKEL	+ 0.35	EFFECT THAN SILICON
	COPPER	+ 0.20 (c)	(c) THE VALUE FALLS TO 0.05 FOR CARBON CONTENT OF OVER ABOUT 3%
	MANGANESE	– 0.25 (d)	(d) FOR MANGANESE CONTENTS BETWEEN 0.8 AND 1.5%. BELOW 0.8% MANGANESE HAS A WEAKER EFFECT ON CARBIDE FORMA-
	MOLYBDENUM	– 0.30 (e)	TION, AND MAY EVEN HAVE A STRONG GRAPHITE-FORMING INFLUENCE BELOW
CARBIDE STABILIZERS	CHROMIUM	– 1.00	0.6% IN THE PRESENCE OF SULFUR.
	VANADIUM	– 2.50 (f)	(e) FOR MOLYBDENUM CONTENTS BETWEEN 0.8 AND 1.5%. BELOW THIS RANGE MOLYBDENUM HAS A WEAKER ACTION OF GRAPHITE-FORMATION; ABOVE IT HAS A STRONGER ACTION.
			(f) MEAN VALUE

ADAPTED FROM GRAY IRON CASTINGS HANDBOOK, 1958 ED.

Effects of some of the major alloying elements used in gray cast iron are as follows:

Silicon is the strongest graphitizer of all alloying elements. It probably forms Fe_3Si leaving free graphite; upper limit of effectiveness is at about 3.0 percent silicon.

Nickel assists in graphitization but is only about half as effective as silicon; promotes density and freedom from porosity by allowing lower silicon content; progressively and uniformly hardens and strengthens the matrix by changing the structure of coarse pearlite to fine pearlite and finally to martensite; helps in refining the grain and promotes dispersion of graphite in a finely divided state, thus improving strength and toughness; is widely used, especially in combinations with chromium or molybdenum to obtain uniformity of castings of varying section thickness.

Chromium forms stable carbides, intensifying the chilling properties of cast iron; promotes the formation of more finely laminated and harder pearlite, thus increasing the strength, hardness, and wear resistance of the iron matrix; in amounts of 1 to 1.5 percent adds resistance to softening and scaling at temperatures of up to 1400°F; tends to prevent excessive graphitization in thick sections; is frequently used in combination with nickel, copper, molybdenum, and vanadium.

Molybdenum decreases rate of austenitic transformation to pearlite during cooling, resulting in a finer pearlite with a consequent increase in strength, hardness, and fatigue life; is the most effective alloying element for increasing strength; promotes structural uniformity in heavy sections and improves machinability; contributes to retention of strength at elevated temperatures.

Copper promotes graphite formation, but has only about one-fifth the graphitizing ability of silicon; is a mild strengthener of the matrix and tends to break up massive cementite spots; reduces hardness of iron having a tendency to chill but in normal gray iron increases hardness, resistance to wear and certain types of corrosion.

Vanadium is a powerful grain refiner, an intense hardener and carbide former; results in greater depth of chill hardness, toughness, and resistance to wear; improves machineability in highly alloyed irons by promoting fine pearlitic structure in compositions that might otherwise be acicular.

11.8 Heat Treatment of Gray Iron. Gray iron castings may be subjected to various heat treatments to improve their hardness, wear resistance and other mechanical properties. The heat treatments usually applied to gray iron includes (1) stress-relief annealing, (2) annealing, (3) hardening and tempering.

Stress-relief annealing of gray cast iron for machine parts is used for the purpose of removing any residual internal stresses present in the castings as they are received from the foundry. Also, the treatment tends to soften any hard spots and hard corners that occur in the castings caused from chills. The stress-relief annealing treatment consists of heating the castings to within a temperature range of 750° to 1250°F for a period of several hours followed by slow cooling to below 700°F. After reaching this low temperature the doors of the furnace may be opened and the castings cooled to room temperature in air. This treatment will remove the internal stresses and allow the castings to be machine finished to accurate dimensions without the fear of warping. Also, this treatment will improve the machinability of castings having hard corners and hard spots by breaking down the combined carbon.

Annealing of gray cast iron increases the softness of the iron for economy in machining. Castings that are hard and difficult to machine because of the amount of combined carbon in their structure can be annealed to graphitize the combined carbon and restore easy machinability. The process consists of controlling the degree of transformation of the combined carbon to graphitic or free carbon. The annealing treatment selected depends upon the structure and composition of the iron being treated and the degree of softening desired. The usual practice is to heat the castings in a furnace to around 1500°-1600°F, hold the castings at this temperature for an hour or more, depending upon the maximum section of the casting, followed by a slow cool to around 700°F, when the doors of the furnace may be opened and the castings removed. The maximum temperature used in this treatment depends upon the composition of the casting. Annealing temperatures from 1200° to 1800°F have been used. The time at heat influences the degree of graphitization and is governed by the required softness.

Hardening and Tempering gray irons of the pearlitic type can be done just as steels can. The pearlite in the structure of the iron can be changed to austenite upon heating to above the critical temperature of the iron, and upon rapid cooling

(quenching) transformed to martensite. The treated casting can be tempered to relieve the quenching stresses, increase the toughness of the iron, and reduce the maximum hardness. The purpose of hardening and tempering gray iron is to strengthen and increase the wear and abrasion resistance of the iron. Ordinary pearlitic gray irons are heated to 1500°F and oil quenched although water is used as a quenching medium to a limited extent. Water quenching will cause more warping and danger from hardening cracks. The hardening of ordinary gray irons by this treatment will increase the Brinell hardness from around 180-200 to 400-500. Following the hardening operation the casting may be tempered from 400° to 800°F, depending upon the desired properties. Fig. 11–16 illustrates the change in tensile strength and hardness that might be expected from hardening a pearlitic iron and tempering within a temperature range of 200° to 1200°F.

Gray iron castings can be locally hardened by either the flame-hardening or induction-hardening method. Local heating of massive castings can be carried out safely by either process. If this is followed by local quench hardening, the heat treater is able to harden surfaces without heating the entire casting to the quenching temperature. This insures a very hard surface without the dan-

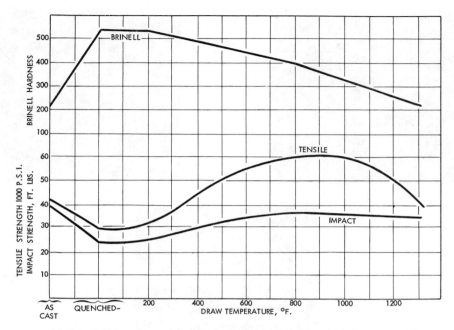

Fig. 11–16. Effects of tempering temperatures on properties of gray cast iron.

ger of cracking and marked distortion that would occur if an attempt should be made to heat and quench the entire casting. Fig. 11–17A shows a fractured section of gray iron that has had one surface hardened by local heating and quenching, using the flame-hardening method. Fig. 11–17B illustrates the type of pearlitic structure in this iron before hardening, and Fig. 11–17C shows the martensitic structure found in the hardened region of the casting after treatment. It will be seen from Fig. 11–17C that the hardening operation did not affect the graphitic

constituent of the casting; it only changed the pearlite to martensite.

11.9 White (Chilled) Cast Iron. Iron castings that are classified as white or chilled iron have practically all of their carbon in the combined condition as cementite. These castings are relatively hard and brittle due to the high carbon content which forms hard and brittle cementite. Cast irons that contain low carbon, 2.0 percent-2.5 percent, and low silicon (See Fig. 11–19, A, B) when cast in sand molds and cooled slowly, solidify and cool to room tempera-

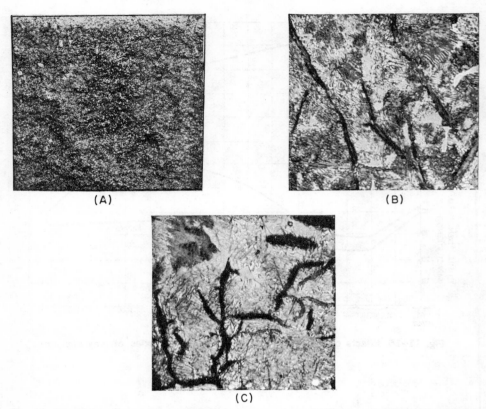

(A)

(B)

(C)

Fig. 11–17. A, fracture of flame hardened gray cast iron. Light edge shows depth of hardened zone (Rockwell C55). B, microstructure before flame hardening shows graphite flakes, pearlite matrix and small amount of ferrite (white). 500X. C, microstructure after hardening shows martensitic structure with graphite flakes. 500X.

ture without graphitizing any of their carbon. These irons are naturally white irons.

A white cast iron that has been slowly cooled in the mold has a structure of pearlite and free cementite similar to that of a high carbon steel except that the white cast iron contains much more free cementite. The microstructure of this type of iron is illustrated in Fig. 11–18. The greatest tonnage of this type of iron is used for the manufacture of malleable iron. The addition of alloys such as nickel, molybdenum, and chromium to white iron

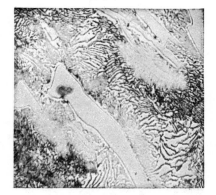

Fig. 11–18. Microstructure of white cast iron. White is cementite, dark is pearlite.

results in the formation of a much harder iron. The type of structure formed by the alloy additions is usually martensite and free cementite with the alloys acting as hardening agents and thus preventing the formation of pearlite from austenite during the cooling cycle. The structure of an alloyed white iron is illustrated in Fig. 11–19A, B. Irons of this type are used for parts of machinery and industrial equipment where extreme hardness and excellent wear resistance are required, such as in crusher jaws and hammers, wearing plates, cams, and balls and liners for ball mills.

White cast iron can also be made by rapid cooling or chilling of an iron which, if cooled slowly, would be graphitic and gray. If cast iron is cooled relatively fast in the mold the carbon does not have an opportunity to graphitize and remains combined.

Also, rapid cooling prevents the formation of a coarse, soft pearlite and adds to the hardness of the casting. If the cooling is fast enough martensite may be formed instead of pearlite.

Local hardening effects may be produced by a local chill in the molds, which results in a hard, white iron surface on a casting that might ordinarily be soft and gray if cooled slowly. Fig. 11–20 illustrates this effect. Factors that influence the depth of the chill include the ratio of carbon to silicon (high silicon content decreases the depth of chill), thickness of the casting, thickness and temperature of the metal placed in the mold to act as a chill, the time that the cast metal is in contact with the metal chill, and the use of alloy additions.

Chilled iron castings are extensively used for railroad car wheels

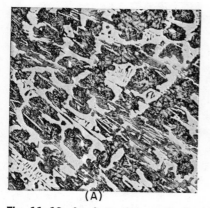
(A)

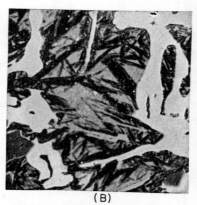
(B)

Fig. 11–19. A, photomicrograph of white cast iron containing molybdenum. Hardness is Rockwell C58. 100X. B, magnified 1000X. Acicular structure is martensite, white is cementite.

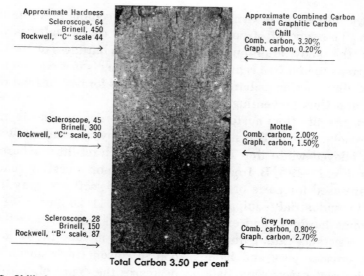

Approximate Hardness
Scleroscope, 64
Brinell, 450
Rockwell, "C" scale 44

Approximate Combined Carbon and Graphitic Carbon
Chill
Comb. carbon, 3.30%
Graph. carbon, 0.20%

Scleroscope, 45
Brinell, 300
Rockwell, "C" scale, 30

Mottle
Comb. carbon, 2.00%
Graph. carbon, 1.50%

Scleroscope, 28
Brinell, 150
Rockwell, "B" scale, 87

Grey Iron
Comb. carbon, 0.80%
Graph. carbon, 2.70%

Total Carbon 3.50 per cent

Fig. 11–20. Chilled casting fracture showing combined carbon, graphitic carbon, and hardness of white, mottled and gray iron in the as-cast condition.

and for many diversified applications where resistance to wear and compressive strength are major requirements. Applications for chilled iron include rolls for crushing grain and ore, rolling mills for shaping metals, farm implements, and cement grinding machinery.

356

11.10. Malleable Cast Iron. Malleable cast iron is made by a process involving the annealing of hard, brittle white iron which, as the name implies, results in an iron that is much more ductile (malleable) than either white or gray cast iron. Malleable cast iron is not malleable in the sense that it is as forgeable as steel or wrought iron, but it does exhibit greater toughness and ductility when compared with other forms of cast iron. Also, malleable iron is softer than gray cast iron and exhibits easier machining characteristics. Because of these characteristics, malleable iron can be used in applications where greater toughness and resistance to shock are required, such as in farm implements, plows, tractors, harrows, and rakes, and many applications in automobile parts, hardware, small tools, and pipe fittings. In spite of its greater cost as compared to gray cast iron, malleable iron finds wide use.

Malleable iron is made from white iron castings by a high-temperature, long-time annealing treatment. The original white iron castings are made of a low carbon, low silicon type of iron—an iron that will solidify in a mold and cool without the formation of graphitic carbon. The iron is usually melted in a reverberatory type furnace, commonly known as an air furnace. Occasional heats of white iron are melted in the electric furnace or the open-hearth furnace.

Some heats are melted in the cupola, but it is difficult to melt irons of this composition due to the high temperatures required for them.

11.11 Annealing or Malleablizing. A white iron casting is first made by casting a controlled composition of metal into a sand mold. This casting is hard and brittle upon cooling, because of its structure of combined carbon (Fe_3C). In the annealing process the castings of white iron are placed in cast iron pots, or rings, and surrounded by a packing material which should be sufficiently refractory in nature so as not to fuse to the castings at the annealing temperatures. Sand is commonly used as a packing material although squeezer slag, crushed blast-furnace slag (used alone or mixed with mill scale), or other forms of iron oxide may also be used. The purpose of the packing material is to protect and support the castings from warping during the annealing cycle.

The object of the annealing cycle is to change the combined carbon or cementite. (Fe_3C) of the white iron to a graphitic carbon (temper carbon) found in malleable iron. The decomposition of the combined carbon to graphitic carbon is as follows:

$Fe_3C = 3 Fe + C$ (graphite)

Cementite (Fe_3C) is unstable at a red heat and decomposes to graphite and ferrite upon heating and slow cooling. The packed castings are placed in a furnace of the box or car type, and a slow fire is started. The temperature is increased at a rate that may require two days to reach an annealing temperature of 1550° to 1600°F. After reaching this temperature range the castings are held there from 48 to 60 hours. The castings are then cooled slowly at a rate of not more than 8° to 10°F per hour until the temperature has dropped to around 1300°F. The castings may be held at this temperature, 1250° to 1300°F for a period up to 24 hours, or the doors of the furnace may be opened after the castings have been slow-cooled to 1250°F and the pots removed and allowed to air-cool. The castings are shaken out as soon as their temperature permits handling.

The annealing cycle should result in all of the combined carbon in the original white iron being completely decomposed to a graphitic or temper carbon condition, and the final structure of the iron should consist of ferrite and graphite. The graphitization of the combined carbon starts as soon as the castings reach a red heat. The initial heating to 1600°F for 48 hours causes the free or excess cementite of the white iron to graphitize, and the slow cooling cycle to 1250°-1300°F allows the graphitization of the cementite that is precipitated from solution during this period. When the temperature of the iron falls below the critical temperature on cooling or the Ar_1 point

(1250°-1300°F), the balance of the cementite dissolved by the iron is precipitated with the formation of pearlite. A soaking period or a very slow cool while this change is taking place will allow the graphitization of the cementite portion of the pearlite. This will complete the graphitization of all the possible cementite contained in the original white iron.

The time required for the complete annealing cycle varies from five to seven days. A so-called accelerated annealing cycle that requires from one-third to one-half the time required by the large annealing ovens in general use has been used successfully. The shorter annealing cycle is accomplished by the use of better designed furnaces of gas or electric type and, because of the rapid temperature changes and the accurate control possible with these furnaces, requires much less total time. Furnaces used in the shorter annealing cycle are the smaller batch-type, continuous car-type, or kiln-type furnaces. Also, if the annealing can be carried out without the use of a packing material, considerable reduction in the annealing time is gained. This is because of more rapid and even heating and cooling of the metal.

11.12 Black-heart Malleable Iron. The ductile nature of malleable iron and the presence of graphitic carbon, will cause a fracture to show dark

Fig. 11–21. Black-heart malleable cast iron showing decarburized outer case. 2½X.

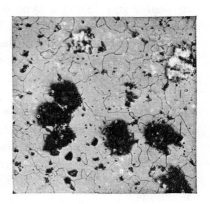

Fig. 11–22. Malleable iron etched to show ferrite grains. Dark spots are graphite (temper) carbon, 100X.

or black with a light decarburized surface; thus the fracture will appear with a white edge and black core. This is known as black-heart malleable iron illustrated in Fig. 11–21. The graphite present in a fully annealed malleable iron differs from the flake-like graphite found in gray cast irons in that it is formed into a nodular shape called temper carbon. The principal constituents of malleable iron are ferrite and nodular graphite, Fig. 11–22. The ferrite matrix of fully annealed malleable iron contains silicon, manganese, and phosphorus in a solid solution condition.

The decarburized or white surface found in malleable iron results from a decarburization or burn-out of the carbon during the annealing cycle. Decarburization can be avoided by the use of a carbonaceous packing material or by annealing in a controlled atmosphere. If annealing is carried out using an iron oxide such as mill scale or iron ore for packing material, it is possible to completely decarburize or burn out all of the carbon in the original iron. This will result in a malleable iron that fractures with a light fracture appearance and is known as white-heart malleable iron. Such a type of iron is seldom manufactured as it is difficult to machine, and its mechanical properties are inferior to those of the black-heart malleable iron.

11.13 Properties of Malleable Iron. The chemical composition of malleable iron is controlled within the limits specified for the various grades. The average chemical composition is as follows:

Carbon	1.00 to 2.00%
Silicon	0.60 to 1.10%
Manganese	under 0.30%
Phosphorus	under 0.20%
Sulfur	0.06 to 0.15%

This composition is only approximate and is changed to suit the requirements of the final product. All of the carbon should be in the graphitic form, with no combined carbon present. The amount of phosphorus and sulfur is not objectionable as

phosphorus does not produce any marked cold-brittleness, and, as the iron is not hot-worked, hot-shortness or brittleness caused by the sulfur has no appreciable effect.

The average properties of malleable iron are as follows:

Tensile strength	54,000 PSI
Yield point	36,000 PSI
Elongation, 2 inches	15% minimum
Brinell hardness	115
Izod impact strength	9.3 foot lbs
Fatigue endurance limit	25,000 PSI

Modifications in composition and heat treatment may alter these properties, and malleable irons that exhibit much higher mechanical properties are often made. One of the modern developments in metallurgy has been the manufacture of high-strength malleable irons known as pearlitic malleable irons, discussed in paragraph 11.15.

11.14 Alloy Malleable Irons. Some producers of straight malleable iron also manufacture a malleable iron to which they add a small amount of copper and molybdenum. These alloy malleable iron castings have numerous applications since they have a yield point that approximates 45,000 psi and an ultimate strength that often exceeds 60,000 psi, accompanied by an elongation in some instances as high as 20 percent or more in two inches. These irons are reported as having excellent machining properties even when the hardness exceeds 200 Brinell, and a very fine surface finish. Castings for use in valve and pump parts render very good service and show excellent resistance to wear and corrosion.

The addition of copper to malleable iron apparently accelerates graphitization during annealing treatments and also strengthens the ferrite, while at the same time greatly improves the iron by increasing its susceptibility to heat treatments following an annealing cycle. Copper additions from 0.70 to over 2.0 percent make possible an improvement in physical properties by a precipitation heat treatment that consists of heating the annealed iron to 1290°-1330°F, quenching and then drawing at about 940°F for three to five hours. The quenching temperature is not high enough to redissolve the graphitic carbon but dissolves much of the copper which is precipitated in a finely divided form by reheating to 940°F. At this temperature copper is relatively insoluble and precipitation occurs at a fast rate. The precipitated copper increases the hardness and strength without much loss in ductility. Precipitation hardening, due to the copper precipitation, also occurs in the regular annealing cycle during the malleabilizing treatment. By controlling the original composition of the casting, marked improvement in physical properties is obtained as compared with the straight malleable iron compositions. The structure of malleable iron alloys with copper, or copper-molybdenum malleable irons, consists of nodular graphitic carbon with a matrix of ferrite and a precipitate of copper.

11.15 Pearlitic Malleable Cast Irons. Straight malleable irons have all of their carbon in the graphitic or temper carbon condition, whereas the so-called pearlite malleable iron retains some of its carbon in the combined condition as cementite, similar to steel and gray cast iron. The structure of the pearlitic malleable iron differs from that of the straight malleable iron in that the matrix consists of a pearlite-like structure of ferrite and cementite. Fig. 11–23A and B illustrate the structure found in this type of iron. It consists of graphite nodules, or temper carbon, with a matrix of spheroidized cementite in ferrite. Pearlitic malleable iron can be made by several different treatments, such as (1) modifying the composition of the original white iron from that used in the manufacture of straight malleable iron, (2) making use of a short annealing cycle, or else (3) subjecting a straight malleable iron casting to a special heat treatment. A brief discussion of these treatments follows:

Modifying Composition of the white iron makes it possible to retard the action of graphitization during the regular annealing cycle so that some combined carbon is retained.

(A)

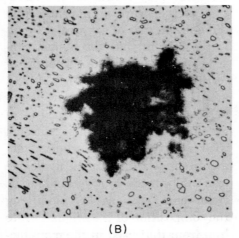

(B)

Fig. 11–23. A, pearlitic malleable cast iron. 100X. B, magnified 750X. Temper carbon (graphite) surrounded by cementite in a ferrite matrix.

Careful control of the silicon-carbon ratio and the additions of manganese, molybdenum, and copper are common practice in securing a retention of combined carbon in the annealed iron. A lower silicon and

carbon content also retards the action of graphitization.

Short-Cycle Annealing is a common practice in making a pearlitic malleable iron. It employs a shorter annealing cycle than that used when complete graphitization is wanted, as in the making of straight malleable iron. Manganese additions can be made to the molten metal in the ladle prior to casting as an aid in retarding complete graphitization. A typical analysis for making a pearlitic malleable iron is:

Total Carbon 2.40%
Silicon 0.92%
Manganese 0.32%
Manganese added to ladle 0.63%

This iron is cast to form a white iron similar to that made for straight malleable iron. The addition of manganese to the ladle helps to retard graphitization during the casting cycle and during the annealing cycle that follows. This requires about 30 hours at 1700°F, followed by cooling to below 1300°F in the annealing furnace, and subsequent reheating to 1300°F for about 30 hours, followed by cooling to normal temperature. The total time consumed by this annealing cycle is about 75 hours, as compared with five to seven days for the annealing of a straight malleable iron. The shorter annealing time prevents complete graphitization and the annealing time at the lower temperature, 1300°F, puts the combined carbon in a spheroidized condition. As a result of this treatment a cast iron is produced containing a matrix of spheroidized cementite and ferrite in which the nodules of temper carbon are present.

Special Heat Treatments makes it possible to obtain a pearlitic type of structure in a straight malleable iron by simply reheating a completely graphitized iron to a temperature high enough to give the ferritic matrix an opportunity to dissolve some of the nodular temper carbon, by heating above the critical temperature of the iron. By controlling the cooling rate from the solution temperature the dissolved carbon may be precipitated as a coarse spheroidized cementite or as a fine laminated pearlite. Accurate control of the complete cycle is necessary in order to obtain the desired amount of combined carbon and type of structure.

11.16 Properties of Pearlitic Malleable Iron. The properties of pearlitic iron depend upon the character of the matrix, that is the amount of combined carbon as cementite and the size, shape, and distribution of the cementite particles. In general, pearlitic malleable iron has a higher yield strength and ultimate strength and lower elongation than normal malleable iron. It machines less readily and has a higher Brinell hardness.

Briefly, the properties may be varied between the following limits:

Ultimate tensile strength 60,000[a] to 90,000[b] psi
Yield point . . . . 43,000[a] to 60,000[b] psi
Elongation in 2 inches. . . 15%[a] to 30%[b]
Brinell hardness . . . . 160[a] to 200[b]

[a]soft pearlitic malleable iron.
[b]hard pearlitic malleable iron.

Applications include use where strength, rigidity and wear resistance are important factors, such as in gears, sprockets, air tools, brake drums, cams, crankshafts, and wearing pads.

11.17 Ductile Irons. Although the manufacture of iron castings is a very old process one new remarkable development has been the discovery of methods to produce a high strength cast iron that can be bent. This new iron is commonly referred to as ductile iron because of its outstanding ductility. It is also referred to as nodular iron or spherulitic graphite cast iron (SG iron) because of its microstructure.

Ductile (nodular) iron consists of graphite nodules dispersed in a metallic matrix. Fig. 11–24 shows a structure which has been only partially nodularized and has both flakes and nodules of graphite. Graphite in the form of spherulites has a minimum influence on the mechanical properties of ductile iron, but in the amount present is significant since the ease with which iron can be melted and cast into complex shapes is dependent upon a relatively high carbon content.

Since the graphite nodules have a

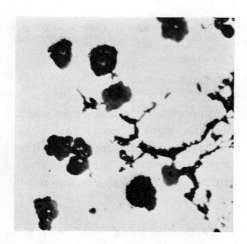

Fig. 11–24. Ductile cast iron partially nodularized. Graphite flakes and nodules in a ferrite matrix. 250. Nital etch.

minimum influence on the properties of ductile iron the main influence is dependent upon the metal matrix surrounding the graphite nodules. Typical matrix structures are ferritic, pearlitic, or acicular (martensitic) shown in Fig. 11–25. This gives possibilities for a wide range of properties. The various types of ductile iron are listed in Table 11–6.

Ductile iron is made by the addition of a small amount of magnesium to the ladle of molten cast iron. The addition of magnesium causes a vigorous mixing reaction resulting in a homogeneous spheroidal structure in the ductile iron casting.

Heat treatments commercially used on ductile iron are stress relief,

annealing, normalizing, quench hardening, austempering, and martempering. Surface hardening treatments commonly employed are induction and flame hardening.

Ductile irons combine the processing advantages of cast iron with the engineering advantages of steel. These advantages include low melting point, good fluidity and castability, excellent machinability, good wear resistance, high strength, toughness, ductility, hot workability, weldability, and hardenability. This combination of properties permits the production of castings and forgings of intricate shape and, if necessary, in light sections, which will withstand severe service conditions. Ductile iron castings can be produced in sizes weighing from a few ounces to over 50 tons with section thicknesses from less than ⅛″ to over 50″. Some typical application of ductile iron are shown in Fig. 11–26.

11.18 Alloy Cast Irons. A number of high alloy cast irons have been developed for increased resistance to corrosion and temperature and for other special purposes. A brief description of each of these alloys will show what can be done to custom make alloys for special applications. These irons may be grouped as follows:

1. Corrosion resistant alloy cast irons,

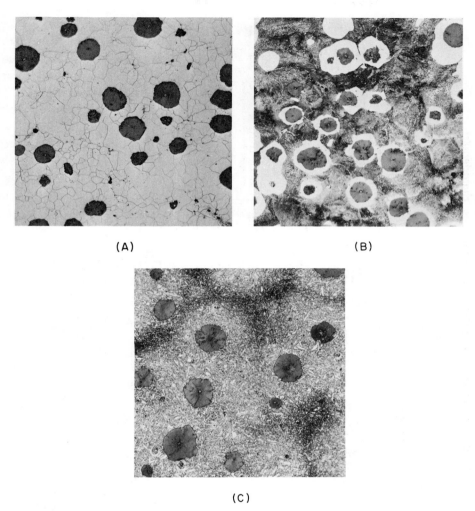

(A)

(B)

(C)

Fig. 11–25. A, ferritic ductile cast iron. B, pearlitic ductile cast iron. C, martensitic ductile cast iron. (International Nickel Co.)

2. Heat resistant alloy cast irons,
3. High strength gray irons,
4. Special purpose alloy cast iron.

11.19 Corrosion Resistant Alloy Cast Irons. These irons provide improvement in resistance to the cor-

Table 11–6. Principal types of ductile (nodular) cast iron: properties, characteristics; applications.

ADAPTED FROM "DUCTILE IRONS THE GRAPHITIC STEELS FOR INDUSTRY" INTERNATIONAL NICKEL CO. N.Y. 1965

Type	Typical Range of Mechanical Properties					Characteristics	Applications
	Tensile Strength psi	Yield Strength psi	Elongation %	Hardness Brinell	Charpy Impact Unnotched (Ft-Lbs)		
60-40-18	60/80,000	40/55,000	18/30	131/207	60/115	Maximum toughness and machinability. Weldability is second only to that of austenitic grade. Structure essentially ferrite. Not as readily flame or induction hardened as next three grades.	Pressure castings, compressor bodies, ingot molds, pipe, pipe fittings, valves, cylinders, pump bodies, connecting rods, shock resistant parts for automotive, agricultural, electrical, railroad, machine, marine and general use; high temperature applications requiring maximum toughness. Applications involving severe thermal and mechanical shock.
65-45-12	65/80,000	45/60,000	12/15 Requires 40,000 psi yield and 15% elongation. Also specific chemical composition and metallurgical structure. One grade, 60-45-15.	131/207	60/115		
80-55-06	80/100,000	55/75,000	6/10 Requires a minimum of 3% elongation.	187/269	15/65	Essentially pearlitic structure with good machinability and toughness. Responds readily to flame or induction hardening. May be cast against a chill to provide a carbidic, abrasion resistant surface. Can be welded.	Gears, cams, bearings, dies, pistons, crankshafts, rolls, sheaves, sprockets, wear and strength applications for automotive, aeronautical, diesel, agricultural, heavy machinery, mining, paper, textile and other related industries.
100-70-03	100/120,000	70/90,000	3/10 Requires 75,000 psi yield and 4% elongation.	217/302	35/50	This grade is usually normalized and tempered although it may be alloyed to produce these properties without heat treatment. Structure is pearlitic. Offers an excellent combination of strength, toughness and wear resistance. Readily flame or induction hardened.	Gears, crankshafts, camshafts, dies, pistons, agricultural implement parts such as bolsters, bolster forks, ratchets, governor weights, track shoes, tractor brake drums.
120-90-02	120/175,000	90/150,000	2/7	269/388	20/40	This grade is usually oil quenched and tempered although sections up to 1.5" thick may be normalized and tempered. Alloying elements such as nickel and molybdenum should be used to obtain adequate hardenability in heavy sections. Readily flame or induction hardened to 58 Rc.	Pinions, gears, cams, dies, machine guides, track rollers, idlers, tractor steering gear arms, drill press columns, pump liners, clutch drums.
Heat Resistant (3)	60/100,000	45/75,000	0/20 Typical composition ranges: T.C. 2.8/3.8%, Si 2.5/6.0%, Mn .20/.60%, P .10% max., Ni 0/1.5%. Although resistance to oxidation increases with silicon content and is excellent at 6% Si, most castings are purchased at 3.0/4.5% Si which provides the best compromise between mechanical properties and oxidation resistance.	143/302	5/115	Maximum resistance to oxidation and growth. Sections less than 1" thick usually annealed for maximum dimensional stability. Higher silicon contents increase hardness and strength but reduce ductility and toughness at room temperature although normal toughness is achieved at elevated temperatures. *This grade is not recommended for applications involving severe thermal shock.*	Furnace doors and frames, grate bars, blast furnace parts, sinter pots, reduction pots, lead pots, aluminum pouring troughs, glass molds and plungers, annealing pots, slag runners, pallets, etc.
Austenitic	55/72,000	32/38,000	4/40	110/240	Vee Notched 10 to 28	Austenitic ductile irons containing 18 to 35% nickel and up to 5% chromium, are available in several types which are especially useful for resistance to corrosion, selected expansivities, high temperature service, resistance to wear and heat and other special industrial services. Structure is austenitic. The reader is referred to "The International Nickel Company's brochure, "Engineering Properties of Ni-Resist Ductile Irons," for detailed information regarding the range of usefulness of each type of austenitic ductile iron.	

The type numbers indicate the minimum tensile strength, yield strength and percent elongation.
The 60-45-10 type has a minimum of 60,000 PSI tensile, 40,000 PSI yield and 10 percent elongation in 2 inches

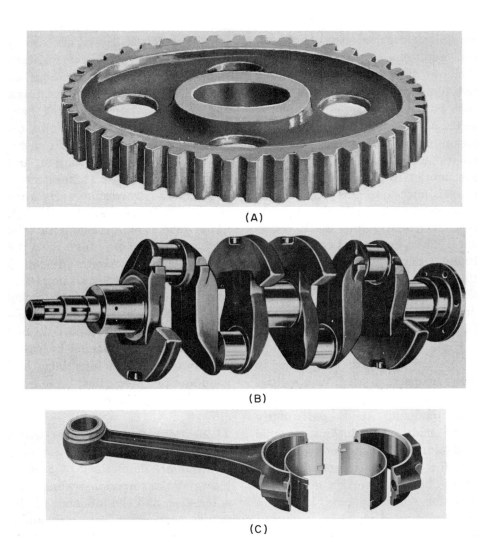

(A)

(B)

(C)

Fig. 11–26. Typical applications of ductile iron castings. A, roller gear for farm implement. B, hollow automotive crankshaft. C, diesel engine connecting rod.

rosive action of acids and caustic alkalis. There are three general classifications of these irons, based on the use of silicon, nickel, chromium, and copper to enhance corrosion resistance.

High silicon irons known as Duriron or Durichlor find wide use for

367

handling corrosive acids. High silicon additions, up to 17 percent, promote the formation of a protective film under oxidizing conditions such as boiling nitric and sulphuric acids. A 3.5 percent molybdenum addition increases resistance to hydrochloric acid. High silicon cast irons are inferior to plain gray cast irons in resistance to alkalis. High silicon irons have poor mechanical properties, particularly low shock resistance. They are very hard and virtually unmachinable. In spite of these disadvantages they are considered the standard material for drain pipe in chemical plants and laboratories.

High chromium irons with 20 to 35 percent chromium are similar in many respects to the high silicon irons, since they give good service with oxidizing acids, particularly nitric acid, salt solutions, organic acid solutions, caustic soda solutions, and for general atmospheric exposure. The lower carbon grades are used for annealing pots, lead, and aluminum melting pots, conveyor links, and other parts subjected to corrosion at high temperatures. Tensile strengths of up to 70,000 psi and hardness of 290 to 340 Brinell may be obtained with chromium cast irons. Regular chromium irons are quite shock resistant and are machinable.

High nickel irons are known commercially by the registered name of Ni-Resist. These alloys are widely used for their corrosion resistant properties in such applications as pump housings, and filter castings, and in applications requiring handling of sulphuric, hydrochloric, organic acids, and caustics in the chemical, food, automotive, marine, and petroleum industries. This cast metal has superior corrosion resistance than 18-8 stainless steels, although not as good as high silicon irons. The austenitic nickel gray irons have the highest impact strength (60-150 ft lbs Charpy) of all the cast irons containing flake graphite. High nickel irons have relatively low tensile strength (20,000-40,000 psi) but since this is accompanied by excellent toughness, machinability, and good foundry properties, these irons find many applications.

11.20 Heat Resistant Cast Irons. The heat resisting properties of a cast iron are determined by growth, intergranular corrosion, scaling characteristics, and the influence of elevated temperature on mechanical properties such as stress rupture, creep, and tensile strength. Growth is a permanent volume increase, occurring at elevated temperatures in some cast irons, produced by the expansion accompanying graphitization, or internal oxidation of the iron. Gases can penetrate around the graphite flakes in hot cast iron and oxidize the graphite as well as the iron and silicon. It is evident, therefore, that white cast irons which do not contain free graphite will be more resistant to growth than the gray irons. White cast irons can be successfully employed at temperatures of up to 1400°F.

The high temperature characteristics of cast irons may be markedly improved by alloying additions. Silicon and chromium are added to increase resistance to heavy scaling by forming a light oxide surface coating that is impervious to oxidizing atmospheres. Molybdenum and nickel increase high temperature strength. Aluminum additions reduce both growth and scaling at elevated temperature although room temperature properties are considerably decreased.

11.21 High Strength Gray Irons. Cast irons whose tensile strength exceeds 40,000 psi have been termed high strength cast iron. Low alloy cast irons have been produced with tensile strengths up to 100,000 psi. In general, these high alloy irons require close metallurgical control and usually have total carbon contents of 3 percent or lower. Special "inoculants" have been developed which, when added to the molten cast iron, modify the structure and thereby change physical properties to an extent not explained on the basis of the change in composition resulting from their use. Alloying elements such as molybdenum, chromium, copper, vanadium, and nickel are frequently employed in various combinations to aid in producing high strength irons. One of the main effects of alloy additions in cast iron is to reduce the influence of section size and thus yield a fine grained, homogeneous structure with a corresponding advantage of high mechanical properties. The high strength irons usually possess an acicular matrix structure shown in Fig. 11–27, which has high strength and toughness while still maintaining some machinability. If sufficient alloy is added to produce a martensitic structure the iron is nonmachinable and is used as a wear and abrasion resisting material.

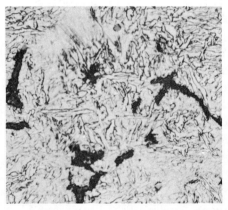

Fig. 11–27. High strength nickel-molybdenum low alloy gray iron with 70,000 psi tensile strength. 500X. (Gray and Ductile Iron Founders' Society)

11.22 Special Purpose Alloy Cast Irons. Some alloy cast irons have been developed for special applications requiring controlled expansion or special electrical or magnetic properties. An austenitic type high nickel cast iron is available which has only ⅕ the coefficient of expansion of steel. This particular alloy also has excellent impact properties at both room temperature and subzero temperatures. This low expansion alloy is used for machine parts which must maintain high accuracy in spite of temperature change. It is also used for instruments, forming dies, and glass molds. Variations in the high nickel type cast iron can produce controlled high expansivity alloys for matching other metals with different expansion rates. Castings can be produced with thermal expansion coefficients ranging from 2.2 to 10.6 × 10^{-6} in./in. deg F by changing the alloy content.

Cast irons with high electrical resistivity and with nonmagnetic properties are very useful for electrical resistance grids and for magnetic clutch and brake parts. Such alloys may be produced from the high nickel austenitic cast irons. Aluminum additions of 18 to 25 percent produce parts with high electrical resistance. Magnetic permeability can be so controlled in one alloy by chromium additions that castings from this alloy are "dead" to the attraction of a strong alnico magnet.

11.23 Wrought Iron. Wrought iron is a mixture of high purity iron and slag (about 2 or 3 percent) with stringers of slag uniformly distributed through the iron matrix, Fig. 11–28. The fibrous structure produces an unusual fracture as shown in Fig. 11–29. An improved process for manufacturing wrought iron has been developed. It is called the Aston-Byers Process. The old process involved the use of a puddling furnace somewhat similar to an open hearth furnace. Cold pig iron was charged into the furnace and melted. Carbon from the molten iron would combine with oxygen from the iron oxide lining of the furnace to yield a low carbon iron. When practically all of the carbon and other impurities were eliminated, the melting point of the molten iron would increase, causing the formation of a pasty mass.

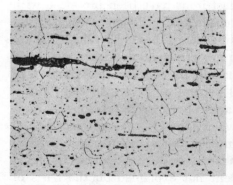

Fig. 11–28. Photomicrograph of wrought iron. Fibrous slag embedded in a matrix of nearly pure iron. 100X.

Fig. 11–29. Wrought iron bars fractured to show characteristic fibrous, hickory-like structure. (A. M. Byers Co.)

The slag and pasty mass was well stirred, formed into a ball, which was then removed from the furnace. The bulk of the slag was then squeezed from the mass in a press after which the iron was rolled into bars.

The bars were cut into lengths and piled, with alternate layers at right angles to each other. The stacks of bars were wired together, heated to a welding heat, and rolled to the desired shapes. The purpose of stacking and rolling was to obtain a fine and uniform distribution of the slag.

The Aston-Byers process gives an iron which cannot be distinguished chemically or metallographically from puddled iron but can be welded somewhat more easily. In this process, pig iron is melted in a cupola furnace and blown to soft steel in a Bessemer converter. The steel is then poured into liquid slag, special precautions being taken to insure thorough mixing. The excess slag is then poured off and squeezed out. After this squeezing, the spongy material is rolled into bars for fabrication into pipe and similar products.

Despite the fact that many different processes have been used in the manufacture of wrought iron, the characteristics and basic principles of metallurgy used in producing it have remained unchanged. The iron silicate or slag was at first considered an undesirable impurity; however, we know now that the slag is responsible for many of the desirable properties of wrought iron, particularly its resistance to fatigue and corrosion.

The principal value of wrought iron lies in its ability to resist corrosion, and fatigue failure. Its corrosion resistance, directly attributable to the slag fibers, is also a result of the purity of the iron base metal and its freedom from segregated impurities. Because of its softness and ductility, wrought iron finds use in the manufacture of bolts, pipe, staybolts, tubing, and nails.

References

Lyman, Taylor, ed. *Metals Handbook, 1948 ed.* Cleveland, O.: American Society for Metals.

Malleable Founders Society. *Malleable Iron Castings.* Cleveland, O. 1960.

Walton, Charles F. *The Gray Iron Castings Handbook.* Cleveland, O.: Gray Iron Founders Society, 1958.

Light Metals and Alloys

12.1 Introduction. High strength, lightweight alloys are of great importance as engineering metals for use in land, sea, air, and space transportation. One pound of metal saved in construction weight can result in savings of many pounds in power plants, transmission systems, and in total fuel requirements.

The metals commonly classed as light metals are those whose density is less than steel (7.8 gm/cc) and which can be used for engineering applications.

The metals in this group are:

	Density (gm/cc)
Magnesium	1.74
Beryllium	1.85
Aluminum	2.71
Titanium	4.5

Probably the greatest application of the light metals is in aircraft construction. Extensive use of light alloys began with the discovery of precipitation hardening for greatly strengthening these alloys.

Aluminum

12.2 Discovery and Uses. Aluminum was discovered in 1825 and introduced for the first time in 1852 at a price of $545 a pound. Despite the fact that the earth's crust contains an abundance of aluminum, a method for economically wresting the metal from nature and separating it from its ore in large quantities was not accomplished until 1886. It was during this year that a young American named Charles Martin Hall and a young Frenchman, Paul L. T. He'-Roult, independently but simultaneously discovered identical processes to produce aluminum inexpensively by electrolysis of a molten bath. A singular fact about these two scientists is that they were both born the same year and both died the same year.

Aluminum is one of our foremost industrial metals and can be manufactured with purity ranging from

99.5 to 99.99 percent. Commercially pure aluminum is used for deoxidizing steel, in the thermit mixture used in welding, in cooking and chemical apparatus, as an electrical conductor, as a pigment in paint, for aluminum foil, and for many other household and chemical applications. However, most of the aluminum produced for industrial applications is used in the alloy form, making a series of alloys with much higher mechanical properties than those found in pure aluminum.

12.3 Properties of Pure Aluminum. Aluminum is a very soft and ductile metal with a tensile strength of 9,000 psi in single crystal form to 24,000 psi in the fully work hardened condition. The excellent ductility and malleability of aluminum permits it to be rolled into thin sheets and foil and to be drawn into wire. Hardness of aluminum varies from 25 Brinell in the annealed condition to 40 Brinell in the fully work hardened state. On a weight basis the electrical conductivity of aluminum is 200 percent of that for copper; whereas on a volume basis it is about 62 percent of that for copper. Aluminum is non-sparking and is non-magnetic. Aluminum containers are non-toxic and will not adversely affect the purity and taste of foods.

Aluminum has nearly five times the thermal conductivity of cast iron. Because of its high reflectivity and low radiating power aluminum is useful for heat and light reflectors and for insulating applications. Aluminum has excellent low temperature properties. As the temperature drops, tensile strength increases as much as 50 percent and impact strength also improves. Pure aluminum has very good resistance to corrosion by weathering. As an example of this, the aluminum ornamental spire on the Washington Monument, Washington, D.C., has successfully withstood many years of weathering. Aluminum's resistance to corrosion is largely due to the formation of a tightly adhering, thin aluminum oxide (Al_2O_3) film which acts as a natural paint in preventing further attack by weathering.

Aluminum alloys are less effectively protected by the natural oxide coating than is pure aluminum and require special treatments which will be discussed later.

Aluminum can be successfully welded by gas, electric arc, and electric resistance methods. Special fluxes and rods are also available for brazing and soldering of aluminum.

12.4 Classification of Aluminum Alloys. Aluminum alloys are broadly classed as (1) casting or (2) wrought alloys. Tables 12–1 through 12–3 list some of the more important alloy compositions for each group. Casting alloys are usually designated by a number assigned to each composition such as 43, 356, and so on. The wrought alloys are designated by a four digit numbering system. The first digit identifies the alloy type. The second digit indicates the alloy modification and the last two digits identify either the specific alloy or the aluminum purity. The alloy types as indicated by the first digit is as follows:

Designation	Alloying Elements
1xxx	Aluminum 99.00% min
2xxx	Copper
3xxx	Manganese
4xxx	Silicon
5xxx	Magnesium
6xxx	Magnesium and Silicon
7xxx	Zinc
8xxx	Other elements

A letter following the alloy designation and separated from it by a hyphen indicates the basic temper treatment employed in hardening, strengthening or softening the alloy. For example: —O indicates an annealing treatment, —H strain hardening, and —W and —T solution heat treatment.

—F as fabricated
—O annealed
—H strain hardened
  —H1 strain hardened only
  —H2 strain hardened followed by partial annealing
  —H3 strain hardened then stabilized
—W Solution heat-treated, unstable temper
—T Solution heat-treated, stable temper
  —T2 annealed (cast shapes only)
  —T3 solution treated followed by cold work
  —T4 solution treated, natural aging at room temperature
  —T5 artificially aged after quenching from a hot working operation such as casting or extrusion
  —T6 solution treated, artificially aged
  —T7 solution treated, stabilized to control growth and distortion
  —T8 solution treated, cold worked, artificially aged
  —T9 solution treated, artificially aged, cold worked.

The number following the basic temper indicates the degree of hardness produced in the alloy by the specific processing operation. Temper hardness ranges from 0 in the fully annealed condition to 8 in the full hard commercial temper. A number "2" represents a quarter hard condition, "4" a half hard condition, and "6" a three-quarter condition. An extra hard temper is designated by the number "9".

12.5 Alloying Elements in Aluminum. The alloying elements commonly used in aluminum include copper, silicon, magnesium, manganese, and occasionally zinc, nickel, and chromium. The use of alloying elements can extend the tensile strength to as high as 125,000 psi in wrought alloys or 50,000 psi in large complex castings. Recently developed alloys prepared from aluminum powder can be used for many low strength applications at temperatures to 1000°F. The overall effect of alloying is to increase tensile strength, yield strength, and hardness with a corresponding reduction in ductility.

Alloying elements such as silicon are added to casting alloys to improve fluidity of the molten metal as well as to improve casting quality and mechanical properties.

In general, aluminum and its alloys may be hardened and strengthened by one of the following three

(Text Cont. on Page 384)

Table 12–1. Non-heat treatable wrought aluminum alloys.

| Alloy | Temper | Typical Tensile Properties | | | | | Typical Physical Properties | | | | |
| | | Strength—ksi[1] | | Elongation % in 2 in. | | Nominal Chemical Composition[2] per cent | Thermal Conductivity at 77°F Eng. Units | Electrical Resistivity at 68°F Ohms/mil-ft | Specific Gravity | Average Coefficient Therm. Exp. 68°-212°F | Melting Range °F |
		Ultimate	Yield—set 0.2%	1/16" thick	1/2" dia.						
EC	O	12	4	…	23[3]	99.45 Al (min.)	1625	17	2.70	13.1	1195-1215
	H19	27	24	…	1.5[4]						
	H26	18	16	…	…						
	H111	11	5	…	…						
	H112	12	6	…	…						
1100	O	13	5	35	45	99.0 Al (min.)	1540	17	2.71	13.1	1190-1215
	H12	16	15	12	25		…	…			
	H14	18	17	9	20		…	…			
	H16	21	20	6	17		…	…			
	H18	24	22	5	15		1510	18			
3003 Alclad 3003	O	16	6	30	40	Core 1.2 Mn Clad 1.0 Zn	1340	21	2.73	12.9	1190-1210
	H12	19	18	10	20		1130	25			
	H14	22	21	8	16		1100	25			
	H16	26	25	5	14						
	H18	29	27	4	10		1070	26			
3004 Alclad 3004	O	26	10	20	25	Core 1.2 Mn 1.0 Mg Clad 1.0 Zn	1130	25	2.72	12.9	1165-1205
	H32	31	25	10	17						
	H34	35	29	9	12						
	H36	38	33	5	9						
	H38	41	36	5	6						
4043	H18	33	…	…	…	5.2 Si					
5005	O	18	6	30		0.8 Mg	1390	20	2.70	13.2	1170-1205
	H12	20	19	10	…						
	H14	23	22	6	…						
	H16	26	25	5	…						
	H18	29	28	4	…						
	H32	20	17	11	…						
	H34	23	20	8	…						
	H36	26	24	6	…						
	H38	29	27	5	…						
5050	O	21	8	24	…	1.3 Mg	1340	21	2.69	13.2	1160-1205
	H32	25	21	9	…						
	H34	28	24	8	…						
	H36	30	26	7	…						
	H38	32	29	6	…						
5052	O	28	13	25	30	2.5 Mg 0.25 Cr	960	30	2.68	13.2	1100-1200
	H32	33	28	12	18						
	H34	38	31	10	14						
	H36	40	35	8	10						
	H38	42	37	7	8						

Table 12-1. (cont.) Non-heat treatable wrought aluminum alloys.

Alloy	Temper	Tensile strength (ksi)[1]	Yield strength (ksi)[1]	Elong. %	Elong. %	Nominal composition[2] (%)			Sp. gr.		Melting range (°F)
5056	O	42	22	...	35	0.1 Mn	810	36	2.64	13.4	1055-1180
	H18	63	59	...	18	5.2 Mg					
	H38	60	50	...	15	0.1 Cr	750	38			
5083	O	42	21	22	...	.6 Mn	810	36	2.66	13.2	1065-1180
	H113	46	33	16	...	4.65 Mg					
						.15 Cr					
5086	O	38	17	22	...	0.5 Mn	870	34	2.66	13.2	1084-1184
	H34	42	30	12	...	4.0 Mg					
	H36	47	37	10	...						
	H38	50	40	9	...						
	H112	39	19	14	...						
5154	O	35	17	27	...	3.5 Mg	870	32	2.66	13.3	1100-1190
	H32	39	30	15	...	0.25 Cr					
	H34	42	33	13	...						
	H36	45	36	12	...						
	H38	48	39	10	...						
	H112	35	17	25	...						
5357 5457	O	19	7	25		0.3 Mn	1160	24	2.70	13.2	1165-1210
	H32	22	19	9		1.0 Mg					
	H25	26	23	12							
	H38	32	30	6							
5454	O	36	17	22		.7 Mn	930	31	2.68	12.1	1115-1195
	H32	40	30	10		2.7 Mg	...	...			
	H34	44	35	10		.12 Cr	...	...			
	H112	36	18	18			...	...			
	H311	38	26	14			...	...			
5456	O	45	23	24		.7 Mn	810	36	2.66	13.3	1060-1180
	H11	45	24	22		5.1 Mg	...	...			
	H31	47	33	18		.12 Cr	...	...			
	H32	51	37	16			...	...			
5657	O	17	6	25		0.8 Mg	1310	21	...	...	...
	H25	23	20	13							
	H38	28	24	7							

1 ksi = 1000 psi.

2 Balance aluminum and impurities.

3 EC wire in 10 inches.

Courtesy: Reynolds Metals Co.

Table 12–2. Heat treatable wrought aluminum alloys.

Alloy	Temper	Strength—ksi① Ultimate	Strength—ksi① Yield—set 0.2%	Elongation % in 2 in. 1/16" thick	Elongation % in 2 in. ½" dia.	Nominal Chemical Composition② per cent	Thermal Conductivity at 77°F Eng. Units	Electrical Resistivity at 68°F Ohms/mil·ft	Specific Gravity	Average Coefficient Therm. Exp. 68°-212°F	Melting Range °F
2011	T3	55	43	..	15	5.5 Cu, 0.5 Pb, 0.5 Bi	990	29	2.82	12.8	995-1190
	T8	59	45		12		1190	23			
2014	O	27	14	..	18	0.8 Si, 4.5 Cu, 0.8 Mn, 0.4 Mg	1340	21	2.80	12.8	950-1180
	T4, T451	62	42		20		840	35		12.8	
	T6, T651	70	60		13		1070	26		12.5	
Alclad 2014	O	25	10	21	..	Clad 0.6 Si / Core 0.8 Si, 4.5 Cu, 0.8 Mn, 0.4 Mg, 1.1 Mg	1340	21	2.80	12.8	950-1180
	T3	63	40	20			840	35		12.8	
	T4, T451	61	37	22			1070	26		12.8	
	T6, T651	68	60	11						12.5	
2017	O	26	10	..	22	4.0 Cu, 0.5 Mn, 0.5 Mg	1190	23	2.79	13.1	955-1185
	T4, T451	62	40		22		840	35			
2024	O	27	11	20	22	4.5 Cu, 0.6 Mn, 1.5 Mg	1310	21	2.77	12.9	935-1180
	T36	70	50	18	..		840	35			
	T4, T351	72	57	13	19		840	35			
		68	47	20			840	35			
Alclad 2024	O	26	11	20	..	Core 4.5 Cu, 0.6 Mn, 1.5 Mg / Clad 99.3 Al (min.)	1310	21	2.77	12.9	935-1180
	T36	65	45	18			840	35			
	T4, T351	67	53	11			840	35			
	T81, T851	64	42	19			840	35			
	T86	65	60	6							
		70	66	6							
2117	T4	43	24	..	27	2.5 Cu, 0.3 Mg	1070	26	2.74	13.0	950-1200
2218	T72	48	37	..	11	4.0 Cu, 1.5 Mg, 2.0 Ni	1070	26	2.81	12.4	940-1175
4032	T6	55	46	..	9	12.2 Si, 0.9 Cu, 1.1 Mg, 0.9 Ni	960	30	2.69	10.8	990-1060
6061	O	18	8	25	30	0.6 Si, 0.25 Cu, 1.0 Mg, 0.25 Cr	1190	23	2.70	13.0	1080-1200
	T4, T451	35	21	22	25		1070	26			
	T6, T651	45	40	12	17		1070	26			
	T93	60	..	..	..						
	T94	54	47								
	T913	68									

Table 12-2. (cont.) Heat treatable wrought aluminum alloys.

Alloy	Temper					Composition			Density		Temp.
Alclad 6061	O	17	7	25	...	Clad 1.0 Zn	1190	23	2.70	13.0	1080-1200
	T4, T451	33	19	22	...	Core 0.6 Si, 0.25 Cu, 1.0 Mg, 0.25 Cr	1070	26			
	T6, T651	42	37	12			1070	26			
6062	O	18	8	...	30	0.6 Si, 0.25 Cu, 1.0 Mg, 0.1 Cr	1190	23	2.70	13.0	1100-1205
	T4, T451	35	21	...	25		1070	26			
	T6, T651	45	40		17		1070	26			
6063	O	13	7	...	...	0.4 Si, 0.7 Mg	...	...	2.70	13.0	1140-1205
	T4	25	13	22	...			21			
	T42	22	13	20	...		1340	20			
	T5	27	21	12	...		1390	20			
	T6	35	31	12	...		1390	...			
	T83	37	35	9	...						
	T831	30	27	10	...						
	T832	42	39	12	...						
6101	H111	14	11	...	...	0.5 Si, 0.6 Mg	1500	19	2.70	13.0	1140-1205
	T6	35	31	12	...		1550	18			
	T61	22	15	...	...		1525	18			
	T62	31	24	...	...						
7075	O	33	15	17	16	1.6 Cu, 2.5 Mg, 0.3 Cr, 5.6 Zn	840	35	2.80	12.9	890-1180
	T6, T651	83	73	11	11						
Alclad 7075	O	32	14	17	...	Clad 1.0 Zn	840	35	2.80	12.9	890-1180
	T6, T651	76	67	11	...	Core 1.6 Cu, 2.5 Mg, 0.3 Cr, 5.6 Zn					
7079	T6, T651	78	68	...	14	0.6 Cu, 0.2 Mn, 3.3 Mg, 0.17 Cr, 4.3 Zn	840	34	2.74	13.0	900-1180
7178	O	33	15	15	16	2.0 Cu, 2.7 Mg, 0.29 Cr, 6.8 Zn	840	34	2.81	13.0	890-1165
	T6, T651	88	78	10	11						
Alclad 7178	O	32	14	16	...	Clad 1.0 Zn	840	34	2.81	13.0	890-1165
	T6, T651	81	71	10	...	Core 2.0 Cu, 2.7 Mg, 0.29 Cr, 6.8 Zn					

1 ksi = 1000 psi.

2 Balance aluminum and impurities.

Courtesy: Reynolds Metals Co.

Table 12–3. Aluminum casting alloys.

SAND CAST ALUMINUM ALLOYS

ALLOY	TEMPER	TYPICAL TENSILE PROPERTIES ksi [1]		Elongation per cent in 2 in.	CHEMICAL COMPOSITION nominal per cent [2]		HARDNESS Brinell 500Kg load 10mm ball	SHEARING STRENGTH ksi	ENDURANCE LIMIT [3] ksi	MODULUS OF ELASTICITY psi x 10⁶
		Ultimate	Yield							
43	F	19	8	8.0	5.5 Si		40	14	8	10.3
108	F	21	14	2.5	4.0 Cu	3.0 Si	55	17	11	..
112	F	24	15	1.5	7.0 Cu		70	20	9	..
113	F	24	15	1.5	2.0 Si	7.0 Cu	70	20	9	..
122	F	24	15	1.5	1.0 Si	10.0 Cu	80	21	9.5	..
	T2	27	20	1.0	1.2 Fe	0.27 Mg				
	T61	41	40	0.5			115	32	8.5	10.7
142	F	31	30	0.5	4.0 Cu	1.5 Mg	70	21	8	10.3
	T21	27	18	1.0		2.0 Ni				10.3
	T571	32	30	0.5			85	26	11	10.3
	T77	30	23	2.0			75	24	10.5	
A142	T75	31	..	2.0	4.1 Cu	0.2 Cr	70	..	..	..
					1.5 Mg	2.0 Ni				
195	T4	32	16	8.5	4.5 Cu	1.0 Si	60	26	7	10.0
	T6	36	24	5.0			75	30	7.5	10.0
	T62	40	34	2.0			90	33	8	10.0
212	F	23	14	2.0	8.0 Cu	1.2 Si	65	20	9	10.3
						1.0 Fe				
214	F	25	12	9.0	4.0 Mg		50	20	7	..
B214	F	20	13	2.0	1.8 Si	4.0 Mg	50	17	8.5	..
F214	F	21	12	3.0	0.5 Si	4.0 Mg	50	17	8	..
A218	F	36	18	9.0	7.0 Mg	0.2 Mn	65	..	..	..
220	T4	46	25	14.0	10.0 Mg		75	33	8	9.5
319	F	27	18	2.0	6.0 Si	4.0 Cu	70	22	10	10.7
	T6	36	24	2.0			80	29	11	10.7
355	T51	28	23	1.5	5.0 Si	1.3 Cu	65	22	8	10.2
	T6	35	25	3.0	0.5 Mg		80	28	9	10.2
	T61	39	35	1.0			90	31	9.5	..
	T7	38	36	0.5			85	28	10	10.2
	T71	35	29	1.5			75	26	10	10.2
	T72	30	24	2.5			80	..	..	10.2
C355	T6	39	29	5.0	Same as 355		85	..	..	..

Table 12-3. (cont.) Aluminum casting alloys.

ALLOY	TEMPER	Ultimate (ksi)	Yield (ksi)	Elongation per cent in 2 in.	Chemical Composition nominal per cent	Hardness Brinell 500 Kg load 10 mm ball	Shearing Strength ksi	Endurance Limit ksi	Modulus of Elasticity psi × 10⁶
356	F T51 T6 T7 T71	24 25 33 34 28	18 20 24 30 21	6.0 2.0 3.5 2.0 3.5	7.0 Si 0.3 Mg 0.2 Fe	60 70 75 60	20 26 24 20	8 8.5 9 8.5	10.5 10.5 10.5 10.5
A356	T6	38	28	6.0	7.0 Si 0.3 Mg 0.1 Fe	70	..	..	..
357	T6	43	36	4.0	7.0 Si 0.5 Mg	80	..	..	..
A347	T6 T6 (Chilled)	46 49	36 36	3.0 8.0	7.0 Si 0.6 Mg	85 85	40 40	12 12	..
TENS-50	T6 T62	46 51	36 40	5.0 3.0	8.0 Si 0.5 Mg	85	41	12.5	..
603	F	35	20	8.5	3.0 Zn 1.6 Mg	65	..	..	..
604	F	36	22	5.0	3.7 Zn 1.8 Mg	74	..	..	..
607	F	37	27	3.0	4.3 Zn 2.2 Mg	85	..	..	..
D612	F	35	25	9.0	5.8 Zn 0.5 Mg	75	26	9	..
613	F	35	25	5.0	7.5 Zn 0.4 Mg 0.7 Cu	74	..	..	9.7
A750	T5	20	11	5.0	2.5 Si 0.5 Ni 1.0 Cu 6.3 Sn	45	14	..	10.3

PERMANENT MOLD CAST ALUMINUM ALLOYS

ALLOY	TEMPER	TYPICAL TENSILE PROPERTIES ksi [1] Ultimate	Yield	Elongation per cent in 2 in.	CHEMICAL COMPOSITION nominal per cent [2]	HARDNESS Brinell 500 Kg load 10 mm ball	SHEARING STRENGTH ksi	ENDURANCE LIMIT [3] ksi	MODULUS OF ELASTICITY psi × 10⁶
43	F	23	9	10.0	5.5 Si	45	16	8	10.3
A108	F	28	16	2.0	5.5 Si 4.5 Cu	70	22	13	..
113	F	28	19	2.0	2.0 Si 7.0 Cu	70	22	9.5	..
C113	F	30	24	1.5	7.0 Cu 4.0 Si 1.4 Fe 2.5 Zn	85	24	9.5	..
122	T52 T551 T65	35 37 48	31 35 36	1.0 0.5 0.5	1.0 Si 10.0 Cu 1.2 Fe 0.5 Mn	100 115 140	25 30 36	8.5 9	..
A132	T551 T65	36 47	28 43	0.5 0.5	12.0 Si 1.0 Cu 1.0 Fe 1.1 Mg 2.5 Ni	105 125	28 36	13.5	10.7

Table 12-3. (cont.) Aluminum casting alloys.

Alloy	Temper				Nominal composition, %				
D132	T5	36	28	1.0	9.5 Si, 3.0 Cu, 1.0 Mg, 1.0 Ni	105	28	13.5	..
138	F	32	24	1.5	10.0 Cu, 4.0 Si, 0.3 Mg	100	22	..	..
142	T571	40	34	1.0	4.0 Cu, 2.0 Ni, 1.5 Mg	105	30	10.5	10.3
	T61	47	42	0.5		110	35	9.5	10.3
B195	T4	37	19	9.0	4.5 Cu, 2.5 Si	75	30	9.5	10.1
	T6	40	26	5.0		90	32	10	10.1
	T7	39	20	4.5		80	30	9	10.1
A214	F	27	16	7.0	4.0 Mg, 1.8 Zn	60	22	..	..
319	F	27	18	2.0	6.3 Si, 3.8 Cu, 0.6 Fe	70	22	10	..
	T6	40	27	3.0		95		..	
333	F	34	19	2.0	9.0 Si, 1.0 Fe, 0.6 Mg, 3.8 Cu, 0.8 Mn	90	27	14.5	10.2
	T5	34	25	1.0		100	27	12	10.2
	T6	42	30	1.5		105	33	15	10.2
	T7	37	28	2.0		90	28	12	
355	T51	30	24	2.0	5.0 Si, 0.5 Mg, 1.3 Cu	75	24	10	10.2
	T6	43	27	4.0		90	34	10	
	T62	45	40	1.5		105	36	10	
	T7	40	30	2.0		85	30		
	T71	36	31	3.0		85	27		
C355	T6	48	28	10.0	5.0 Si, 0.5 Mg, 0.1 Fe, 1.3 Cu	90	32	14	10.2
	T61	46	34	6.0		100			
356	F	26	18	5.0	7.0 Si, 0.3 Mg, 0.2 Fe	..	..	13	10.5
	T51	27	20	2.0		..	..	11	10.5
	T6	40	27	5.0		90	32		
	T7	33	24	5.0		70	25		
A356	T6	41	28	12.0	7.0 Si, 0.3 Mg, 0.1 Fe	80	28	..	..
	T61	37		5.0					
357	F	28	15	6.0	7.0 Si, 0.5 Mg	..	..	..	..
	T51	29	22	4.0		..			
	T6	49	36	8.0		85			
A357	T6	50	41	10.0	7.0 Si, 0.6 Mg	85	43	16	..
TENS-50	T6	50	44	6.0	8.0 Si, 0.5 Mg	90	44	16	..
	T62	53	45	5.5			45		
750	T5	23	11	12.0	1.0 Cu, 6.3 Sn, 1.0 Ni	45	15	9	10.3
A750	T5	20	11	5.0	2.5 Si, 0.5 Ni, 1.0 Cu, 6.3 Sn	45	14	9	10.3
B750	T5	32	23	5.0	2.0 Cu, 1.2 Ni, 0.8 Mg, 6.3 Sn	70	21	11	10.3

Table 12-3. (cont.) Aluminum casting alloys.

DIE CAST ALUMINUM ALLOYS

Alloy	Temper				Composition				
13	F	43	21	2.5	12.0 Si, 1.0 Fe	80	28	19	10.3
Al3	F	35	16	3.5	12.0 Si, 0.5 Fe	80	:	:	:
43	F	33	16	9.0	5.5 Si	50	21	17	10.3
85	F	40	24	5.0	4.0 Cu, 5.0 Si, 2.3 Fe	75	23	22	:
L214	F	41	:	10.0	0.75 Si, 0.5 Mn, 3.0 Mg	:	:	:	:
218	F	45	27	8.0	8.0 Mg	80	30	23	:
360	F	47	25	3.0	9.5 Si, 0.55 Mg, 1.0 Fe	75	30	19	10.3
A360	F	46	24	5.0	9.5 Si, 0.55 Mg, 0.5 Fe	75	29	18	:
380	F	48	24	3.0	8.5 Si, 3.5 Cu, 0.8 Fe	80	31	21	10.3
A380	F	47	23	4.0	8.5 Si, 3.5 Cu	80	30	20	:
384	F	47	25	1.0	12.0 Si, 0.9 Fe, 3.5 Cu	:	30	21	10.3

1 ksi = 1000 psi.

2 Balance aluminum and impurities.

3 Endurance limit—500,000,000 cycles of completely reversed stress.

Courtesy: Reynolds Metals Co.

383

methods: (1) strain hardening by cold working, (2) solid solution strengthening by alloying, (3) precipitation hardening treatments.

The highest strength alloys may employ a combination of all three methods for obtaining maximum properties. The effect of common alloying elements in aluminum are generally as follows:

Copper has been the principal alloying element in aluminum for many years. It is used in amounts of up to 5.5 percent in wrought alloys and up to 10 percent in casting alloys. Copper provides a basis for precipitation hardening alloys by the formation of submicroscopic particles of copper aluminide. Solubility of copper in aluminum is 5.65 percent at the eutectic temperature and decreases to less than 0.25 percent at room temperature. Copper additions tend to decrease shrinkage and hot shortness in aluminum alloys.

Silicon is second to copper in importance as an alloying element and is used principally in casting alloys. It is used in amounts ranging from 1 to 12 percent as a primary or secondary alloying element. Silicon improves casting qualities by improving fluidity of the molten metal with freedom from hot shortness.

It also provides increased corrosion resistance, low thermal expansion, high thermal conductivity, and low specific gravity. Casting alloys containing silicon are noted for pressure tightness and good impact toughness. Silicon usually is present as either an alloying element or an impurity in all commercial alloys.

Magnesium is alloyed with aluminum in amounts ranging from less than 1 to 10 percent. These alloys are light, possess good mechanical properties, and are easily machined. The alloys containing high percentages of magnesium have high corrosion resistance to both salt water and alkaline solutions. Aluminum-magnesium alloys are not particularly difficult to cast as they have good fluidity and are quite free from cracking even in large castings. Pressure tight castings are less easily obtained than with aluminum-silicon alloys, however.

Zinc additions in percentages of up to 7.5 percent improve mechanical properties through the formation of hard intermetallic phases such as Mg_2Zn. Zinc is usually added in conjunction with other elements such as magnesium and copper.

Other Elements are manganese and chromium. In small amounts they tend to improve corrosion resistance of aluminum alloys. Nickel additions improve strength at elevated temperature but with some loss in corrosion resistance. Iron and copper, when added to alloys containing more than 3 percent silicon, result in alloys usually free from hot shortness. Lead and bismuth additions to aluminum alloys provide free machining alloys. Small additions of titanium and columbium have been used as grain refining elements in certain alloys.

12.6 Aluminum Casting Alloys.

The nominal composition and mechanical properties of common aluminum casting alloys are given in Table 12–3. Casting alloys have been grouped according to their suitability for either sand casting, permanent mold casting, or die casting. Some alloys such as 43, 122, and 356 are included in more than one group because of their suitability for use by more than one casting process.

The original binary aluminum-copper sand casting alloy with 8 percent copper has been largely replaced by alloys such as 112, 113 and 212. These alloys contain elements to improve casting and machining properties while still maintaining practically the same mechanical properties.

Aluminum-silicon casting alloys such as 43 and 356, because of their fluidity, are useful for thin walled pressure tight castings with intricate detail. This property, together with their good resistance to atmospheric corrosion make them particularly suitable for architectural and ornamental castings. Aluminum-silicon alloys have greater ductility and toughness than the aluminum-copper alloys and often find uses in aircraft and marine fittings, cooking utensils, and intricate heat treated castings.

The properties of certain aluminum-silicon alloys may be improved by inoculation with metallic sodium just prior to casting. The addition of a small amount of sodium to the ladle of molten metal causes the castings to freeze with a much finer and tougher grain structure. A comparison of the microstructures obtained by this modified freezing practice with conventional practices is shown in Fig. 12–1A and B.

Aluminum-magnesium alloys such as 214 and 220 have even better resistance to salt water corrosion than aluminum-silicon alloys and develop higher mechanical properties.

Aluminum-copper-silicon alloys such as 108 and 319 develop desirable properties such as weldability, good casting properties, moderate strength and pressure tightness. These alloys are used for manifolds, valve bodies, and automotive cylinder heads. Additions of iron and magnesium provide wear resistance required for piston alloys.

Certain casting alloys such as 195 and 356 develop improved mechanical properties through heat treatment. The precipitation hardening treatment consists of heating the alloy to 980-1000°F and holding for a period of time followed by quenching and aging.

Aluminum-zinc alloys are less expensive than aluminum-copper al-

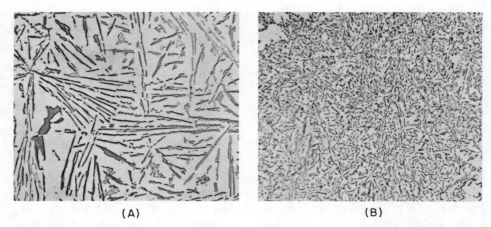

(A) (B)

Fig. 12–1. A, microstructure of unmodified aluminum-silicon casting. 250X. ½% HF etch. B, inoculated with metallic sodium to refine grain structure.

loys with approximately the same properties. However, aluminum-zinc alloys are heavier than the aluminum-copper alloys and have been reported to be less resistant to corrosion.

12.7 Aluminum Permanent Mold Casting Alloys. The low melting temperatures of aluminum alloys make it possible to cast them into gravity fed permanent molds of cast iron or steel. These permanent mold castings have the advantage of excellent surface finish, close dimensional tolerances, and high mechanical properties in the as-cast and heat treated conditions. Solidification is rapid when the alloys are cast in metal molds producing a fine grained strong casting. In general, perma-

nent mold castings must be relatively small and simple in shape although some rather complex shapes and castings weighing several hundred pounds have been produced by the permanent mold process. The relative casting characteristics for several common permanent mold casting alloys are given in Table 12–4.

As with sand castings and die castings the alloy for permanent mold castings should be selected after consideration of the casting characteristics as well as consideration of physical and mechanical properties.

12.8 Aluminum Die Casting Alloys. Die casting is similar to permanent mold casting except that the molten

Table 12–4. Relative casting characteristics of aluminum permanent mold alloys.

ALLOY	LACK OF HOT SHORTNESS	PRESSURE TIGHTNESS	LACK OF SHRINKAGE TENDENCY
43	1	1	1
A108	2	3	1
C113	3	2	1
122	3	3	3
A132	1	1	4
142	5	4	4
B195	4	4	3
A214	5	5	5
355	1	1	1
356	1	1	1

(NUMBERED IN ORDER OF PREFERENCE, 1 BEST)

Table 12–5. Relative casting characteristics of aluminum die casting alloys.

ALLOY	CASTING CHARACTERISTICS	
	LACK OF HOT SHORTNESS	MOLD FILLING CAPACITY
13	1	1
43	2	3
85	3	2
218	4	4
360	1	1
380	2	2

(NUMBERED IN ORDER OF PREFERENCE, 1 BEST)

alloy is forced into metal molds at pressures of 750 to 20,000 psi. Die cast parts have been produced which weigh from a fraction of an ounce to engine block castings weighing over 100 lbs. Because of the nature of the process, sections can be made that are somewhat thinner than can be made with sand or permanent mold processes. Small castings can be produced with sections of only .045″ by die casting whereas about ⅛″ is the thinnest section that can be produced by other methods. The surface smoothness and detail of die cast alloys is superior to that of other forms of aluminum castings. Tolerances as low as $\pm$ .0015 in. per lineal in. can be achieved by die casting. In Table 12–5 are given the relative casting characteristics of common die casting alloys.

Not all casting alloys are suitable for die casting. Many have hot shortness or do not have satisfactory mold filling characteristics; others introduce operational problems which limit casting complexity and intricacy. Operating speeds of up to 150 cycles per hour make die casting a very competitive process.

12.9 Aluminum Sand Casting Alloys.

Sand casting, because of its low cost and adaptability, is frequently used for production of small quantities of parts requiring intricate coring or large parts. Castings weighing from a few ounces to as much as 7000 lbs have been produced by the sand casting process. Expendable sand molds are produced from a suitable pattern. The pattern is made larger than the finished part to allow for shrinkage and subsequent finishing. Smooth finishes on small castings may be obtained by use of fine sand, although coarser sand is desired on large castings. The coarse sand allows the escape of gases from the mold cavity and makes for castings with greater soundness. The sand casting process permits the use of a number of different aluminum alloys with a wide variety of properties. Alloy selection should take into consideration casting characteristics (Table 12–6) in addition to mechanical and physical properties. These vary enough to have an important influence on the cost and usefulness of the part.

In melting all aluminum alloys for casting it is necessary to avoid overheating for two main reasons: First, overheated metal will require a longer cooling time resulting in a coarser grain and weaker structure. Second, aluminum at high temperature readily reacts with water vapor forming aluminum oxide and hydrogen. The hydrogen gas is avidly dissolved by the molten metal. As the temperature drops solubility decreases resulting in the formation of entrapped gas bubbles which produce a porous casting. Various fluxes are used to minimize hydrogen pick up. The practice of flushing the molten alloy with nitrogen or chlorine gas may also be used to remove dissolved hydrogen.

Although overheating is undesirable the liquid must be sufficiently hot to fill the mold before solidification begins. Cold shuts and misruns result from streams of metal coming from different directions. These streams make physical contact, but do not completely fuse because of heavy oxide films or lack of fluidity due to a low pouring temperature.

Table 12–6. Relative casting characteristics of aluminum sand casting alloys.

ALLOY	CASTING CHARACTERISTICS			
	LACK OF HOT SHORTNESS	PRESSURE TIGHTNESS	LACK OF SHRINK-AGE TENDENCY	LACK OF GAS ABSORPTION
43	1	1	1	1
108	2	1	1	3
113	3	2	1	3
122	3	2	3	5
142	4	4	3	4
195	5	3	3	3
214	3	5	4	3
220	3	5	5	3
355	1	1	1	2
356	1	1	1	2

(NUMBERED IN ORDER OF PREFERENCE, 1 BEST)

12.10 Wrought Aluminum Alloys. Wrought alloys make up by far the greatest tonnages of commercially produced aluminum alloys. These alloys are those which have been shaped by plastically deforming the crystal structure of the metal. The high mechanical properties developed in these alloys are due to the grain refinement, homogenization, and work hardening effects obtained from the mechanical treatment of rolling, forging, and extruding.

The action of breaking down the cast structure by a mechanical working process such as forging always results in the development of properties superior to those of a casting. In general, lower percentages of alloying elements are used in wrought alloys as indicated in Tables 12–1 and 12–2. The wrought alloys are classed as (1) non-heat treatable, meaning that they may only be hardened by cold working, and (2) heat treatable, indicating that they may be precipitation hardened.

The production of wrought aluminum alloys involves three steps: first, casting of an ingot; second, hot working (and often cold working) or shaping; and third, heat

389

treating. There are two principal requirements for producing a successful ingot. First, the molten metal must be introduced quietly and uniformly into the mold without the inclusion of dross or the formation of cold shuts. Second, heat must be withdrawn from the ingot at a sufficient rate and in a proper direction to produce a fine grained structure without surface cracking. Both tilting-mold and continuous-cast ingots are used for the production of wrought alloys.

There are two principal reasons for hot working of aluminum alloys. The first reason is to deform plastically the as-cast structure, converting it into a more workable wrought structure for subsequent hot or cold working operations. The second reason for hot working is because of the increased plasticity of the alloy at elevated temperatures; this increased plasticity permits more drastic reductions in size, and the manufacture of intricate parts with thin sections which could not be made by cold working. The general availability of wrought aluminum products is given in Table 12–7.

12.11 Non-Heat Treatable Aluminum Alloys.

The non-heat treatable aluminum alloys may be hardened and strengthened by cold working. Shaping by cold rolling, drawing, or by any of the other common forming methods may be used for the double purpose of obtaining the desired shape while at the same time causing an improvement in the mechanical properties. Cold working of an aluminum alloy begins where the hot working stops although an annealing operation may precede the cold working.

Aluminum finished by cold rolling is available with varying amounts of cold reduction and therefore with varying degrees of hardness. In Fig. 12–2 A through G are shown photomicrographs of the grain structure of 99.99% aluminum subjected to different degrees of cold reduction.

The strain hardenable alloys include such grades as EC, 1100, 3003, 3004, 4043, 5005, 5052, 5056, and 5456. These alloys vary in hardness from about Brinell 23 (500 kg load) for 1100-0 in the annealed condition to about Brinell 100 for 5056-H38 in the fully strain hardened condition. The modulus of elasticity of both cast and wrought aluminum alloys varies from 10×10^6 to 11.4×10^6 lb/in.^{2} with a value of 10.3×10^6 lb/in.^{2} usually being taken for most calculations. Grade EC is a high purity aluminum used for electrical conductors. Most strain hardenable alloys are employed for formed parts not requiring high strength. Typical applications include railway and bus bodies, airplanes, shipping containers, cooking utensils, architectural uses, and forgings not subjected to reverse stress conditions.

Table 12–7. General availability of wrought aluminum products.

Alloy	Foil	Sheet	Plate	Tube Drawn	Tube Extruded	Pipe	Structural Shapes[1]	Extruded Shapes	Rod	Bar	Wire	Rivets	Forgings & Forging Stock
EC	....	●	●	●	●	⊙	●		●	●	●	●	....
1100	●	●	●	●	●	....	....	....	●	●	●	●	●
1235	●	....	....	....	●	....	....	....	....	....	....	....	....
1145	●	....	....	....	....	....	....	....	....	....	....	....	....
1345	....	....	....	....	....	....	....	....	....	....	●	....	....
1060	....	●	●	●	●	....	....	....	....	....	....	....	....
2011	....	....	....	....	....	....	....		●	●	●	....	●
2014	....	●	●	●	●	....	....	●	●	●	●	....	●
Alclad 2014	....	●	●	....	....	....	....		....	....	....	....	....
2017	....	....	....	....	....	....	....		●	●	....	●	....
2117	....	....	....	....	....	....	....		....	....	●	●	....
2018	....	....	....	....	....	....	....		....	....	....	....	●
2218	....	....	....	....	....	....	....		....	....	....	....	●
2618	....	....	....	....	....	....	....		....	....	....	....	●
2219	....	●	●	....	●	....	....	●	●	....	●	....	●
2024	....	●	●	●	●	....	....	●	●	●	●	....	●
Alclad 2024	....	●	●	....	....	....	....		....	....	....	....	....
2025	....	....	....	....	....	....	....		....	....	....	....	●
3003	●	●	●	●	●	●	....	●	●	●	●	....	....
Alclad 3003	....	●	●	●	●	....	....		....	....	....	....	....
3004	....	●	●	●	●	....	....		....	....	....	....	....
Alclad 3004	....	●	●	....	....	....	....		....	....	....	....	....
4032	....	....	....	....	....	....	....		....	....	....	....	●
4043	....	....	....	....	....	....	....		....	....	●	....	....
5005	....	●	●	....	....	....	....		●	●	●	....	....
5050	....	●	●	●	....	....	....		●	●	●	●	....
Alclad 5050	....	●	●	....	....	....	....		....	....	....	....	....
5052	●	●	●	●	....	....	....		●	●	●	●	....
5252	....	●	....	....	....	....	....		●	●	●	....	....
5652	....	●	●	....	....	....	....		●	●	●	....	....
5154	....	●	●	●	●	....	....	●	●	●	●	....	....
5254	....	●	●	●	....	....	....	●	●	●	●	....	....
5454	....	●	●	....	●	....	....	●	●	....	....	....	....
Alclad 5155	....	●	●	....	....	....	....		....	....	....	....	....
5056	....	....	....	....	....	....	....		....	●	●	●	....
Alclad 5056	....	....	....	....	....	....	....	....	....	●	●	●	....
5456	....	●	●	....	●	....	....	●	....	●	....	....	....
5257	....	●	....	....	....	....	....		....	....	....	....	....
5457	....	●	....	....	....	....	....		....	....	....	....	....
5557	....	●	....	....	....	....	....		....	....	....	....	....
5657	....	●	....	....	....	....	....		....	....	....	....	....
5083	....	●	●	....	●	....	....		●	....	●	....	....
5086	....	●	●	●	●	....	....	●	●	●	●	....	....
6101	....	....	....	●	●	●	●		●	....	●	....	....
6201	....	....	....	....	....	....	....		....	....	●	....	....
6011	....	●	....	....	....	....	....		....	....	....	....	....
6151	....	....	....	....	....	....	....		....	....	....	....	●
6053	....	....	....	....	....	....	....		●	....	●	●	●
6061	....	●	●	●	●	●	●	●	●	●	●	....	●
Alclad 6061	....	●	●	....	....	....	....		....	....	....	....	....
6262	....	....	....	●	●	....	....		●	●	●	....	....
6063	....	....	....	●	●	●	....		●	●	....	....	....
6463	....	....	....	●	....	....	....		●	....	....	....	....
6066	....	....	....	....	●	....	....		●	●	●	....	●
7001	....	....	....	●	....	....	....		●	●	●	....	●
7039	....	●	●	....	....	....	....		....	....	....	....	....
7072	....	....	....	....	....	....	....		....	....	●	....	....
7075	....	●	●	●	●	....	....	●	●	●	●	●	●
Alclad 7075	....	●	●	....	....	....	....		....	....	....	....	●
7076	....	....	....	....	....	....	....		....	....	....	....	●
7277	....	....	....	....	....	....	....		....	....	....	●	●
7178	....	●	●	....	....	....	....		●	●	●	....	....
Alclad 7178	....	●	●	....	....	....	....		....	....	....	....	....
7079	....	●	●	....	....	....	....		●	....	....	....	●
Alclad 7079	....	●	....	....	....	....	....		....	....	....	....	....

●Indicates the products in which the alloy is normally produced.
[1]Rolled or Extruded.

*"1966 Standards for Aluminum Mill Products," The Aluminum Association.

12.12 Heat Treatment of Aluminum and Its Alloys.

The main types of heat treatments employed with aluminum and its alloys include annealing, stress relieving, solution heat treatment, and precipitation hardening.

Annealing of cold worked commercially pure aluminum such as 1100 consists of heating the metal

391

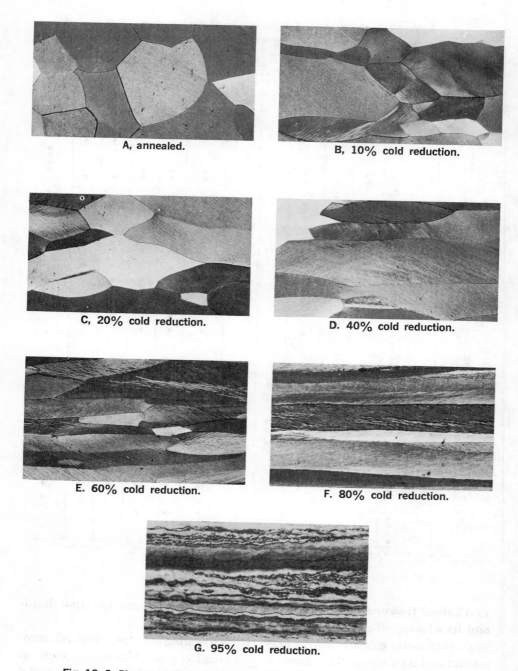

A, annealed.

B, 10% cold reduction.

C, 20% cold reduction.

D. 40% cold reduction.

E. 60% cold reduction.

F. 80% cold reduction.

G. 95% cold reduction.

Fig. 12–2. Photomicrographs of 3003 aluminum alloy. (Reynolds Metal Co.)

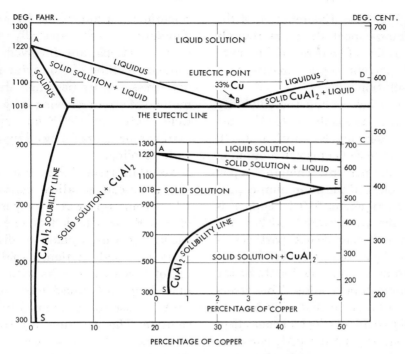

Fig. 12–3. Aluminum end of aluminum-copper diagram.

slightly in excess of 650°F, where complete softening is almost instantaneous, followed by cooling to room temperature. The exact temperature used and the time at this temperature, along with cooling rate, is not very important provided the annealing temperature is above the recrystallization temperature. The annealing heat treatment permits the strain hardened structure to recrystallize into new stress-free grains. As these new born grains grow they remove all of the strain hardened condition

and restore the original soft, ductile structure.

Wrought aluminum alloys may be annealed by subjecting them to heat treatments similar to those for pure aluminum. The main difference is in the rate of cooling which may be employed. Aluminum alloys of the precipitation hardening grades must not be cooled at a rate greater than 50°F per hour down to 500°F; the subsequent rate is unimportant. The reason for this may be obtained from a study of the aluminum-copper dia-

393

gram, Fig. 12–3. Examination of the aluminum-copper diagram reveals that an alloy of 95 percent aluminum and 5 percent copper freezes as a solid solution alloy.

Upon slow cooling the solubility of copper (probably as $CuAl_2$) decreases along line E-S until at 300°F the solubility of copper in aluminum is only 0.5 percent. Therefore, upon slow cooling of the aluminum-copper alloy from about 970°F, copper is gradually precipitated within the grains as rather coarse particles of $CuAl_2$. This condition may be considered as the annealed state of the aluminum-copper alloy. This structure is in a state that may be subjected to cold forging and consists of the relatively hard $CuAl_2$ constituent embedded in a soft aluminum matrix. The alloy may be annealed following cold working to remove any strain hardening effects. An annealing temperature of 650°F will recrystallize the cold worked aluminum phase without dissolving the $CuAl_2$ phase. The rate of cooling from the annealing temperature is not important if the $CuAl_2$ phase has not been dissolved; however, if this $CuAl_2$ phase has dissolved, rapid cooling will keep it in solution and the alloy will not be in its best annealed condition. The annealing treatments for several wrought aluminum alloys are given in Table 12–8.

Stress Relieving Treatments are sometimes recommended for welded aluminum alloys, especially where heavy sections are involved. One common method of modifying residual stresses is by peening. Another method of stress relieving is by postweld heating at 450°F for approximately four hours followed by air or water cooling.

Solution Heat Treatment of an aluminum-copper alloy containing 4.5 percent copper at a temperature of 950°F for a period of one hour causes the $CuAl_2$ phase to dissolve or go into solid solution. Rapid cooling of this structure, as with water quenching, will cause most of the $CuAl_2$ phase to be retained in solid solution. The structure of a copper-aluminum alloy is shown in Fig. 12–4 A prior to solution treatment and in Fig. 12–4 B after solution treatment. Following solution treatment and quenching examination of the microstructure will reveal very little of the $CuAl_2$ phase. Apparently, the balance has been retained in solution by the severe quench. This treatment has been referred to as a *solution treatment*. Improvement in corrosion resistance is obtained by this treatment. Cold working may be carried out after the solution treatment to increase strength without adverse effect upon corrosion resistance. However, after about 30 minutes to one hour at room temperature, the solution treated alloy may begin to spontaneously increase

(A)

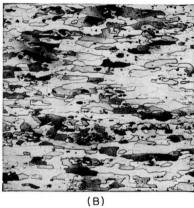

(B)

Fig. 12–4. A, annealed aluminum-copper alloy prior to heat treatment. Particles of $CuAl_2$ present. B, after solution heat treatment and quench. $CuAl_2$ dissolved and in solution. 100K.

caused by the gradual precipitation of the $CuAl_2$ phase in very fine submicroscopic particles. This precipitation takes place at the aluminum grain boundaries and along crystallographic planes within the aluminum crystals. The very fine particles of $CuAl_2$ precipitate act as keys and build up resistance to slip, thereby greatly increasing the strength of the alloy. After the solution heat treatment, the strength of the 2017-O alloy is about 44,000 lbs per sq in., with an elongation of 20 percent in 2 in. Upon aging at room temperature the strength increases rapidly within the first hour, and after 4 to 5 days the strength has increased to approximately 62,000 lbs per sq in. A very singular fact is that the ductility, as measured by elongation, does not change during the aging period, but remains about 20 percent in 2 in. This treatment produces an alloy that is almost as strong and as ductile as structural steel, with about one-third the weight.

Instead of aging the quenched alloy at room temperature in order to obtain the precipitation hardening, the alloy may be caused to precipitate the $CuAl_2$ phase more rapidly by heating somewhat above room temperature. However, rapid precipitation does not develop as high a tensile strength. Heating to above room temperature, say 375°F, is called *artificial aging*. The precipita-

in hardness and strength. This phenomenon is known as *age hardening* or *precipitation hardening*.

Precipitation Hardening causes the increase of strength that occurs with the passage of time at room temperature in a solution-heat-treated alloy. It is thought to be

395

Table 12–8. Recommended heat treatments for wrought aluminum alloys.

Alloy	Annealing Treatment			Solution Heat Treatment[1][2]		Precipitation Heat Treatment		
	Metal Temperature (°F)	Approx. Time of Heating (hours)	Temper Designation	Metal Temperature (°F)[3]	Temper Designation[4]	Metal Temperature (°F)[3]	Approx. Time of Heating (hours)[5]	Temper Designation[4]
1100	650	(6)	-O	...	...	...	...	...
2011	775[7]	2-3	-O	950	-T4	320	12-16	-T6
2014[8]	775[7]	2-3	-O	940	-T4[14]	340	8-12	-T6[9]
2017	775[7]	2-3	-O	940	-T4	...	...	...
2117	775[7]	2-3	-O	940	-T4	...	...	...
2018	775[7]	2-3	-O	950	-T4	340	8-12	-T61
2218	775[7]	2-3	-O	950	-T4	460	5-8	-T72
2024[8]	775[7]	2-3	-O	920	-T4[14]	375[10] / 375[10]	11-13[15] / 7-9	-T81 / -T86
2025	775[7]	2-3	-O	960	-T4	340	8-12	-T6
3003	775	(6)	-O	...	...	...	...	...
3004	650	(6)	-O	...	...	...	...	...
4032	775[7]	2-3	-O	950	-T4	340	8-12	-T6
4043	650	(6)	-O	...	...	...	...	...
5005	650	(6)	-O	...	...	...	...	...
5050	650	(6)	-O	...	...	...	...	...
5052	650	(6)	-O	...	...	...	...	...
5652	650	(6)	-O	...	...	...	...	...
5154	650	(6)	-O	...	...	...	...	...
5254	650	(6)	-O	...	...	...	...	...
5454	650	(6)	-O	...	...	...	...	...
5056	650	(6)	-O	...	...	...	...	...
5456	650	(6)	-O	...	...	...	...	...
5457	650	(6)	-O	...	...	...	...	...
5557	650	(6)	-O	...	...	...	...	...
5083	650	(6)	-O	...	...	...	...	...
5086	650	(6)	-O	...	...	...	...	...
6151	775[7]	2-3	-O	960	-T4	340	8-12	-T6
6053	775[7]	2-3	-O	970 / 940	-T4 / -T41	450 / 350 / 350	1-2 / 6-8 / 6-10	-T5 / -T6 / -T61

Table 12-8. (cont.) Recommended heat treatments for wrought aluminum alloys.

Alloy	Temp, °F	Time, hr	Temper	Temp, °F	Temper	Temp, °F	Time, hr	Temper
6061	775[7]	2-3	-0	970	-T4[14]	{320 / 350}	{16-20 / 6-10}	-T6
6063	775[7]	2-3	-0	....	....	450 / 365 / 350	1-2 / 4-6 / 6-8	-T5 / -T6
6463	775[7]	2-3	-0	....	....	450 / 365 / 350	1-2 / 4-6 / 6-8	-T5 / -T6
6066	775[7]	2-3	-0	970	-T4	{320 / 350}	{16-20 / 6-10}	-T6
7001	775[11]	2-3	-0	870[12]	-W	250[13]	24-28	-T6
7039	775	2(min.)[8]	-0	750-850	-T4	[18] / [18]	[18] / [18]	-T61[18] / -T64[18]
7072	650	[6]	-0					
7075[8]	775[11]	2-3	-0	870[12]	-W	250[13]	24-28	-T6
7277			....	890	-W	[17]	[17]	-T6
7178[8]	775[7]	2-3	-0	870[12]	-W	250	24-28	-T6
7079	775[7]	2-3	-0	830	-W	250	24-28	-T6

(1) The time of heating varies with the product, the type of furnace and the size of load. For sheet heat-treated in a bath of molten salt, the time may range from 10 minutes for thin material to 60 minutes for thick material. Time of several hours may be required in air furnaces because the metal comes to temperature less rapidly. A minimum of 4 hours is suggested for average forgings. When heat-treating in air furnaces, it may be desirable to use a protective compound to prevent oxidation along the grain boundaries in certain alloys.

(2) The material should be quenched from the solution heat-treating temperature as rapidly as possible and with a minimum delay after removal from the furnace. Quenching in cold water is preferred although less drastic chilling (hot or boiling water, air blast) is sometimes employed for bulky sections, such as forgings, to minimize quenching stresses.

(3) The temperature specified should be attained by all portions of the load as rapidly as possible and should be maintained, with as little variation as possible, during the recommended time at temperature. Furnaces capable of maintaining the temperature well within plus or minus 10 F of that desired are readily available.

(4) These designations apply to material heat-treated by the producer. (See notes 9 and 14 for exceptions.) A different designation may apply to material heat-treated by the user.

(5) The rate of cooling from the precipitation heat treatment is unimportant but should not be unduly slow.

(6) Time in the furnace need not be longer than is necessary to bring all parts of the load to the annealing temperature. Cooling rate is unimportant.

(7) This treatment is intended to remove the effect of heat treatment and includes cooling at a rate of about 50 F per hour from the annealing temperature to 500 F. The rate of subsequent cooling is unimportant. The treatment recommended for 1100 can be used to remove the effects of cold work or partially to remove the effect of heat treatment if a completely annealed material is not required.

(8) Alclad sheet is heat treated under the same conditions as the core alloy but the shortest heat treatment time consistent with securing the required properties should be used, and repeated reheat treatment should be avoided. Prolonged heating, or repeated reheat treatments cause diffusion of alloying elements into the coating and impair the resistance to corrosion.

(9) For extrusions this treatment also applies to temper -T62.

(10) Cold working subsequent to the solution heat treatment and prior to the precipitation treatment is necessary to secure the required properties.

(11) Should be followed by heating for about 6 hours at about 450 F if material is to be stored for an extended period of time before use.

(12) Sheet may also be heat-treated at higher temperatures (up to 925 F) if desired.

(13) Two-stage treatments comprising 4 to 6 hours at 210 F followed by 8 to 10 hours at 315 F or 2 to 4 hours at 250 F followed by 2½ to 3½ hours at 325 F are recommended for sheet and cold drawn wire.

(14) For some products the correct temper designation is -T42 when material is heat-treated by the user.

(15) For extrusions, a time of 6 hours should be used.

(16) Two-stage treatment comprising 5 days of room temperature, and then 48 hours at 230-250 F.

(17) Two-stage treatment comprising 4 hours at 210 F, and then 8 hours at 315 F.

(18) Thermal treatments for artificially aging 7039 to the -T61 and -T64 tempers are patented processes. For information on these processes, contact Kaiser Aluminum & Chemical Corporation.

* Standards for Aluminum Mill Products," The Aluminum Association. To locate a specific alloy, see basic alloy group, as identified by first digit. Then note sequence of last two digits, which identify the specific alloy or its aluminum purity. The second digit indicates a modification of an alloy whose other three digits are identical to one or more others in the same group.

tion hardening that takes place at room temperature may be prevented by cooling solution heat treated alloys to below 0°F, and holding them at this low temperature. The heat treatments for several precipitation hardening aluminum alloys are given in Table 12–8.

12.13 Protective and Decorative Coatings.

Aluminum forms its own natural protective oxide coat to resist corrosion, but the alloys of aluminum are less effectively protected than the pure metal. Consequently, many of the aluminum alloys are specially treated to increase their corrosion resistance. Some treatments can be also coupled with decorative effects to enhance sales appeal. Two widely used methods for improving corrosion resistance for aluminum alloys are through the use of *Alclad* and *anodizing* treatments.

The *Alclad* treatment consists of coating an aluminum alloy ingot with high purity aluminum and reducing the duplex by rolling and shaping to the desired dimensions.

The Alclad coat shown in Fig. 12–5 has a cladding thickness of between 1½ and 15 percent per side of the composite thickness. The bond between the high purity aluminum and the alloy is extremely good. The coating offers protection even at sheared edges because of the electronegative electrode potential. Aircraft sheets, for example, made from

Fig. 12–5. Photomicrograph of Alclad 2-24-T6 aluminum sheet. 125X. (Research Lab., Aluminum Co. of America)

strong wrought alloys such as 2017 and 2024 are coated on both sides, although some alloys may be obtained with cladding on one side only.

Anodizing treatments, which were discussed briefly in Chapter 8, consist of producing an oxide coating ranging in thickness from .0001 to .0005″. The oxide coatings are produced by either an oxidizing chemical solution such as sulfuric acid or chromic acid, or by making the metal surface anodic in a suitable electrolyte whereby nascent oxygen liberated by the passage of current reacts chemically with the metal surface to form this coating.

The anodized surface is hard, abrasion resistant, and corrosion resistant, but somewhat porous, thus producing a suitable surface for paints or organic dyes. Anodized sur-

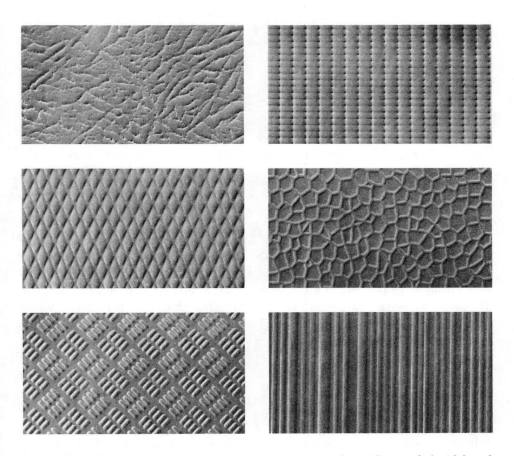

Fig. 12–6. Embossed patterns in aluminum decorative panels for appliances, industrial, and transportation equipment.

faces which have been dyed are usually sealed by immersion in hot water containing nickel acetate. Other protective coating methods such as electroplating, metallizing, porcelainizing, painting, and vinyl coating are among the ones commonly used for finishing aluminum and its alloys.

Fig. 12–6 shows several attractive embossed patterns which have many uses in modern appliances, door panels, industrial packaging, automobiles, buses and aircraft.

399

Magnesium

12.14 Properties of Magnesium.

Magnesium, with a specific gravity of 1.74 has long been recognized as the lightest structural metal. The chief source of magnesium is from sea water which contains approximately 9 billion *pounds* of magnesium per cubic mile. Extraction of magnesium from sea water is by electrolytic reduction of magnesium chloride. In the annealed condition wrought magnesium has a tensile strength of about 27,000 psi and an elongation of 15 percent in 2 inches. It can be work hardened to a strength of about 37,000 psi, but is cold formed with difficulty. The melting point of magnesium is 1202°F and the low heat content (18.8 btu/cu-in.) permits high casting rates. At 650°F magnesium is plastic enough to be hot worked by either rolling, forging, or extrusion. The metal, with its hexagonal close packed structure, work hardens rapidly upon cold rolling with slip occurring along its basal planes. It requires frequent annealing to maintain its plastic and workable nature. Above 400°F additional slip planes become active and plasticity is improved. The desirable low specific gravity of magnesium is offset somewhat by its lack of stiffness, and low resistance to corrosion, particularly in marine atmospheres.

The modulus of elasticity of magnesium is only 6.5×10^6 psi, compared with 10.3×10^6 and 29.5×10^6 for aluminum and steel respectively. Improvements in stiffness and strength with little increase in weight can be obtained through small increases in thickness, since these properties are proportional to the cube or square of section thickness.

12.15 Alloying Elements in Magnesium.

Although the mechanical properties of magnesium are relatively low, additions of alloying elements such as aluminum, zinc, manganese, and for special purposes, tin, cerium, thorium, beryllium, zirconium, and lithium, greatly improve these properties. Copper, iron, and nickel are considered impurities and must be kept to a minimum to insure the best corrosion resistance. Some of the most common types of magnesium alloy combinations are listed in Tables 12–9 and 12–10 along with their properties and designations.

Aluminum in amounts ranging from 3 to 10 percent is the principal alloying element in most magnesium alloys. It is used to improve strength, hardness, and castability. Magnesium will dissolve a maximum of 12 percent aluminum at 817°F as seen in the magnesium-aluminum diagram, Fig. 12–7. This solubility diminishes to about 2 percent at room temperature. This change in solid solubility with temperature makes it possible to subject these alloys to

Table 12–9. Composition and properties of magnesium casting alloys.

ALLOY AND TEMPER	Al	Mn	Zn	Zr	RARE EARTHS	Th	YIELD 2% OFFSET	ULTIMATE	ELONGATION % IN 2 IN.	IMPACT STRENGTH UNNOTCHED, PER LBS.
SAND AND PERMANENT MOLD CASTINGS										
AZ63A–T4	6.0	0.23	3.0	--	--	--	14	40	12	--
AZ81A–T4	7.6	0.13	0.7	--	--	--	14	40	12	13–20
AZ91C–T4	8.7	0.20	0.7	--	--	--	14	40	11	5–7
AZ91C–T6	8.7	0.20	0.7	--	--	--	19	40	5	--
AZ92A–T6	9.0	0.15	2.0	--	--	--	23	40	2	--
EK35A–T5	--	--	2.7	0.7	3	--	16	23	3	--
HK31A–T6	--	--	--	0.7	--	3	16	30	3	--
HZ32A–T5	--	--	2.5	0.7	--	3	14	29	7	--
DIE CASTINGS										
AZ91A–F	9.0	0.20	0.6	--	--	--	22	33	3	--
AZ91B–F	9.0	0.20	0.6	--	--	--	22	33	3	--

TEMPER DESIGNATIONS:

O = FULLY ANNEALED
F = AS EXTRUDED
T4 = SOLUTION HEAT TREATED
T5 = PRECIPITATION HARDENED
T6 = FULLY HEAT TREATED

(1) BALANCE MAGNESIUM

precipitation hardening treatments comparable to those given to aluminum alloys. The solution heat treatment usually requires heating to 760-780°F for 16-18 hours followed by water quenching. The aging or precipitation treatment is carried out at 350-360°F for 4-18 hours. The precipitation hardening results in increased strength and hardness but lower ductility and toughness.

Zinc in amounts of up to about 3 percent is used in combination with aluminum to increase resistance to salt water corrosion, to offset harmful effects of iron and copper impurities, and to improve casting properties.

Manganese has very limited solubility in magnesium. Less than about 0.50 percent is used in magnesium-aluminum and magnesium-aluminum-zinc alloys for the purpose of improving corrosion resistance and weldability without impairing strength properties. Up to 1.20 percent manganese is used in a binary magnesium-manganese alloy to promote weldability and hot forming characteristics, but with some sacrifice of strength.

Silicon which is not soluble in

Table 12–10. Composition and properties of wrought magnesium alloys.

ALLOY AND TEMPER	NOMINAL COMPOSITION, PERCENT (1)						STRENGTH, KSI		ELONGATION % IN 2 IN.	SHEAR STRENGTH KSI	BRINELL HARDNESS NO.
	Al	Mn	Zn	Zr	RARE EARTHS	Th	YIELD 2% OFFSET	ULTIMATE			
EXTRUSIONS, BAR, ROD, SHAPES											
AZ31V-F	3.0	0.45	1.0	--	--	--	29	39	15	19	51
AZ31C-F	3.0	0.45	1.0	--	--	--	29	38	15	19	49
AZ61A-F	6.5	0.30	1.0	--	--	--	33	45	16	23	60
AZ80A-T5	8.5	0.25	0.5	--	--	--	40	55	7	24	82
AZ80A-F	8.5	0.25	0.5	--	--	--	36	49	11	22	60
HM31A-T5	--	1.2	--	--	--	3.0	39	44	10	27	--
ZK60A-T5	--	--	5.7	0.55	--	--	44	53	11	26	82
AZ60A-F	--	--	5.7	0.55	--	--	38	49	14	24	75
SHEET AND PLATE											
AZ31B-H24	3	0.45	1.0	--	--	--	32	42	15	29	--
AZ31B-0	3	0.45	1.0	--	--	--	22	37	21	26	--
HK31A-H24	--	--	--	0.7	--	3	29	37	8	26	--
HK31A-T6	--	--	--	0.7	--	3	21	37	14	--	--
AZ31B-F(2)	3	0.45	1.0	--	--	--	19	35	12	19	--
AZ31C-F(3)	3	0.45	1.0	--	--	--	19	35	14	19	49

TEMPER. DESIGNATIONS:

O = FULLY ANNEALED
F = AS EXTRUDED
T5 = PRECIPITATION HARDENED
T6 = FULLY HEAT TREATED
H24 = STRAIN HARDENED THEN PARTIALLY ANNEALED

(1) BALANCE MAGNESIUM
(2) TOOLING PLATE
(3) TREAD PLATE

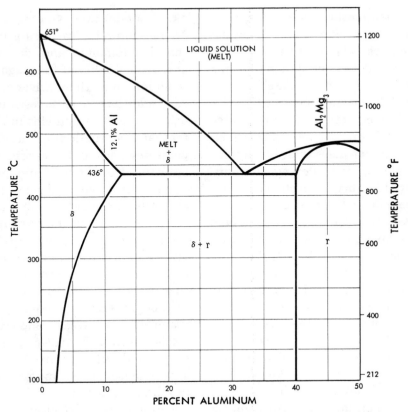

Fig. 12–7. Magnesium end of the magnesium-aluminum alloy diagram.

magnesium, forms a compound Mg₂Si, which increases hardness of the alloy. Additions are usually kept below 0.30 percent to avoid extreme brittleness.

Tin additions of 5 percent to magnesium-aluminum-manganese alloys promote hammer forging properties.

Thorium-Zirconium-and Thorium-Manganese additions to magnesium preserves useable strength properties at elevated temperature. Rare earth additions with zinc and silver produce special properties in casting alloys. Magnesium-lithium alloys with a specific gravity of 1.35 are the lightest commercial magnesium alloys.

The magnesium alloys have many desirable characteristics and are second in importance to aluminum alloys as light structural metals.

12.16 Magnesium Alloy Designations. Magnesium alloys may be classed as (1) casting and (2) wrought alloys. The designation for alloys in each of these classifications follows the recommendations for the American Society for Testing Materials, B275-61. These designations consist of not more than two letters representing the alloying elements followed by their respective percentage rounded off to the nearest whole number. The letters used to represent alloying elements are listed as follows:

A — Aluminum	M — Manganese
B — Bismuth	N — Nickel
C — Copper	P — Lead
D — Cadmium	Q — Silver
E — Rare earths	R — Chromium
F — Iron	S — Silicon
G — Magnesium	T — Tin
H — Thorium	Y — Antimony
K — Zirconium	Z — Zinc
L — Beryllium	

As an example of the above designation for magnesium alloy AZ92A, the "A" represents aluminum and the "Z" zinc. The aluminum percentage of "9" indicates that it is somewhere between 8.6 and 9.4 and the "2" indicates that the zinc composition lies somewhere between 1.5 and 2.5 percent. The final letter "A" signifies that this is the first alloy whose composition received assignment of this AZ92 designation.

12.17 Magnesium Casting Alloys. Magnesium alloy castings generally contain aluminum and zinc to increase the fluidity of the magnesium to fill in thin sections. Silver and calcium additions increase resistance to oxidation and cerium aids in the retention of strength at elevated temperatures. Most of the casting alloys are subject to age hardening with a number of them achieving tensile strengths as high as 40,000 psi.

The composition and properties of casting alloys are given in Table 12–9. Sand, permanent mold, and die casting alloys have been developed to give the best properties from each method of casting. A photomicrograph of a typical magnesium cast alloy is shown in Fig. 12–8. It should be pointed out that magnesium alloy castings require some special care in foundry practice in order to produce the most satisfactory castings. A protective flux of magnesium chloride must be used during melting and pouring to prevent flash fires; a specially treated molding sand must be used to prevent reaction between the sand and metal, and owing to the large shrinkage ratio, large gates and risers must be used.

Magnesium casting alloys are used in the manufacture of household appliances, foundry equipment, portable power tools, aircraft components, high speed reciprocating and rotating machinery, and numerous other applications.

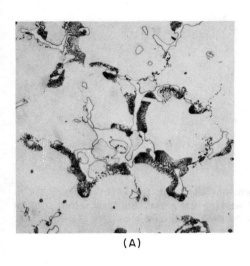

(A)

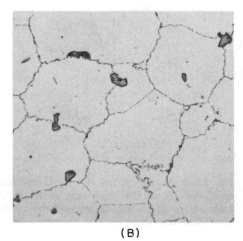

(B)

(C)

Fig. 12–8. A, microstructure of sand cast magnesium alloy with 9% aluminum and 2% zinc. 250X. B, after solution heat treatment. Note distinct grain boundaries after treatment. 250X. C, after solution heat treatment and artificial aging. Aging causes compound to precipitate causing dark areas. 250X. (Dow Chemical Co.)

12.18 Magnesium Wrought Alloys. Alloys of magnesium containing aluminum, manganese, zinc, zirconium, and rare earths (see Table 12–10), are now available in many of the standard wrought structural shapes such as bars, rods, tubing, sheets, and angles. Although these alloys do not lend themselves readily to cold forming or forging methods they may be satisfactorily shaped by hot methods such as extrusion, press forging, bending, and drawing. They work harden so rapidly upon cold forming that only small amounts of cold work can be carried out without annealing. However, at a temperature range of 550° to 600°F most of

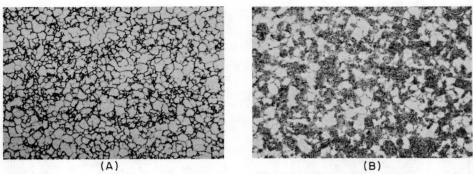

(A)   (B)

Fig. 12–9. A, microstructure of extruded aluminum alloy with 8% aluminum and 0.5% zinc. Grain is smaller than cast as result of working the metal. 250X. B, after artificial aging. Note dark areas of precipitation similar to aged sand cast alloy (Fig. 12–8). 250X. (Dow Chemical Co.)

the alloys may be successfully shaped by the usual mechanical methods. If only slight or light working is required to shape a part, temperatures as low as 400°F may be used. In Fig. 12–9 is a photomicrograph of an extruded magnesium alloy showing the fine and uniform grain size. Mechanical properties of magnesium alloys are greatly improved when grain size can be maintained at about 0.0003 in. For most of the wrought alloys the maximum properties are obtained in the extruded form. Wrought alloys containing 9 percent or more of aluminum can be age hardened, but very little heat treating is done on wrought alloys.

12.19 General Characteristics of Magnesium Alloys. The general characteristics of magnesium alloys can be summarized under seven major topics.

Corrosion Resistance was seriously questioned for a long time. The chemical stability of magnesium was in doubt due to its affinity for oxygen and its early use in pyrotechnics and as a chemical reagent. Unfortunately, many of the early tests on magnesium involved subjecting the material to salt water solutions. These tests had little significance for normal uses of the metal. Corrosion resistance to ordinary outdoor atmosphere has been substantially improved through use of higher purity magnesium and by modifications of the foundry fluxing treatments. Although commercial alloys are reasonably stable in inland atmosphere, protective treatment such as anodizing and painting are frequently employed. Sea coast locations involving contact with salt-bearing atmospheres are definitely corrosive to magnesium alloys and protective chemical treatment and painting must be employed as preventatives.

Formability of magnesium alloys is good. They can be readily fabricated by most metal working processes. Workability at elevated temperatures is excellent, although at room temperature only mild deformations such as bends around generous radii can be made. When the temperature approaches 600°F deep draws with up to 70 percent reduction in blank diameter can be made in a single draw.

Notch sensitivity is shown in stress concentrations such as occur at sharp corners, abrupt section changes, deep machine marks, sharp discontinuities, and some weldments on heavy or rigid structures. Particular care should be taken in the design of stressed parts to avoid abrupt changes in cross section and to provide ample fillets and smooth surfaces.

Modulus of Elasticity for magnesium is 6,500,000 psi as compared with 10,300,000 psi for aluminum and 29,500,000 psi for steel. Thus a magnesium bar will deflect over four times as much as steel and 50 percent more than aluminum. However, since the relative densities are 1.74, 2.71, and 7.8, equal weights of the three metals would show them to be of nearly equal stiffness. The extreme lightness of magnesium, therefore, compensates for its lack of rigidity. The added bulk or thickness of magnesium alloys over aluminum alloys of equal strength may eliminate the need for stiffening ribs thus simplifying design and lowering fabrication and assembly costs.

Machinability of magnesium allows greater speeds than any other structural material. Many magnesium alloys have a machinability rating of at least five times that of free machining brass. Low cutting pressures and high thermal conductivity, with rapid dissipation of heat produces long tool life, high dimensional accuracy, and excellent surface finish. Tools must be kept sharp and high rake angles are desirable. Magnesium chips and dust, if properly cared for and not allowed to accumulate, are not a serious fire hazard.

Weldability of magnesium alloys is good. Joints have high efficiency. Common welding methods used for joining magnesium include gas tungsten-arc, gas metal-arc, and electric resistance processes. In addition brazing, soldering, adhesive bonding, bolting, and riveting may be used for assembly.

Heat Treatment followed by water quenching of cast magnesium alloys may cause cracking if the quenching temperature is too high or if the water is too cold. Solution treatment and precipitation hardening temperature and time is given in Table 12–11.

407

Table 12–11. Heat treatments for magnesium casting alloys.

TYPE OF TREATMENT	ALLOY	CONDITIONS		TYPICAL TENSILE PROPERTIES			
		TIME (HR)	TEMP.(°F)	AS CAST		HEAT TREATED	
				U.T.S. KSI	%EL. ON Q	U.T.S. KSI	%EL. ON Q
ANNEALING	EZ33A-T5	2	650	22.4	4	22.4	4
	HZ32A-T5	6	420	30	5.5	33.6	7
PRECIPITATION	ZK51	10-16	340-350	33.6	10	38	7
	ZE41	24	480	25	5	31.4	4
	ZH62	{ 2 { 10-16	625 340-350	38	17	30	10
SOLUTION	AZ81	18	770	22.4	4	35.8	13
FULL, WITH AIR COOLING	AZ91	{ 8 { 16 { 10	720-735 770-790 390	22.4	2	33.6	7
	AZ92	{ 18 { 16	760 340	24	--	40	--
	AZ63	{ 12 { 5	725 425	29	--	40	--
	ZK61	{ 2 { 58	930 265	38	--	43	--

Titanium

12.20 Properties and Uses of Titanium. Titanium is a relatively new addition to the family of engineering metals. Initially, it was used in aircraft, aerospace, and military applications and was a very rare, expensive metal. The U.S. Air Force claims that a 1 lb weight reduction in jet engine weight results in an 8-10 lb reduction in air frame weight. Thus, the use of titanium in place of stainless steel saves 500 lbs in each of the eight jet engines of a heavy bomber and makes possible a weight reduction of 40,000 lbs. Titanium is the main structural material for supersonic transports with as much as 130,000 lbs. required for each aircraft.

Today, titanium is used in many structural, chemical processing, transportation, sporting goods, surgical implants, fasteners, and corrosion resisting applications. Its price has decreased by 70 percent over a recent ten year period with processing and fabrication costs now only slightly higher than for conventional stainless steels.

Titanium's most important properties are its superior strength-to-weight ratio at temperatures ranging up to 800°F. It has strength comparable to stainless steel but with only half the weight. Ultimate strength value for titanium-base alloys start at about 30,000 psi and reach 200,-000 psi. Another important characteristic is its exceptional resistance to corrosion and erosion in marine environments. In fact, titanium is the only known structural metal having corrosion-fatigue properties in salt water that are practically identical to those in air. Titanium is virtually immune to atmospheric corrosion and is highly resistant to metal salts, chlorides, hydroxides, nitric and chromic acids, and organic acids such as acetic and formic.

Titanium weighs 0.16 lbs/cu in., is non-magnetic, and has excellent ductility. It has useful properties over a wide range of temperatures (-423 to 1000°F). The modulus of elasticity in tension is 16×10^6 psi which is about 55 percent that of steel. Titanium generally has excellent welding characteristics with the advantage of equal corrosion resistance in the weld deposit, heat affected zone, and parent metal. The endurance limit of titanium is well above that for heat treated alloy steels and other non-ferrous metals; it varies

between $\frac{1}{3}$ and $\frac{1}{2}$ of its high tensile strength. However, titanium appears to be somewhat notch sensitive and special precautions should be taken to prevent sharp notches and marks.

12.21 Commercially Pure Titanium. Titanium is a silver-gray metal whose density is 4.5 gm/cc with a melting point of 3140°F. Above 1200°F it readily absorbs oxygen and nitrogen from the air embrittling the metal. Titanium is allotropic in nature and may exist in either the hexagonal close packed (alpha) or body centered cubic (beta) structure. The alpha phase is stable from room temperature to approximately 1625°F at which time transformation to the beta phase occurs.

Titanium with 20 specific commercial alloy compositions provides a wide range of physical and mechanical properties. Alloying elements are used to effect the transformation temperature of titanium in the following ways.

Raise the transformation temperature by certain alloying additions, notably oxygen, nitrogen, and aluminum. They tend to stabilize the alpha phase thus raising the temperature at which transformation to the beta phase will occur. The transformation temperature is referred to as the beta transus temperature. Oxygen and nitrogen also increase hardness and strength but reduce ductility.

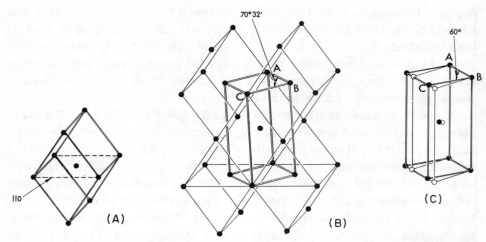

Fig. 12–10. Diffusionless transformation of B titanium to A titanium. A, B.C.C. cell of B phase. B, an array of B.C.C. cells shown with potential C.P.H. cell in bold lines. C, final transformation by shear into C.P.H. cell of A titanium. (By permission: *Structure and Properties of Alloys*, McGraw-Hill Book Co.)

Lower the transformation temperature by most alloying additions including chromium, columbium, copper, iron, manganese, molybdenum, tantalum, and vanadium. These act to stabilize the beta phase by lowering the transformation temperature. Columbium and tantalum also improve strength and prevent embrittlement from titanium-aluminum compounds.

Little change on transformation temperature by alloying elements such as tin and zirconium. These are soluble in both the alpha and beta phase and act as strengtheners of the alpha phase at room temperature.

Upon cooling the B.C.C. beta (β)

alloy to just below the transformation temperature, a C.P.H. alpha (α) alloy slowly forms. At lower temperatures the transformation rate increases. Even by very rapid quenching it is not possible to retain the beta (β) phase. The explanation for this fact is readily explained if it is assumed that the transformation is diffusionless and does not require large atom movement. The mechanism of diffusionless transformation shown in Fig. 12–10 was deduced by W. G. Burgers. One B.C.C. cell is shown at A standing on edge with the (110) plane horizontal. In B is shown an assembly of several of these cells with the partially transformed C.P.H. cell shown by bold lines. The angle ABC which is 70° 32' must become 60° in order to complete the transformation to a hexagonal cell. The mechanism for reducing angle ABC is shown at C in which shear movements occur. An additional displacement of the center atom followed by slight expansion and contraction of cell edges produces the close packed hexagonal structure of the alpha phase.

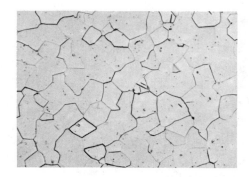

Fig. 12–11. Titanium alpha alloy in annealed condition. 500X. (Reactive Metals, Inc.)

12.22 Titanium Alloys. Titanium alloys can be conveniently classified into three groups by means of the crystal structure which is stable at room temperature. Three groups are: alpha alloys, beta alloys, and alpha-beta alloys.

Alpha alloys are single phase alloys in which the room temperature stable phase is of a hexagonal close packed structure. This is the same microstructure which typifies pure titanium and is shown in Fig. 12–11. These alloys usually contain aluminum which acts as a stabilizer, and which has a strong solution hardening effect. The additions further strengthen the phase. Alpha alloys are characterized by excellent strength characteristics and oxidation resistance in the 600°-1100°F range. They have good ductility and good weldability. Alpha alloys are single phase alloys and cannot be hardened by heat treatment. These alloys may be hardened by cold working but may develop directional properties.

Beta alloys have a room temperature stable phase of the body centered cubic structure. A high percentage of beta stabilizing elements such as vanadium, chromium, and aluminum results in a microstructure that is substantially a beta phase, Fig. 12–12. Alloys which contain both iron and chromium in small percentages form metastable alloys which are susceptible to heat treatment to high strength levels. Beta alloys have inherent superiority with their B.C.C. structure over the C.P.H. alpha alloy structure with regard to bending and cold forming operations. Chemical composition and properties of typical titanium alloys are included in Table 12–12.

Alpha-beta alloys are formed by the addition of controlled amounts

411

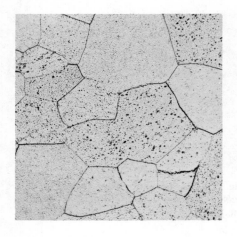

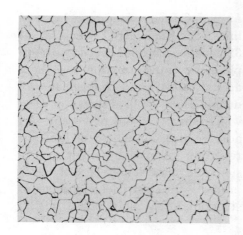

Fig. 12–12. Titanium beta alloy in annealed condition. 500X. (Reactive Metals, Inc.)

of beta stabilizing elements to titanium which causes the beta phase to persist below the beta transus temperature, down to room temperature. This results in an alpha-beta two phase system of body centered and close-packed hexagonal crystal structures. Such two phase alloys can be significantly strengthened by quenching from some temperature in the alpha-beta range followed by an aging cycle at somewhat lower temperature. Quenching supresses the beta phase, forming a metastable structure. During the aging cycle fine alpha particles are precipitated

Table 12–12. Composition and properties of typical titanium alloys.

COMMERCIALLY PURE TITANIUM GRADES

COMPOSITION OF GRADE	TEMP.	TYPICAL TENSILE PROPERTIES				MECHANICAL PROPERTIES, ROOM TEMP.	HEAT TREATMENT
		U.T.S.	Y.S.	%El.	%R$_A$		
GRADE 30 (99.6%)	R.T.	40,000	30,000	25	35	BEND RADIUS: 1.5–2.0 THICKNESS	BETA TRANSUS: 1630°F.
	600°F.	26,000	15,000	32	70/80	HARDNESS: R$_B$70	ANNEALING, FULL 1300°F/2 HRS; AIR COOL
	800°F	20,000	13,000	26	70/80	WELDABILITY: GOOD	STRESS RELIEF: 1000–1100/30 MIN; AIR COOL
	1000°F	---	---	---	---	CHARPY IMPACTS: ---	FORGING: 1600–1700°F.
GRADE 40 (99.6%)	R.T.	50,000	40,000	22	35	BEND RADIUS: 2.0 X THICKNESS	BETA TRANSUS: 1675°F.
	600°F	30,000	18,000	37	70/80	HARDNESS R$_B$80	ANNEALING, FULL: 1300°F/2HR; AIR COOL
	800°F	26,000	15,000	25	70/80	WELDABILITY: GOOD	STRESS RELIEF: 1000–1300/30 MIN; AIR COOL
	1000°F	19,000	11,000	32	70/80	CHARPY IMPACT: 25–40 FT LBS	FORGING: 1600–1700°F.
GRADE 55 (99.6%)	R.T.	65,000	55,000	20	35	BEND RADIUS: 2.0 X THICKNESS	BETA TRANSUS: 1690°F.
	600°F	32,000	20,000	35	70/80	HARDNESS: R$_B$90	ANNEALING, FULL: 1300°F/2 HR.; AIR COOL
	800°F	29,000	17,000	18	70/80	WELDABILITY: GOOD	STRESS RELIEF: 1000–1100/30 MIN.; AIR COOL
	1000°F	22,000	13,000	33	70/80	CHARPY IMPACT: 20–40 FT LBS	FORGING: 1600–1700°F.
GRADE 70 (99.6%)	R.T.	80,000	70,000	15	35	BEND RADIUS: 2.5 X THICKNESS	BETA TRANSUS: 1740°F.
	600°F.	44,000	24,000	18	70/80	HARDNESS: R$_B$ 100	ANNEALING, FULL 1300°F/2 HRS. AIR COOL
	800°F	34,000	21,000	22	70/80.	WELDABILITY: GOOD	STRESS RELIEF: 1000–1100/30 MIN; AIR COOL
	1000°F.	27,000	16,000	29	70/80	CHARPY IMPACTS: ---	FORGING: 1650–1700°F.

Table 12-12. (cont.) Composition and properties of typical titanium alloys.

ALPHA BETA TITANIUM ALLOYS

COMPOSITION OF GRADE	TYPICAL TENSILE PROPERTIES					MECHANICAL PROPERTIES, ROOM TEMP.	HEAT TREATMENT
	TEMP.	U.T.S.	Y.S.	%El.	%RA		
8Mn	R.T.	120,000	110,000	10	---	BEND RADIUS: 3 X THICKNESS	BETA TRANSUS: 1475°F.
	600°F.	104,000	82,000	18	---	HARDNESS: R_c 35	ANNEALING, FULL: 1300/1 HR.; FURN. COOL TO 1000°F., AIR COOL
	800°F.	87,000	66,000	20	---	WELDABILITY: NOT RECOMMENDED	STRESS RELIEF: 900-1100/30 MIN.- 2 HRS.; AIR COOL
	1000°F.	55,000	30,000	32	---	CHARPY IMPACT: ----	FORGING: NOT A FORGING ALLOY
7Al – 4Mo	R.T.	145,000	135,000	10	25	BEND RADIUS: ----	BETA TRANSUS: 1840°F.
	600°F.	130,000	110,000	15	40	HARDNESS: Rc38	ANNEALING, FULL: 1500/1 HR.; FURN. COOL TO 1050°; AIR COOL
	800°F.	126,000	106,000	16	45	WELDABILITY: NOT RECOMMENDED	STRESS RELIEF: 1300/1 HR. AIR COOL
	1000°F.	105,000	88,000	20	50	CHARPY IMPACT: 18 FT LBS	FORGING: 1800-1850°F.
6Al – 6V – 2Sn	R.T.	125,000	115,000	10	---	BEND RADIUS: 3.5 X THICKNESS	BETA TRANSUS: 1755°F.
	600°F.	92,000	82,000	18	---	HARDNESS: R_c 32-38	ANNEALING, FULL: 1225/4 HRS.; FURN. COOL TO 1050; AIR COOL
	800°F.	84,000	74,000	19	---	WELDABILITY: LIMITED	STRESS RELIEF: 1000-1100 1 HR.; AIR COOL
	1000°F.	65,000	55,000	35	---	CHARPY IMPACT: ----	FORGING: 1750-1800°F.
1Al – 8V – 5Fe	R.T.	210,000	200,000	8	15	BEND RADIUS ----	BETA TRANSUS: 1525°F.
	600°F.	128,000	115,000	19	59°	HARDNESS: R_c 35-37	ANNEALING, FULL: 1250/4 HRS.; FURN. COOL TO 950°F.; AIR COOL
	800°F.	108,000	85,000	32	72	WELDABILITY: NOT RECOMMENDED	STRESS RELIEF: 1000-1100 1 HR.; AIR COOL
	1000°F.	---	---	--	---	CHARPY IMPACT: ----	FORGING: 1500

Table 12-12. (cont.) Composition and properties of typical titanium alloys.

COMPOSITION OF GRADE	TEMP.	TYPICAL TENSILE PROPERTIES				MECHANICAL PROPERTIES, ROOM TEMP.	HEAT TREATMENT
		U.T.S.	Y.S.	%El.	%RA		
ALPHA-BETA TITANIUM ALLOYS (CONT'D)							
6Al – 4 V	R.T.	130,000	120,000	10	25	BEND RADIUS: 4.5 X THICKNESS	BETA TRANSUS: 1820°F.
	600°F.	105,000	93,000	17	52	HARDNESS: R_c 36	ANNEALING, FULL: 1300–1525/1-2 HR. FURNACE COOL TO 1100; AIR COOL
	800°F.	97,000	85,000	18	53	WELDABILITY: FAIR	STRESS RELIEF: 900–1200/1-4 HRS. AIR COOL
	1000°F.	80,000	60,000	30	68	CHARPY IMPACT: 18 FT. LBS.	FORGE: 1800–1850°F.
ALPHA TITANIUM ALLOYS							
5 Al – 2.5 Sn	R.T.	120,000	115,000	10	15	BEND RADIUS: 4 X THICKNESS	BETA TRANSUS: 1900°F.
	600°F.	82,000	65,000	18	45	HARDNESS: R_c 36	ANNEALING, FULL 1325–1550/15 MIN.-4 HRS; AIR COOL
	800°F.	78,000	59,000	18	45	WELDABILITY: GOOD	STRESS RELIEF: 1000–1100 15 MIN. – 1 HR.; AIR COOL
	1000°F.	67,000	55,000	19	47	CHARPY IMPACT: 19 FT. LBS	FORGING: 1850 – 1900°F.
4Al – 2.5 Sn, EXTRA LOW INTERSTITIAL	R.T.	100,000	90,000	10	- -	BEND RADIUS: 4 X THICKNESS	BETA TRANSUS: 1900°F.
	-320°F.	180,000	168,000	16	- -	HARDNESS: R_c 33	ANNEALING, FULL: 1325–1550/15 MIN.-4 HRS;AIR COOL
	-423°F.	229,000	206,000	15	- -	WELDABILITY: GOOD CHARPY IMPACT: 19 FT LBS.	STRESS RELIEF: 1000–1200/15 MIN. 1 HR.; AIR COOL FORGING: 1850–1900°F.
7Al – 2Cb – 1Ta	R.T.	115,000	110,000	10	20	BEND RADIUS: 4 X THICKNESS HARDNESS: R_c 37 WELDABILITY: GOOD	ANNEALING, FULL: 1650/ 1HR
	600°F.	90,000	75,000	20	30		BETA TRANSUS: 1880°F.
	800°F.	83,000	68,000	25	40	CHARPY IMPACT: 28 FT LBS	STRESS RELIEF: 1000–1200/15 MIN. – 1 HR.; AIR COOL
	1000°F.	70,000	55,000	35	50		FORGING: 1850–1900°F.

Table 12-12. (cont.) Composition and properties of typical titanium alloys.

COMPOSITION OF GRADE	TEMP.	TYPICAL TENSILE PROPERTIES				MECHANICAL PROPERTIES, ROOM TEMP.	HEAT TREATMENT
		U.T.S.	Y.S.	%El.	%R_A		
ALPHA TITANIUM ALLOYS (CONT'D)							
7Al – 12Zr	R.T.	130,000	120,000	10	20	BEND RADIUS: ----	BETA TRANSUS: 1825°F.
	600°F.	110,000	90,000	28	40	HARDNESS: ----	ANNEALING, FULL: 1650/30 MIN.; AIR COOL
	800°F.	100,000	82,000	25	45	WELDABILITY: GOOD	STRESS RELIEF: 1000-1200/15 MIN-1 HR.; AIR COOL
	1000°F.	95,000	75,000	23	48	CHARPY IMPACT: ----	FORGING: 1850-1900°F.
BETA TITANIUM ALLOY							
13V – 11Cr – 3Al	R.T.	130,000	125,000	10	25	BEND RADIUS: 3 X THICKNESS	BETA TRANSUS: 1325°F.
	600°F.	118,000	108,000	20	--	HARDNESS: R_c 32-36	ANNEALING, FULL: ----
	800°F.	112,000	102,000	18	--	WELDABILITY: FAIR	STRESS RELIEF: ----
	1000°F.	100,000	75,000	35	--	CHARPY IMPACT: ----	FORGING: 1800-2050°F.

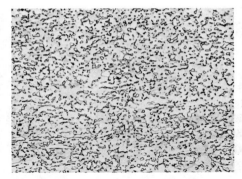

Fig. 12–13. Annealed alpha-beta titanium alloy. 500X. (Reactive Metals, Inc.)

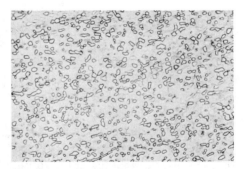

Fig. 12–14. Alpha-beta titanium alloy solution treated at 1500°F and water quenched. Structure is entirely beta phase. 300X.

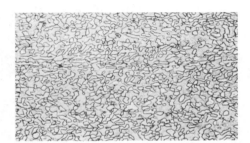

Fig. 12–15. Alloy in Fig. 12–14 after aging at 900°F for 24 hours. Microstructure is beta plus dark network of alpha phase.

from the metastable beta phase, thus distorting the lattice structure and hardening and strengthening the alloy. The aging time and temperature should be chosen to avoid overaging. While the properties of titanium alpha-beta alloys can be significantly improved by heat treatment and aging, the relative strengthening is much less than that obtained by heat treatment of steel and aluminum alloys. In Fig. 12–13 is shown an annealed alpha-beta titanium alloy while in 12–14 and 12–15 are shown solution treated and aged alloys.

The majority of alpha-beta alloys are weldable but with some sacrifice in joint ductility. A few alpha-beta alloys are now weldable by fusion welding although flash or spot welding may be used on some.

Titanium alloys are finding new uses daily in pressure vessels, honeycomb panels, missile castings, and other applications requiring high strength-to-weight ratio, outstanding corrosion resistance, and retention of strength at elevated temperatures.

Beryllium

12.23 Properties. Beryllium is a high strength, light weight metal which is finding increasing use as a structural material in aerospace vehicles. It has density of only 0.066 lb/cu in. which is very near that of

417

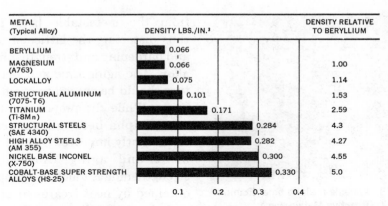

METAL (Typical Alloy)	DENSITY LBS./IN.³	DENSITY RELATIVE TO BERYLLIUM
BERYLLIUM	0.066	
MAGNESIUM (A763)	0.066	1.00
LOCKALLOY	0.075	1.14
STRUCTURAL ALUMINUM (7075-T6)	0.101	1.53
TITANIUM (Ti-8Mn)	0.171	2.59
STRUCTURAL STEELS (SAE 4340)	0.284	4.3
HIGH ALLOY STEELS (AM 355)	0.282	4.27
NICKEL BASE INCONEL (X-750)	0.300	4.55
COBALT-BASE SUPER STRENGTH ALLOYS (HS-25)	0.330	5.0

Fig. 12–16. Density of typical materials used in aerospace structural applications. (The Beryllium Corp.)

magnesium. In Fig. 12–16 is a comparison of the density of beryllium as compared with other aerospace alloys. With new hot pressing techniques parts have been produced with minimum strengths of 70,000 psi, while beryllium wire develops strengths of over 200,000 psi. The high strength of beryllium coupled with its light weight gives outstanding strength-to-weight ratios. Fig. 12–17 shows that a forged beryllium alloy had a strength of 128,000 psi which on a weight basis is equivalent to 555,000 psi in steel. Beryllium maintains its strength up to temperatures of about 1100°F.

The high elastic modulus of beryllium (44×10^6 psi) makes possible the design of light weight, thin members with high rigidity. A beryllium column in pure buckling applica-

tions will have greater load carrying capacity and will be lower in weight than any other metal of equal length and size. The elastic modulus of typical aerospace materials is shown in Fig. 12–18. It should be observed that the elastic modulus of beryllium is nearly three times that of titanium, four times that of aluminum, and over six times that of magnesium.

Other properties of beryllium which are of interest are its high specific heat (0.445 Btu/lb/°F. This property makes beryllium attractive for heat sinks, aircraft brakes, and re-entry heat shields. A pound of beryllium will absorb as much heat as 2 lb of aluminum or 5 lb of copper in changing temperature one degree.

An adherent, protective oxide film forms on the surface of beryllium

418

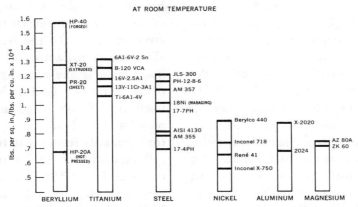

Fig. 12–17. Ultimate tensile strength to weight ratios of some structural materials. (The Beryllium Corp.)

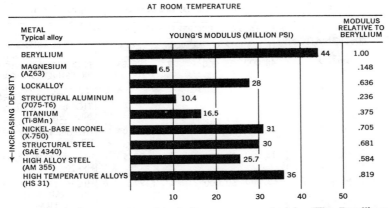

Fig. 12–18. Elastic moduli of typical aerospace materials. (The Beryllium Corp.)

providing excellent corrosion resistance in the atmosphere. It is also resistant to halogens, pure water, and oxygen-free liquid metals. Beryllium is the lightest available metal which will not vaporize appreciably in the vacuum of space.

Beryllium sheet has higher notch sensitivity than many structural materials at room temperature but above 300°F sensitivity to notches is greatly diminished. The coefficient of thermal expansion for beryllium is very near that of stainless steel, nickel, and cobalt alloys thus minimizing design problems.

419

12.24 Fabrication of Beryllium.

Beryllium (like magnesium, titanium, and zirconium) has a hexagonal lattice structure. At room temperature beryllium plastically deforms by slip on the basal or (0001) plane. At elevated temperatures slip may occur on a number of 1010 pyramidal planes. Mechanical twinning occurs on the 1012 planes. The plastic deformation system utilized depends upon the direction of strain. Very high elongation (140 percent) has been recorded in a single crystal of zone refined beryllium when stressed at 45° to the basal plane. However a tensile stress at 90° to the basal plane is likely to initiate basal plane cleavage, with brittle fracture resulting. Three disadvantages of beryllium are its high cost, potential toxic hazards from inhalation of beryllium dust from machining, and its tendency for brittleness at room temperature. High purity beryllium from electron beam zone refining has

Table 12–13. Beryllium products, properties and composition.

PRODUCT	DESIGNATION	MECHANICAL PROPERTIES			BERYLLIUM ASSAY.	DESCRIPTION
		U.T.S.	Y.S. 2%	% EL.		
BLOCK	HP – 20	48,000	35,000	2	98.0%	PRODUCED BY HOT PRESSING OF POWDER; MOST COMMONLY USED GRADE FOR INSTRUMENTATION, OPTICAL AND NUCLEAR USES.
	HP – 8	37,000	27,000	3	99.0%	HIGHER PURITY THAN HP–20
	HP – 12	38,000	30,000	3	98.5	HIGHER PURITY THAN HP–20
	HP – 40	65,000	60,000	1	92.0	HIGH STRENGTH AND DIMENSIONAL PROPERTIES
	C – 10	20,000	---	---	99.0	CAST, FINE GRAINED, HIGH PURITY
SHEET AND PLATE	PR – 2	77,000	55,000	9	98.00	MOST COMMONLY SPECIFIED SHEET, MODEST FORMABILITY AT 1350°F.
	IS – 2	45,000	31,100	---	99.0	HIGH PURITY SHEET; HIGH FORMABILITY AT 700°F.
EXTRUSIONS	XT – 20	77,000	40,000	7	98.0	MOST COMMONLY SPECIFIED; GOOD MECHANICAL PROPERTIES
	XT – 40	90,000	52,000	5	92.0	HIGH STRENGTH ALLOY
	LOCKALLOY	60,000	50,000	---	38% Al 62% Be	NEW BERYLLIUM-ALUMINUM ALLOY FOR UP TO 600-800°F. SERVICE

been found to be significantly more ductile than arc melted beryllium. Wrought beryllium is anisotropic and may show excellent ductility in one direction, but brittle behavior in another direction.

Machining operations can cause surface twinning and microcracking. Chemical etching away of about .001 inch of a previously machined surface removes incipient notches and microcracks.

Machining operations produce a discontinuous type chip similar to that formed in machining cast iron. This may be a potential health hazard unless proper ventilation is provided. Solid beryllium is completely non-hazardous although dust from machining, or dusts, fumes or vapors generated by any process, that can be inhaled can be toxic.

Beryllium can be forged, extruded, compacted from powdered metals, drawn, and rolled. Forming and forging is usually done in the 1200° to 1450°F range. Forged parts are stress relieved at 1300° to 1450°F. Beryllium parts may be satisfactorily joined by Tig and Mig welding processes.

Typical beryllium products along with composition and properties are shown in Table 12–13.

References

Abkowitz, Stanley, Burke, John J. and Hiltz, Ralph, Jr. *Titanium in Industry.* Princeton, N.J.: D. Van Nostrand Co. 1955.

Aluminum Co. of America. *Alcoa Aluminum Handbook.* Pittsburg, Pa., 1967.

American Society for Metals. *Magnesium.* Cleveland, O., 1961.

American Society for Testing and Materials. *Symposium on Titanium.* Philadelphia, 1957.

Beryllium Corp. *Beryllium Properties and Products.* Reading, Pa., 1965.

Lyman, Taylor, ed. *Metals Handbook, 1948 ed.* Cleveland, O.: American Society for Metals.

Reactive Metals, Inc. *Basic Design Facts About Titanium.* Niles, O., 1965.

Reynolds Metals Co. *Reynolds Aluminum Data Book.* Richmond, Va.

Von Zeerleder, A. *Technology of Light Metals.* New York: American Elsevier, 1959.

White, D. W., Jr. and Burke, J. W. *The Metal Beryllium.* Cleveland, O.; American Society for Metals, 1955.

Chapter
13

Copper, Nickel and Alloys

Copper

13.1 Copper. Where copper was first produced is not known. It was known to the Romans as the metal from Cyprus, from which the name is derived. Many of the ancient peoples used it. The Chaldeans had developed the art of working copper as early as 4500 B.C. The mines of Cyprus yielded copper to the ancient Egyptians and from Egypt the use of copper spread into Europe. In America copper was used by the pre-Columbian peoples and the Indians long before the arrival of the Europeans. The native copper was found on the earth's surface, distributed, no doubt, by glaciers; it is doubtful if the Indians did any mining of native copper, although pre-Columbians probably did.

Copper seems to have been used first for ornaments; later it was used for tools and arms. Discovery of the hardening effects of tin combined with copper to produce bronze, gave various peoples, in succession, important advantages in war. Where, when, and how brass was discovered is not known, but the early and long used method was by direct reduction from a mixture of copper, or some of its ores, and calamine (zinc carbonate). It was only in 1781 that Emerson invented the process of direct fusion of the metals to produce a copper-zinc alloy we call brass. In America, the first rolling of brass sheet was done in Waterbury, Connecticut, and Connecticut has continued to be the center of American brass manufacture.

Copper is used extensively as an alloying element for precious metals. Most of the gold and silver on the market is alloyed with copper. Copper is also added to some steels to improve corrosion resistance. When copper is added to aluminum an age hardenable alloy results. Babbitts (tin or lead base alloys) often contain copper. Nickel alloys containing copper exhibit outstanding corrosion resistance. Another extensive use of copper is in the electroplating industry where copper is used as an undercoat for nickel and chromium plating.

13.2 Properties of Copper. Although most metals find their major industrial use in the form of alloys, this is not necessarily true of copper. It is used extensively in its unalloyed form for electrical and heat conductors. Copper ranks second to silver in electrical conductivity. It exhibits great resistance to weathering and has good mechanical properties. Copper with a face centered cubic structure, has high plasticity and is easily rolled or drawn into wire. It is surpassed in ductility only by gold and silver. Its high ductility allows shaping and forming without trouble from cracking and with relatively small deformation forces. Cast copper has a tensile strength of 24,000 psi. Copper is a soft metal with a Brinell hardness of 35 in the as-cast condition. The hardness may be increased to approximately 100 Brinell in cold working operations. It was believed by many that in ancient times copper was made approximately as hard as hardened tool steel by some method of heat treatment, a treatment that has often been referred to as the lost art of hardening copper. Present knowl-

edge leads us to believe that some of the ancient tools made from copper may have contained some impurities which contributed to its hardness and, coupled with cold forging, resulted in a harder variety of copper. We know that the only way to harden commercially pure copper is by cold work hardening methods.

Copper is an extremely tough metal that has a remarkable resistance to fracture from sudden shock loads. The fact that its elastic limit upon loading is only 50 percent of its ultimate strength may be taken as an indication of its great ability to deform without rupture when loaded beyond its elastic limit.

13.3 Commercial Grades of Copper. Three commercial classes of copper are given in Table 13–1. These grades include the so called tough pitch, the oxygen free, and the deoxidized coppers.

Tough pitch copper is metal that has been remelted under conditions producing a final oxygen content of 0.02 to 0.05 percent and which exhibits a high degree of toughness. The oxygen present combines with copper to form cuprous oxide (Cu_2O). If this type of copper is heated in a reducing atmosphere above 750°F, the reducing gases react with the oxide particles at the grain boundaries resulting in a brittle structure. The electrical conductivity of electrolytic tough pitch copper is approximately

423

Table 13–1. Classification of copper alloys.

ASTM DESIGNATION	TYPE	TYPICAL USES
	TOUGH PITCH COPPERS	
E T P	ELECTROLYTIC TOUGH PITCH	BUS BARS, BRAZING ROD, WIRE ANODES, FORGINGS
F R H C	FIRE REFINED, HIGH CONDUCTIVITY TOUGH PITCH	MECHANICAL APPLICATIONS
F R T P	FIRE REFINED TOUGH PITCH	
A T P	ARSENICAL, TOUGH PITCH	SHEETS, STRIP, PLATE
S T P	SILVER–BEARING, TOUGH PITCH	ROOFING, RADIATOR CORES,
S A T P	SILVER BEARING ARSENICAL TOUGH PITCH	PANS, PRINTING ROLLS, FASTENERS
	OXYGEN FREE COPPERS	
O F H C	OXYGEN FREE (WITHOUT RESIDUAL DEOXIDANTS)	TUBING WAVE GUIDES, STARTING ANODES, WIRE
O F P	OXYGEN-FREE, PHOSPHORUS BEARING	WELDING RODS, FORGING
O F T P E	OXYGEN-FREE, PHOSPHOROUS AND TELLURIUM BEARING	
O F S	OXYGEN-FREE, SILVER BEARING	PLATE, SHEET, RODS, FORGING BARS
O F T E	OXYGEN-FREE, TELLURIUM BEARING	FREE MACHINING
	DEOXIDIZED COPPERS	
D H P	PHOSPHORIZED, HIGH RESIDUAL PHOSPHOROUS	TUBES, PIPE, ANODES, PROJECTILE ROTATING BANDS
D L P	PHOSPHORIZED, LOW RESIDUAL PHOSPHOROUS	TUBES, WAVE GUIDES, GENERAL USE
D P S	PHOSPHORIZED, SILVER BEARING	HEAT EXCHANGES, STEAM LINES
D P A	PHOSPHORIZED, ARSENICAL	CONDENSER TUBES, TUBES FOR GENERAL USE
D P T E	PHOSPHORIZED, TELLURIUM BEARING	FREE MACHINING

(FROM ASTM DESIGNATIONS B224--58)

98 percent that of pure copper and is used extensively for electrical purposes. The presence of small amounts of oxygen is essential to sound castings and results in good mechanical properties. The structure of this type of copper is illustrated in Fig. 13–1. Fire-refined tough pitch copper containing a maximum of 0.10 percent arsenic is used for mechanical purposes. Arsenical tough pitch copper containing from 0.25 to 0.50 percent arsenic is used where its high recrystallization temperature is an advantage. Silver bearing tough pitch coppers have increased recrystallization and annealing temperatures. Its uses in-

Fig. 13–1. Electrolytic tough-pitch copper, 99.90% pure, annealed, 75X.

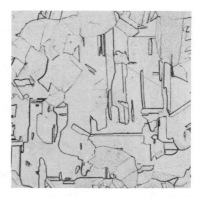

Fig. 13–2. Oxygen-free high conductivity copper (OFHC), annealed, 75X.

clude down spouts, gutters, roofing, screening, radiator cores, bus bars, switches, terminals, kettles, pans, printing rolls, vats and fasteners.

Oxygen-free copper is a special type of copper in which the last traces of cuprous oxide (Cu_2O) are removed during melting by means of a charcoal treatment. Cathode copper is melted, deoxidized, and cast under a protective atmosphere to exclude oxygen producing a metal with less than 0.000 percent oxygen. Oxygen-free high conductivity copper (O.F.H.C.), Fig. 13–2, possesses exceptional plasticity and good welding properties and is not subject to embrittlement resulting from exposure to a reducing atmosphere. Typical uses include copper tubing, wave guides, welding rods, bus bars, starting anodes, wire, and forgings.

Deoxidized copper has been treated with about 0.02 percent phophorus just before casting. The last traces of cuprous oxide are removed by this treatment. Deoxidized copper has better welding and brazing characteristics than the other grades. Uses for deoxidized copper include copper tubing for air, gasoline, and hydraulic fluids, heat exchangers, steam and water lines.

13.4 Impurities in Copper. Some of the more common impurities found in copper are oxygen (as Cu_2O), sulfur, bismuth, antimony, arsenic, iron, lead, silver, cadmium, phosphorus, and others. Certain of these impurities such as arsenic and phosphorus have a marked effect of lowering electrical conductivity. Cuprous oxide in the concentration

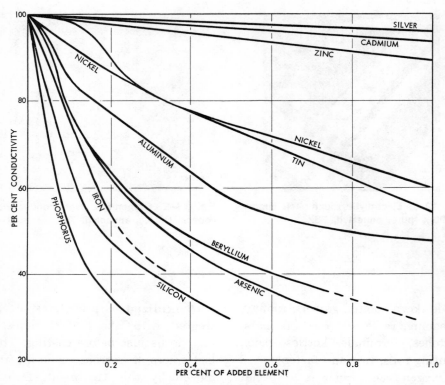

Fig. 13–3. Influence of elements on conductivity of copper. (By permission: *Copper Data, 1936 ed.,* British Copper Devel. Assn.)

found in commercial copper has little effect on the mechanical properties. Silver is one of the most common impurities found in copper. It has little effect upon the properties of copper with the exception of raising the recrystallization and annealing temperatures.

Lead in amounts over 0.005 percent reduces the hot working properties, has little effect on plasticity at room temperature, but does improve machinability. Tellurium and selenium, while not normally impurities, have largely superseded lead as elements for improving machinability. In Fig. 13–3 is shown the influence of some elements on electrical conductivity.

13.5 Temper Designations. A certain minimum deformation is required to transform a cast structure into a wrought one. The degree of deformation may be reported as percent reduction in thickness in rolling, or by the percent reduction in cross-section by drawing and forging. For most soft metals a total reduction of 50 percent is considered to be about the minimum required to produce properties characteristic of a wrought metal. The different cold worked tempers for copper and copper alloys are shown in Table 13–2. The tempers are obtained by cold working an annealed metal the amount indicated. The percent reduction of the strip is based on thickness whereas the reduction of the wire is based on cross sectional area.

Individual grains in a homogeneous metal are altered in shape in a cold deformed metal in about the same manner as the bulk part is altered. Soft inclusions deform in about the same manner as the matrix. Hard inclusions deform less and may even fragment as the matrix becomes work hardened.

Very heavy cold reductions produce elongated grains which are difficult to resolve microscopically because of slight variations in etching characteristics. Careful examination, however, does reveal boundaries of the original grains which persist after cold deformation. The principal direction of elongation, which determines fiber direction, may be clearly seen in cold worked metals (See Fig. 7–11). Hot worked or annealed metals also may exhibit a

Table 13–2. Temper designations for cold worked copper.

| TEMPER | APPROXIMATE PER CENT REDUCTION BY COLD WORKING | |
	STRIP	WIRE
QUARTER HARD, 1/4 H	10	20
HALF HARD, 1/2 H	20	37
THREE-QUARTER HARD 3/4 H	30	50
HARD, H	37	60
EXTRA HARD E H	50	75
SPRING, S	60	85
EXTRA SPRING, E S	68	90

fibrous structure because of heterogeneous inclusions or incomplete annealing.

13.6 Grain Size in Copper Alloys. Grain size of hot worked products is principally a function of the temperature at which hot working is completed. When hot working is carried out at the recrystallization temperature, new grains may form from this deformed metal but little coalescence of the newly formed grains is possible. This results in a fine grained metal. Grain size in steel and other polymorphic metals* can be altered by heat treatment without prior plastic deformation. However, metals which do not exhibit polymorphic change, including copper alloys, cannot obtain grain refinement by recrystallization unless plastically deformed by more than a threshold amount.

The grain size of copper alloys may be estimated by direct comparison of a projected image, or of a photomicrograph of a representative area of the test specimen with standard grain size photomicrographs (See Fig. 13–4). The comparison is preferably done at a specimen magnification of 75X. It should be recognized in making the estimation that a metal structure is an aggregate of three dimensional grains of varying sizes and shapes. Even if all

grains were of identical size and shape a cross section through such a structure would have grain areas varying statistically from zero to a maximum depending upon where the plane cut each individual crystal. All that is being estimated is an average size of grains as observed on this particular plane. Mixed grain sizes are sometimes encountered in hot worked metal. In this case an estimated percentage of the area occupied by each of the two sizes of grains is given. For example 50%— 0.015 mm and 50%—0.070 mm.

The grain size of a copper alloy is important in determining its suitability for a particular application. In Table 13–3, for example, are given recommended grain sizes for various types of drawing operations. The fine grained alloys are used for shallow drawing or forming operations whereas coarse grained metals have longer slip planes and can be used for deep drawing operations. The grain size best suited for a particular application depends upon the depth of draw, material thickness, and the type of finish required after the draw.

13.7 Annealing of Copper. Annealing is the only method of heat treatment used on pure copper although other heat treatments are used in connection with some of the copper alloys. The purpose of annealing copper is to restore the orig-

---

*Polymorphic metals are those which can exist in more than one lattice structure, with each structure stable in a separate temperature range.

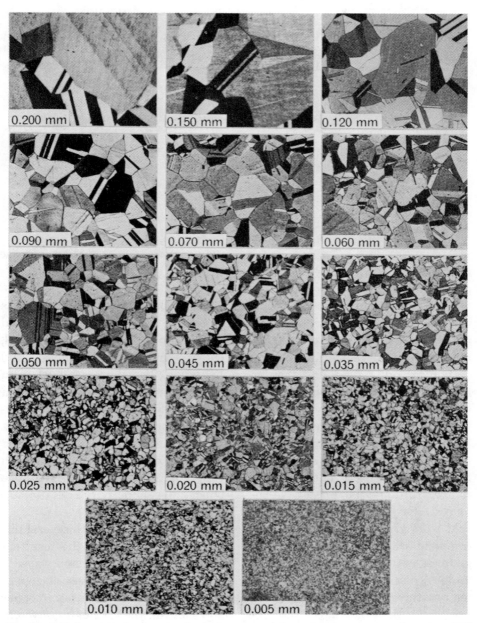

Fig. 13–4. Twinned grain size standards for copper alloys showing average diameters, 75X.

Table 13–3. Common grain sizes for drawing operations with copper alloys.

GRAIN SIZE MILLIMETERS	RECOMMENDED FORMING APPLICATIONS
0.015	SLIGHT FORMING OPERATIONS; BEST POLISHING
0.025	EASY DRAWING, GOOD POLISHING ON ITEMS SUCH AS HUBCAPS
0.035	GOOD DRAWING AND POLISHING AS FOR HEADLIGHT REFLECTORS
0.050	HEAVY DRAWING AND SPINNING; MORE DIFFICULT TO POLISH
0.100	SEVERE DRAWS ON HEAVY MATERIAL

inal ductility and softness to work hardened copper resulting from any cold working operations.

The process of annealing cold work hardened copper involves heating to a temperature of 1100°F, holding at this temperature for a certain period of time, and then allowing the metal to cool to room temperature. The rate of cooling from annealing temperature of 1100°F is without effect; therefore, fast cooling rates such as a water quench may be used. The annealing temperature of 1100°F is above the recrystallization temperature of the cold worked state and results in a complete change in the original grain structure without promoting undue grain growth. Annealing at temperatures higher than 1100°F has little material effect on the strength of copper but the ductility

increases somewhat due to increase in size of grains.

13.8 Copper Alloys. The elements that are most commonly alloyed with copper are zinc, tin, nickel, silicon, aluminum, cadmium, and beryllium. The elements which form solid solutions with copper have a marked effect in decreasing electrical conductivity (Fig. 13–3). Elements such as lead, selenium, and tellurium are relatively insoluble in copper and consequently show little effect on conductivity, but do improve machinability.

Copper is alloyed with several different metals because the resulting alloys are superior in many ways.

Strength and hardness of copper alloys are greater than pure copper and they may be further improved in their mechanical properties by cold working and, in some instances, by heat treatments.

Castability of commercial grades of copper is unsatisfactory. Alloying of copper improves the casting characteristics, and these alloys, such as brass and bronze, are used to make castings.

Machinability of the alloys of copper, in general, is much greater than commercial copper, which is too soft and tough for easy machining.

Corrosion resistance of many of the copper alloys is superior to that of commercial copper.

Lower cost of alloys of copper-zinc (brass) is due to the low cost of zinc, with the alloys being superior to copper in many respects.

Improved elasticity of the copper alloys makes them superior to commercial copper, the latter being almost totally lacking in elastic properties unless it has been subjected to severe cold work hardening.

13.9 Brass—General. The alloys of copper and zinc are commonly classified as *brasses;* however, the term *commercial bronze* may also be used for some compositions of copper and zinc. Since the terms used in the brass industry may be misleading to the student the alloys of copper will be discussed from the standpoint of both composition and name rather than from name alone. The brasses may be classified as (1) alpha brasses or (2) alpha-beta brasses.

Alpha brasses containing from 5 to 20 percent zinc are called red brasses because of their copper color while those containing 20 to 36 percent zinc are called yellow alpha brasses. Alpha-beta brasses contain 36 to 45 percent zinc. The copper-zinc alloys are the most important of the copper alloys due to their desirable properties and relatively low cost.

Zinc readily dissolves in copper in both the liquid and solid states, forming a series of solid solutions; with less than 36 percent zinc, a solid solution is formed that is referred to as alpha (α) solid solution. The alpha phase is a strong and very ductile structure. If the combination contains above 36 percent zinc, a solid solution known as beta (β) is formed, which is relatively hard and much less ductile than the alpha brass. When zinc exceeds 50 percent in the alloy, a gamma solid solution is formed which is hard and brittle and of no value industrially except for decoration.

A study of the copper-zinc alloy diagram, Fig. 13–5 illustrates the areas where the alpha (α), beta (β), and gamma (γ), solid solutions exist. The copper-zinc diagram may be readily understood and interpreted if the student will refer back to Chapter 5 dealing with the types of alloy systems.

Brasses containing over 62 percent copper consist of only one phase, the alpha solid solution,

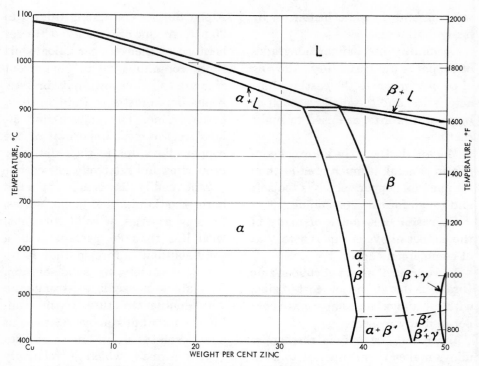

Fig. 13–5. Copper-rich portion of the copper-zinc phase diagram. (By permission: *Metals Handbook, 1948 ed.,* American Society for Metals)

which is very ductile and has a face centered cubic type of crystal structure. The alpha solid solution brasses are used mostly where the parts are wrought to shape. Many of the commercially important brasses are of this type. The mechnical properties of these alpha brasses depend largely upon the zinc content and the degree of cold working they receive. Fig. 13–6 illustrates the effect of composition upon the tensile strength and ductility of alloys con-

taining up to 50 percent zinc. It will be seen from Fig. 13–6 that the tensile strength and ductility are both improved with additions of zinc up to approximately 30 percent, above which the tensile strength continues to increase to about 45 percent zinc, with a marked drop in the ductility when the zinc content exceeds 30 percent.

When the percentage of zinc exceeds approximately 36 percent the structure of the alpha solid solution

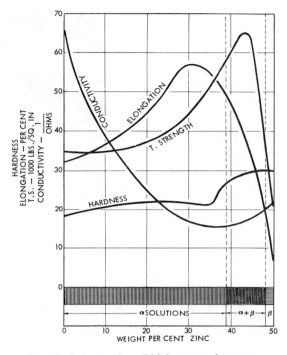

Fig. 13–6. Properties of high copper brasses.

changes to the beta phase. The beta phase is much harder and stronger but less ductile than the alpha phase. Also, the beta phase differs from the alpha phase in that it is a body centered cubic crystal structure, and upon cooling to approximately 880°F, the beta phase undergoes a structural transformation changing the beta (β) to a phase designated as β'. The two beta phases differ in that the β has its solute atoms haphazardly arranged in the solvent lattice, where the β' phase has the solute atoms orderly arranged in its atom structure of solvent atoms.

The color of brass varies as its composition; it changes from a copper red for the high copper alloys to a yellow color at about 36 percent zinc. The color changes and is slightly more reddish in the β' phase.

13.10 Red Brasses. There are four types of wrought brasses in this group: gilding metal—95–5, commercial bronze—90–10, rich low brass—85–15, and low brass—80–20. These brasses are very workable both hot and cold. The lower the zinc content, the greater the plasticity and workability. These brasses are superior to the yellow brasses or high brasses for corrosion resistance and show practically no bad effects from dezincification or season cracking. Due to their low zinc content, they are more expensive and are used primarily when their color or greater corrosion resistance or workability are distinct advantages. The applications for red brass include valves, fittings, rivets, radiator cores, detonator fuse caps, primer caps, plumbing pipe, bellows and flexible hose, stamped hardware, caskets, and screen cloth. These alloys may be shaped by stamping, drawing, forging, spinning, and other processes. They have good casting and machining characteristics and are weldable.

13.11 Yellow or Cartridge Brass.

The yellow alpha brass is the most ductile of all the brasses, and its ductility allows the use of this alloy for jobs requiring the most severe cold forging operations such as deep drawing, stamping, and spinning. Fig. 13–7 illustrates the structure of annealed yellow or cartridge brass, which is a copper containing from 28 to 32 percent zinc. This alloy is used for the manufacture of sheet metal, rods, wire, tubes, cartridge cases (from which it gets its name) and many other industrial shapes.

Fig. 13–7. Annealed cartridge brass. Copper 70%, zinc 30%, 75X.

Unless the brass is exceptionally free from lead it may not respond well to hot working, thus fabrication should be done cold. Lead additions may be used to improve the cold working characteristics of cartridge brass.

A rolled and annealed cartridge brass has a tensile strength of about 48,000 psi. By cold rolling, however, its tensile strength may be increased to 100,000 psi. Brass is obtainable in varying degrees of hardness by cold rolling, usually designated as *quarter-hard, half-hard,* and *hard.* The hard brass is brass with the maximum degree of cold rolling that may be considered practical for a given thickness of section and from a workability viewpoint.

The effect of cold working upon the mechanical properties of this alloy is illustrated in Fig. 13–8. The

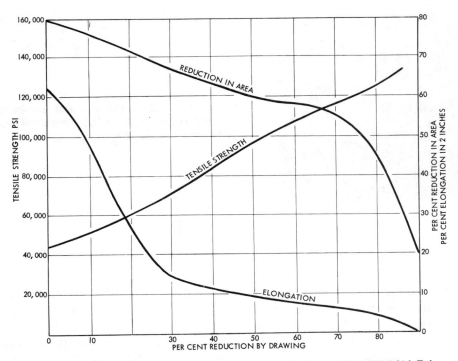

Fig. 13–8. Effect of cold drawing on yellow brass properties (66% Cu 34% Zn).

only heat treatment given to this alloy is annealing to remove any cold work hardening and to remove residual stresses as the result of cold-working operations.

13.12 Special Yellow Brasses. Two variations of yellow brasses have been developed for special applications. Admiralty metal is a copper-zinc alloy containing approximately 1.0 percent tin and 0.03 percent arsenic. Its resistance to corrosion is superior to that of ordinary brasses

that are free from tin. The structure of this type of alloy is shown in Fig. 13–9. This composition is used for tubing condensers, preheaters, evaporators, and heat exchangers in contact with fresh and salt water, oil, steam, and other liquids at temperatures below 500°F. Aluminum-brass (76 Cu, 22 Zn, 2Al) has better corrosion resistance because of a tenacious, self-healing film which protects tubing against high velocity cooling water in marine and land power stations.

Fig. 13–9. Annealed arsenical admiralty metal, copper 70%, tin 1%, arsenic 0.03%, balance is zinc, 75X.

13.13 Annealing of Brass. The only heat treatment applicable to alpha brass is that of annealing after cold working. Cold working, as we have seen, produces a distortion of the crystal state, which is accompanied by an increase of hardness and strength and a loss of plasticity. The cold worked condition, with its greater strength and lower ductility, renders the metal less workable and may result in failure due to rupture of the less plastic condition of the crystal state. Restoration of the original properties may be accomplished by an annealing operation; heating of the cold-worked brass to above its recrystallization temperature. During the heating of the cold worked brass any internal stresses which may be present in the brass from the cold forming operations may be removed. Three changes take place during the heating of the cold-worked brass in an annealing operation; (1) the relief of internal stresses, (2) the recrystallization of the cold worked crystal structure into new and, at first, very fine grains, and (3) the growth of the fine new-formed grains into larger and fewer grains or crystals. The annealing operation completely removes any trace of the original cold worked state of the brass, and restores the original ductility, with a lowering of the tensile strength and hardness to the normal values.

The effect of annealing temperatures ranging from room temperature up to 1472°F on the properties of cold work hardened brass of the yellow (high brass) composition is illustrated by Fig. 13–10.

Annealing may be carried out by heating the cold worked brass to

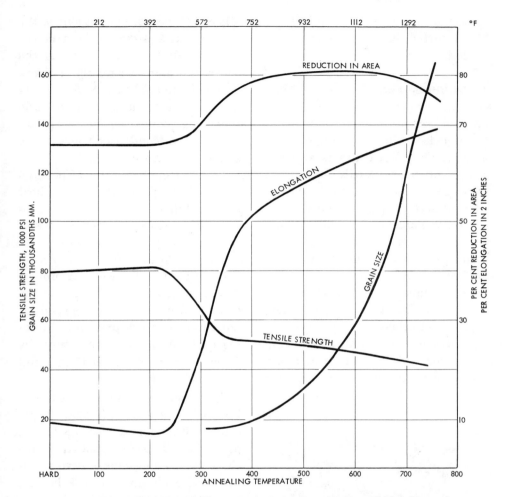

Fig. 13–10. Effect of annealing temperature on properties of high brass.

a temperature range of 1100 to 1200°F, followed by cooling at any convenient rate. The rate of heating and cooling is almost without effect on the size of the newly formed crystals. The size of the annealed crystals is influenced largely by the amount of cold work received prior to annealing and by the maximum annealing temperature. The atmosphere in the annealing furnace may often be controlled to prevent excessive oxidation or even discoloration of the surface of the brass to be annealed.

13.14 Alpha-Beta Brasses. When the content of zinc in brass is increased from 40 to 45 percent the alloy is called Muntz metal. (60 Cu, 40 Zn). This alloy contains both the alpha and beta constitutents in its structure. Muntz metal may be hot worked even when it contains a high percentage of lead. Such brass is accordingly useful for screws and machine parts where ease of working and particularly ease of machining are more important than strength.

A photomicrograph of this alloy, Fig. 13–11 illustrates the type of structure found in Muntz metal. As stated previously, both the alpha and beta phases are present in this alloy. The lighter constituent is the harder and less ductile beta phase, and the darker constituent is the softer and more plastic alpha phase. Some of the applications for Muntz metal include sheet form for ship sheathing, condenser heads, perforated metal, condenser tubes, valve stems, and brazing rods.

Free cutting brass (61.5 Cu—35.5 Zn—3 Pb) which has the highest machinability rating of any brass combined with good mechanical and corrosion-resistant properties.

Forging brass (60 Cu—38 Zn—2 Pb) has the best hot working properties of any brass when heated to the single phase β region and is used for hot forgings, hardware, and plumbing parts. At room temperature the β' phase becomes harder than α and more difficult to work.

Architectural bronze (57 Cu—40 Zn—3 Pb) is really a brass with excellent forging and free machining properties. Typical applications include handrailings, decorative moldings and grills, hinges, lock bodies, and industrial forgings.

Naval brass or **Tobin bronze** (60 Cu—39.25 Zn—0.05 Sn) has increased resistance to salt water spray and is used for condenser plates, welding rod, propeller shafts, and marine hardware.

Manganese bronze (58.5 Cu—39 Zn—1.4 Fe—1 Sn—0.1 Mn) is really a brass and is well known as an alloy for ship propellers. It has high

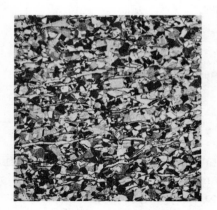

Fig. 13–11. Muntz metal annealed. Copper 60%, bal. zinc. Alpha (dark) and beta (light) structure.

strength and excellent wear resistance. Typical applications are clutch discs, extrusions, forgings, pump rods, valve stems, and welding rods.

13.15 Season Cracking and Dezincification. Highly stressed brass and bronze may be sensitive to failure by cracking in service under conditions of a corrosive nature. This type of failure is called season cracking or stress-corrosion cracking. It is due to the high residual stresses resulting from severe cold working which make the metal more susceptible to intergranular corrosion (See Fig. 13–12). This failure is spontaneous and may occur without warning and without any added strain from service.

The conditions leading to season cracking may be eliminated by careful control of the cold working operations, or by a 500-800°F stress relief annealing treatment. Substitution of a less susceptible copper alloy will also minimize the danger of stress-corrosion cracking.

Sensitivity of a copper alloy to season cracking may be determined by exposure to a mercurous nitrate solution as discussed in paragraph 2.11. Satisfactory brass will show no evidence of a crack after 30 minutes immersion in the solution.

Some brass alloys, notably, yellow brasses, are subject to a pitting corrosion called dezincification. This type of corrosion usually occurs when the brass is in contact with water having high percentages of dissolved oxygen and carbon dioxide.

Dezincification involves a loss of zinc by dissolution and a subsequent

Fig. 13–12. Photomicrograph through crack in 70/30 alpha brass nozzle. Note intergranular path of crack which originated at the surface. 25X, etchant: ammonium hydroxide-hydrogen peroxide. (Revere Copper and Brass Co.)

surface deposit of one or more less active components, usually porous, non-adherent copper. Action of this kind, unless stopped will eventually lead to penetration of the cross section of the metal with resulting leakage. Small amounts of tin or antimony alloying elements minimize dezincification of yellow brasses.

13.16 Bronzes—General. The term bronze was originally used to describe copper-tin alloys. The term is now used for any copper alloy, excepting copper-zinc alloys, which contain up to 25 percent of the principal alloying element. The name bronze conveys the idea of a higher class alloy than brass, although it has been incorrectly applied to some alloys which are really special brasses. Commercially used bronze alloys are alloys of copper with tin, aluminum, silicon, or beryllium. Bronzes may also contain phosphorus, lead, zinc, and nickel.

Bronzes may be classed as: tin bronzes, aluminum bronzes, silicon bronzes, beryllium bronzes.

13.17 Copper-tin Bronzes. Alloys containing principally copper and tin are referred to as the true bronzes although other elements are frequently present to improve characteristics of the plain two component system.

Tin increases the hardness of copper and its resistance to wear to a

much greater degree than does zinc. Tin also improves copper salt water corrosion resistance. Additions of tin up to about 16 percent form a homogeneous face centered cubic solid solution with copper. At room temperature the solubility of tin in copper is restricted as shown by the equilibrium diagram (see Fig. 13–13). However, in commercial operations, the diminution of solubility rarely occurs. In all but alloys annealed for extremely long times after severe cold working, the solubility below 968°F can be considered as constant at 15.8 percent. Similarly, the decomposition of ϵ (Cu_3Sn), below 662°F occurs extremely slowly. Thus, for most purposes, the ($\alpha + \epsilon$) region can be considered as extending vertically downward to 15.8 percent and 32.4 percent as shown by the dotted lines. The copper-tin alloys may be divided into four groups depending upon the percentage of tin present.

1. Up to 8 percent tin; used for sheets, wire, coins; readily cold worked.

2. Between 8 and 12 percent tin; used for gears, machine parts, bearings and marine fittings.

3. Between 12 and 20 percent tin; used considerably for bearings.

4. Between 20 and 25 percent tin; used primarily for bells; are very hard and brittle and used in the cast condition.

Copper-tin alloys are generally

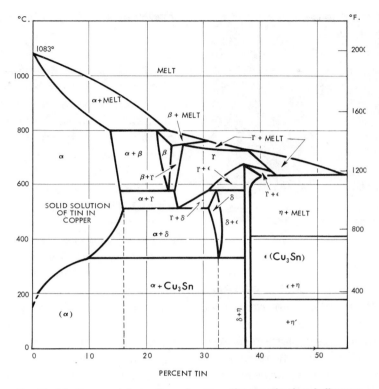

Fig 13–13. Copper-rich portion of copper-tin constitutional diagram. (By permission: *Metals Handbook, 1939 ed.*, American Society for Metals)

more expensive than copper-zinc brasses but are also stronger and more resistant to corrosion. Typical parts include high strength springs, clips, snap switches, electric sockets and plug contacts, fuse clips, flexible tubing, hardware, bushings, bearings, and welding rods.

Two modifications of copper-tin bronze are the phosphor bronzes and the leaded bronze.

Phosphor bronze has a small amount of phosphorus added during melting and casting for deoxidation purposes. The phosphorus also increases fluidity of the molten metal thereby increasing the ease of casting and aiding in the production of sounder castings. The structure of a cast phophorus-treated bronze is shown in Fig. 13–14. The alpha constituent appears as dendritic crys-

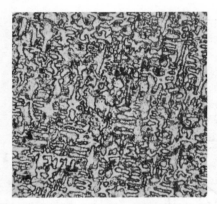

Fig. 13–14. Phosphor bronze as cast. Alloy is 92% copper, 8% tin, 75X.

tals in a matrix of alpha phase and Cu$_3$Sn. These alloys are often used in gears and bearings.

Leaded bronze is a term used to describe a mechanical mixture of lead in a copper alloy. Lead does not alloy with copper but may be mixed with molten copper and under suitable conditions, may be cast into a mold. The solidified alloy has lead well distributed throughout the casting in small particles. Lead may be added to both bronze and brass alloys to improve machinability and bearing properties.

The lead particles, with their softness and low shear strength, reduce friction in parts subjected to sliding wear, such as in a bearing, or machining tool wear as at the tool-chip interface. Lead is a source of weakness and is usually kept below 2 percent although some bearing bronzes may contain a much higher percentage.

13.18 Aluminum Bronzes. Alloys of copper containing between 5 and 11 percent aluminum are known as aluminum bronzes. They may also contain other elements such as iron, silicon, manganese, and nickel, for increased strength. The choice of aluminum bronze over tin bronze is usually based on properties other than corrosion resistance such as superior wear resistance and mechanical properties at room and elevated temperatures. The hardness and strength of aluminum bronzes may be markedly improved by heat treatment. Reference to the copper-rich end of the copper-aluminum equilibrium diagram Fig. 13–15, shows how the heat treatment is accomplished. An alloy containing approximately 10 percent aluminum may be solution treated by heating it to 1650°F followed by water quenching. During the solution treatment the alpha phase is transformed to the beta phase. The beta phase is retained, in part, by the rapid cooling. (Fig. 13–16). A subsequent aging treatment at a temperature of 700-1100°F causes the beta phase to become unstable; it then undergoes a transformation to a fine alpha and gamma crystal form. Such a transformation causes an increase in tensile strength of from 80,000 psi in the annealed condition

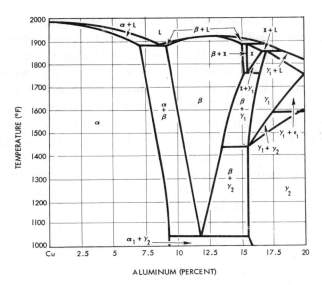

Fig. 13–15. Copper rich end of copper-aluminum equilibrium diagram. (By permission: *Physical Metallurgy for Engineers*, D. Van Nostrand Co.)

to 100,000 psi in the solution treated and aged condition. Hardness is increased from RB90 to RC25, while elongation is decreased from 22 percent to about 2-6 percent.

Heat treated aluminum bronzes may be used for hand tools, such as chisels, where non-sparking characteristics are essential as in explosive environments. Other typical uses are in valve seats and guides, bushing, bearing plates, decorative purposes, grills, imitation gold jewelry, and paint pigment.

With aluminum content below 10

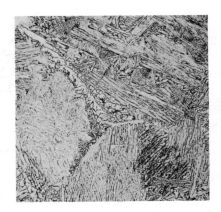

Fig. 13–16. Aluminum-bronze, heated to 1700°F and quenched in water. Structure is alpha and beta phases. (Revere Copper and Brass Co.)

443

percent, the copper dissolves the aluminum and forms solid solutions, α, β and γ phases—not unlike those found in copper-zinc and copper-tin alloys. Aluminum bronze in the soft condition is nearly as ductile as brass and has twice the ductility of copper-tin bronze. However, it is difficult to cast because of its marked shrinkage during solidification, and its tendency to form shrinkage cavities. Alpha phase alloys may be satisfactorily hot or cold worked but are somewhat difficult to machine.

13.19 Silicon Bronzes. A portion of the copper-silicon equilibrium diagram is shown in Fig. 13–17. Silicon

is soluble in copper up to about 5.3 percent at a temperature of 1550°F. This solubility decreases to less than 4 percent at room temperature. Although these alloys are not hardenable by heat treatment they may be cold worked to achieve strengths as high as 145,000 psi, although the usual strength of castings is in the range of 40,000 to 70,000 psi with 20 to 35 percent elongation in 2 inches. Iron and other elements are frequently added for the purpose of further strengthening, grain refinement, or avoidance of patent infringement.

In addition to high capacity for

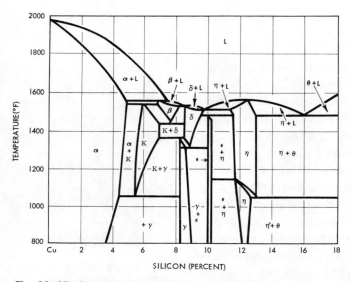

Fig. 13–17. Copper rich portion of the copper-silicon equilibrium diagram. (By permission: *Physical Metallurgy for Engineers,* D. Van Nostrand Co.)

work hardening silicon bronzes exhibit excellent resistance to corrosion from organic acids, sugar solutions, sulfite solutions, and others. Silicon bronzes find application in electrical fittings, marine hardware, boilers, pumps, shafting, and as low-cost substitutes for tin bronzes. These alloys are cast and worked hot or cold. They are noted for their ease of welding, combined with the strength of mild steel and the corrosion resistance of copper.

13.20 Beryllium-Copper Alloys. The answer to the "lost art of hardening copper" may be found in some of the modern copper alloys such as the newer beryllium-copper alloys containing 1 to 2.25 percent beryllium. Although beryllium makes a very costly alloy its high initial cost is offset by the remarkable properties developed by this new alloy. It may be cast, worked hot and cold, welded, and given heat treatments similar to those given the aluminum-copper alloys.

As may be seen from the alloy diagram of beryllium-copper, Fig. 13–18, the solubility of beryllium in copper increases with temperature from less than 1 percent beryllium at room temperature to more than 2 percent at 1500°F. Also, as the beryllium content in copper increases, the alpha solid solution, which forms first, changes to a eutectoid structure consisting of a mixture of alpha

and gamma phases. Any gamma phase present in the alloy at room temperature changes to a beta phase upon heating to above 1050°F. Due to these changes in solubility and the change from one phase to another, the beryllium alloys are susceptible to heat treatments which greatly alter their mechanical properties with improvement in both strength and hardness. Heat treatments include annealing, heating, and quenching, known as solution heat treatment, and aging or precipitation hardening treatments.

In the annealed condition, obtained by quenching the alloy from a temperature of 1472°F, an alpha solid solution structure is developed. The tensile strength of this quenched alloy is about 60,000 psi. After cold working the alloy may be subjected to a precipitation hardening treatment by heating to a temperature between 475 and 575°F. This treatment precipitates a hard gamma crystal phase of the alloy rendering it much harder and less ductile. By means of these treatments the hardness may be varied from 215 Brinell to 400 Brinell, with a tensile strength in the higher Brinell hardness range of up to 200,000 psi. These alloys exhibit good electrical properties and excellent corrosion resistance. Because they possess a combination of high Brinell hardness and tensile strength, these alloys find many applications as

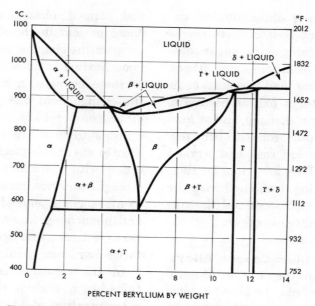

Fig. 13—18. Copper rich part of the copper-beryllium equilibrium diagram. (By permission: *Metals Handbook, 1939 ed.*, American Society for Metals)

springs, tools, surgical and dental instruments, watch parts, firing pins, and welding electrodes. At the present time, the main objection to these alloys is high cost; and due to the difficulty in extracting beryllium from its ores it is doubtful if the price of this alloy will be reduced to that of other copper alloys.

A substitute for the beryllium alloys has been developed that contains approximately 22-24 percent manganese, 22-24 percent nickel, and the balance copper. This alloy exhibits age hardening effects resulting from heat treatments and may be cast and wrought to shape. Alloys of this type may be greatly improved in strength and hardness by quenching from 1200°F followed by tempering or aging at 660 to 840°F for approximately 24 hours. With proper heat treatments hardness values of Rockwell C54 and higher are obtainable with tensile strength values in excess of 200,000 psi. These particular alloys may be used in applications similar to the beryllium-copper and the aluminum bronzes and are used where hardness and non-sparking characteristics are needed, such as some applications for hand tools.

13.21 Copper-Nickel Alloys. Copper and nickel form a very simple alloy as both metals dissolve in each other in all percentages to form a solid solution. Nickel added to brasses and bronzes has been found to improve their electrical properties and toughness. A familiar copper-nickel alloy is found in the United States five-cent piece, which is coined from alloy sheets containing 75 percent copper and 25 percent nickel. All alloys of copper and nickel with more than 20 percent nickel are white in color. Alloys containing 30 percent nickel known as cupro-nickel are much used in marine service where strength and resistance to corrosion are important factors in the selection of an alloy. These alloys are used for condenser tubes, evaporator and heat-exchanger tubes and ferrules.

Nickel-silver, also called German silver, is a brass alloy containing 10 percent nickel and 5 to 50 percent zinc, the balance copper. These alloys are very white, similar to silver, hence the name. They are resistant to atmospheric corrosion and to the acids of food stuffs. Their use depends largely upon their resistance to corrosion and pleasing appearance (See Fig. 13–19).

Nickel

13.22 Nickel. Nickel is one of the most important industrial metals. It has very good resistance to corrosion and is often used in electroplating as a base coating for other metals. Nickel-clad steel is used in the construction of heavy kettles and tanks. It is also used in the electronics industry for vacuum tubes and in the chemical industry as a catalyst for producing caustic soda. The greatest quantity of nickel, however, is used in alloying with other metals, particularly steel.

Nickel in commercially pure form

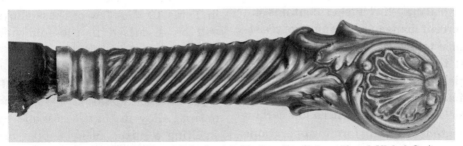

Fig. 13–19. Pre-Civil War nickel silver knife handle. (International Nickel Co.)

may contain a very slight amount of sulfur resulting from sulfur containing fuel used in the melting furnace. The sulfur may form a continuous sulfide envelope around each grain at the grain boundaries embrittling the structure. The addition of a very small amount of manganese combines with the sulfur forming an innocuous dispersion of MnS particles permitting the nickel to display its inherent plasticity. Pure nickel (electrolytic) of 99.95 percent purity has a tensile strength of 46,000 psi in the annealed condition but an extremely low yield strength of 8,500 psi. It is very ductile and has an elongation of 30 percent in the annealed condition. Nickel is of the face centered cubic crystal structure and has an atomic diameter very close to that of iron (Ni—2.49 Å, Fe—2.48 Å). Commercially pure nickel is available in several grades, depending upon its impurities or minor alloying elements. "A" nickel is a commercially pure nickel (99.4 percent) with the balance being principally cobalt. Its use is mostly in the chemical industry and in electroplating. "D" nickel contains 4.5 percent manganese to improve resistance to attack by sulfur compounds at temperatures up to 1000°F. "E" nickel, which contains 2 percent manganese is very similar to "D" nickel. "L" nickel, with a low carbon content is used in forming operations requiring large plastic deformation.

Another nickel alloy called "Z" nickel contains about 4.5 percent aluminum and is subject to precipitation hardening. Maximum yield strength of cold worked and age hardened strip approaches 230,000 psi with a hardness of Rockwell C46. This alloy is used principally for pump rods, springs, and shafts that demand high strength and high resistance to corrosion.

13.23 Nickel-Base Alloys. Alloys which contain a substantial portion of nickel are of great commercial importance. The corrosion resisting stainless steels that contain nickel were discussed in Chapter 9. Aside from the nickel containing stainless steels, there are other nickel alloys which have special properties such as high magnetic properties, magnetic shielding properties, low coefficient of expansion, outstanding heat resistance, and corrosion resistance. Nickel alloys which contain greater than 50 percent nickel are classed as nickel-base alloys. The nominal composition and typical uses of nickel-base alloys are shown in Table 13–4. A few of these alloys will be discussed in the following sections.

13.24 Nickel-Copper (Monel) Alloys. Monel metal is a so-called natural alloy in that it is made by smelting a mixed nickel-copper ore mined in Sudbury, Ontario. It contains approximately 65 percent nickel, 28 percent copper, the balance consisting chiefly of iron, manganese, and cobalt as impurities. Monel metal is harder and stronger than either copper or nickel in the pure form; it is also cheaper than pure nickel and will serve for certain high-grade purposes to better advantage than the pure metal. Monel is a substitute for steel where resistance to corrosion is a prime requisite. Monel metal can be cast, hot and cold worked, and welded successfully. In the forged and annealed condition it has a tensile strength of 80,000 psi with an elongation of 45 percent in 2 inches, and is classified as a tough alloy. In the cold worked condition it develops a tensile strength of 100,000 psi with an elongation of 25 percent in 2 inches. With excellent resistance to corrosion, pleasing appearance, good strength at elevated temperatures, Monel metal finds many uses in sheet, rod, wire, and cast forms. Included in the applications for Monel metal are sinks and other household equipment, containers, valves, pumps, and many parts of equipment used in the food, textile, and chemical industries, "R" Monel is a free machining grade for automatic screw machining operations.

A Monel metal containing 3 to 5 percent aluminum, known as "K" Monel, is of interest. It can be made to develop very high properties by means of heat treatment as a substitute for cold work hardening as required in the other nickel-copper alloys. A Brinell hardness of 275 to 350 Brinell may be obtained by means of a precipitation heat treatment. This hardness may be further increased by cold work hardening. The "KR" grade of Monel is the free machining variety of the age hardenable alloy. Mechanical properties of a number of Monel group alloys are given in Table 13–5.

13.25 Nickel-Iron Alloys. Nickel and iron are completely soluble in the liquid state and solidify as solid solutions. Nickel additions retard transformation of gamma iron to alpha iron during cooling. Additions of nickel up to 6 percent produce a ferritic alloy. Alloys with 6 to 30 percent nickel have a tendency to air harden on slow cooling. Rapid cooling produces a martensitic structure which transforms into α and γ phases upon reheating. Alloys containing greater than 34 percent nickel are entirely austenitic, nonmagnetic, and will not change to the ferritic structure even at extremely low temperatures. Alloys of nickel and iron have rather unique properties of thermal expansion, magnetism, thermal conductivity, and modulus of elasticity. The above properties can be varied considerably by changes in the percentage of nickel present in the alloy. One

Table 13—4. Nominal composition and uses of nickel and nickel-base alloys.

ALLOY	Ni	C	Mn	Fe	S	Si	Cu	Cr	Al	Ti	OTHER	TYPICAL PROPERTIES AND USES
1. COMMERCIALLY PURE NICKEL												
"A" NICKEL	99.4	0.1	0.2	0.15	0.005	0.05	0.1					CHEMICAL INDUSTRY, ELECTROPLATING
"D" NICKEL	95.2	0.1	4.5	0.15	0.005	0.05	0.05					RESISTANCE TO SULFUR ATTACK
"E" NICKEL	97.7	0.1	2.0	0.10	___	0.05	0.05					SIMILAR TO "D" NICKEL
"L" NICKEL	99.4	0.02	0.2	0.15	0.005	0.05	0.1					SEVERE PLASTIC FORMING OPERATIONS
"Z" NICKEL	94.0	0.16	0.25	0.25	0.005	0.40	0.05			0.33		AGE HARDENABLE, SPRINGS, PUMP RODS, SHAFTS
2. NICKEL AND COPPER ALLOYS												
MONEL	66.0	0.12	0.90	1.35	0.005	0.15	31.50					CORROSION RESISTANCE, TOUGHNESS, HIGH STRENGTH
R-MONEL	66.0	0.18	0.90	1.35	0.050	0.15	31.50					FREE MACHINING
K-MONEL	65.3	0.15	0.60	1.00	0.005	0.15	29.50		2.80	0.50		AGE-HARDENABLE NON-MAGNETIC
H-MONEL	65.0	0.1	0.9	1.5	0.015	3.0	29.50					AGE-HARDENABLE, MACHINABLE
S-MONEL	63.0	0.1	0.9	2.0	0.015	4.0	30.0					AGE-HARDENABLE, ANTI-GALLING, IMPELLERS, PUMP LINERS
3. NICKEL-IRON ALLOYS												
INVAR	36.0	*	*	63.0	*							VERY LOW EXPANSION COEFFICIENT LENGTH STANDARDS
PLATINITE	46.0			54.0								GLASS SEAL
PERMALLOY	78.5			21.5								HIGH MAGNETIC PERMEABILITY, SUBMARINE TELEGRAPH CABLES

Table 13-4. (cont.) Nominal composition and uses of nickel and nickel-base alloys.

ALLOY	Ni	C	Mn	Fe	S	Si	Cu	Cr	Al	Ti	OTHER	
4. NICKEL–CHROMIUM AND NICKEL–IRON–CHROMIUM ALLOYS												
INCONEL	76.4	0.04	0.20	7.20	0.007	0.2	0.10	15.85				HEAT AND COR-ROSION RESIST-ING, NON-MAGNETIC
INCONEL 702	78.0	0.02	0.05	0.30	0.007	0.15	0.05	15.85	3.0	0.60		AGE–HARDENABLE, OPERATES UP TO 2400°F.
S-INCONEL	68.0	0.20	1.0	8.5	0.008	5.0	0.5	15.5				AGE-HARDENABLE, ANTI-GALLING
INCONEL-X-750	72.9	0.04	0.70	6.8	0.007	0.30	0.05	15.0	0.80	2.5	Cb-0.9	AGE-HARDENABLE, USED UP TO 1500°F.
NIMONIC 80A	74.5	0.05	0.55	0.55	0.007	0.20	0.05	20.0	1.30	2.50		AGE-HARDENABLE, SIMILAR TO INCONEL X
NIMONIC 90	57.0	0.05	0.50	9.50	0.007	0.20	0.05	20.45	1.65	2.60	Co-17.00	SIMILAR TO NIMONIC 80A
NICHROME	80.0	0.05	0.1	0.5	--	0.20	--	20.0				HEATING ·ELEMENT
CHROMEL	90.0							10.00				HEATING ELEMENTS, THERMO COUPLES
INCOLOY 901	37.35	0.04	0.45	33.75	0.007	0.30	0.10	13.50	0.25	2.50	Mo-5.90	AGE HARDENABLE, GAS TURBINE WHEELS
Ni-O-NEL	41.35	0.03	0.65	31.65	0.007	0.35	1.80	20.2	0.40	0.50	Mo-310	RESISTANT TO CERTAIN HOT ACIDS
60-12 Ni-Cr	62.45	0.55		25.0				12.0				RESISTANCE HEATING ELEMENT
65-15 Ni-Cr	65.0			24.00	0.007			15.00				RESISTANCE HEATING ELEMENT
67-17 Ni-Cr	72.45	0.55		10.0				17.00				HEAT RESISTING ALLOY

*COMBINED AMOUNT OF SILICON, MANGANESE AND CARBON
TOTAL LESS THAN 1 PERCENT

Table 13-4. (cont.) Nominal composition and uses of nickel and nickel-base alloys.

ALLOY	Ni	C	Mn	Fe	S	Si	Cu	Cr	Al	Ti	OTHER	
5. OTHER HIGH NICKEL ALLOYS (HIGH-STRENGTH, HIGH-TEMPERATURE APPLICATIONS)												
HASTELLOY A	57.0			20.0							Mo-20	CORROSION RE-SISTANT AT ELEVATED TEMPERATURE
HASTELLOY B	62.0			5.0							Mo-30	HEAT AND CORROSION RESISTANT AT ELEV. TEMP.
HASTELLOY X	47.0	0.10		18.0				22.0			Co-1.5 Mo-9.0 W-0.6	HEAT RESISTING ALLOY UP TO 2000°F
ILLIUM G	58.0			6.0			6.0	22.0			Mo 6.0	CORROSION RESISTING ALLOY
INCO 700	49.00	0.1	0.05	0.5	0.2			15.0	3.0	2.0	Co-28.0 Mo-3.0	HEAT RESISTING
NICROTUNG	61.85	0.10						12.0	4.0	4.0	Co-10.0 W-8.0 B-0.5 Zr-0.05	HIGH TEMP. ALLOY
REFRACTALOY 26	38.77	0.03	0.80	16.0		1.00		18.00	0.2	2.0	Co-20.0 Mo-3.2	HIGH STRENGTH TO 1600°F
RENE' 41	55.3	.09						19.0	1.5	3.1	Co-11.0 Mo-10.0 B-.005	HIGH STRENGTH TO 1800°F
UNITEMP	51.6	0.24		9.5				16.3	1.9	3.2	Co-7.2 Mo-1.6 W-8.4 B-.008 Zr-.06	HIGH STRENGTH TO 2000°
UDIMET 500	53.57	0.80						18.0	2.9	2.9	Co-18.5 Mo-4.0 B-.006 Zr-.05	HIGH STRENGTH TO 2000 °F
ALUMEL	94.5		2.5			1.0		2.0				THERMOCOUPLE WIRE

*COMBINED AMOUNTS OF SILICON, MANGANESE AND CARBON TOTAL LESS THAN 1 PERCENT

Table 13—5. Mechanical properties of monel (nickel-copper) alloys.

ALLOY	CONDITION	TENSILE STRENGTH	YIELD STRENGTH	PER CENT ELONGATION 2-INCHES	HARDNESS-ROCKWELL
MONEL	ANNEALED	70-95,000	30-60,000	45-25	B60 MAX.
	QUARTER HARD	95-120,000	65-45,000	25-15	B74-82
	HALF HARD	104-130,000	80-110,000	15-8	B83-89
	THREE-QUARTER HARD	115-145,000	95-130,000	8-5	B90-93
	FULL HARD (SPRING)	130-160,000	120-150,000	5-2	B98 MIN.
MONEL "K"	ANNEALED	80-105,000	35-65,000	40-20	B85 MAX.
	ANNEALED-AGE HARDENED	120-150,000	90-110,000	25-10	C20-25
	SPRING-AGE HARDENED	160-200,000	14-190,000	8-3	C33-40

COMPILED FROM "A HANDBOOK ON WIRE ROD AND STRIP," ALLOY METAL CO., DIV. OF H. K. PORTER, PROSPECT PARK, PENNSYLVANIA

nickel-iron alloy known as *Invar* has an extremely low coefficient of thermal expansion (0.64×10^{-6}) between the temperature range of 32-212°F when quenched from 1526°F. This low coefficient of expansion is applicable over a rather narrow range of temperatures. The stability of quenched Invar may be improved by a 200 to 300°F anneal followed by slow cooling to room temperature over a period of months. Invar is used in applications such as absolute standards of length, compensating balance wheels in clocks, tuning forks, and variable condensers.

The alloy of nickel and iron containing 46 percent nickel is known as "platinite" since it has the same coefficient of expansion of glass and may be used in place of platinum for metal-to-glass seals.

An alloy of 79 percent nickel in iron has very high magnetic permeabilities at low magnetic fields and has been successfully used in the construction of underwater transoceanic telegraph cables.

Since the nickel-iron alloys are austenitic, they can be hardened only by cold working. Recent findings show that additions of titanium and aluminum make these alloys precipitation hardenable. Free machining properties are imparted to nickel-iron alloys by the addition of 0.25 percent selenium.

13.26 Nickel-Iron-Chromium Alloys.

Nickel-iron-chromium heat and corrosion resisting alloys are able to withstand high temperatures without scaling. These alloys have high creep resistance and good resistance to chemical corrosion and oxidation at temperatures up to 2200°F. Alloys containing aluminum and titanium are age hardenable. Inconel was first used in food processing equipment such as that used for pasteurizing milk. One of its outstanding properties is that of withstanding repeated heating and cooling (0° to 1600°F) without becoming embrittled. It is this property coupled with its excellent corrosion resistance which has encouraged its use as airplane exhaust manifolds, and heaters. Inconel is also used extensively for nitriding containers, carburizing boxes, muffles, and thermocouple protection tubes.

Nimonic is a high temperature-high strength alloy used in gas turbines. It maintains its strength above 1700°F and has good resistance to high temperature creep.

Nichrome is a popular and excellent metal for heating elements in household appliances, industrial furnaces, and rheostats for electronic equipment. Nichrome is also used for dipping baskets in acid pickling.

Many of the nickel-iron-chromium alloys give good resistance to oxidation, heat fatigue, and carburizing gases.

13.27 Nickel-Base Super-Alloys.

Most of the alloys listed in Section 5 of Table 13–4 have been developed for high strength, high temperature applications. These metals, often referred to as heat-resisting super-alloys were developed for application at temperatures from 1000 to 2000°F. High nickel super-alloys are of the Al-Ti age hardenable type. In these alloys, chromium provides oxidation and corrosion resistance along with some auxiliary strengthening. Columbium, molybdenum, tungsten, and tantalum additions provide solid solution strengthening of the matrix. However, the major part of the strengthening at high temperatures is due to the precipitation of the gamma prime phase, $Ni_3(Al, Ti)$. In some superalloys cobalt is often present replacing substantial amounts of nickel. Other alloys have iron as the base with chromium and nickel additions. Boron and zirconium additions impart improvement in high temperature creep properties, and increased malleability. Because of the outstanding high temperature strength of the superalloys they are inherently difficult to deform during hot working and are often sensitive to cracking. Likewise cold working is difficult.

These alloys are more difficult to machine than regular austenitic stainless steels and require rigid machines and slow cutting speeds.

Although superalloys were originally developed for jet engine parts, they are now finding application in other areas requiring outstanding high temperature strength. Such applications include automotive and auxiliary gas turbines, air frames, rocket engines, missiles, and various hot die applications.

References

Bunn, E. S. and Wilkins, R. A. *Copper and Copper-Base Alloys.* New York: McGraw-Hill, 1943.

Ellis, O. W. *Copper and Copper Alloys.* Cleveland, O.; American Society for Metals, 1948.

Flinn, Richard A. *Fundamentals of Metal Castings.* Reading, Mass.: Addison-Wesley Pubg. Co., 1963.

Lyman, Taylor, ed. *Metals Handbook, 1961 ed.* Cleveland, O.; American Society for Metals.

International Nickel Co. *Nickel and its Alloys.* New York.

———. *Nickel and Nickel-Base Alloys.*

———. *Age-Hardening INCO Nickel Alloys.*

———. *Engineering Properties of Duranickel.*

———. *Engineering Properties of Inconel.*

———. *Engineering Properties of Monel.*

———. *Engineering Properties of Nickel.*

———. *Copper-Nickel Alloys.*

Lead, Tin and Zinc

Lead

14.1 Introduction. Lead is probably one of the oldest metals known to man. The Egyptian Pharoahs, and Assyrians, and Babylonians had many uses for lead in ornamental objects and for structural purposes. The Chinese made money from lead about 2000 B.C. The bronze coinage of ancient Greece and Rome contained up to 30 percent lead. Perhaps the most interesting historical application of lead was its use by the Romans for water pipes. The pipes were made in 15 standard sizes and 10 foot lengths. Many of these lead pipes, in almost perfect condition, have been dug up in modern times from the ruins of Pompeii. The old Roman baths at Bath, England, still use the same lead pipe installed by the Romans 1900 years ago. Roman lead pipe was made by folding heavy sheets of cast lead together and fusing the seams. Many

magnificient buildings erected in Europe during the fifteenth and sixteenth centuries still stand under their original lead roofs. Lead ornaments, statues, cisterns, and other architectural treasures have remained through the centuries in a state of excellent preservation.

14.2 Properties and Uses of Lead. Both lead and the alloys in which it is a major constituent are characterized by high density, softness, malleability, low melting point, low strength, and high corrosion resistance. Less well known properties are its low electrical conductivity, its lubricating properties and its high coefficient of thermal expansion. The melting point of lead is 621°F. Its density is 11.34 grams per cubic centimeter. It has a face centered cubic structure and alloys well with many metals including antimony, tin, bismuth, and cadmium. The high malleability of lead permits it to be rolled into foil as thin as 0.0005 in. or less. Since the recrystallization point of pure lead is below room temperature it cannot be work hardened by usual methods.

Lead and its alloys can be extruded, drawn, rolled, cast, stamped, or spun. Furthermore, it can be applied as a coating to other metals by hot dipping, electroplating, or spraying.

As for applications, lead probably has more varied fields of application than any other metal. The greatest tonnage of lead used is in the metallic form in either the commercially pure form or as a lead alloy. A large quantity of lead is also used in chemical compound forms as white lead, red lead, lead chromate, and tetraethyl lead, to mention only a few.

Metallic lead is used for x-ray shielding, type metal, storage batteries, cable coverings, ammunition, bearings, calking, collapsible tubes, solder, tank linings, coatings, and as an alloy addition to steel, aluminum, bronze, and other metals to improve machinability.

14.3 Lead Alloys. Antimony and tin are the metals most commonly alloyed with lead. Antimony is used to increase hardness and strength in storage battery plates, type metal, sheet, pipe, and castings. Antimony additions also increase the recrystallization temperature. Tin likewise increases hardness and strength but is mostly used in soft solder because of its melting characteristics and be-

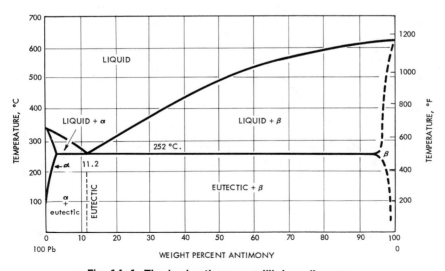

Fig. 14–1. The lead-antimony equilibrium diagram.

cause it imparts the ability to the solder for bonding with metals like steel and copper.

The alloy diagram for lead-antimony alloys is shown in Fig. 14–1. This is a simple eutectic system with the eutectic composition at 11.2 percent antimony. The solid solubility of antimony in lead decreases with decreasing temperature below the eutectic temperature. This permits age hardening of rapidly cooled lead-antimony and lead-antimony-tin alloys. If type metal alloy containing 1 percent tin and 2 percent antimony is quenched from a temperature slightly below the eutectic temperature, age hardening proceeds rather slowly and may require several days to achieve maximum hardness. The age hardening time can be greatly shortened by aging for one minute at 212°F or for one hour at 185°F. Hardness in the aged condition may reach 18-21 Brinell whereas in the unaged condition its hardness may be as low as 7-8 BHN.

The properties of cast lead-antimony alloys are shown in Fig. 14–2. A maximum tensile strength of 7,670 psi is achieved with a 10 percent antimony addition.

The equilibrium diagram for lead-tin alloys is shown in Fig. 14–3. The eutectic composition consists of 61.9 percent tin and has a solidus temperature of 361°F. The eutectic composition is used as a "fine solder" where temperature requirements are critical. The usual soft solder composi-

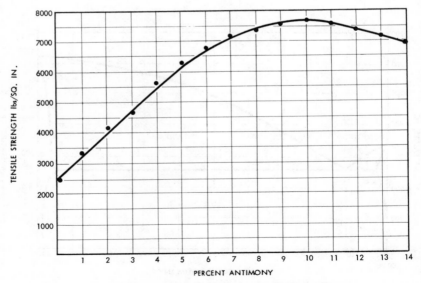

Fig. 14–2. Properties of cast lead-antimony alloys.

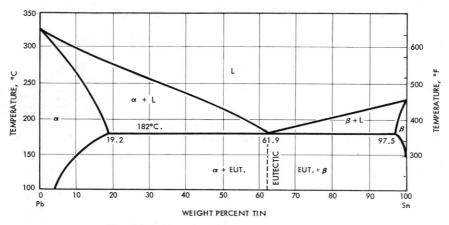

Fig. 14–3. The tin-lead equilibrium diagram.

tion is reduced to 50 percent tin and lead is increased to 50 percent. A few other solders have been developed as substitutes to reduce the use of tin.

Terne metal is a lead-tin alloy containing from 10 to 25 percent tin. This alloy has been used to coat steel sheets for such applications as roofing and automotive fuel tanks.

Type metals used in the printing industry are generally alloys of lead, antimony, and tin, with small amounts of copper in the hardest alloys. The microstructure of a linotype metal is shown in Fig. 14–4. These alloys have high fluidity and give sharp definition of printing characters. The fine detail from type metals has been erroneously ascribed to slight expansion during solidification. In reality these metals gen-

erally contract somewhere between 2-3 percent. Most type metals may be age-hardened to improve wear resistance. Grain size in cast type

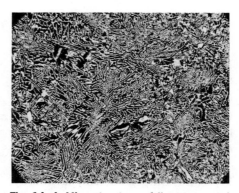

Fig. 14–4. Microstructure of linotype metal. Composition is 12% antimony, 4% tin and 84% lead. The ternary eutectic structure consists of alpha and beta phases, 300X. (American Smelting and Refining Co.)

459

Table 14–1. Composition and mechanical properties of lead alloys.

ALLOY	COMPOSITION	TENSILE STRENGTH PSI	YIELD STRENGTH PSI	ELONGATION %	BHN	TYPICAL USES
CHEMICAL LEAD	99.90 + Pb	2,385	1,180	29		CHEMICAL STOR-AGE TANKS
CORRODING LEAD	99.73 + Pb	1,700	800	30	3.2–4.5	STORAGE BAT-TERIES, CABLE SHEATHING, PAINT CALKING, ANTI-KNOCK FLUID
ANTIMONIAL LEAD	99 Pb – 1 Sb	6,100	4,800	60	14.5	COATINGS, BODY SOLDER
HARD LEAD (AGED)	94 Pb – 4 Sb	11,670	----	6.3	24	SHEET, PIPE
SOFT SOLDERS	95 Pb – 5 Sn	3,400	1,500	50	8	COATING AND JOINING
	80 Pb – 20 Sn	5,800	3,650	16	11.3	COATING AND JOINING, BOBY SOLDER
	50 Pb – 50 Sn	6,100	4,800	60	14.5	GENERAL PURPOSE SOLDER
LEAD BASE BAB-BITTS (CHILL CAST)	85 Pb – 10 Sb–5 Sn	10,000	----	5	19	LIGHT LOADS, SHAFTING
	75 Pb – 15 Sb–10 Sn	10,500	----	4	22	MODERATE LOADS, PUMPS, BLOWERS
	83 Pb–15 Sb–1S +1 As	10,350	----	2	20	HIGH LOADS, DIESEL ENGINES
	83.5 Pb–12.75 Sb, 3 As–0.75 Sn	9,800	----	1.5	22	ELEVATED TEM-PERATURE BEAR-INGS, TRUCKS

COMPILED FROM "METALS HANDBOOK," 1948 EDITION
AMERICAN SOCIETY FOR METALS

metal depends upon cooling rate during solidification. Small alloy additions such as calcium, lithium, or magnesium may also be used to produce desirable fine grains.

Babbitt is a lead-base bearing alloy containing lead, tin, antimony, and usually arsenic. These are used for bearings in automotive engine connecting rods, diesel engine bearings, railroad car journal bearings, pumps, and construction equipment.

The nominal composition and typical mechanical properties of some lead alloys are given in Table 14–1.

Wood's Metal, which melts at 154°F, has a nominal composition of 33.1 percent Pb, 14.3 percent Cd, 19 percent Sn, 33.6 percent Bi. This metal is used in fuse links for heat actuated water spray fire extinguishing systems.

Tin

14.4 Properties of Tin and Its Uses.
Tin is a soft, white metal with good corrosion and lubricating properties. Tin exists as tetragonal structure from its melting point (450°F) down to 55.8°F when it undergoes a polymorphic transformation from its normal tetragonal structure (white tin) to a cubic form (gray tin). This transformation results in disintegration of the metal to a coarse powder, and is known as *Tin Pest*. The change is accompanied by a reduction of density from 7.3 for the white tin to 5.7 for the gray tin. The transformation is very sluggish and requires considerable undercooling to initiate it. Impurities in tin or alloy additions, such as 0.1 percent bismuth, will prevent this transformation from occurring.

During the plastic deformation of a single crystal of tin, the gliding or shearing action occurs by distinct movement on certain planes which make a distinctive audible sound known as "tin cry." Chill cast tin has an elongation of 69 percent and a tensile strength of 2100 psi. At 100°F the elongation of tin exceeds 80 percent. Hardness of tin is 3-4 Brinell at room temperature.

Although the production of tin is relatively small in tonnage its uses are widely spread industrially. The major application of tin is as an electrolytically plated coating approximately 15 millionths of an inch thick on cold-rolled low-carbon steel sheet. This tin plate is subsequently fused to brighten the surface coat and then made into containers for perishable foods. A second major application of tin is as an alloying element. Tin is alloyed with copper to produce various tin brasses and bronzes; with lead to produce solder; with antimony, copper or lead to form babbitts, type metals and cast alloys. Other uses of tin are in production of foil, tinned ware, dental amalgams, fusible alloys, pharmaceutical, and chemical applications.

14.5 Tin Alloys. Tin is alloyed with antimony, silver, or lead to produce soft solders. The tin base solders have higher strength than lead base alloys. The solders containing 5 percent antimony or silver are preferred for electrical equipment because of their good electrical conductivity. Copper joints soldered with tin-based solders have strengths ranging between 14,000-29,000 psi in tension and 8,000-11,000 psi in shear.

Tin babbitts are noted for their anti-galling characteristics and corrosion resistance as bearing alloys. Lead in amounts up to 30 percent may be added to reduce cost. Lead in babbitts is reduced to 0.35 percent when high temperatures are to be encountered because of the formation of a tin-lead eutectic which

461

Fig. 14—5. Tin base hard babbitt of 84% tin, 7% copper, and 9% antimony, 50X. Needle-like crystals of CuSn and rectangular SnSb are contained in a ductile ternary eutectic matrix.

Table 14—2. Composition, properties, and uses of tin alloys.

ALLOY	NOMINAL COMPOSITION PERCENT	CONDITION	TENSILE STRENGTH PSI	ELONGA- TION PERCENT	BHN	TYPICAL USES
TIN (GRADE A)	99.8 Sn (MIN.)	CAST	2,100	55	5.3	ELECTROTIN- NING, ALLOY- ING, POWDER
HARD TIN	99.6 Sn-0.4 Cu	80% RED.	4,000	---	---	COLLAPSIBLE TUBE AND FOIL
ANTIMONIAL SOLDER	95 Sn – 5 Sb	CAST	5,900	38	---	SOLDER FOR ELECTRICAL EQUIPMENT
TIN-SILVER SOLDER	95 Sn – 5 Ag	SHEET	4,600	49	---	SOLDER FOR ELECTRICAL EQUIPMENT
SOFT SOLDER	70 Sn – 30 Pb	CAST	6,800	---	12	JOINING AND COATING
	63 Sn – 37 Pb	CAST	7,500	32	14	ELECTRICAL WORK
TIN BABBITT	91 Sn – 4.5 Sb-4.5 Cu	CHILL CAST	9,300	---	17	BEARINGS, DIE CASTINGS
	89 Sn-7.5 Sb-3.5 Cu	CHILL CAST	12,600	---	24	BEARINGS
	83.4 Sn-8.3 Sb-8.3 Cu	CHILL CAST	10,000	1	27	BEARINGS
	65 Sn-18 Pb-15 Sb – 2 Cu	CHILL CAST	7,800	1.5	225	BEARINGS; DIE CASTINGS
TIN FOIL	92 Sn-8 Zn	SHEET	8,700	40	---	PACKAGING, GASKETS
WHITE METAL	92 Sn-8 Sb	CHILL CAST (ANNEALED)	6,500	50	17	COSTUME JEWELRY
PEWTER	91 Sn-7 Sb-2 Cu	SHEET	7,600	50	8	VASES, CAND- LESTICKS, BOOK ENDS

COMPILED FROM DATA IN "METALS HANDBOOK" 1948 EDITION
AMERICAN SOCIETY FOR METALS

melts at 361°F. Copper additions of 2-8 percent and 10-15 of antimony result in the formation of a dark eutectic matrix containing embedded needle-like crystals of CuSn and cube-like crystals of SbSn (Fig. 14–5). The hard crystals of CuSn and SbSn increase the hardness and wear resistance of the bearing alloy.

White metal is a tin casting alloy containing 92 percent Sn and 8 percent Sb. This alloy has been used for costume jewelry and other small castings.

Pewter is a tin alloy which contains from 90 to 95 percent tin and 1 to 3 percent copper and the remainder antimony. Pewter has been used for articles such as vases and hollow slush cast book ends. Pewter is readily solderable and may be formed by hammering, spinning, and casting. The composition and uses of some tin alloys are given in Table 14–2.

Zinc

14.6 Properties of Zinc and Its Uses.
Zinc coatings are well known for their ability to prevent corrosion of steel. Zinc is more highly anodic than steel and in corrosive atmospheres the zinc is consumed while protecting the steel from any attack. Thus, the life of the zinc coating is proportional to its thickness.

Zinc has a tensile strength of about 16,000 psi. Wrought zinc has the characteristic of flowing under loads that are substantially below the ultimate strength. These creep characteristics for zinc make it difficult to establish elastic moduli or yield strength. Standard engineering formula based on elastic materials cannot be used for zinc alloys. Approximate formulas have been developed to determine the elongation that will occur in a definite period of time under a given load.

Zinc has a melting point of 787°F, a close packed hexagonal lattice structure, and a density of 7.14 at room temperature. Zinc and zinc alloys may be shaped by rolling, drawing, extruding, and casting. The recrystallization temperature of pure zinc (and soft grades) is below room temperature. It cannot be work-hardened.

Zinc and its alloys find wide applications as metallic coatings for corrugated and flat sheet, chains, fencing, hardware, pipe, electrical conduit, screws, tanks, wire, and wire cloth. Zinc die castings include automotive parts, household utensils, office equipment, hardware, padlocks, toys, and novelties.

Galvanized coatings are applied either by dipping, electroplating, mechanical plating, or flame spraying.

Zinc sheet is used for battery cans, eyelets, grommets, address plates, photoengraver's, and lithographer's sheet.

Zinc oxide is used in the manufacture of dental cement, paint, floor tile, matches, pottery, and rubber goods.

14.7 Zinc Alloys. Commercial wrought zinc contains small amounts of impurities such as lead, iron, and cadmium. Commercial alloys contain the above impurities and small percentages of copper, magnesium, aluminum, and tin.

For deep drawing purposes a rel-atively pure zinc should be used. Additions of lead or cadmium results in higher hardness, higher stiffness, and more uniform etching quality. Copper additions of 0.85 to 1.25 percent improve stiffness, creep resistance, and raise the recrystallization temperature to the point where the alloy can be work hardened.

Copper and titanium additions produce outstanding creep resistance in zinc alloys. Typical wrought

Table 14–3. Composition and mechanical properties of wrought zinc alloys.

ALLOY AND COMPOSITION	CONDITION	TENSILE STRENGTH PSI (1)　　　(2)	ELONGATION PERCENT (1)　(2)	BHN	CHARACTERSTICS
COMMERCIAL ROLLED ZINC Zn–0.08 Pb	H.R.	19,500–23,000	65 – 50	38	DEEP DRAWING QUALITIES
	C.R.	21,000–27,000	50–40		CREEPS UNDER LOAD, PARTICULARLY ABOVE 100°F
COMMERCIAL ROLLED ZINC Zn–0.06 Pb–0.06 Cd	H.R.	21,000–25,000	52–30	43	DRAWING AND SPINNING QUALITIES WITH SOME RIGIDITY, CREEPS ABOVE 100°F
	C.R.	22,000–29,000	40–30		
COMMERCIAL ROLLED ZINC Zn–0.3 Pb–0.3 Cd	H.R.	23,000–29,000	50–32	47	HIGH HARDNESS AND STIFFNESS; UNIFORM ETCHING QUALITY, CAN BE WORK HARDENED
	C.R.	25,000–31,000	45–28		
COPPER HARDENED ROLLED ZINC	H.R.	24,000–32,000	20–15	52	HIGH HARDNESS AND STIFFNESS; EASILY WORK HARDENED; GOOD DUCTILITY AND CREEP RESISTANT
	C. R.	32,000–36,000	5–3		
ROLLED ZINC ALLOY	H.R.	28,000–36,000	20–10	61	MAXIMUM HARDNESS, STIFFNESS AND CREEP RESISTANCE. CAN BE SEVERELY WORK HARDENED
	C,R.	37,000–48,000	20–2		

COMPILED FROM "METALS HANDBOOK," 1948 ED., AMERICAN SOCIETY FOR METALS.
(1) PARALLEL TO ROLLING DIRECTION

(2) PERPENDICULAR TO ROLLING DIRECTION

zinc alloys along with their mechanical properties are given in Table 14–3.

The major use of zinc as a structural material is in the form of zinc die castings, examples of which are shown in Fig. 14–6. Zinc die casting alloys are low in cost, and have good strength. They can be cast to close dimensional limits, require minimum machining and possess good resistance to surface corrosion. Zinc

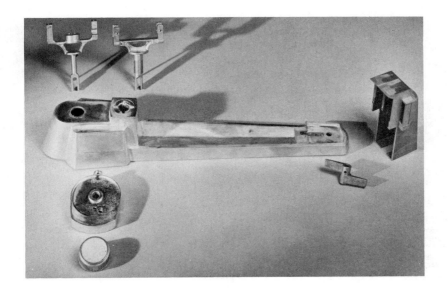

Fig. 14–6. Typical zinc die casting. (American Zinc Institute)

Table 14—4. Composition and mechanical properties of casting zinc alloys.

ALLOY[1] AND COMPOSITION	CONDITION	TENSILE STRENGTH, PSI	ELONGATION PERCENT	BHN	CHARPY IMPACT FT. LBS	TYPICAL USES
ZAMAK – 3 Zn–4AL–0.04 Mg	DIE CAST	41,000	10	82	43	AUTOMOTIVE PARTS, UTENSILE, OFFICE EQUIPMENT, HARDWARE
ZAMAK–5 Zn–4 A1–1 Cu–0.04 Mg	DIE CAST	47,000	7	91	48	AUTOMOTIVE PARTS, PADLOCKS, TOYS
ZAMAK–2 Zn4 A1–3Cu–0.03 Mg	SAND CAST	20,000–30,000	-----	70-100	---	DROP HAMMER DIES
ZAMAK–5 Zn–4 Al–1 Cu–0.04 Mg	SAND CAST	20,000–30,000	-----	70-100	---	DROP HAMMER DIES
ZINC–BASE SLUSH CASTING ALLOY Zn–4.75 A1–0.25 Cu	CHILL CAST	28,000	-----	---	3	LIGHTING FIXTURES
ILZRO 12 (PROTOTYPE ALLOY)	GRAVITY CAST	46,400	3	101	18	PROTOTYPE DEVELOPMENT; RELATIVELY INSENSITIVE TO COOLING RATE
ILZRO 14 (NEW DEVELOPMENT)	COLD CHAMBER CAST	NOT AVAILABLE	NOT AVAILABLE	---	---	CREEP RESISTANT; STABLE PHYSICAL PROPERTIES

[1]SPECIAL HIGH GRADE ZINC (99.99 % + PURE) MUST BE USED AS THE BASIC MATERIAL IN MAKING THESE ALLOYS

die castings are limited to maximum operating temperatures of 200°F because of their rapid loss in strength and hardness above this temperature.

The commonly used zinc casting alloys are given in Table 14—4.

The casting alloys are subject to intergranular corrosion, warping, and cracking if the impurity limits are exceeded. Typical zinc castings have a shrinkage of 0.14 in/ft during solidification and cooling. Some of the zinc base casting alloys are subject to embrittlement and growth on aging and mechanical properties such as impact strength or tensile strength after aging 20 years may be reduced as much as 25 percent or more.

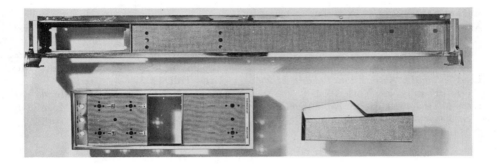

Fig. 14—7. New finishes on zinc die castings include anodizing and clear polymer coatings. (American Zinc Institute)

Recent developments in finishing of die cast zinc alloys include anodizing processes, clear polymer coatings, and improved electroplating. Examples of new finishes on die castings are shown in Fig. 14—7.

467

References

American Zinc Institute. *Wrought and Cast Zinc Alloy Characteristics.* New York, 1967.

————. *Zinc Die Castings.* 1962.

Lead Industries Association. *Lead in Modern Industry.* New York, 1952.

Lyman, Taylor, ed. *Metals Handbook, 1961 ed.* Cleveland, O.: American Society for Metals.

Hoyt, Samuel L. *Metal Data.* New York: Reinhold Publishing, 1952.

Mondolfo, L. F. and Zmeskal, Otto. *Engineering Metallurgy.* New York: McGraw-Hill, 1955.

Samans, Carl H. *Engineering Metals and Their Alloys.* New York: Macmillan, 1949.

Precious and Specialty Metals

15.1 Introduction. Of all the metals found in nature, eight of these are known as "noble" or "precious" metals. These metals are precious because of their high price and noble because of their resistance to oxidation and to solution by inorganic acids. The precious metals include silver, gold, and the six platinum group metals—platinum, palladium, rhodium, ruthenium, iridium, and osmium. This group of metals is characterized by softness in the pure state, high ductility and malleability, good electrical conductivity, heat resistance, high reflectivity, and excellent corrosion resistance.

Specialty metals, which will be discussed at the end of the chapter include germanium, selenium, tellurium, and silicon for use in solid-state electronic devices; uranium, thorium, and zirconium for nuclear applications; and high-strength, high-temperature refractory alloys.

Precious Metals

15.2 Silver and Silver Alloys. Silver and gold were discovered and used in early Biblical times. These metals were present in their native state in small quantities in the regions first civilized. Silver and gold were rare but could be worked readily using crude equipment and could be melted without too much difficulty. Since these metals remained tarnish free, and were rather precious because of their limited availability, they became articles of adornment and coinage.

It is believed that prior to the use of coal, very little sulfur was present in the atmosphere and silver remained bright and untarnished over long periods of time.

Silver is the least expensive of the precious metals and because of this fact, coupled with its desirable properties, it is widely used industrially.

Pure silver has the highest electrical conductivity of any known metal. It is also an excellent conductor of heat. These two properties have made silver useful in electrical conductors, contacts, and chemical equipment. The photosensitivity of silver halide salts is the basis of photography. The high reflectivity of silver in the visible region and the ease with which it may be electroplated makes silver attractive for reflectors, silverware, and jewelry. Silver is also used in bearings and in brazing alloys.

Metallic silver has the following characteristics:

Density	10.5 grams/cm³
Melting point	1760.9°F
Electrical resistivity	1.59 microhm-Cm (68°F)
Crystal structure	Face centered cubic
Tensile strength	18,200 psi
Yield strength	7,900 psi
Elongation	54 percent

Silver is a soft, white metal that is extremely ductile. It may be formed with high reductions by both hot and cold forming methods. The annealing temperature is approximately 570°F for silver cold worked at least 50 percent.

Silver-copper alloys have been used for thousands of years for silver coins, but only recently was it discovered that these alloys could be age hardened to improve strength and resistance to denting. Silver and copper form a simple eutectic alloy with the eutectic composition being at 28.1 percent copper and the eutectic temperature at 435°F. The maximum solubility of copper in silver is 8.8 percent at the eutectic temperature, and diminishes to less than 1 percent at room temperature.

The most important of the silver-copper alloys are those used for coining and jewelry; these contain between 5 and 10 percent copper. *Sterling silver*, the British coining metal, contains 7.5 percent copper. The eutectic composition containing 28 percent copper is used as a solder. However, the silver solders frequently contain small amounts of zinc, cadmium, phosphorus, or tin. These alloys are used for many applications in the joining of ferrous and non-ferrous materials. Caution should be exercised in the use of silver solders containing cadmium as the fumes are toxic.

Silver-base alloys containing mercury, tin, copper, and zinc are used as dental amalgams. Alloys of silver with tungsten, molybdenum, graphite, nickel, and cadmium, or lead oxide are used in electrical contacts requiring resistance to arcing, and minimum welding of contacts. The above metals are immiscible with liquid silver and hence are made by the methods of powder metallurgy discussed in Chapter 17.

15.3 Gold and Gold Alloys. For many centuries gold has been the world's standard of value. Gold was found in small quantities as a native metal by early man and became a natural medium of exchange. The exceptional malleability of gold has led to its use in the form of beaten leaf for achitectural ornamentation. Ancient historical and religious records have also been found in the Middle East and in South America inscribed on thin gold sheets. Gold is a face centered cubic, soft, yellow metal, non-tarnishable and exceptionally corrosion resistant. The melting point of gold is 1945.4°F. Its tensile strength is 18,000 psi in the cast condition and 32,000 psi after 60 percent reduction. Annealed gold has 45 percent elongation and a Brinell hardness value of 25. Gold can be worked at any temperature below the melting point and may be joined by nearly all welding processes. Unless one is licensed it is illegal to deal in gold.

The high cost for precious metals (Table 15–1) has led to the development of extremely thin electroplated surface coatings. These coatings on either metal, glass, or ceramic substrates provide the corrosion resistance and reflectivity of the pure metals while requiring a minimum quantity of metal. Pure gold leaf is used to a limited extent in the form of very thin foil (0.000005 in.) for lettering on books. The high malleability of gold may be demonstrated by the fact that gold leaf has been produced thin enough so that under high illumination it will transmit a small amount of greenish colored

Table 15–1. Prices of precious metals.

METAL	PRICE* PER TROY OUNCE, DOLLARS	CALCULATED COST (PER CUBIC INCH) DOLLARS
OSMIUM	230	2739
IRIDIUM	165	1968
RHODIUM	197	1288
PLATINUM	100	1130
RUTHENIUM	55	361
GOLD	35	356
PALLADIUM	33	209
SILVER	1.29	7

*SOURCE: AMERICAN METAL MARKETS. SEPT. 1, 1966 AS QUOTED FROM "CHARACTERISTICS OF PRECIOUS METAL ELECTRODEPOSITS," INTERNATIONAL NICKEL COMPANY

light and objects can actually be seen through it.

Gold plating and gold cladding are used extensively to produce cheaper jewelry with the appearance and surface properties of gold. Clad products have a relatively thick coating of gold which is bonded by either heat and pressure, solder, or welding. For gold-clad (gold-filled) stock, the ratio by weight of gold to the total weight of the composite article is specified. For example, a $\frac{1}{10}$ 12-carat gold filled composite material would require a sheet of 12-carat gold equal to 10 percent of the composite assembly weight.

Gold alloys used for jewelry are described by carat and color; in this system 100 percent gold corresponds to 24 carats, an alloy containing 50 percent gold corresponds to 12 carats, and so on. Gold color is determined by the alloying metals used and the amount present. Gold is commonly alloyed with copper and silver to increase its hardness and strength and to minimize cost. Gold is completely miscible in both of these metals in the liquid and solid states. Other less common alloying elements are lead, platinum, nickel, palladium, and zinc.

White gold is gold alloyed with nickel, copper, and zinc. It is used primarily as a substitute for platinum in jewelry. *Green, yellow,* and *red* golds are alloys of gold, copper, and silver. These alloys are used in jewelry and dental castings. *Pink* or *suntan golds* are age hardenable alloys containing gold, copper, nickel, and zinc. Tensile strengths range from 60,000 to 90,000 psi in the annealed state to 120,000 psi in the 50 percent cold worked condition. These golds are susceptible to stress corrosion-cracking under certain conditions and may darken when exposed to sulfur containing or oxidizing atmospheres. Certain of these alloys have been used for jewelry but are principally used for eyeglass frames.

Some of the more recent uses for gold have been for printed circuit boards, commutator and encoder discs, and sealed-reed electronic relays.

15.4 Platinum and Platinum Alloys. According to anthropologists who worked along the Esmeraldas coast of Ecuador native Indians made ornaments of platinum prior to the discovery of America by Columbus. One story is that the difficulty of separating the gold and platinum or "platina" as obtained from the South American mines caused King Charles III of Spain to issue a special order. His order was that the "platina" found with gold in Central and South American rivers be replaced in the waters, hoping that by further immersion, it might "ripen" into gold. Specimens of "platina" are known to have reached

England as early as 1741. Here, this heavy metal attracted the attention of several prominent scientists. They discovered that they could not get sufficient heat to melt platinum but that it could be forged at a white heat. The resistance of platinum to high temperature oxidation and its immunity to attack by most acids led to its use as laboratory equipment which use continues to this day. Platinum has a melting point of 3216°F, and is resistant to many corrosive agents such as boiling sulfuric acid and molten chloride solutions.

Platinum is one of six elements known as the *Platinum Group*. This group of transition metals in the Periodic Table, Fig. 15–1, has distinctive family properties and similarities. This family in one way resembles two sets of triplets in that ruthenium, rhodium, and palladium have densities of around 12 while the densities of osmium, iridium, and platinum are near 22. In other ways, the family resembles three sets of twins. For example, ruthenium and osmium have hexagonal close packed structures, and are quite hard. Rhodium and iridium, with their face centered cubic structure are more ductile and have the lowest electrical resistivity of the platinum metals. Palladium and platinum have the lowest annealed hardness and are easily worked.

AT. NO. 44	M.P. 4190°F	AT. NO. 45	M.P. 3560°F	AT. NO 46	M.P. 2825 °F
Ru RUTHENIUM HCP		Rh RHODIUM FCC		Pd PALLADIUM FCC	
DENSITY 12.45	AT. WT. 101.07	DENSITY 12.41	AT.WT. 102.905	DENSITY 12.02	AT WT. 106.4
AT. NO. 76	M.P. 5522°F	AT. NO 77	M.P. 4430°F	AT. NO. 78	M.P. 3216°F
Os OSMIUM HCP		Ir IRIDIUM FCC		Pt PLATINUM FCC	
DENSITY 22.61	AT.WT. 190.2	DENSITY 22.65	AT.WT. 192.2	DENSITY 21.65	AT.WT. 195.09

Fig. 15–1. The platinum group metals. (American Metal Market, Sept. 1, 1966)

Table 15–2. Physical and mechanical properties of the platinum group metals.

PROPERTY	RUTHENIUM	RHODIUM	PALLADIUM	OSMIUM	IRIDIUM	PLATINUM
ATOMIC NO	44	45	46	76	77	78
ATOMIC WT	101.07	102.905	106.4	190.2	192.2	195.09
CRYSTAL STRUCTURE	HCP	FCC	FCC	HCP	FCC	FCC
LATTICE CONSTANT AT 20°C, KX UNITS*	a 2.7001 c/a 1.5820	3.7963	3.8829	a 2.7286 c/a 1.5800	3.8317	3.9152
DENSITY AT 68°F gm/cm^3	12.45	12.41	12.02	22.61	22.65	21.45
MELTING POINT °F	4190	3560	2825	5522	4430	3216
BOILING POINT °F	7412	6692	6252	9032	8132	6872
YOUNGS' MODULUS— STATIC PSI	60x10^6	46.2x10^6	16.75x10^6	81x10^6	75x10^6	24.8x10^6
TENSILE STRENGTH ANNEALED, psi	78,000	100,000	25,000	-----	155,000	18,000
HARDNESS DPN— ANNEALED-ELECTRO- DEPOSITED	250–450 900–1300	120 800–900	37–40 200–400	500 -----	200–240 900	37–42 200–400

*1 KX UNIT = 1.00202 A° (ANGSTROM)

BASED ON: THE PLATINUM GROUP METALS IN INDUSTRY, THE INTERNATIONAL NICKEL CO., INC. NEW YORK.   10005

Physical and mechanical properties of the platinum group metals are given in Table 15–2. Platinum metals are used primarily for industrial applications, although about 30 percent of the annual production is for jewelry. Speaking of annual production, it was not long ago that the annual world production of platinum metals could be stored in a cubical bin slightly over 3 feet on a side. Important sources of platinum metals include Canada, Russia, South Africa, Colombia, and Alaska.

Platinum is the most noble of metals and is used for chemical laboratory equipment (Fig. 15–2), thermocouple wire, as a catalyst in the manufacture of gasoline, bushings for extrusion of molten glass fibers and rayon fibers, containers for x-ray therapy, and humidity control devices.

15.5 Palladium. Palladium is a widely useful silver-white noble metal resembling platinum in many respects. It is extremely ductile and

Fig. 15–2. Platinum and platinum alloy equipment used in chemical laboratories. (Engelhard Minerals and Chemical Co.)

can be worked hot or cold. Cold working will increase tensile strength from about 30,000 psi to about 55,000 psi with a 50 percent reduction, and to 65,000 psi with 75 percent reduction. Like gold, palladium can be beaten into leaf as thin as 1/250,000 of an inch. Palladium is the lightest of the platinum metals and also has the lowest melting point. Palladium is generally resistant to corrosion by single acids, alkalies, and simple salt solutions. It is attacked, however, by nitric and hot sulfuric acid, ferric chloride, chlorine, bromine, and iodine.

Palladium is used extensively in automatic dial telephone contacts, with an average of five thousand contacts required during the connection of one phone with a dialed number. A useful life of one billion operations may be achieved with a 0.010″ thick contact. Palladium contacts resist atmospheric corrosion, and spark erosion assuring quiet circuits and long service. It is outstanding in its ability to form extensive ductile solid solutions with other metals. One alloy of palladium-platinum-gold is used in dentistry for the structural framework of plates,

475

partial dentures, pins, bands and wires used in straightening of teeth. Controllable hardness, resiliency, high strength, and resistance to oral fluids are among its properties.

Palladium coatings may be successfully applied by electroplating, cladding, electroforming, and other common coating methods. Palladium is used in film resistors, thermocouples, brazing alloys, selectively permeable membranes; catalysts are also numbered among the important uses of palladium and its alloys.

15.6 Iridium. Typical uses of iridium include electrical contacts, catalysts, aircraft sparkplug electrodes, and high temperature crucibles. It is used as an oxidation resistant coating for refractory metals and graphite to be operated at very high temperatures, exceeding 3600°F, as a hardening element in platinum; and for applications requiring chemical inertness and high temperature strength. Iridium has a number of interesting features including the distinction of being the heaviest element known with a specific gravity of 22.65. Osmium is a close second to iridium with a specific gravity of 22.61. The melting point of iridium is 4430°F—second highest of the platinum group metals. The modulus of elasticity is iridium is one of the highest known for an element at 75,000,000 psi. Iridium shows a high degree of work

hardening capability for a face centered cubic metal. Following a 20 percent cold reduction, there is an increase of hardness of 250 DPH (Diamond pyramid hardness) for iridium as compared with an increase of 30 DPH for pure platinum worked a similar amount. The excellent high temperature strength properties of iridium shown in Fig. 15-3, place it in the category of the refractory metals—tungsten, tantalum, molybdenum, and columbium. The tensile strength of wrought iridium rod is about 90,000 psi at room temperature and 22,600 psi at 2730°F.

Iridium is the most corrosion resistant metal known. It is unattacked by common mineral acids, boiling aqua regia, and many molten salts and molten metals such as lead, tin, silver, gold, and cadmium. It is, however, readily attacked by molten aluminum, copper, zinc, and magnesium. Small additions of iridium (0.1 percent) increase the corrosion resistance of titanium to non-oxidizing acids a hundred-fold. Similar improvements in the corrosion resistance of chromium have been reported with 0.5 percent iridium.

Iridium additions of 10 percent increase the tensile strength of platinum from 19,000 psi to 60,000 psi while a 35 percent iridium increases tensile strength to 140,000 psi. Although iridium is a less potent hardener of platinum than ruthenium, iridium may be added in signifi-

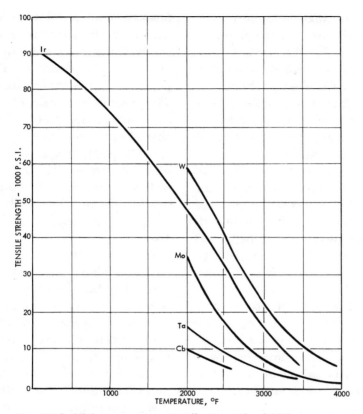

Fig. 15–3. High temperature tensile strength of iridium and refractory metals.

cantly higher concentration without
unduly impairing workability and
ductility of platinum.

Some intermetallic compounds of
iridium with columbium, thorium,
titanium, and zirconium are super-
conductors. Iridium-platinum alloys
are used in the international standards of weight and length.

15.7 Osmium. The naturally occurring alloy "osmiridium" (30 to
65 percent osmium with the balance
iridium) has been widely used for
wear- and corrosion-resisting foun-
tain pen nibs. It has also been used
for instrument pivots, record player
needles, and similar applications requiring high hardness and extreme

resistance to wear and corrosion. Osmium is the rarest of the platinum metals; it is also the hardest (600 BHN) and has the highest melting point (5522°F). One drawback of osmium is that it is completely unworkable, and must be either melted and cast or compacted and sintered to achieve the desired shape.

The resistance to rubbing wear of high osmium alloys is an important asset; this is not solely the result of high hardness as other metals of greater hardness are inferior in this respect. Osmium is resistant to sulfuric, hydrofluoric, hydrobromic, and hydrochloric acids at room temperature. It is, however, attacked by nitric acid and by aqua regia. The modulus of elasticity of the hexagonal close packed osmium has been estimated to be as high as 81,000,000 psi; the highest of all elements.

15.8 Rhodium. The pure metal is used principally as a hard, non-tarnishing electroplate for searchlight and projector mirrors, jewelry, silver cutlery, hollowware, and sliding contact (commutator) surfaces of printed circuit boards. Rhodium is outstanding in that it is not attacked by boiling aqua regia, and single acids such as hydrochloric and concentrated nitric acid. It is slowly attacked by solutions of sodium hypochlorites, hydrobromic acid, and hot sulfuric acid, but is resistant to many molten salts. Rhodium has a

face centered cubic lattice structure, has the highest electrical and heat conductivity of the platinum metals, and has high reflectivity, (78 to 80 percent) in the visible light range. The tensile strength of annealed rhodium is 100,000-110,000 psi while that of drawn wire is in the 200,000-220,000 psi range. Rhodium alloys with all of the platinum metals as well as a number of other metals. The hardening effect of rhodium on the platinum metals is given in Fig. 15–4.

Commercially the most important rhodium alloys are those formed with platinum. Rhodium and platinum are completely miscible and form solid solutions in all proportions. Stress-rupture strength becomes less marked as rhodium content is increased above 20 percent and is counteracted by a decrease in creep ductility. Platinum-rhodium alloys may be fabricated into sheet and wire from either the sintered compact or the cast ingot. Although rhodium work hardens rapidly, thinner gages can be cold worked up to 40-50 percent between stress relief anneals. Relatively large quantities of platinum-rhodium alloys are used in the glass industry for bushings used in the manufacture of glass fiber. Other applications include crucibles for growing laser crystals, aircraft turbine glow plugs, thermocouples, and ignition coils for kitchen range broilers. Rhodium metals and

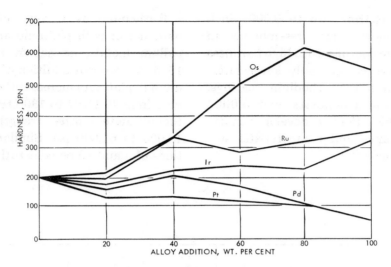

Fig. 15-4. Hardness of cast rhodium with platinum group metal additions.

compounds are widely used in the chemical industry as catalysts to control various chemical reactions.

15.9 Ruthenium. In relation to other metals of the platinum group, ruthenium with its close packed hexagonal structure bears close resemblance to osmium. Wrought polycrystalline ruthenium rod has a tensile strength of 73,000 psi, yield strength of 54,000 psi and 3 percent elongation. The hardness and tensile properties of ruthenium varies according to the orientation of the hexagonal lattice. Thus, on single crystals, the hardness on the basal (0001) plane is 200 DPN; tensile strength parallel to the hexagonal

axis is 40,000 psi, with negligible elongation. In contrast, the hardness on the prismatic (1010) plane is 480 DPN, tensile strength is 22,000 with elongations up to 30 percent.

Ruthenium is resistant to attack of hot and cold acid solutions including aqua regia. It exhibits good resistance to attack by molten lithium, sodium, potassium, copper, silver, and gold when contained in an argon atmosphere. It is also resistant to attack of molten lead and bismuth at temperature up to 1290°F. Ruthenium may be electrodeposited to produce electrical contacts with low electrical resistance and good resistance to welding for operation at temperatures up to 1100°F. Thin,

479

bright coatings up to 0.0001 inch thick show general resemblance to rhodium coatings with high hardness but have considerably lower internal stress than rhodium coatings. Ruthenium deposits over 0.0002 inch thick tend to develop a matte appearance and characteristic surface cracks.

Ruthenium is a very effective hardener of both platinum and palladium metals, as shown in Fig. 15–5. A 5 percent addition of ruthenium to platinum increases the hardness from 40 DPN to 130 DPN and the ultimate tensile strength from 18,000 to 61,000 psi. Similar additions of 5 and 10 percent ruthenium

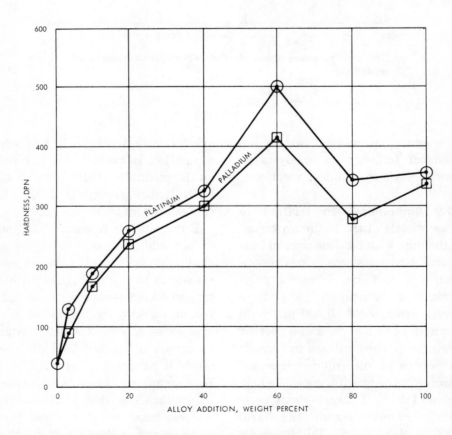

Fig. 15–5. Hardening effect of ruthenium addition on platinum and palladium metals. (After Jaffee et al.)

to palladium increase hardness to 90 DPN and 170 DPN respectively as compared with 40 DPN for pure annealed palladium.

Platinum-ruthenium alloys are used for electrical contacts in voltage regulators, thermostats, relays, and other applications where high wear resistance, oxidation, and corrosion resistance is required. Palladium-ruthenium alloys are also used in electrical contacts, jewelry, and dentistry. Ruthenium alone is used as a versatile catalyst in numerous chemical reactions.

Specialty Metals

15.10 Metals for Solid-State Electronics. A number of metals have found special application in the development of solid state electronic devices. One of these devices, the semi-conductor, as first used in the old crystal set radio, utilized a galena crystal and a thin wire called a cat whisker to achieve rectification of electrical signals. Later, during the early 1930's the selenium rectifier was developed and was the main solid state or metal rectifier for about 20 years until germanium and silicon rectifiers were developed. Selenium has also found application in solar cells for converting light energy into electrical energy. Tellurium has likewise found application in certain thermoelectric devices. The silicon rectifier, the more recently devel-

oped transistor, and the still more recent integrated circuits are some of the extraordinary, and important applications of metals in solid state electronics.

15.10.1. Germanium is a rare grayish-white metal, rather hard and brittle. It has a melting point of 1748°F, density of 5.36 and is a poor conductor of heat and electricity. It was discovered by the German, Clemens Winkler in 1886, although its discovery had been predicted 15 years earlier by Dmitri Mendeleev who noticed a vacancy in the periodic chart of the elements corresponding to this element. Germanium atoms are arranged in a modified diamond structure. It has been used for tarnish resistant mirrors, precision gold-germanium dental castings, and as filters in infrared spectrometers passing wavelengths greater than 1.8μ and absorbing shorter wavelengths.

Its chief use, however, is in the manufacture of transistors. This electronic device, which performs many of the functions of the vacuum tube and some far more efficiently, was developed in 1948 at the Bell Telephone Laboratories. The inventors, William Shockley, John Bardeen, and Walter H. Brattain received the 1956 Nobel Prize in physics for their discovery. The transistor is shock resistant, requires little power (as little as one millionth of a watt), results in extreme miniaturization of electronic equipment and requires no warm-up time. It is perhaps one of the most important inventions of the twentieth century.

Since germanium cannot be used in ambient temperatures much above 212°F, it was supplanted in many military applications by silicon.

15.10.2. Silicon, which is abundant in nature, possesses a gray metallic lustre and fair electrical conductivity. However, it too has a drawback in that it has certain severe upper frequency limitations.

A search for new semi-conductor materials which might be used at high temperatures and high frequencies lead to a number of new materials.

15.10.3. Gallium is silver-white with a slightly blue lustre in incandescent light. It has a melting point of 85.5°F (see Fig. 15–6) and a boiling point of 4357°F which gives it an extremely high liquid temperature range. Gallium has an orthorhombic crystal structure and in the single crystal form shows good malleability. However, in the polycrystalline form, it shatters

Fig. 15–6. Gallium metal melts slightly below body temperature. (Aluminum Co. of America)

readily. Gallium lies between aluminum and indium in Group IIIb of the periodic system and has properties similar to these neighboring metals. Gallium, like aluminum, forms an oxide film upon exposure to air. It alloys readily with tin, zinc, cadmium, aluminum, silver, magnesium, copper, and other metals. Like bismuth and germanium, gallium expands upon freezing. Gallium is extremely anisotropic. The variation in electrical conductivity as measured in different directions through the crystal is thought to be greater than for any other metal with a 1:3 to 2:7 range. It also has definite differences in coefficient of thermal expansion and varies from 31:16:11 along different crystallographic axes.

Some applications of gallium as a metal alloy are in dental alloys and as a sealant for glass joints in laboratory equipment. Although gallium metal was first discovered in 1875, only recently has it become available in quantities and at extreme purity.

Perhaps the most interesting of new materials is a compound of gallium metal and arsenic, gallium arsenide. Both gallium and arsenic are available in the extreme purities required for semi-conductor applications; and the resulting compound can be used in transistors at high frequencies and at high temperatures. Gallium arsenide is now used in tunnel diodes, in oscillators, switching circuits, FM transmitters, and amplifiers. Gallium arsenide is also used in increasing quantities for solar cells in orbiting satellites.

By far the largest current application of gallium and certain of its compounds is in semiconductors.

15.10.4. Selenium was used for years in rectifiers and photo-cells. There is very little precise knowledge about its fundamental characteristics. Selenium has a number of allotropic forms which may coexist, making it difficult to obtain homogeneous samples; it is also very difficult to purify. The hexagonal rhombohedral form of selenium is known as metallic or gray selenium. The electrical conductivity of this gray form is poor, but is greatly increased by exposure to light. This property makes selenium useful in the construction of apparatus for measuring light intensity. Selenium is also used in making of ruby red colored glass. Some selenium has been used in place of sulfur in the vulcanizing of certain types of rubber goods. Small additions of selenium to copper alloys and to stainless steel improves machinability characteristics of those metals.

15.10.5. Tellurium was called the useless metal; but is now finding important uses in improving machinability of steel and copper, improving corrosion resistance of lead, and in solid state devices. As a semiconductor, it acts very similarly to selenium; unlike selenium, however,

483

it behaves like a semi-conductor even above its melting point, 845°F. Tellurium crystallizes into a silver white, semi-metal having a rombohedral crystal structure. It is the poorest electrical conductor of any of the metals. It is easier to purify and work with than selenium since it appears to exist in only a single phase. Fairly large quantities of tellurium are present in combination with gold as gold telluride; it may also be recovered from electrolytic copper refining operations.

15.11 Metals Used in Nuclear Work. The great developments in physics that have resulted in the production of electrical power through controlled nuclear reactions has placed a tremendous demand upon metallurgists for the development of suitable reactor materials. Some of these materials must be employed in the fission zone where conditions are probably more severe than in any

engineering application hitherto encountered. Other materials must be suitable for tubing to conduct liquid-metal coolants such as molten sodium. The metals employed for these and other equally severe applications must withstand the effects of irradiation, high temperatures, and corrosive environments.

During the nuclear reaction, certain isotopes such as Uranium 235 split into approximately two equal parts when bombarded by neutrons, Fig. 15–7. When the split occurs, either two or three neutrons are emitted. During the fission or splitting of each uranium 235 atom, approximately 200 MEV (million electron volts) of energy is released. If properly controlled, this energy is available for utilization as a power source. In order to produce power continuously it is necessary to have a chain reaction occur. If the three newly emitted neutrons of the original reaction are used to bombard

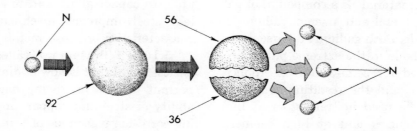

Fig. 15–7. Fission or splitting of Uranium 235 atom by bombardment with neutron. Neutrons produced split other atoms and start a chain reaction.

other U-235 atoms a chain reaction starts—after one second, two million split, 1/1000 second later 40 billion split, then 900 trillion, and the reaction proceeds faster and faster. In order to control this chain reaction and not let it proceed too rapidly, causing an explosion, it is necessary to remove some of the emission neutrons and prevent them from splitting other atoms. This control is accomplished by containing the fission reaction in a special furnace known as a reactor, Fig. 15–8. The uranium fuel rods are separated by graphite to slow down the neutrons and control the reaction. Adjustable control rods of boron or cadmium which absorb neutrons in large amounts are inserted into reactor to provide the fine adjustment for a self-sustaining chain reaction. Heat is produced from the energy liberated during the fission reaction. This heat is re-

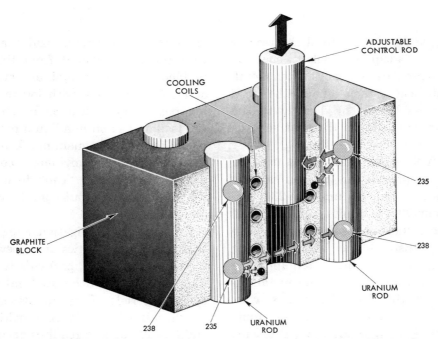

Fig. 15–8. Simplified reactor with uranium rods separated by graphite to slow the neutrons. Moveable control rods absorb neutrons to retard the reaction. During fission uranium U-235 splits and bombards U-238 with neutrons converting it to plutonium. Fission is accompanied by intense heat which is carried away through the cooling coils to a heat exchanger.

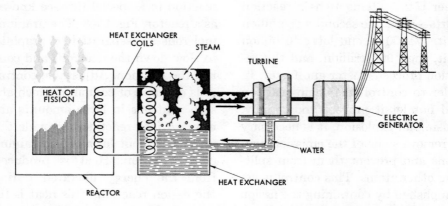

Fig. 15–9. Atomic power plant to harness the tremendous heat energy from fission and generate electrical power.

moved from the core of the reactor by an adequate coolant such as high pressure steam or molten metal. This heat is then taken either to a heat exchanger or directly to the unit that will use the heat such as steam turbine, Fig. 15–9.

Fuels which have been used or contemplated for use in atomic reactors include uranium-235, uranium-238, plutonium, and thorium. These fuel materials must be able to withstand the high operating temperatures, which often approach the melting point of the metal encasing the fuel. Difficulties with uranium stem from the anisotropic nature of alpha uranium. Problems of growth, wrinkling, and swelling under irradition and thermal cycling arise from this anisotropic factor. Interest in alloy additions and use

of refractory uranium oxides and carbides have resulted from these problems. Plutonium fuels are under active consideration with low-melting point alloys such as bismuth, tin, lead and so on in a liquid metal fueled reactor. The main problem to be overcome is in providing a compatible container or "can" to hold the liquid fuel without contaminating the fuel material.

Preliminary investigations suggest that thorium properties are superior to those of uranium, particularly in its compatibility with structural reactor materials. Thorium presents a radioactive hazard during melting and fabricating and requires special precautions.

Virtually all of the reactors which have been built thus far are those in which the energy of the neutrons is reduced by elastic collision between the neutrons and the nuclei of the *moderator*. Graphite has been the most popular moderator because of its good moderating properties. High temperature properties of graphite are outstanding. Limitations in the usage of graphite arose primarily from its low oxidation resistance at high temperatures and its low impact strength. The effects of irradiation on graphite are notable; damage to crystal structure by bombardment of fast neutrons brings about marked changes in many of the physical properties of carbon. At higher temperatures these changes tend to anneal and reduce the effect of the damage.

The *cladding* or can which separates the fuel from the coolant is subject to intensely strenuous requirements. Several materials including aluminum, magnesium, stainless steel, beryllium, and zirconium have been used for encasing. The cladding material must resist corrosion, intense irradiation, be a good thermal conductor, allow free travel of neutrons, have sufficient strength within the temperature range encountered, be weldable and formable to permit fabrication. Beryllium metal is unique in that it has the lowest neutral absorption of all metals, being only one-sixth of its nearest rival, magnesium, and shows promise as a cladding material as fabrication problems are solved.

The primary function of the control rods is that of capturing neutrons. These rods are automatically moved into or out of the reactor core in order to produce the stable condition known as the *critical condition*. The neutron absorbing capacity of a material is indicated by its nuclear *"cross-section"*. When a neutron enters a one centimeter cube perpendicular to one face, the chance of collision is the ratio of area of the nuclei to the total area. For the sake of simplicity, the cross section has been designated as a "barn" which is equal to 10^{-24} cm^2. Natural cadmium has a cross section of 3000 barns, boron[10] 3990 barns, natural boron 750 barns, stainless steel 2.75 barns and aluminum only 0.22 barns. The general tendency is to use cadmium for control rods for the capture of thermal neutrons.

One of the problems encountered with boron[10] is its transmutation to lithium[7] plus an alpha particle which picks up an electron and becomes helium gas. The high internal pressures created by the gas and possibilities for mechanical difficulties reduce its attractiveness as a control rod material even though it has a very high cross section.

15.12 Refractory Metals. The demand for materials to operate at temperatures exceeding 1600°F in aerospace and other environments has directed the attention of the me-

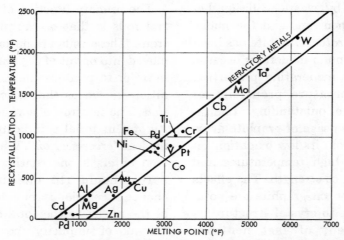

Fig. 15–10. Recrystallization temperature and absolute melting point for cold worked pure metals.

tallurgist to the refractory metals and their alloys. The maximum temperature at which a metal exhibits satisfactory strength properties is directly related to its recrystallization temperature. The recrystallization temperature for pure metals is approximately 40 percent of the absolute melting temperature as shown in Fig. 15–10. However, the recrystallization temperature may be substantially increased by the addition of alloying elements. For example, cobalt which has a recrystallization temperature of 810°F can be alloyed to provide useful properties at temperatures of up to about 1700°F. The strategic position of the refractory metals, niobium (columbium), molybdenum, tantalum, and tung-

sten is also shown in Fig. 15–10. Until recently, the preparation of the refractory metal parts was limited to small sizes that could be produced by compacting and sintering of the powdered metals. The early production of incandescent lamp filament wire from sintered compacts by swagging and drawing is an example of this method. One of the difficulties of melting and casting the refractory metals is that of providing a crucible to withstand the very high melting temperatures. Some of the refractory metals including ductile molybdenum are now produced by the arc-cast process.

15.12.1. Niobium (Columbium) was first tried as a base metal by the desire to take advantage of its

488

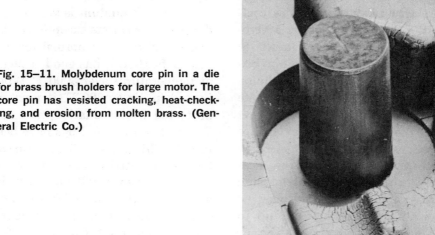

Fig. 15–11. Molybdenum core pin in a die for brass brush holders for large motor. The core pin has resisted cracking, heat-checking, and erosion from molten brass. (General Electric Co.)

low neutron absorption cross section (1.1 barns) and its high temperature strength as a fuel element container and structural material in nuclear reactors. Later this interest was rapidly extended to the use of columbium base alloys for high temperature applications in gas turbines, ramjet components, and other aerospace applications. Columbium has excellent fabricability in the pure state and can be forged, drawn, stamped, and machined by existing commercial techniques. It reacts readily with air at elevated temperatures and requires a protective coating. Both ceramic and zinc coatings have shown promise for protecting columbium and its alloys during extended high temperature service. Alloying of columbium with zirconium, vanadium, titanium, hafnium, and the other refractory metals improves its oxidation resistance more than fifty times. The high temperature tensile properties are likewise increased, although not so dramatically, by a factor of 4 or 5. These improvements are gained at some expense to fabricability because of their greater strength at high temperatures.

15.12.2. Molybdenum has the physical properties of high melting point, high thermal conductivity, high elastic modulus, and low coefficient of thermal expansion. Molybdenum has excellent resistance to liquid metals such as molten bismuth, sodium, potassium, lithium, and molten glass. Its resistance to reaction with mercury, even at elevated temperatures, has led to its use as electrodes in mercury switches. At temperatures over 1000°F unprotected molybdenum oxidizes rapidly in air and must be protected. For temperatures up to 2200°F, nickel-base cladding or sprayed coatings are suitable. For temperatures up to 2800°F, or for short periods at higher temperatures, modified chromized coatings and sprayed aluminum-chromium-silicon coatings have been used. Still higher temperatures require siliconizing or ceramic coating.

The three most prominent molybdenum-base alloys are (1) molybdenum-titanium, (2) molybdenum-titanium-zirconium and (3) molybdenum-tungsten.

The main advantages of the alloys over pure metal are their higher recrystallization temperatures, and higher hot strength. Applications are generally those that include extrusion dies, die casting cores (Fig. 15–11), piercer points, and structural parts in aerospace vehicles. The molybdenum-tungsten alloy has

excellent resistance to molten zinc and has been used in pumps for liquid metal handling.

15.12.3. Tantalum is very dense, ductile and exhibits exceptional corrosion resistance at normal temperatures, further, it has good strength at temperatures over 3000°F. When heated in air at 500°F, it must be protected from oxidation, and tantalum-aluminide coatings have been used for this purpose. Tantalum will become embrittled when heated in hydrogen; also it will form nitrides or carbides when heated in the presence of nitrogen or hydrocarbons. At temperatures between 70-300°F tantalum possesses negligible corrosion rates in most chemical environments except fluorine, hydrofluoric acid, potassium, and sodium hydroxide solutions and fuming sulfuric acid. Arc-melted or electron-beam melted tantalum ingots are readily cold worked. Annealing must be carried out in a vacuum or an inert gas atmosphere. This annealing (recrystallization) temperature is from 1800-2500°F depending upon purity and the degree of prior cold-work.

Although tantalum has a body centered cubic structure, it remains ductile down to liquid helium temperatures in the pure form; even the relatively impure metal is ductile at room temperatures. Pure tantalum was produced in 1905, and used as filament material in millions of incandescent lamps before tungsten wire replaced it as a filament material. Tantalum is used extensively in the chemical industry for acid resistant heat exchangers, condensers, duct work, and chemical lines. It is also used in medical and dental applications for sutures, gauze, pins, and plate. The unique electrical properties of tantalum have also led to its use in rectifiers, compact capacitors, electronic tubes, and lightning and high voltage arresters. It is added to stainless steel to prevent intergranular corrosion, and to tungsten carbide cutting tools to improve resistance to cratering.

15.12.4. Tungsten has the property of its high melting point—6170°F—highest of all metals. Tungsten also has the highest ductile-to-brittle transition temperature of the body centered cubic system. The high transition temperature makes it necessary to conduct most cold forming operations at 1290°F or over. Stress relieving treatments vary from 1830 to 3180°F. Hot working temperatures start initially at 2550°F but can be progressively lowered to about 2010°F as the amount of work increases. Hard drawn tungsten wire has an amazing tensile strength of 600,000 psi and a yield strength of 540,000 psi. The high strength properties require consideration of equipment capacity during hot and cold working operations.

The oxidation resistance of tungsten is better than that of molyb-denum but still requires protection from the atmosphere when used at elevated temperatures. A 1 percent thoria addition strengthens tungsten for high temperature service from 6000 psi to nearly 20,000 psi.

Although the Mushet self-hardening tungsten containing tool steels were patented in 1868, it was not until 1898 that F. W. Taylor and M. White developed a tungsten containing tool steel that created a sensation at the 1900 Paris Exposition by operating cutting tools at speeds unheard of with conventional tools.

The use of tungsten in electric lamp filaments was suggested as early as 1904, but its use awaited the development of the ductile filament in 1907. Tungsten carbide cutting tools were developed in Germany by Baumhaver and Schroeter in 1926.

The major uses of tungsten are: carbides (35%), ferrous alloys (37%), metal (14%), non-ferrous alloys (12%), and chemicals (2%). Tungsten metal is used for lamp filaments, x-ray tubes, radio tubes, surgical wire, thermocouples, crucibles, and laboratory apparatus.

Tungsten alloys find use in hard-facing materials, tool steels, tools and dies, boring bars, electrical contacts, welding electrodes. Tungsten chemicals are used in textile dyes, inks, paints, enamels, in glass, and as coatings in phosphorescent TV tubes and in fluorescent lighting.

Metallurgy Theory and Practice

References

Aluminum Co. of America. *Gallium and Gallium Compounds*. Pittsburg, Pa.: Chemicals Div.

American Society for Metals. *The Metal Molybdenum*. Cleveland, O., 1959.

Clough. *Reactive Metals*. New York: John Wiley & Sons, 1959.

International Nickel Co. *Characteristics of Precious Metal Electrodeposits for Industrial* Use. New York.

———. *Iridium*.

———. *Palladium*.

———. *Rhodium*.

———. *Ruthenium*.

———. *The Platinum Group Metals in Industry*.

———. *Six Precious Metals*.

Raudebaugh, R. J. *Nonferrous Physical Metallurgy*. New York; Pitman Publishing, 1952.

Samans, Carl H. *Engineering Metals and Their Alloys*. New York: Macmillan, 1949.

Semchysen-Harwood. *Refractory Metals and Alloys*. New York: Interscience Publishers, 1966.

Plastic Deformation of Metals

16.1 Introduction. The ability of metals to be plastically deformed without rupture is one of their most valuable characteristics. Plasticity in metals allows the manufacture of sheet, plates, strip, rods, tubes, structural shapes, and the forging and forming of metal.

During plastic deformation at room temperature (cold working) most metals undergo a noticeable increase in strength and hardness. Thus, many metals whose properties cannot be improved by heat treatment can be modified by cold working. The increase in strength resulting from plastic deformation requires the application of higher and higher stresses during the metal forming operation if plastic deformation is to continue. In many cases, however, the metal being deformed becomes excessively hard and strong before the final shape is reached. Thus, it must be annealed and made

suitable for further plastic working. During the annealing operation, the structural disturbances within the metal are relieved, softening it and increasing its ductility for further processing. The fundamental principles of plastic deformation of single crystals discussed in Chapter 3 will be reviewed with the emphasis now placed on deformation of polycrystalline metals. Both hot and cold working operations will be covered.

16.2 The Nature of Plastic Deformation. The ability of metals to deform permanently without rupture may be attributed to their unique atomic structure. As discussed in Chapter 3 the metallic bond in metal consists of atomic nuclei surrounded by an electron cloud. This structure allows shifting of atom bonds and facilitates interchange and movement of electrons. An in-

493

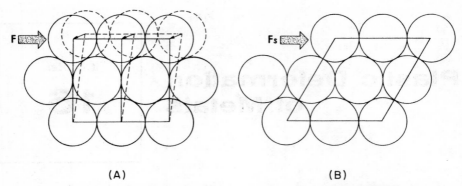

Fig. 16–1. A, elastic deformation caused by small force. F, Arrows indicate the tendency for atoms to return to original positions after the force is removed. B, plastic deforma-mation caused by force F_s which caused the crystal to permanently slip by one atomic distance.

dividual atom nucleus is not rigidly attached to any particular electrons so that electron-to-nucleus bonds may be shifted during movement of a nucleus along crystallographic planes (slip planes) without breaking the bonds. Thus an entire section of a metal crystal is able to move in relation to the remainder of the crystal without producing fracture in the region of deformation. This type of deformation is shown in Fig. 16–1A. For low loads the deformation is elastic. If the applied load exceeds a certain limit known as the *elastic limit,* the atoms are moved to new positions in the lattice. Since the atoms are at equilibrium in the new positions, this type of deformation does not disappear when the load is released and is called *plastic* or *permanent* deformation.

16.3 Plastic Deformation by Slip.
The movement of atoms depicted in Fig. 16–1B requires large amounts of energy. Nevertheless, the calculated theoretical strength should be 100 to 1000 times greater than the actual observed shear strength as measured by tensile tests. This discrepancy is much too large to be explained away by experimental errors but must be explained by the existence of certain defects in the lattice.

Investigation of this discrepancy has led to the widely accepted dislocation theory. In essence, this theory states that the movement of one plane of atoms with respect to another can be accomplished with minium energy by movement of vacant atom sites (dislocations) along the slip plane, Fig. 16–2. One analogy used to describe the movement of a

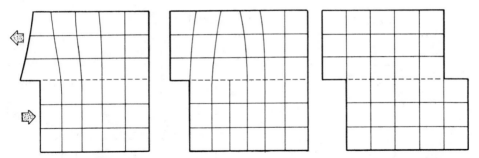

Fig. 16–2. Edge dislocation through a crystal resulting in plastic deformation.

dislocation and the lower stresses associated with it is the movement of an earthworm in which a hump starting at the tail moves toward the head. A second analogy is that of moving a large rug by forming a hump (or ruck) at one end and moving it toward the other. There is movement of the rug with respect to the floor but sliding has not occurred. In both cases, movement (slip) is achieved with relatively small force as compared with moving the bodies all at once.

As discussed earlier, the most important planes related to plastic deformation are those of high atomic density and greatest interplanar spacing. In the ductile body centered cubic metals, slip may occur in 4 possible directions in each of 12 possible slip planes, making 48 possible slip systems. Face centered cubic metals have 4 diagonal planes

with 3 slip directions in each plane for a total of 12 possible slip systems. In the less ductile hexagonal close packed metals, slip occurs on the basal plane along the diagonals making 3 possible slip systems; however, at elevated temperatures, other slip systems become active.

Deformation of metallic crystals takes place by shear even though they are stretched by tensile forces or squeezed by compressive forces. As pointed out in paragraph 3.6, deformation in a crystal takes place on planes which are most favorably oriented and on which the applied force produces the greatest shearing stress. Experiments show that fresh dislocations are continually generated during plastic flow allowing slip on any given plane to continue almost indefinitely. If it were not for this fact, it might be supposed that slip would have to stop after all of

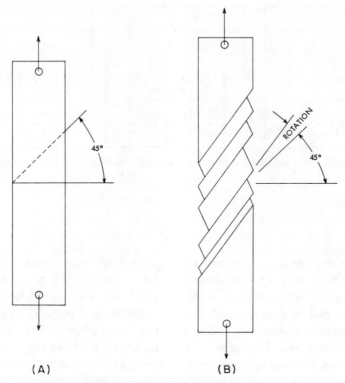

Fig. 16–3. A, maximum resolved shear stress is 45° to axis of deformation during tensile testing. B, rotation or tilting of slip planes during plastic deformation of a crystal under tensile load with constrained ends.

the dislocations were used up by being run out to the surface of the crystal. During plastic deformation by tensile forces, the sliding of portions of the crystal relative to other portions of the crystal causes a tilting or rotation of the active slip plane, Fig. 16–3. The direction of tilting is such that the slip planes tend to become more nearly parallel to the line of applied force. This tilting action brings new slip systems into more favorable orientation thus facilitating slip on secondary slip planes. In polycrystalline metals the interference of adjacent crystals causes the greatest slip on the most favorably oriented crystals, continuing on to other planes as the first planes become strengthened by plastic deformation. This interference with slip and crystal rotation in polycrystalline metals accounts for the increased strength of fine grained metals. The tendency for crystals to align themselves parallel to the direction of forming leads to preferred orientation of grains and consequent directional properties in the metal known as anisotrophy.

16.4 Plastic Deformation by Twinning.

There is another mechanism of deformation, not involving slip, which is important in some metals. This mechanism, called twinning, is often an important mode of deformation in the H.C.P. metals such as magnesium, zinc, and zirconium, when the orientation is not favorable to slip; and in the B.C.C. metals at low temperatures including tin, brass, and iron. Mechanical twinning is not readily obtained in F.C.C. metals. Twinning results when the deformation of the metal is such that adjacent planes of atoms are realigned as a reoriented crystal lattice, Fig. 16–4. The difference between slip and twinning should be carefully recognized since both mechanisms involve shear. In slip,

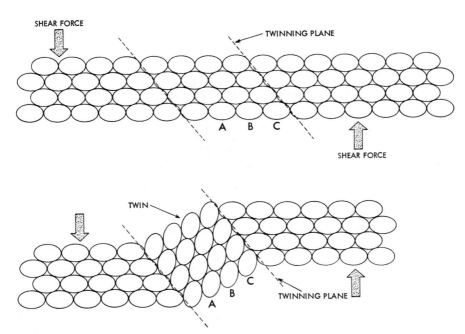

Fig. 16–4. Schematic showing a twin produced during shearing by simple movement of atoms. Atoms A, B, and C shift very little in position during twinning.

497

deformation occurs on individual lattice planes. The shear associated with twinning, on the other hand, is uniformly distributed over a discrete number of slip planes. Shear movement in slip may be many times larger than the lattice spacing but in twinning movement is only a fraction of an interatomic spacing. The total shear movement due to twinning is small, so that slip is a much more important primary mode of deformation. Some metals deform by both slip and twinning as shown by the discontinuous strain increments in the stress strain diagram for a single crystal, Fig. 16–5. Twins may

grow in size by increasing both their length and thickness. In some instances, twin formation is extremely rapid, setting up audible shock waves that may be heard as audible clicks.

The familiar "tin cry" produced when a bar of tin is bent is due to shock waves from deformation twinning. Twinning results in a lattice having a mirror image of the parent lattice. Under certain conditions the lattice within the twin may become favorably aligned with respect to the applied force allowing deformation to proceed by means of the slip mechanism.

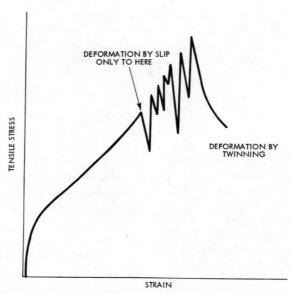

Fig. 16–5. Stress-strain diagram for a single crystal during deformation by both slip and twinning.

16.5 Strain Hardening. Plastic deformation at a temperature and rate which causes an increase in the hardness and strength of a metal is known as cold working. The increase in hardness due to cold working is called strain hardening. The probable reasons for the occurrence of strain hardening in a metal whose lattice is apparently unaltered after deformation was discussed in paragraph 3.8. One practical aspect of strain hardening is that it can be used to strengthen metals. The relationship of strain hardening to other observable events resulting from cold working is significant and should be well understood by the student. Fig. 16–6 is a typical stress-strain diagram showing the increase

in strength due to the effects of cold working. Severe cold working may increase strength of some metals as much as 50 to 100 percent. The slope of the stress strain curve, Fig. 16–6, beyond the yield point "A" shows the rate of increase of stress with increased plastic deformation or strain. The steeper this curve, the greater is the hardening and strengthening effect of plastic deformation. While this section of the stress-strain diagram is ordinarily curved, it becomes a straight line when plotted on log-log graph paper. The slope of the plot, *n*, is known as the strain hardening exponent. This useful dimensionless quantity is numerically equal to the strain at maximum load, point "B",

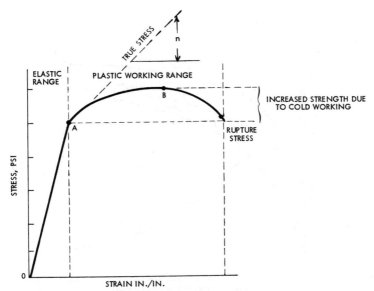

Fig. 16–6. Stress-strain diagram showing increase in strength due to cold working.

Fig. 16–6. Metals with a low strain hardening rate show little elongation and start to neck locally early in the tensile test. Metals with a very high strain hardening rate tend to have a large amount of uniform elongation and no localized necking under tension. The usual effect of cold deformation on the strength, hardness, and ductility of metals is shown in Fig. 16–7.

One interesting effect that is frequently present when sheet metal with large grain size (above about 0.05 mm. in diameter) is deformed, is the appearance of a roughened objectional surface. This phenomenon, called the *orange peel* effect (Fig. 16–8A) may be readily observed with the unaided eye. The uneven nature of the plastic strains and subsequent strain hardening of planes with particular orientations are evidenced by the non-homogeneous nature of the deformed surface. Materials in which the grain size is nonuniform and which varies over wide limits is especially prone to poor surface finish. Metals with fine grain size are stronger and tougher than the coarse grained metals. However, they also have lower ductility and consequently, may not be subjected to as severe deformation as the more coarse grained metals. A compromise between grain size and formability must be made which will still give suitable results.

One other desirable property of deep drawing sheet metal is freedom

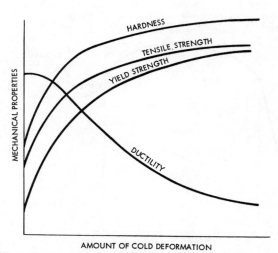

Fig. 16–7. Effects of cold deformation on strength, hardness, and ductility.

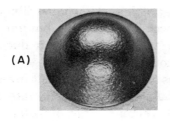

(A)

(B)

Fig. 16–8. Problems revealed by cupping tests. A, orange peel effect in coarse grained brass, 2X. B, earing tendency in sheet copper caused by preferred orientation of grain structure. (Chase Brass and Copper Co.)

TENSILE FORCE

TENSILE FORCE

Fig. 16-9. Stretcher strains in rolled and aged low carbon steel. The aging process prior to forming restores characteristic yield point resulting in surface irregularities shown. 2X. (U.S. Steel Corp.)

from preferred orientation. Metals having grains with preferred orientation are particularly sensitive to earing, (Fig. 16–8B), or to metal fracture in severe deforming operations.

In Fig. 16–9 is shown another surface defect which is sometimes encountered during the deformation of soft, low carbon steels which have a sharply defined yield point. This surface irregularity, called stretcher strains or *Lüders bands,* occurs at approximately 45° angles to the stress axis. Lüders bands should not be confused with slip lines as the bands are hundreds of crystal planes in width. This effect may be eliminated by subjecting the material to 1 to 2 percent reduction by cold rolling. This process, called *temper rolling* evenly strains the entire piece of metal and removes the sharply defined yield point jog. Stretcher strains are non-existent when there is no sharply defined yield point. Another method of eliminating the problem of stretcher strains if the deformation is not to be too drastic is to use sheet steel with higher carbon content. Higher carbon steels do not exhibit a sharply defined yield point, but unfortunately do not have the high ductility of lower carbon steels.

Recent investigations of the effect of strain rate on mechanical properties of metals has led to some interesting conclusions. In general, as the strain rate (plastic deformation) is increased from normal testing speeds of 10^{-3} in./in./sec to 10^2 in./in./sec, the strength increases significantly; sometimes as much as 50 percent (see Fig. 16–10). In some materials such as aluminum, titanium, tantalum, and uranium, the strain rate effect is predominantly noticeable at elevated temperatures. Thus, as metals are heated to increase formability at high strain

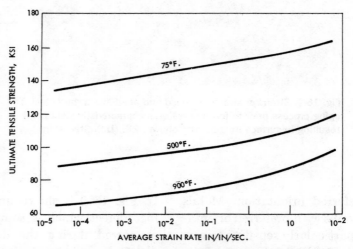

Fig. 16–10. Effect of strain rate on tensile strength of Ti-5 Al-2.5 Sn alloy.

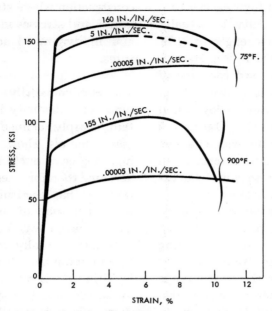

Fig. 16–11. Stress-strain diagrams for Ti-5 Al-2.5 Sn at 75° and 900°F at strain rates indicated.

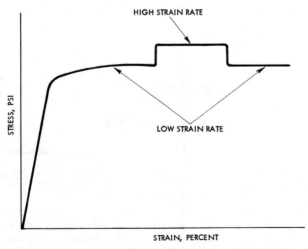

Fig. 16–12. Effect of changing testing speed during a tensile test of a titanium alloy.

rates such as with high energy rate forming, the plasticity is actually reduced. Compare strength at 75°F with strength at 900°F, Fig. 16–11.) In Fig. 16–12 is shown the type of result that occurs in a stress-strain diagram for a titanium alloy tested at different strain rates. During the test the strain rate was changed showing the obvious increase in stress at higher strain rates. One explanation for this increase in stress is that at higher strain rates the energy requirement to move dislocations is increased, thus requiring higher stresses. The strain rate effect is not operative in the elastic range of the diagram since there is no dislocation movement. Practical applications for an understanding of the strain rate effect are in explosive forging and forming, nuclear reactor construction where starting temperatures and stresses increase rapidly, and in dynamic loading of structures.

16.6 Superplasticity. The idea behind superplasticity is that, at certain sharply defined strain rates, many metals show an unusually high degree of ductility. These metals are therefore less difficult to form at that particular rate of deformation than at any other rate. This behavior was found previously only in structurally weak alloys such as aluminum-zinc, tin-bismuth, and lead-tin, but is now found to occur in certain stainless steels and titanium alloys especially at elevated temperatures.

A study of magnesium alloy shows that it has superplasticity at about

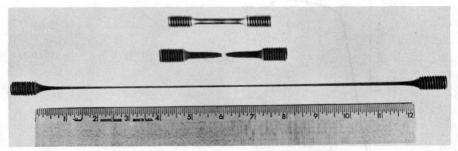

Fig. 16–13. Superplasticity of IN744X stainless alloy at critical strain rate when tested at 1800°F. Top: standard test specimen. Middle: alloy broken in hot tensile test. Bottom: superplastic deformation of IN744X at 1800°F. Specimen is about four times longer. (International Nickel Co.)

10^{-2} in./in./sec then loses that superplasticity as the strain rate is increased. Fig. 16–13 shows the dramatic elongation of a 26 percent chromium and 6.5 percent nickel alloy designated IN744X. This stainless steel alloy has about double the yield strength and twice the fatigue limit of typical ferritic and austenitic stainless steels at room temperature. Yield strength of the new alloy is 75,000 psi as compared with 45,000 psi for ferritic stainless steel.

16.7 Distinction Between Cold Working and Hot Working. The relief of internal stress by annealing was discussed in Chapter 3, paragraph 3.9. As a metal is plastically deformed at a temperature above its recrystallization temperature, little, if any strain hardening will occur. Recrystallization, the formation of new, unstrained crystals, takes place during the plastic deformation process, thus eliminating any strain hardening effect. Under this condition of continuous annealing, plastic deformation of metals is relatively easy. Because of this difference in behavior of materials above and below the recrystallization temperature, consideration will be given to both hot and cold working as typical processes are discussed.

16.8 Cold Working of Metals. Cold working is defined as plastic deformation of metal below the recrystallization temperature. Working of magnesium at 200°F would be defined as cold working since its lowest recrystallization temperature is approximately 300°F.

Although cold working operations involve higher forces than hot working, they do offer the following advantages over hot working: (1) closer dimensional control, (2) better surface finish, (3) grain refinement, and (4) increased mechanical properties. In cold working the metal temperature is approximately room temperature and it is not necessary to correct measurements for thermal contraction. Oxidation, which is appreciable at hot working temperatures, is negligible in cold working with resulting clean, scale-free surfaces. Pronounced grain refinement of most steels can be achieved by cold working followed by annealing; in this case a cold working finishing operation is used to improve surface finish and dimensional tolerances. Stainless steels and other metals which cannot be normalized to achieve grain refinement, are often subjected to extensive cold working followed by proper annealing to control grain size. During cold working the lattice is strained resulting in markedly changed mechanical properties, including increased tensile and yield strength, and increased hardness.

After a metal has been subjected to severe cold working operations, further cold deformation may cause a decrease in strength. When this occurs, the metal is said to be overworked. The loss of strength is due to internal failure of the metal structure. Most metals, including stainless steel, show somewhat lower resistance to corrosion after being subjected to cold working operations. The effect depends upon composition, environment, and the extent and uniformity of cold work.

The cold working processes which will be discussed and which are typical of the many cold forming processes available include rolling, shearing, extruding, drawing, and forming.

16.9 Hot Working of Metals. Hot working involves plastic deformation of metals at temperatures such that they are continually annealed during working, and strain hardening does not occur. Hot working of

Table 16–1. Hot working temperatures for various metals and alloys.

MATERIAL	TEMPERATURE RANGE DEG. F
ALUMINUM	650–900
ALUMINUM ALLOYS	750–900
BERYLLIUM	700–1300
BRASSES	1200–1475
COPPER	1200–1650
HIGH SPEED STEELS	1900–2200
INCONEL	1850–2350
MAGNESIUM ALLOYS	400–750
MONEL	1850–2150
NICKEL	1600–2300
REFRACTORY METALS & ALLOYS	1800–3000
STEEL: CARBON	1900–2400
LOW ALLOY	1800–2300
STAINLESS	1900–2200
TITANIUM	1400–1800
ZINC ALLOYS	425–550

metals is done primarily for two reasons. (1) to plastically mold the metal into the desired shape, thus minimizing requirements for machining, (2) to improve the properties of the metal as compared to the as-cast condition.

In all plastic deformation processes a major part of the energy employed to change the shape of the metal is transformed into heat. This is also true in hot working. Large sections of hot worked metal with a relatively small surface-to-volume ratio may become appreciably hotter during deformation and may even partially melt if deformed too rapidly. This factor must be considered when establishing the maximum hot working temperature. The

hot working temperature, it should be noted, is a function of the composition of the metal, its grain size, its previous degree of cold work, and other factors, including strain rate. For example, working of high purity lead at $-10°F$ is still hot working, while working steel at $700°F$ is cold working. Some typical hot working temperatures are given in Table 16–1.

16.10 Grain Refinement. During the hot working operation there is a continual refining of the grain structure as illustrated in Fig. 16–14. Here, the coarse cast grain structure is broken up and recrystallizes into new, smaller stress-free grains. Repeated deformation at elevated tem-

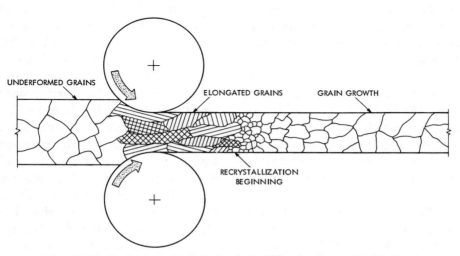

Fig. 16–14. Grain refinement during hot rolling due to recrystallization.

507

peratures can produce a desirable very fine grain structure in metals. This is one definite advantage of hot working. In addition to breaking up the large grains typical of a cast structure, mechanical working also reduces other internal defects within the metal. Typical defects in cast metals which are minimized in hot worked metals include:

1. Large grain size (due to slow cooling),

2. Porosity (voids due to shrinkage),

3. Blow holes (due to gas evolution during solidification),

4. Segregation (due to limited solubility in the solid state),

5. Dirt and slag inclusions (due to oxides, dirt, etc.),

6. Poor surface condition (due to oxides, and scale).

As mentioned, hot working breaks up large grains. It also reduces the effect of segregation by breaking up eutectic and dendritic structures. Porosity and blow holes are often eliminated during hot working due to the internal welding that occurs under the high pressure and at the high temperature. Slag inclusions may be dispersed throughout the metal as flow lines, fiber, or banded structure with little effect on mechanical properties. No matter how well the hot working process may reduce the major defects noted, there is still the problem of heavy oxide surface scale on some metals

during hot working. Care must be exercised in this case to prevent imbedding of these oxides into the surface of the formed part.

The hot working processes discussed in this section will be principally those which produce finished products rather than a discussion of primary metal working processes as used in working ingots, blooms, billets, or the like.

16.11 Rolling. Plastic deformation of a metal by passing it between driven cylindrical rolls is known as *rolling* and is perhaps the most widely used hot deformation process. The major reasons for hot rolling are refinement of the original cast metal structure and the rapid production of useful shapes. Rolling is a high production metal forming process in which control of the grain structure is possible. Fig. 16–15 is a schematic drawing illustrating the deformation which takes place during rolling. As the metal passes through the rolls it is reduced to a thinner section, elongated proportionally in length, but is spread laterally only a small amount. Metal is drawn into the rolls by frictional force. Friction between the rolls and the workpiece also produces movement of the workpiece surface layers relative to the interior. This effect is shown by the cross-hatched element emerging from the rolls in Fig. 16–15. The reason for this move-

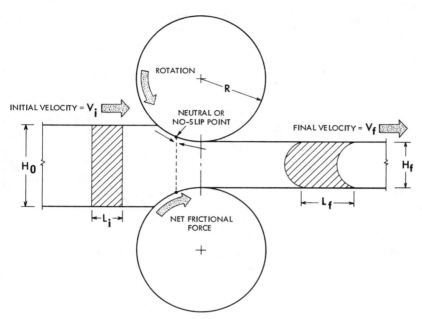

Fig. 16–15. Deformation during a typical rolling operation.

ment may be explained by the following: During a 50 percent reduction in thickness, for example, the work will leave the rolls at approximately twice the speed of the entering work, and the peripheral speed of the rolls will be an intermediate value.

It is thus evident that the work is accelerated while moving through the rolls in order to keep the volume rate of flow constant. At only one point will the work have the same velocity as the rolls. This point is called the *neutral point* or no-slip point. On both sides of this point, there is relative velocity as indicated by the direction of the arrows. Rolling speeds range up to 5000 ft/min.

Rolling temperatures are similar to those given in Table 16–1. The reduction in thickness per pass depends upon the coefficient of friction between the work and the roll, the roll diameter, and possible applications of front tension to the emerging workpiece. In hot rolling the maximum reduction per pass has been reported to be about 13 percent of the roll diameter. For cold rolling the maximum reduction is usually less because of lower coefficient of friction and larger separating forces. Very high unit pressures are required to roll thin sections; for hot rolling the minimum thickness that can be rolled is about $\frac{1}{50}$ of the roll diameter unless front tension is

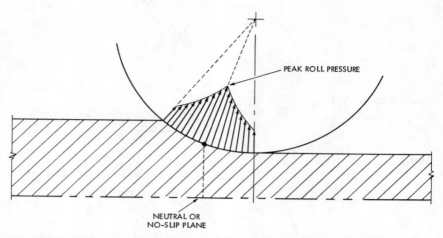

Fig. 16–16. Normal stress distribution during rolling. Pressure at exit is greater than at entry because of strain hardening effects.

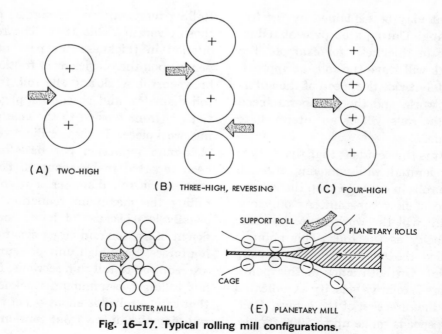

Fig. 16–17. Typical rolling mill configurations.

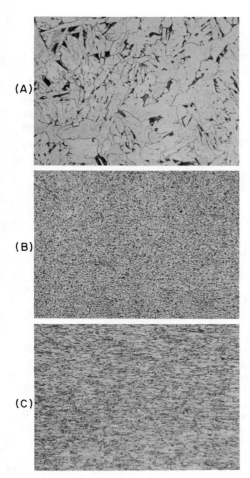

(A)

(B)

(C)

Fig. 16–18. Microstructure of low carbon cast steel. Top, as cast showing dendritic structure. Center, after hot rolling, note reduced grain size. Bottom, after temper rolling the grains become elongated and give directional properties to the steel. (Columbia-Geneva Div., U.S. Steel Corp.)

applied to the work. For cold rolling this thickness is about $\frac{1}{200}$ of the roll diameter. The reason for this minimum thickness is due to the fact that as the metal thickness is

reduced below a certain point, the elastic spring of the rolls becomes so great that they flatten out and further reduction cannot be effected. The normal pressure distributions in rolling is shown in Fig. 16–16. Typical rolling mill configurations are shown in Fig. 16–17. Fig. 16–18 shows the microstructure of a cast structure A before rolling, B after hot rolling, and C after cold rolling.

16.12 Forging. Forging is basically a hot working process in which metal is made to flow under high compressive stresses. Frequently, the metal must flow into a die cavity during forging. The relative ease with which metals may be made to flow plastically at temperatures above the recrystallization temperature is associated with the number of available slip systems, decrease in yield strength at elevated temperature, and the freedom from strain hardening due to continuous self-annealing.

The simplest forging operation is that of upsetting as shown in Fig. 16–19. This process is generally limited to cylindrical workpieces whose length is not usually more than about three times their diameter because of the tendency to buckle. Friction between the workpiece and die results in barreling. This effect is shown in Fig. 16–20. Lubrication, increased or intermittent loading speed, or vibrating of the dies can re-

Fig. 16–19. Deformation during upset forging. Barreling is due to friction between dies and workpiece. (F. M. Kulkarni, Illinois Institute of Technology)

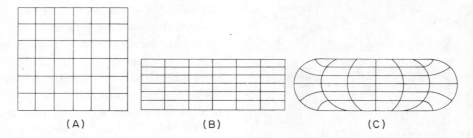

(A)   (B)   (C)

Fig. 16–20. Grid deformation in a compression specimen. A, before deformation. B, after deformation without friction between workpiece and die. C, after deformation with constraining forces resulting in barreling.

sult in reduction of frictional forces during forging. The stress distribution during forging with flat dies is shown in Fig. 16–21A. We notice from this distribution that the normal stress between the die and workpiece increases exponentially and is a maximum at the center of the workpiece. The maximum value of these stresses is a function of the coefficient of friction and the diameter or width to height ratio, a/h, where a is the width or diameter, and h is the workpiece height. Thus, for very small values of h the forces required for upsetting become extremely high. The barreling effect, shown in Fig. 16–21B resulted in cracking from what is known as secondary tensile stresses on the cir-

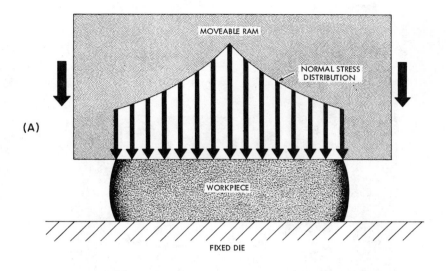

(A)

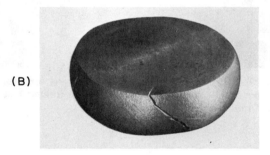

(B)

Fig. 16–21. A, stress distribution with flat dies, considering plane strain and sliding friction. B, cracks along circumference of aluminum cylinder upset with flat dies.

cumference of the specimen. For simple cases it is possible to carry out an upsetting test with controlled temperature and deformation rate and from it determine the maximum deformation or forgeability of the metal prior to cracking. Although no standard forgeability test has yet been devised, it is generally agreed that forgeability is related to three basic quantities: flow stress, ductility, and coefficient of friction.

Some tests which have been used for measuring forgeability include

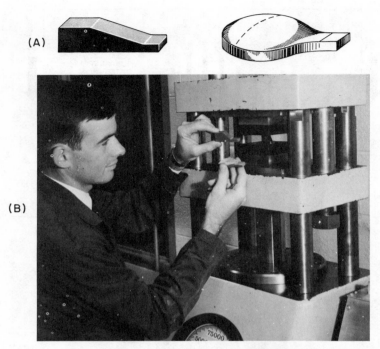

Fig. 16–22. A, wedge test specimen for determining forgeability. B, multipurpose laboratory press (75 tons) with forging test attachment. (Omnipress, Inc.)

the hot torsion test for steels, upsetting tests, and the wedge test shown in Fig. 16–22. In addition, hot tensile-impact tests have been used for determining optimum forging temperatures. The mechanical properties of many metals are sensitive to strain rate especially at elevated temperatures. High deformation rates cause an exponential increase in the upper yield point. It is believed that dislocations may become blocked from movement by segregate atoms or precipitate particles. Thus plastic flow will not occur until mobile dislocations are found or generated. Consequently, a finite delay is encountered before plastic flow is observable, resulting in an increase in the upper yield stress. High strain rates (up to 1000 feet per second) as used in high energy rate forging (HERF) presses may cause a considerable difference in forging characteristics from those encountered at slower deformation rates.

In high velocity forging presses

the stored energy of compressed gas is used to drive the ram at high speed as it impacts the workpiece. This method has been successful in forming almost every type of metal including cast iron, refractories, and other metals which have hitherto been unforgeable. Draft allowances can be reduced or eliminated and closer tolerances can be held than with conventional hydraulic press forging or drop hammer forging. Examination of the microstructure of high velocity forged parts reveals a finer grain size and better controlled grain flow. Because of the high speed metal flow, die wear may become excessive. Therefore, whenever possible, parts should be designed without sharp corners but with maximum radii to give less resistance to metal flow. High production rates possible with this method call for rapid heating and special lubricating and handling equipment to take advantage of high velocity forging.

One distinguishing characteristic of forging is the formation of flow lines or fiber caused by the elongation of slag particles or non-metallic inclusions. This fiber is shown in Fig. 16–23 as revealed by macro-etching. In this same figure is illustrated a very important principle in forging; that of the difficulty of completely converting the cast structure into the wrought structure. Forged products are generally considered to have substantially higher properties than

Fig. 16–23. Flow lines in upset forging of a 1.5″ dia AISI 1045 steel specimen at 1800°F. Non-uniform deformation is apparent (John A. Schey, Illinois Institute of Technology)

cast products, and they usually do. However, some caution must be exercised in advancing claims for all forgings, and especially those structural parts which are stressed in three directions. In a cast structure the strength is approximately the same in all three directions, whereas a forged structure may have different strength values in each of the three directions. It was reported by Sachs some years ago that without exception the mechanical properties of forgings were inferior in the transverse direction to those in the longitudinal direction of fibering. In general, the harder the alloy chosen for forging, the greater is the scatter in ultimate tensile strength and elongation values, and the greater the difference between these properties in the longitudinal and transverse directions. Soft, ductile alloys are more easily converted from the cast structure into the stronger wrought structure, with a disappearance in the difference between the properties in the longitudinal and transverse directions. With high strength aluminum alloys, it is necessary to effect an 80 to 90 percent reduction to obtain the fully wrought structure with high mechanical properties. In steels at least 75 percent reduction in cross section of the cast structure is required. Thus, the general practice for obtaining high strength forg-ings is to start with previously forged stock possessing the desired strength.

16.13 Extrusion. Extrusion is a process which permits the production of shapes which would be almost impossible by any other manufacturing method. Some of the typical extruded rods, tubing, and special

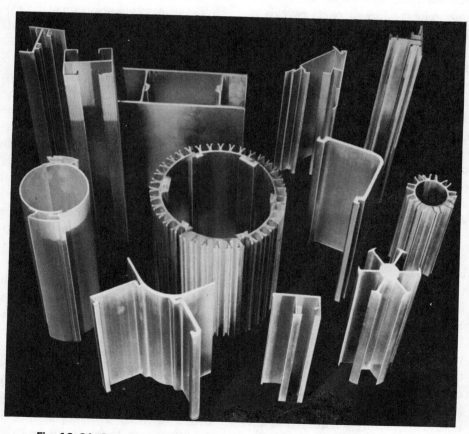

Fig. 16–24. Samples of extruded aluminum shapes. (Reynolds Metal Co.)

shapes are shown in Fig. 16–24. In extrusion, a billet of circular cross section is inserted in a heavy walled cylinder and pressed by a ram. The metal is thus forced through a machined die opening and emerges in the desired shape. Many of the nonferrous metals are extruded including aluminum, magnesium, copper, brass, lead, zinc, and tin. Some steels are also extruded but require special die and liner materials because of the high temperatures required. The hot steel is often rolled in powdered glass prior to extrusion, allowing the glass to melt and act as a lubricant to reduce die wear. In Fig. 16–25 is shown the various extrusion processes. During direct extrusion the billet slides relative to the cylinder wall. The resulting frictional force increases the required ram pressure very considerably. A steel dummy block is used to protect the ram against damage from pressure or heat. Indirect extrusion requires less force than direct extrusion because the billet is not required to slide and

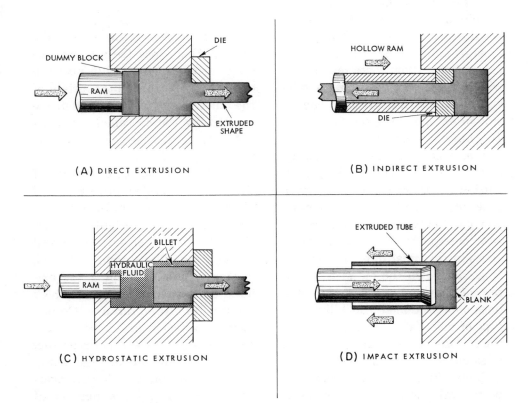

Fig. 16–25. Common methods of extrusion. A, direct. B, indirect. C, hydrostatic. D, impact.

much of the sidewall friction is eliminated. A practical limitation of indirect extrusion is the requirement for hollow rams to allow movement of the product through the ram.

In hydrostatic extrusion the sidewall friction is eliminated by separating the billet from both the cylinder wall and the ram, and applying force through a hydraulic fluid. Impact extrusion, which is normally done cold, is used for the low cost production of thin wall containers such as toothpaste tubes, and the like. The metal slug, normally soft aluminum, tin, or lead, is loaded in the die and impacted with a rapidly moving punch. The metal fills the die and then flows between the die wall and the punch forming a thin-walled tube. Cold extrusion can result in excellent surface finish (30-100 micro-inches, rms), accurate size, and improved mechanical properties. The severe cold working of extrusion results in a fine, dense grain structure with a high degree of grain flow parallel to the axis of extrusion and consequently high tensile properties in this axial direction. In Fig. 16–26 is shown the metal flow pattern of extrusion as revealed by flow lines on a sectioned lead billet. The type of material being extruded and the lubrication both play significant roles in the type of metal flow. With high wall friction the deformation is particularly non-uniform. In direct extrusion the center of the billet

Fig. 16–26. Deformation flow lines in a lead specimen extruded from a powdered lead compact.

moves much faster toward the die opening than does the material toward the sides. As a result, near the end of extrusion the outer surface of the billet begins to flow through the die causing the so called *extrusion defect*. This is somewhat analogous to pipe in an ingot.

Fig. 16–27 shows a modification of the direct extrusion process called direct sleeve extrusion. In this process a dummy block slightly smaller than the container bore is forced through the billet during extrusion, leaving a shell containing the oxidized surface layer about $\frac{1}{8}''$ thick around its circumference. Consequently, when inversion begins the metal is all clean and homogeneous and "pipe" and "core" are eliminated. Extrusion pressure, temperature and speed for a number of commonly extruded materials is

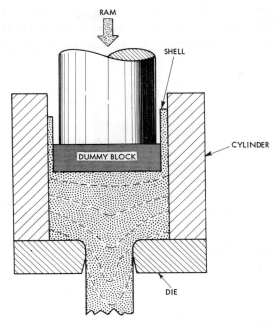

Fig. 16–27. Direct sleeve extrusion.

Table 16–2. Extrusion conditions for typical metals.

METAL	PRESSURE TONS/SQ-IN	TEMPERATURE, DEG. F.	EXTRUSION SPEED FEET/MIN.
ALUMINUM (EX-TRUSION GRADE)	40	700–900	1 1/2–300
BRASS (SOFT)	30	1200–1400	NOT CRITICAL
COPPER (SOFT)	25	1500–1600	NOT CRITICAL
LEAD	20	400–500	----
MAGNESIUM, M1	6–10	780–820	20–100
NICKEL	50–60	2000–2300	400–730
STEEL (C1010 EXTRUSION GRADE)	50	2000–2300	400–730
STEEL (C1020 SPHEROIDIZED)	60	2000–2300	400–730
TIN	20	140–150	10–30

given in Table 16–2. It should be mentioned that extrusion pressure is very much related to the percent reduction in area, (sometimes called extrusion ratio) part thickness, work hardening, extrusion rate, and die

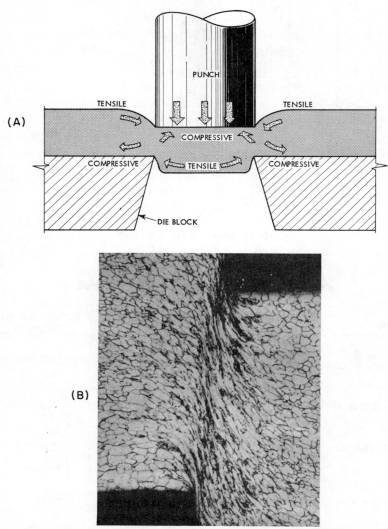

Fig. 16–28. A, deformation stresses in the shearing zone of low carbon steel sheet. B, photomicrograph showing deformation in shear zone of low carbon steel sheet, 100X. (General Motors Institute)

lubricant. Although several of the metals given in Table 16–2 may be extruded at room temperature a reduction in pressure can be achieved by extruding at elevated temperatures.

16.14 Shearing. Shearing is a process for separating a metal by forcing two opposing and slightly offset blades against it with sufficient force to cause fracture. As shown in Fig. 16–28A, the shearing force causes both tensile and compressive forces within the workpiece. During initial stages of the shearing process the metal is elastically, then plastically deformed; during plastic deformation there is a reduction in area, followed by fracturing along cleavage planes in the reduced area. Sheared edges are not perfectly square but may show both burnished, and rough or torn surfaces. Fig. 16–28B is a photomicrograph showing deformation in the shear zone of a low carbon steel sheet. The clearance between the blades is very important in providing clean sheared surfaces. In Fig. 16–29 A, B, and C are shown the effects of excessive, insufficient, and proper clearance. When clearance is excessive as shown at A, the cut edges are distorted and the cut surface is rough and may have a sharp feather edge. In B is shown the result of insufficient clearance. As in A the two fractures forming on opposite sides of the sheet do not meet. Cutting forces are high, tool life is low, and cut edge may appear highly burnished. In C is shown proper

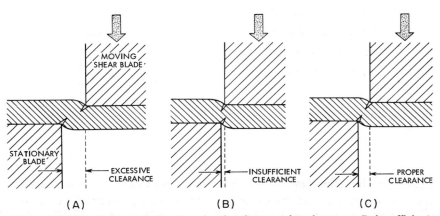

Fig. 16–29. Mechanism of shearing showing A, excessive clearance, B, insufficient clearance, and C, proper clearance.

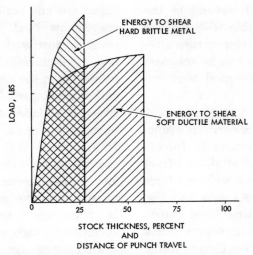

Fig. 16–30. Comparison of force and energy to shear hard and soft metals.

Table 16–3. Average shear strength for various metals and alloys.

METAL	SHEAR STRENGTH, PSI
ALUMINUM (2024–T)	40,000
COPPER	26,000
MAGNESIUM, M1	17,000
MONEL	43,000
STEEL, AISI 1020, ANNEALED	42,000
AISI 1030, ANNEALED	52,000
AISI 1030, ANNEALED	80,000
AISI 1095, ANNEALED	110,000

clearance so that the cracks which form at the sharp edges of the blades and continue through the metal, meet near the center of the work-piece providing a clean fracture. As the shearing edges of the blades become dulled, the shearing force is spread over a larger area and greater plastic deformation is required before shearing begins. This requires greater shearing force and results in greater distortion of the workpiece. Hard, brittle materials tend to crack when plastically deformed only a small amount and thus require only small die clearance. Soft, ductile materials, on the other hand, require greater plastic deformation before fracture and thus require a correspondingly increased die clearance. Although the maximum shear force is higher for harder metals than it is for soft, ductile metals, the total energy required for cutting is less as illustrated by Fig. 16–30. Hard brittle metals are cut with less energy and result in smoother and less distorted edges than softer and weaker metals. Under ideal shearing conditions, the punch should penetrate the material to a depth equal to about one-third of the sheet thickness before fracture occurs. The shear strength values of various metals is given in Table 16–3.

16.15 Forming. Metal forming operations such as deep drawing in-

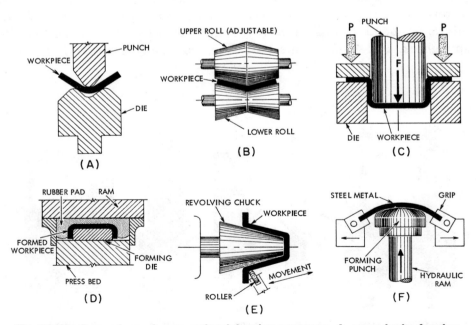

Fig. 16–31. Comparison of conventional forming processes. A, press brake forming, B, roll forming, C, deep drawing, D, rubber pad forming, E, shear spinning, and F, stretch forming.

523

volve stretching and drawing of flat sheets to produce shapes with compound surfaces. Simple forming operations, such as roll forming or press brake forming, involve only bending whereas other forming operations such as shear spinning require considerable metal flow and change of metal thickness. Fig. 16–31 shows a comparison of some conventional forming processes.

16.15.1. Press brake forming is used to produce parts such as in Fig. 16–32. These parts have simple curves involving only bending.

During bending the metal workpiece is subjected to both elastic and plastic deformation. Two important factors affecting bending are (1) minimum bend radius, and (2) springback. Fig. 16–33 shows that

bending a metal results in tensile stresses on the outside of the bend where the metal must be stretched, and in compressive stresses on the inside of the bend where the metal is compressed. The neutral axis designates a line along which the metal is neither stretched nor compressed. As the bend radius is diminished or as stock thickness is increased for a given bend radius, the metal at the extreme fibre is subjected to increased tensile stress. Since tensile stresses rather than compressive stresses are usually the ones which cause failure, tensile stresses in the outer fibre must be kept within an allowable range. The minimum bend radius determines the limit of forming after which the material cracks and is usually expressed in terms of

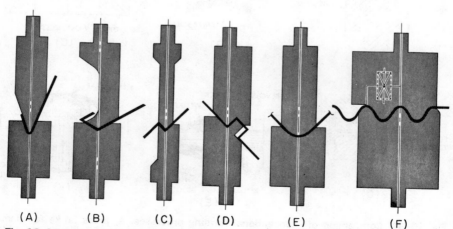

Fig. 16–32. Press brake forming dies and parts. A, acute angle bend, B, obtuse gooseneck die, C, v-bend die, D, channel die, E, radius die, and F, self-gaging corrugating die.

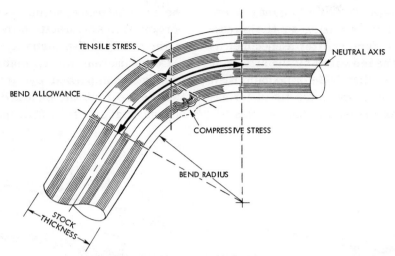

TENSILE STRESS

NEUTRAL AXIS

BEND ALLOWANCE

COMPRESSIVE STRESS

BEND RADIUS

STOCK THICKNESS

Fig. 16–33. Bend terminology.

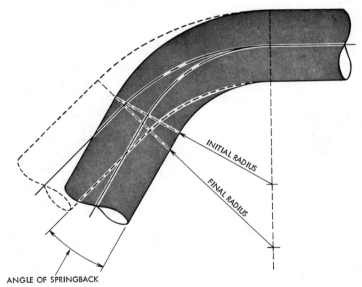

INITIAL RADIUS

FINAL RADIUS

ANGLE OF SPRINGBACK

Fig. 16–34. Spring-back due to elastic recovery of deformed fibers.

material thickness at $2t$, $3t$, $4t$, and so on.

Springback is the elastic move-ment of a bent material during re-lease of the forming pressure. It re-sults in a decrease in the bend angle.

525

Fig. 16–34. The situation may be explained by referring to the stress-strain diagram. After a specimen is loaded as indicated by the arrows, it is first elastically deformed and then finally undergoes plastic deformation. As the load is finally released, the material recovers from its elastic portion of the deformation resulting in the so called "springback" effect.

Compensation for springback is usually accomplished by designing the dies so that the included angle is somewhat less than that required

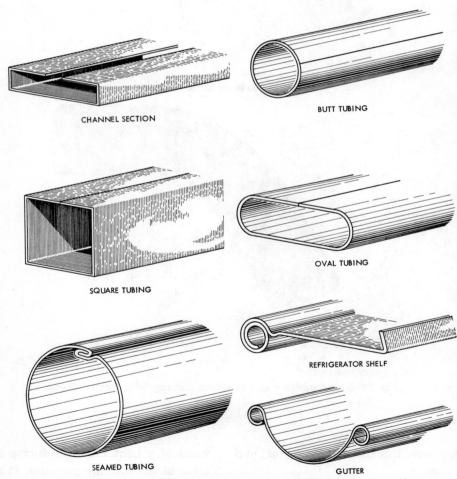

CHANNEL SECTION

BUTT TUBING

SQUARE TUBING

OVAL TUBING

REFRIGERATOR SHELF

SEAMED TUBING

GUTTER

Fig. 16–35. Shapes produced by continuous roll forming.

in the part. This causes the part to be bent through a greater angle than is required but allows it to spring back to the desired angle. Spring-back varies in steel from about ½ to 5° depending upon its hardness. Phosphor bronze, for example, may spring back from 10 to 15° while soft copper has virtually no springback.

16.15.2. Continuous roll forming is a method of forming strips into shapes such as those in Fig. 16–35. The process is practically automatic once the tooling has been built and proper setup has been made. Continuous roll forming reduces handling, power, and labor costs considerably under the cost of press brake forming. Because of the expensive tooling, however, roll forming is not economical except where the required production is greater than about 25,000 lineal feet. During continuous roll forming, the strip must pass through a number of rolls, each of which plays an important role in producing the final shape. In order to maintain a slight tension on the strip, the roll diameters, which are usually around 5 inches, are increased ½ to 1 percent as the strip progresses. Springback is handled by overbending and then bending back to the desired shape. The rolls should be designed to avoid excessive work hardening of the strip at any one pass. On aluminum, for example, angle bending should not exceed 22½° at any one pair of rolls.

Typical continuous rolled parts include seamed tubing, window and screen frame members, bicycle wheel rims, garage door trolley rails, metal molding, siding, and trim.

16.15.3. Drawing is a process used in the production of sheet metal vessels and compound curved sheet metal parts. Examples include seamless pots, pans, tubs, cans, automobile panels and tops, cartridge cases, and parabolic reflectors. The sheet metal is stretched in at least one direction and compressed in other directions. Fig. 16–36A & B shows the basic principles involved in drawing a round cup. The sheet metal blank must be of sufficient diameter to produce the desired depth of cup. The *drawing ratio* is the ratio of maximum blank diameter that can be successfully drawn, to the punch diameter. There is a maximum limit to the blank diameter after which the punch will pierce a hole instead of drawing the blank. This drawing ratio, which ranges from about 1.6 to 2.3, is dependent upon many factors including punch and die radii, and die clearance. It does not seem to depend upon tensile strength or elongation of the material. As the blank is drawn down into the die (Fig. 16–36C), the walls of the blank are under tension. Compressive stresses are set up in the flange as it is drawn into a smaller and smaller diameter. If the flange is thin (less than about 2 per-

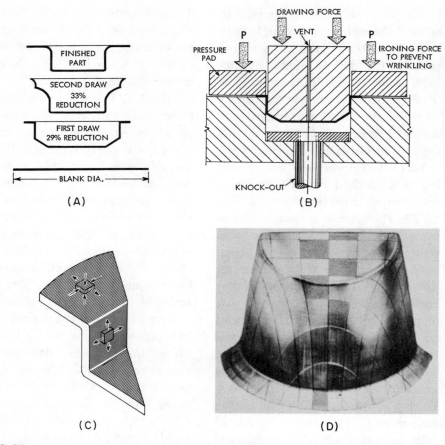

Fig. 16–36. Principles in drawing a sheet metal cup. A, blank for wide flange part, B, drawing and clamping for 1st die, C, stresses on wall and flange element, and D, flow lines on finished cup. (Aluminum Co. of America)

cent of the cup diameter) it will buckle unless pressure is applied to the flange by a pressure pad. The blank is usually lubricated to permit it to slide under the pressure pad and over the edge of the die. Flow lines of a finished cup are shown in Fig.

16–36D. The earing tendencies are due to directional properties in the metal; the cup is normally made higher than needed and the excess trimmed away. The cup may be redrawn in other operations to produce a longer cup of a smaller diameter. Practical reductions for the 1st, 2nd, 3rd, and 4th draws are reported to be 30, 25, 16, and 13 percent.

16.15.4. Flexible die forming and drawing operations, Fig. 16–37, use either a punch or a die, but not both.

Rubber pad forming, Fig. 16–37A, holds the workpiece between the lower die and a laminated rubber pad held in a container attached to the upper ram. Under pressures of 1,000 to 2,000 psi, the rubber flows readily, applying hydrodynamic force forming the blank around the die.

Marforming, Fig. 16–37B, is used for deeper draws than the rubber pad process and with less wrinkling. Blank holder pressure is regulated automatically by hydraulic fluid.

Hydroform process, Fig. 16–37C, is particularly suitable for deep drawing of parts with sharp detail. The process employs a flexible rubber diaphragm which is backed up by oil pressure. The dome is lowered until the diaphragm covers the blank, and the initial oil pressure is applied. As the punch is raised, the oil pressure forms the metal.

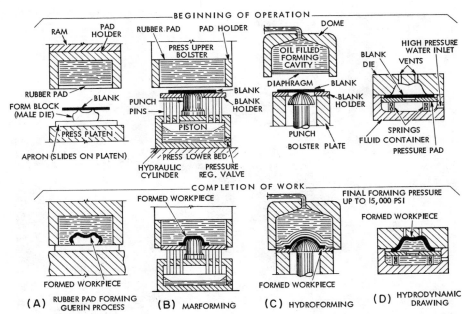

Fig. 16–37. Flexible die forming processes. A, rubber pad forming, B, marforming, C, hydroforming, D, hydrodynamic drawing. (By permission: *Manufacturing Processes and Materials for Engineers,* Prentice-Hall)

Hydrodynamic forming, along with marforming and hydroforming are patented processes. Hydrodynamic forming (Fig. 16–37D) utilizes water or oil pressure to force the blank into the die cavity. The process is limited to forming of shallow pieces, but is able to finish in one operation what may require several steps by other methods.

16.15.5. Stretch forming, Fig. 16–38, gives shallow shapes of large areas. Drawing alone does not strain the metal sufficiently to stress it beyond its elastic limit and give it the

desired permanent set. This difficulty is overcome by stretching the metal beyond its elastic limit by means of hydraulically actuated grips while at the same time forcing the forming punch into the metal. Stretch forming results in minimum springback. Dies may be of wood, zinc alloy, masonite, cast iron, or other easily worked materials.

16.15.6. High-velocity forming has been employed very successfully in recent years for difficult-to-form metals, large shapes beyond the capabilities of even large presses, and

Fig. 16–38. Stretch forming. Pre-stretch blank holders hydraulically grip sheet, stretch it and pull it over male die. Mating ram presses the piece. (Cyril Bath Co.)

for improvement of part tolerances and mechanical properties. Three high velocity metal forming processes are illustrated in Fig. 16–39 including explosive forming, electromagnetic forming, and electrohydraulic forming.

Explosive forming utilizes the energy stored in chemical explosives. In the process illustrated an explosive charge is detonated in a water tank containing the workpiece and die. Shock waves from the explosion propagate throughout the liquid and impact the workpiece with sufficient energy to drive it into the female die (Fig. 16–39A).

Electromagnetic forming uses the force produced on the workpiece by a rapidly collapsing magnetic field. Electrical energy stored in a capacitor bank is discharged into an electromagnetic coil producing a magnetic field around the work coil. This changing magnetic field induces eddy currents in the workpiece material placed in the field. The eddy currents produce a secondary magnetic field which resists the initial field and thus produces a force on the workpiece. For a magnetic field strength of 500,000 Gauss,* a pressure of about 90,000 psi is developed on the workpiece (Fig. 16–39B).

Electrohydraulic forming is similar in many respects to explosive forming except that the energy is derived from the discharge of a high-

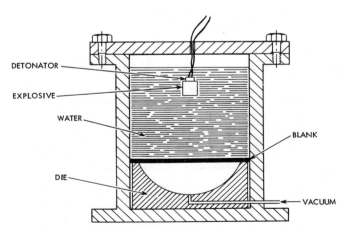

Fig. 16–39A. High velocity forming processes, explosive.

*A Gauss is defined as one line of magnetic induction running perpendicular to an area of one square centimeter, one line of flux per sq-cm.

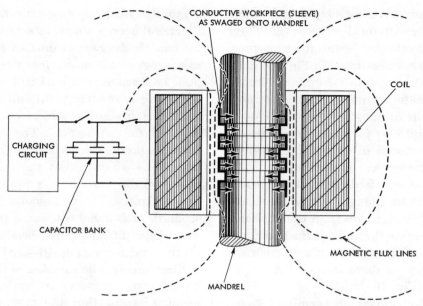

Fig. 16–39B. High velocity forming processes, electromagnetic.

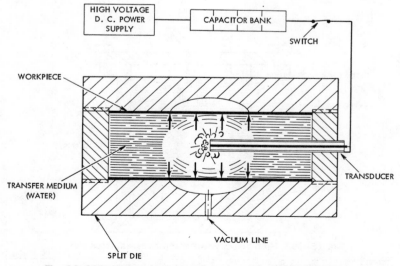

Fig. 16–39C. High velocity forming processes, electrohydraulic.

voltage capacitor bank under water. The discharge is between two electrodes, which may be in some instances spanned by a wire. The wire is vaporized, or the water ionized during the discharge. The expansion of the metal vapor or ionized liquid creates a strong pressure pulse which moves radially outward, deforming the workpiece (Fig. 16–39C).

The high velocity metal forming processes just described produce strain rates of 50 to 1,000 feet per second. Under these high deformation rates many metals show increased formability, minimum springback, and more uniform deformation of the workpiece. As deformation rate exceeds a certain critical value, the ductility drops rapidly to

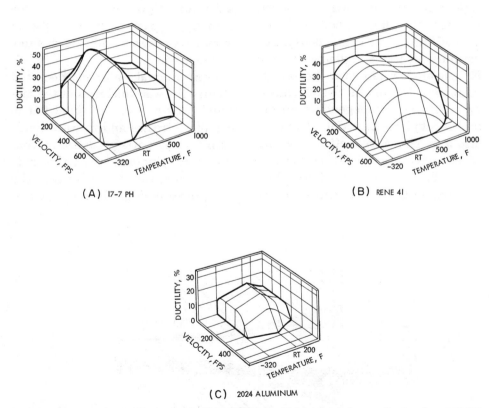

(A) 17-7 PH

(B) RENE 41

(C) 2024 ALUMINUM

Fig. 16–40. Combined effect of velocity and temperature on maximum strain of aerospace alloys. A, 17-7 precipitation hardening stainless, B, Rene '41 high strength alloy, C, 2024 aluminum age hardening alloy. (By permission: *Product Engineering,* **Sept. 30, 1963, McGraw-Hill)**

an insignificant value. The temperatures at which the deformation is carried out also has an important bearing on formability. The combined effect of velocity and temperature on maximum strain is shown in Fig. 16–40 for selected aerospace alloys.

16.16 Die Drawing. Die drawing is similar to the extrusion process except that the material is pulled through the die instead of being pushed through it (Fig. 16–41). For certain reductions in diameter there is a die angle where drawing force is minimum. Die angles usually range up to 15 percent depending on die and workpiece material. There is also a maximum reduction (approx. 45 percent) after which the rod will fail under the drawing force. The grid deformation of a strip with dif-

ferent die profiles is shown in Fig. 16–42. The residual stress distribution in a drawn rod shows tensile forces at the surface and compressive forces at the center. The magnitude of the residual stresses is dependent upon the reduction, die geometry, and temperature.

16.17 Spinning. In Fig. 16–43 are shown two spinning processes, A, conventional spinning, and B, shear spinning. Spinning involves the forming of parts having rotational symmetry over a mandrel with a tool or roller.

Conventional spinning does not change the wall thickness of the material. The material is subjected to complex stresses and to work hardening. Many parts produced by conventional spinning may also be made by drawing; the choice between the two processes depends upon the ma-

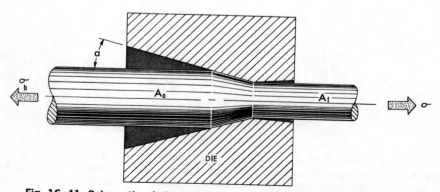

Fig. 16–41. Schematic of die drawing for reducing rod and wire diameter.

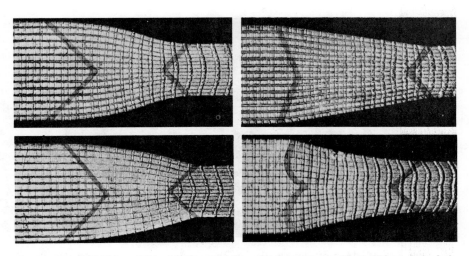

Fig. 16–42. Grid deformation in strip drawing with four die profiles. (Edgar Bain Laboratory, U.S. Steel Corp.)

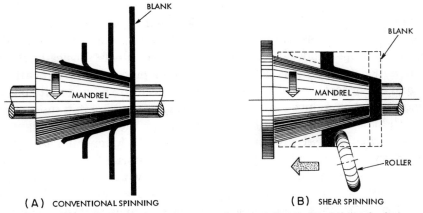

(A) CONVENTIONAL SPINNING   (B) SHEAR SPINNING

Fig. 16–43. Metal spinning processes. A, conventional, B, shear spinning.

terial properties, quantity of production, tool costs, and surface finish.

Shear spinning uses the techniques of spinning but also involves drastic squeezing and extruding of the metal. Blank thickness reductions of up to 90 percent in low carbon steel have been reported. Reduction in thickness is accompanied by cold working and changes in some

Fig. 16–44. Grid deformation in shear spun copper cone. (The Cincinnati Milling Machine Co.)

properties of the metal. Fig. 16–44 shows the grid deformation in a shear spun copper cone. The main advantage of this process is that large and heavy parts with rotational symmetry can be produced in a short time with little metal waste and with improved mechanical properties. Some of the major applications of shear spinning are in the production of aerospace components such as rocket motor casings and missile nose cones.

16-18 Ausforming. One new hot working operation which holds great promise for the production of high strength steel parts is that of ausforming or austenitic forming. Ausforming is a combination of hot working and heat treating processes designed to improve toughness, tensile strength, ductility, and hardness of certain steels. Basically the ausforming process involves heating the workpiece to the austenitizing temperature, working it in the austenitic phase to the desired configuration, followed by quenching to complete the transformation to martensite. Generally the initial ausforming temperature is around 1000°F. Since the metal cools while being deformed all working operations must be completed before the martensite transformation begins. Metalworking processes which have been used in ausforming operations include rolling, extrusion, shear spinning, explosive forming, and others.

Steels which can be successfully ausformed are those having an I-T

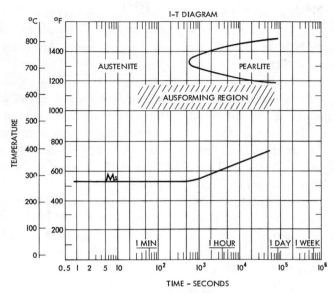

Fig. 16–45. Isothermal transformation diagram showing ausforming region in type H-11 hot work steel.

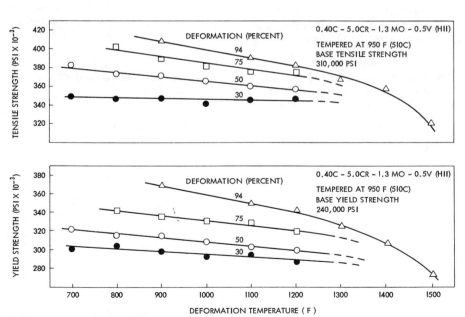

Fig. 16–46. Effect of temperature and deformation on yield and tensile strength of ausformed type H-11 hot work tool steel.

537

diagram with an austenitic bay region as shown in Fig. 16–45. Generally the steel to be ausformed must have a minimum of 0.10 percent carbon. Closely related to the chemistry of the steel are the parameters of time and temperature. As indicated in Fig. 16–45, the transformation from austenite to pearlite occurs relatively rapidly above 1100°F. Thus, the hot working operation must be completed rather quickly. The ausforming temperature is selected upon the basis of yield strength and tensile strength ratios which in turn are related to temperature. Higher temperatures reduce hot working forces and also reduce the amount of time allowed for completion of the operation. The most dramatic strength increases in steel occur when they are subjected to high deformation at the lower ausforming temperatures. An increase in yield strength of up to 125,000 psi for ausformed H-11 steel is shown in Fig. 16–46.

Ausforming is finding applications for parts having high strength-to-weight ratings such as bolts, engine mount components, and automotive leaf springs.

References

Clark, Donald S. and Varney, Wilbur R. *Physical Metallurgy for Engineers. 2nd ed.* Princeton, N. J.: D. Van Nostrand, 1966.

Hodge, K. G. *The Effect of Strain Rate on Mechanical Properties of Some Widely Used Engineering Metals.* Livermore, Cal.: U. of Cal. E. O. Lawrence Radiation Lab. Pub. No. UCRL-14599, 1965.

Kalpakjian, Serope. *Mechanical Processing of Materials.* Princeton, N. J.; D. Van Nostrand, 1967.

Wilson, Frank W. *High-Velocity Forming of Metals.* Englewood Cliffs, N. J. Prentice-Hall, 1964.

Powder Metallurgy

17.1 Introduction. Powder metallurgy is the process of molding metal parts from metal powders to precise shapes through the application of high pressures. Compaction of the metal powders is followed by a high temperature sintering operation in an atmosphere controlled furnace in which the cold welded metal compact is bonded into a strong homogeneous structure. Physical properties of the sintered compact are similar to those of the parent metal. Sintering is usually accomplished at about 80 percent of the melting point of the principal constituent to allow particle bonding and recrystallization across the powder particle interface. Compacting of the powdered metals in the die is accomplished by simultaneously pressing on it with the upper and lower punches at pressures of about 30 tons per square inch. Typical parts produced by P/M, the accepted acronym for powder metallurgy, include cutting tools, machine components, automotive parts, self-lubricating bearings, porous filters, magnetic materials, and composite metal-nonmetals. The P/M technique is also useful in forming of metals with low ductility and extremely high melting points such as tungsten, and some refractory metals.

The major applications of powder metallurgy today may be grouped into five areas:

Alloying of essentially non-alloyable metals. A number of metals which are not mutually soluble in either the liquid or solid states can be combined by powder metallurgy. This is particularly true for the electrical industry where motor brushes and contact points are made from powders of copper and graphite, and where contact parts are made from silver and nickel, or tungsten and copper. These metals form a mechanical mixture as shown in Fig. 17–1.

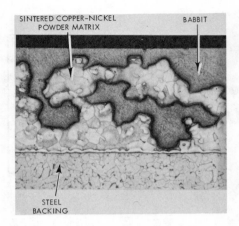

Fig. 17–1. Powder metal part of two non-alloyable metals, copper-lead forming a mechanical mixture, used for bearing material. (General Motors Res. Lab.)

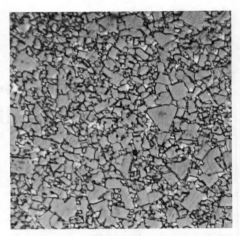

Fig. 17–2. Cemented tungsten carbide cutting material. Tungsten carbide in matrix of cobalt binder. 1500X. (Metallurgical Products Depart., General Electric Co.)

Combining metals and non-metallic materials. Examples of metal-nonmetal combinations include friction materials made from copper, iron, and asbestos; non-lubricated bearings made from iron and teflon, and heat resistant metal-ceramics combinations such as $Al_2 O_3$ and nickel. The metal-ceramic combination is frequently referred to as a cermet and finds applications in corrosion resistant chemical apparatus, pumps for severe service, and nuclear energy equipment.

Unitizing metals with melting points too high for casting. The melting points of the refractory metals such as tungsten, tantalum, and molybdenum are so high that it is difficult to melt and cast them, whereas with powder metal tech-

niques, they are sintered below their melting points. Powder metal compacting and sintered is the only feasible method of forming superhard tool materials such as cemented carbides and sintered oxides. The micrograph of a carbide cutting tool in Fig. 17–2 shows tungsten carbide particles held in a cobalt matrix. The cobalt binds the hard, wear-resisting carbide particles together.

Fabrication of metals for unique structural properties. A feature unique to P/M is the ability to produce self lubricating bearings containing a network of small interconnected pores filled with lubricant. Porous filters for diffusion, separation, and regulation of fluid flow may

540

Fig. 17–3. Permanent porous filter element of sintered bronze. Replaces resin impregnated pleated paper cartridges. (Pacific Sintered Metals, Inc.)

also be made by P/M. In Fig. 17–3 are porous metal filter elements of sintered bronze. Vibration damping is another unusual and often desirable property peculiar to the P/M part.

Economical production of precision parts. Powder metallurgy is a competitive process to die casting and injection molding of precision parts. When large quantities of high strength parts of iron, steel, or nickel are wanted, P/M offers an excellent solution for economical production of bushings, cams, gears, and other parts shown in Fig. 17–4.

Even with all of his important discoveries, the modern metallurgist cannot claim credit for the discovery of the powder metallurgy process. Metallic powders and solids from metallic powders have been made for many years. Powder metallurgy was used in Europe at the end of the 18th century for working platinum into useful forms. Platinum was infusible at that time. However, it is known that the early Incas in Ecuador manufactured shapes from platinum powders long before Columbus made his famous voyage. At the present time, the powder metallurgy process is used for the manufacture of shapes of up to about 350 pounds in weight, but some 1600 years ago, the famous iron pillar in Delhi, India, weighing

541

Fig. 17–4. Parts made by powder metal compacting requires little or no machining. (Pacific Sintered Metals, Inc.)

6½ tons, was made from iron particles or sponge iron similar to an iron used in the modern process. Wollaston, by 1829, had developed a technique that proved very successful in the manufacture of a malleable platinum from platinum powder that permitted forging of the resultant solid like any other metal. His published paper on the process is considered the first scientific work in powder metallurgy and laid the foundation for modern techniques.

The first modern application of the powder metallurgy process was the making of filaments for incandescent lamps. The first successful metal filament was made from tantalum, but with the discovery in 1909 by Coolidge that tungsten sintered from tungsten powder could be worked within a certain temperature range and then retain its ductility at room temperature, tungsten became the most important filament material. Finely divided tungsten powder was compressed into small briquettes which were sintered below the melting point of tungsten. Those sintered briquettes were brittle at room temperature but could be worked at 4530-4890°F. Subsequent hot working improved ductility until a stage was reached where the metal was ductile at room temperature and could be drawn into fine wire with tensile strengths approaching 600,000 psi.

17.2 Metal Powders. Metal powders are now available for many of the common metals and some alloys. Metals and alloys such as aluminum, antimony, brass, bronze, cadmium, cobalt, columbium, copper, gold, iron, lead, manganese, molybdenum, nickel, palladium, platinum, silicon, silver, tantalum, tin, titanium, tungsten, vanadium, zinc, carbides, boron, and tungsten, have been successfully produced in powder form. The amount of metal produced in the powdered form is an insignificant but rapidly increasing part of the total tonnage used by the metal working industry. However, the importance of metal powders and shapes produced from the powders is very great as compared with its tonnage. Perhaps the greatest growth in the use of P/M parts has been in the automotive industry, where as little as 10 years ago the average American car contained only about 4 lbs of P/M parts. Today the same vehicle may contain up to twice that amount with the trend sharply upward.

Metal powders should generally be considered as raw materials for the fabrication of metal shapes and not as end products in themselves.

Most of the metal powders are made in a wide variety of grades to satisfy various manufacturing and physical property requirements. Different production processes also impart different properties to the powdered metal. For example, electrolytic iron is characterized by high purity which in turn aids in obtaining exceptional compactability or unusual magnetic properties. Carbonyl iron powder is composed of small spherical particles which make

it desirable for the production of certain electronic cores; reduced oxide powders have distinguishing properties such as uniform size and porosity; atomized molten iron particles likewise have certain unique properties such as a fine dendritic cast structure and irregular to spherical shapes. Thus, the powder for a given application should be carefully selected in order to satisfy the end use requirements. The selection should consider (1) the alloy required in the finished part, and (2) the physical characteristics needed in the powder. Many of the metal powders are made for specific applications and the producers should be contacted for their recommendations as to the most suitable powders. The manufacturer will also likely run his own tests to evaluate both the powders and the production pressure.

Powders may be distinguished from one another and classified by a study of the following characteristics: (a) particle size and distribution, (b) particle shape, (c) surface profile, (d) solid, porous, or spongy nature, (e) internal grain size, (f) lattice distortion within each particle, and (g) the impurities present, their location, and whether or not they are in solid solution or exist as large inclusions or as surface and grain boundary films.

The shape of the metallic powder particles requires study to determine whether or not the shape is angular, dendritic, or fernlike, ragged and irregular, or smooth and rounded. It is obvious that such a study of all these factors relating to metal powder is relatively arduous and requires skill and special technique. The manufacturer of metallic powders tries to control all of these characteristics and improve techniques so as to manufacture a uniform product in each batch.

The characteristics of any metal powder should receive close study because to develop intelligent specifications for any given metal powder, one must understand what makes the powder behave as it does. At the present time, there are no hard and fast specifications for metal powders, as most of the specifications that are being used by the different manufacturers have been developed by purely empirical means. Some of the factors that are important to the behavior of any powder will be reviewed briefly.

Chemical composition of metal powders is not as important as their physical and mechanical properties. However, composition, and particularly impurities, will greatly influence the characteristics of any powder. The most important factor is the amount of oxygen that may be found in the powder as oxides. Oxides may be present in the powder from the method of refining and may form on the surfaces of powder that are exposed to air, as in storage or handling. The oxides found in the metal powder from the manufacturer may greatly weaken the final product. The thin oxide films that form during storage or handling may not have any detrimental effect because the powder may be cold weldable and the thin films can be reduced by the atmosphere used during the sintering operation. However, it may be found that any oxide film will reduce the ability of the powder to be cold pressed into a satisfactory shape; accordingly cracks and ruptures will occur in the cold pressed shape. Also, in iron powders, the presence of carbon (as cementite, Fe_3C), silicon, sulfur, and manganese may greatly influence the plasticity of the iron powders, making them difficult to cold press. Carbon in the form of graphite may be desirable from the viewpoint of lubrication during cold pressing, but its presence during sintering may result in changing the iron to a product similar to steel. Graphitic carbon is often added to iron powders for the purpose of making a harder iron similar to steel.

Structure of the powders greatly influences such characteristics as plasticity and ability to be cold pressed or made into briquettes. It influences pressures required in pressing, flow characteristics, and the strength of the final product. The study of the powder will reveal whether or not the powder is angular in shape, solid, porous and spongy,

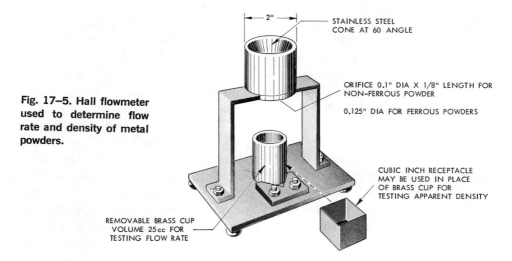

Fig. 17–5. Hall flowmeter used to determine flow rate and density of metal powders.

2"

STAINLESS STEEL CONE AT 60 ANGLE

ORIFICE 0.1" DIA X 1/8" LENGTH FOR NON-FERROUS POWDER

0.125" DIA FOR FERROUS POWDERS

CUBIC INCH RECEPTACLE MAY BE USED IN PLACE OF BRASS CUP FOR TESTING APPARENT DENSITY

REMOVABLE BRASS CUP VOLUME 25cc FOR TESTING FLOW RATE

crystalline, dendritic or fern-like, or ragged and irregular in shape. The type of structure that is desired in any given powder is usually determined from experience by the metallurgist.

Flow characteristics include the ability to flow uniformly, freely, and rapidly into the die cavity by gravity. If the flow is poor it becomes necessary to slow down the process of cold pressing in order to get sufficient powder into the die. The flow rate of any powder is largely influenced by the size of the powder particles, distribution of size, particle shape, and freedom from even minute amounts of absorbed moisture. Flow rate and apparent density may be determined by means of the Hall Flowmeter shown in Fig. 17–5.

Apparent density is the weight of a known volume of unpacked powder. It is usually expressed as grams per cubic centimeter. The apparent density is a very important factor affecting the compression ratio that is required in order to press the powder to a given density. If the apparent density of a given powder is one-third that of the required density, then three times the volume of powder is required to produce a given volume; that is the loose powder in a die cavity should be compressed from 3 inches in height to 1 inch, or the stroke of the punches used in the dies should close up 2 inches. It is obvious that a bulky or low apparent density in the powder will require a longer compression stroke and relatively deep dies to produce a cold pressed piece of a given size and density.

Particle size is one of the most important characteristics of any powder. The most commonly used method of measuring the size of any

545

metal powder particle is to pass the powder through screens having a definite number of meshes to the lineal inch. In the screening method, the size of the particle is measured by a square-mesh screen of standard weave which will just retain the particle. However, most powders are composed of non-spherical particles, and therefore particle size is not a concise method of measurement. Most frequently the powders used are made up of various sizes of particles. Their size is commonly reported by screening out the coarse, then the medium, and then the fine particles and reporting thus: 66 percent -100 to 200 mesh; 17 percent -200 to 325 mesh; and 17 percent -325 mesh. This means that 66 percent of the powder by weight will pass the 100-mesh screen but not the 200-mesh screen, etc. Other methods used for sizing powders include microscopic measurement and sedimentation, both being used to determine the size of particles when they are smaller than the finest mesh size of screens.

17.3 Production of Metal Powders.

The primary methods used in the production of metallic powders include mechanical pulverizing by ball milling, atomization, electrolytic deposition, and gaseous reduction. Other less common methods involve vapor condensation, the carbonyl process, shotting, chemical precipitation, machining, and granulation by vigorous stirring during solidification. The primary methods of metal powder production along with typical characteristics of powders produced by these methods will be discussed.

Mechanical pulverization such as milling, or grinding, in ball mills is a means of producing metal powders of almost any degree of fineness from friable or malleable metals. Friable metals yield angular, jagged particles of irregular shape (Fig. 17–6A). Tungsten carbide particles are pulverized in this way. Some malleable metals such as aluminum and bronze are milled with a lubricant to form flakes. These flakes are not generally suitable for molding but are used for parts and pigments. The lubricant prevents the flakes from welding together. Prolonged grinding in a ball mill may aid compacting because of the fine powder produced, but may also necessitate annealing before compacting because of the resulting work hardening.

Atomization is a method used most frequently for metals having low melting points such as tin, lead, cadmium, zinc, brass, bronze, and aluminum. This process has also been used for high purity iron powders. During atomization the molten metal is forced through a small orifice and broken up in a stream of compressed air, steam, or inert gas. This process permits the production of metal powders with ragged, tear-

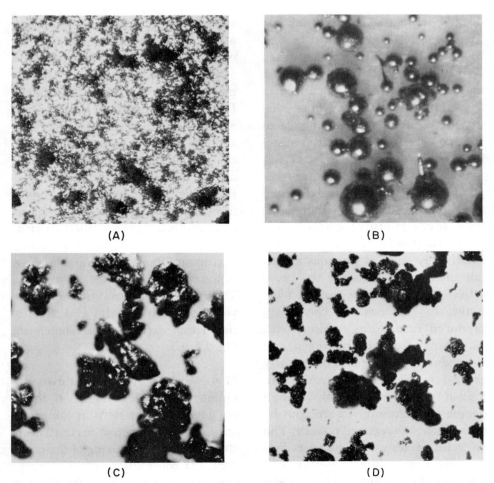

Fig. 17–6. Shapes of metal powders produced by various methods. A, ball milling gives fine acicular shapes to tungsten carbide. 125X. (Easton Metal Powder Co.) B, atomization produces spherical shape in a nickel brazing alloy. 108X. (SCM Corp.) C, gaseous reduction gives sponge iron from oxide. 125X. (Easton Metal Powder Co.) D, electrolytic deposition produces a dendritic copper. 125X. (Easton Metal Powder Co.)

drop shapes, or spherical particles, depending upon the surface tension of the material. Particle size and uniformity varies over a considerable range and is dependent upon orifice design, temperature, viscosity, and flow rate of the metal and pressure and temperature of the gas. The atomized particles are collected by means of a suction system and de-

547

posited in bags. The particles are then annealed and screened to size. For structural parts the atomized powders of copper, nickel, and iron may be oxidized then pulverized to the desired fineness and finally reduced to metal in reducing gases such as hydrogen. A spherical atomized powder is shown in Fig. 17–6B.

Gaseous reduction of metal compounds is one of the most flexible and economical methods of making powders for structural parts. This method of powder production is used extensively since it provides a flexible means of controlling the properties of the powder over a wide range. In this process, chemically produced oxides are reduced with carbon monoxide or hydrogen and the reduced powder is subsequently ground to size. If the oxides are graded prior to reduction, a high degree of uniformity in size can be obtained in the reduced powder. The particles produced by oxide reduction are sponge-like in nature and are ideal for molding. Their shape is generally jagged and irregular as shown in Fig. 17–6C.

Electrolytic deposition is used for the production of extremely pure powders, principally copper and iron. The electrolytic process is essentially an adaptation of electroplating and begins with the production of an anode—by casting in the case of copper, or rolling in the case of iron. The current density, temperature,

type of electrolyte and circulation affect the type of metal deposited on the cathode. The deposit may be a soft, spongy substance, or the deposit may be a hard, brittle metal. Powders obtained from hard, brittle electro-deposition are generally not suitable for molding purposes. After the porous metal deposit has built up on the cathode to the desired thickness, the deposit is stripped from the starting plate, washed free of electrolyte, dried, milled to size, and given a reduction-annealing treatment. Screening and blending of the powder completes the process. The resulting electrolytic powder is very pure and soft and has a dendritic particle shape shown in Fig. 17–6D.

17.4 Mixing of Metal Powders. Following powder preparation, there are generally four steps in the preparation of compacted parts. These steps include: (1) mixing of the powders, (2) compacting, (3) sintering, and (4) supplementary treatments such as sizing, heat treatment, oil impregnations, or surface treatments. The first of these steps, mixing of the powders, has several purposes. It is used for blending of several grades or sizes of the same powder, for blending of alloys, for adding of friction reducing lubricants, and for blending of additives to control porosity of the sintered product. Conditioning of the powders for processing is as important as the pressing and sintering operations. Mixing time may vary from a few minutes to several hours. A blending time of 5 minutes is better than 30 minutes if uniformity is achieved with the shorter time. Over-mixing should be avoided as it may decrease particle size and work harden the metal particles. Various types of mills and blenders are used for mix-ing of metal powders. Double cone blenders, Fig. 17–7 are widely used for blending soft metal powders. Ball and rod mills are used extensively for mixing hard metals such as carbides but are less suitable for soft powders where the danger of particle de-formation and work hardening is greater.

To prevent powder segregation during mixing of powders of greatly

Fig. 17–7. Laboratory cone blender for metal powders. Lucite model is used for studying powder flow. (Patterson-Kelley Co.)

differing particle sizes, a fraction of a percent of camphor, lauryl alcohol, or a similar organic substance with adhesive quality is added. Usually the heavier material in the mix is coated with this additive. In addition to preventing segregation of the particles, the additive also aids bonding of the particles during pressing.

Segregation of powders is also common when mixing materials with different particles such as atomized powders with electrolytic or reduced powders. Additives such as graphite tend to equalize the different shapes and therefore produce a more homogeneous mixture.

Die lubricants such as graphite, stearic acid, zinc stearate, and others, are widely used to reduce friction between metal powder particles and the faces of punches and dies. Friction between powder particles and die walls results in parts without uniform density and creates problems in filling of complicated dies, and lower tool life. Interparticle lubrication is of great importance in reducing these problems. Additions of non-metallic lubricant to the metal powder mix is usually less than 1 percent. When the additives are used to increase porosity of the sintered product they are used in larger amounts.

17.5 Compacting of Metal Powders. Compacting of the metal powder is a very important step in the production of powder metal parts. In addition to producing the required shape the compacting operation also influences the subsequent sintering operation, and to a large degree, governs the properties of the final product. The various compacting methods now in use include both cold and hot die compacting, isostatic pressing, slip casting, centrifugal casting, and rolling. Cold (room temperature) compacting, which is widely used for high production parts, usually with a minimum lot quantity of 20,000 will be discussed in some detail. The other processes are for special applications.

By far, most compacting is performed at room temperature in hardened steel or tungsten carbide dies. The die is filled with the mixed and blended powdered metal after which pressure is applied. Fig. 17–8. The most important effects of compacting are:

1. Increased density by reduction of the voids between powder particles.

2. Cold welding, mechanical locking, and adhesion of the particles, to produce sufficient "green strength" to permit handling of the compact.

3. Plastic deformation of the powder particles to induce recrystallization during sintering.

4. Increased contact area between the particles by plastically deforming the powder.

Filling of the die cavity is done by

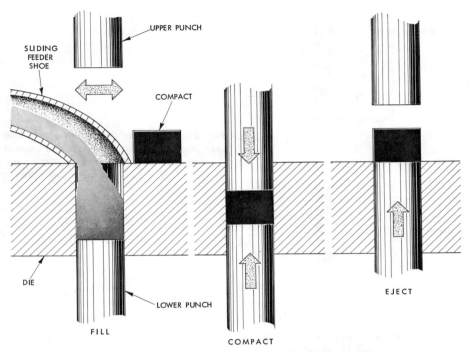

Fig. 17–8. Schematic of steps in compacting powdered metal parts.

introducing a measured volume or weight of powder through a sliding feeder shoe (Fig. 17–8) which also moves the ejected compact off the lower punch. Movement of the feeder helps to distribute powder evenly over the entire die area and also to scrape off excess powder. A non-uniform fill will result in variations in density in the compact. Metal powders do not usually flow readily around corners and into recesses like fluids. Neither do they transmit pressure uniformly from particle to par-

ticle. High friction between adjacent particles and between particles and the die wall occurs during compaction. This results in inability of powder to effectively transmit uniform pressure at right angles to the direction of applied force. Thus, the powder is not compressed to uniform density throughout the compact, as shown in Fig. 17–9, and consequently does not react uniformly to subsequent sintering.

In Fig. 17–10 is illustrated the plastic deformation of the powder

551

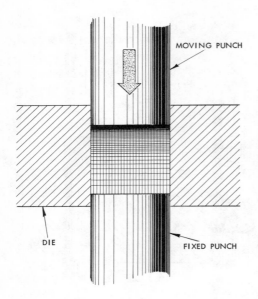

MOVING PUNCH

DIE

FIXED PUNCH

Fig. 17–9. Uneven density in single end pressing caused by side wall and powder friction. Compact becomes harder, more dense, and stronger near the moving punch.

GRAIN BOUNDARIES

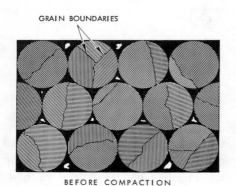

BEFORE COMPACTION

COLD WELDS

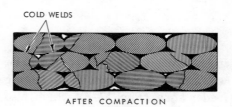

AFTER COMPACTION

Fig. 17–10. Schematic of plastic deformation and cold welding that occurs during compaction of powders.

particles and the cold welding which occurs during compaction. As the adjacent particles are brought into contact under pressure, and as they rub against one another, the surface oxide film is pushed aside. The intimate contact of adjacent particles initiates high interatomic attraction, resulting in a cold weld. The total cold weld area is rather small, and

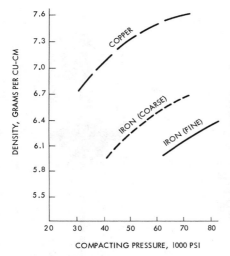

Fig. 17–11. Relationship between compacting pressure and density on copper and iron powders.

the ejected compact is not strong. The green compact is, however, strong enough to be carefully handled but will normally break if dropped. A subsequent sintering operation will be used to strengthen the compact. Severe plastic deformation of the individual powder particles produces a hard and rather brittle structure, which will recrystallize under the elevated temperature used for sintering. The relationship between compacting pressure, and density values for iron and copper powders is shown in Fig. 17–11.

The type of powder, particle size, distribution, shape, and composition, all influence the behavior of powder

under pressure. For example, the compacting pressures for fine powders are usually greater than those required for coarse powders. Smooth powders such as spherical shapes pack more readily than those of irregular shape but also require higher deformation pressures than do irregular shaped powders; better green strength is usually obtained with particles of irregular shape. Hard metals require higher pressures than soft metals of the same structure to achieve satisfactory density throughout a compact.

Compacting pressures generally range from 20,000 psi to 200,000 psi with 100,000 psi being the range in which most work is done in actual practice. Some reports of high frequency vibration of the die while compacting indicate a significant reduction of frictional forces and a consequent reduction in the maximum compacting pressures required.

Excessive compacting pressures may produce cleavage fractures or slip cracks, especially in parts with thick and thin sections in the direction of pressing. Insufficient pressure, on the other hand, produces fragile parts that will not hold together during handling. Horizontal laminations may result from expansion of the upper part during ejection while the lower part is still constrained. A slight taper of the die above the compressing position will reduce the sharp transition stresses

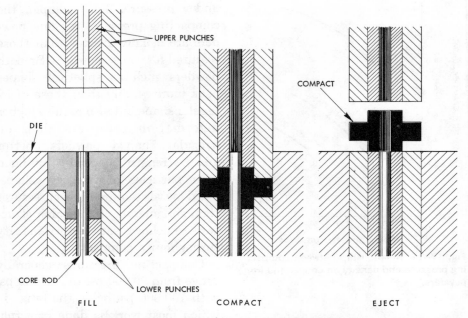

Fig. 17–12. Multiple punches in compacting achieve uniform pressures. Inner punch rises from fill position to push powder into inner section of upper punch. All four punches squeeze and at ejection the two lower punches rise.

during ejection and minimize this problem. Entrapped air may also produce lamination as the compacting pressure is released. This problem may be alleviated by changing powder particle size to a coarser one or reducing pressing speed to facilitate escape of air.

One serious handicap of powder metal compacting is the problem of pressure gradient in thick or irregularly shaped parts. This problem may be minimized by use of multiple punches illustrated in Fig. 17–12 and by use of graphite and metallic

stearates for die wall lubrication. The usual limitation on height for parts of less than 10 square inches in cross section is that the height be not more than the square root of the cross sectional area. Thus, for an area of 4 sq in. the part height should not be greater than about 2 inches.

Compacting is done on either mechanical or hydraulic presses. Mechanical presses, Fig. 17–13, are preferred for high production rates. They usually range in sizes from 5 to 150 tons capacity. Production output for a mechanical press with dies

554

Fig. 17–13. Dual motion press for production of powdered metal gears, rings, discs, and bushings at rates up to 55 parts per minute. (Stokes Powdered Metal Press Dept., Pennsalt Equipment Div.)

mounted in a rotary table may reach 1000 parts per minute. Hydraulic presses (Fig. 17–14) range in size from 5 to 2000 tons and over and are used almost exclusively for the production of large parts where pressures above 150 tons are required. These presses are relatively slow except in the smaller size ranges.

17.6 Sintering. Sintering involves heating the compact to a high temperature in a controlled atmosphere furnace to increase bond strength between the powder particles and to promote alloying by solid state diffusion. Methods of both solid phase and liquid phase sintering will be discussed.

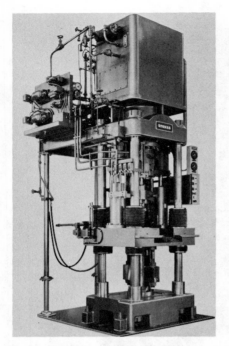

Fig. 17–14. Hydraulic compacting press, 300 ton capacity, for large metal compacts. (Stokes Powdered Metal Press Dept., Pennsalt Equipment Div.)

17.6.1. Solid phase is the most widely used method of sintering in that no melting of the particles occurs. The sintering temperature is usually 60-80 percent of the melting temperature of the lowest melting point constituent.

Sintering is generally carried out in controlled atmosphere furnaces, or occasionally in vacuum furnaces, to prevent or reduce surface oxidation of the large number of metal particles. Oxidation of individual metal particles weakens the sintered bond or may even prevent bonding. A dry hydrogen atmosphere is used in the sintering of refractory carbides and electrical contacts, but most commercial sintering atmospheres are produced by partial combustion of hydrocarbons such as natural gas or propane or by dissociation of ammonia.

The mechanism by which sintered metal particles are joined into a coherent structure having increased density and strength, has been the subject of many investigations and much discussion. Sintering is probably the most complicated of all heat treating operations because it involves not only the problems involved in heat treatment of solid metals but also peculiar problems of dimensional growth or shrinkage, internal and external gaseous reactions, and solid phase alloying. The mechanism by which cold welded metal particles are joined into a coherent mass having increased strength and density is thought to include principally the following: (1) diffusion, (2) recrystallization, (3) grain growth, and (4) densification.

The increased strength developed in the powder compact during sintering is due principally to the disappearance of the individual particle boundaries (Fig. 17–15) through solid state diffusion and recrystalli-

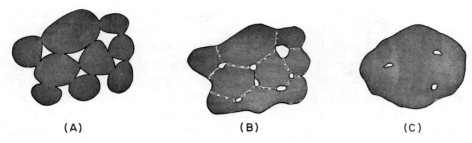

Fig. 17–15. Three stages during sintering. A, initial cold weld contact areas. B, enlargement of contact areas, rounding of interstices, size of holes reduced. C, density increased and impurities and oxide films pushed into internal holes.

zation. Diffusion rate is dependent upon the particle size, the percentage of each alloy present, and upon the temperature. According to Rhinds, most diffusion controlled processes follow the rule of doubling their rates for every 18°F temperature increment up to the melting point where a very large increase occurs. Thus the diffusion at 1420°F should be approximately 1000 times than at 1247°F. High temperatures are applicable in sintering because of the oxidation protection afforded by the controlled atmospheres and because of minimum problems of grain growth due to the initially small powder size.

The powder particles are usually deformed an amount greater than their critical strain during compacting so that recrystallization and coalescence of the grains within the particle are common during sintering. The speed of diffusion is increased by internal stresses, dislocations and atomic voids arising from severe plastic deformation of the powder particles during compacting. The sintering process may increase rapidly above the recrystallization temperature with an increase in strength and density. Sintering of compacts whose powders were subject to less than the critical strain precludes recrystallization and bonding will occur only by diffusion and coalescence. Under certain conditions recrystallization in powder compacts leads to development of grains whose size is larger than the initial grain size within the powder particles. Grain growth may occur within the particles themselves, or across particle boundaries. Grain growth is somewhat inhibited by pores or inclusions but in some pure metal compacts becomes excessive. Additional study is required to more clearly understand this aspect of sintering.

Ordinarily, the density of the compact is increased by sintering and shrinkage will occur. Shrinkage is associated with increase in areas of contact of the particle and reduction of the size of holes in the compact. In some instances of sintering, however, there is a growth in the pressed compact. This growth is thought to be caused by liberation of dissolved or entrapped gases or by chemical reactions within the compact. Slow heating or degasing can be used to minimize problems associated with expansion.

17.6.2. Liquid phase sintering is used in some cases. This is carried out at a temperature at which a liquid phase exists. It may be above the melting point of one of the alloy constituents or above the melting point of an alloy formed during sintering. In the former case liquid exists during the entire sintering operation, but in the latter case liquid is initially formed but disappears prior to completion of sintering. Liquid phase sintering may be successful as long as the essential part of the compact is not molten. The liquid phase may be either soluble or non-soluble with the matrix. A non-soluble molten phase located in the grain boundaries hinders diffusion between the particles of the non-melting phase. When solubility exists between the non-melting

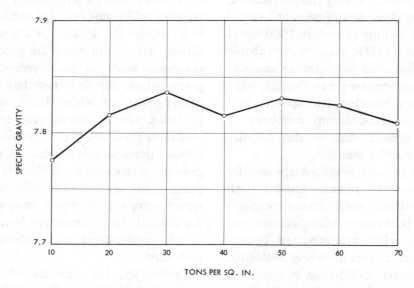

Fig. 17–16. Effect of compacting pressure on density of sintered iron compacts.

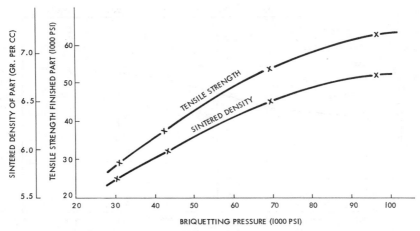

Fig. 17–17. Relationship between compacting pressure, sintered density, and tensile strength for carbon-iron powder sintered at 2030° for 30 minutes.

phase and the liquid one, a very good contact between the particles occurs and high density is achieved as with the tungsten carbide-cobalt and tungsten-nickel-copper systems.

The density, strength and other properties of sintered compacts are dependent upon a number of factors including compacting pressure, sintering temperature, and sintering time. The effect of compacting pressure on density of iron compacts is shown in Fig. 17–16. There seems to be an optimum compacting pressure range to achieve the highest density of the sintered compact for a given sintering temperature. Excessive compacting pressures seem to produce cleavage fractures and other problems. The relationship of compacting pressure, sintered density,

and tensile strength for a carbon-iron powder is shown in Fig. 17–17. Fig. 17–18 shows the effect of sintering time on density. Temperature is the most important factor in bringing about the maximum sintered density. Fig. 17–19 gives the effect of sintering time at a particular sintering temperature. Sintering time and sintering temperature values for a number of metals are given in Table 17–1. Fig. 17–20 is a typical controlled atmosphere sintering furnace; a purging and preheating chamber and a cooling chamber are shown as extensions on the furnace.

17.7 Supplementary Operations. A number of finishing or supplementary operations have been designed to impart specific properties to sin-

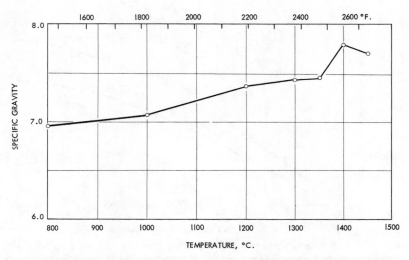

Fig. 17–18. Effect of sintering temperature on density of sintered iron compacts.

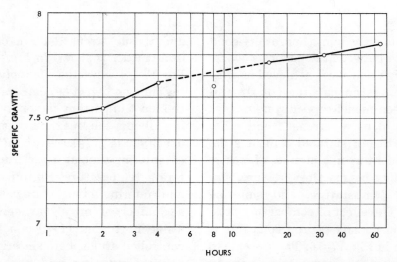

Fig. 17–19. Effect of sintering time on density of iron compacts. Compacted at 60,000 psi, sintered at 2600°F.

tered metal parts. These operations, shown schematically in Fig. 17–21 include infiltration, heat treatment, impregnation, sizing, machining, and extruding. Infiltration, for example, is done by placing a copper sheet on

Table 17–1. Sintering temperature and time for selected powdered metals and alloys.

MATERIAL	TEMPERATURE	TIME, MINUTES
BRONZE	1400–1600	10–20
COPPER	1550–1650	30–45
BRASS (80–20)	1550–1650	30–45
IRON	1850–2100	30–45
NICKEL	1850–2100	30–45
STAINLESS STEEL	2150–2300	30–45
ALNICO MAGNETS	2200–2375	120–150
TUNGSTEN CARBIDE	2600–2700	20–30

DATA COURTESY OF LINDBERG ENGINEERING CO.

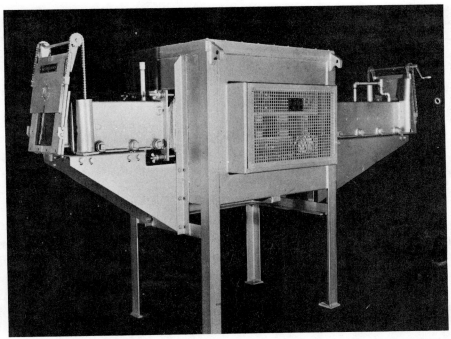

Fig. 17–20. Atmosphere controlled sintering furnace. Purging and preheating vestibule is the left extension, and cooling chamber is the right.

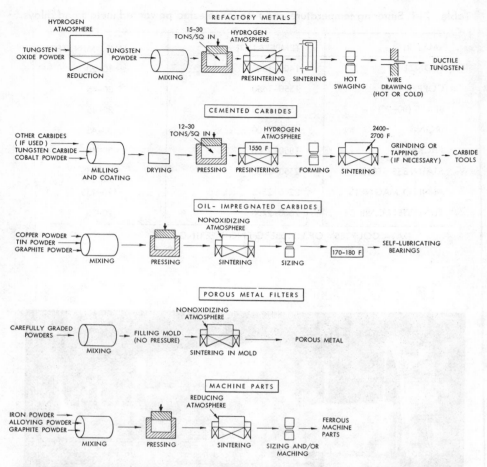

Fig. 17–21. Schematic of powder metallurgy processes. (By permission: *Metals Handbook, 1948 ed.*, American Society for Metals)

a presintered iron part and heating it until the copper melts and is drawn down into the pores of the part by capillary action. This infiltration increases strength of this part by 70 to 100 percent but reduces toughness. Iron parts may be given a carburizing heat treatment to increase strength and hardness. Sintered bronze bearing materials are frequently impregnated with oil (up to 20 percent by volume) by immersion in hot oil or by drawing oil into them with a vacuum.

Powdered metal parts are sometimes sized or coined to improve dimensional accuracy by forcing them into a mold under pressure. Coining and forging also increases density and strength of the material. Machining operations such as drilling, tapping, turning undercuts or close tolerances are sometimes required of powdered metal parts. Tools should be sharp, speeds high, and feeds light. Water-base cutting fluids should not be used as they tend to contaminate oil impregnated bushings and to corrode iron compacts.

17.8 Applications of Powder Metallurgy. Some of the many shapes that have been successfully produced by the powder metallurgy process are shown in Fig. 17–22. The manufacture of porous metal bushings from iron, brass, bronze, and aluminum alloys has proved to be one of the major applications of the process. Millions of gears, levers, and cams for desk calculators are also products of the powder metallurgy process. Gun parts, locks, precision gears, sprockets, hydraulic

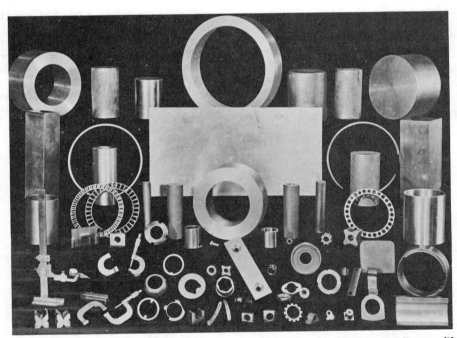

Fig. 17–22. Bearings, tools, and machine parts produced by powder metallurgy with little machining required. (Chrysler Corp.)

pump rotors, ferrite computer core doughnuts, and automotive shock absorber components should be added to this list.

Better understanding of the advantages and limitations of powder metallurgy by designers and producers has opened many new fields of application for P/M parts. Relatively large and complex parts may now be made by special P/M techniques such as isostatic pressing and high velocity forming. Tensile strength properties as high as 180,000 psi and more have been achieved with carbon and alloy steels. The most ideally suited parts for P/M, however, are those which are used in relatively large quantities (20,000 and up) and which would otherwise require extensive machining. The shapes most suited to P/M are cylindrical, rectangular, or irregular shapes which do not have large variations in cross-sectional dimensions. Indentation, or projections on the top or bottom surfaces can be easily produced. Splines, gear teeth, axial holes, counterbores, straight knurls, slots and keyways present no problem for the P/M process.

References

Avner, Sidney H. *Introduction to Physical Metallurgy.* New York: McGraw-Hill, 1964.

Burton, Malcolm S. *Applied Metallurgy for Engineers.* New York: McGraw-Hill, 1956.

DeGroat, George H. *Tooling for Metal Powder Parts.* New York: McGraw-Hill, 1958.

Doyle, Lawrence E. et al. *Manufacturing Processes and Materials for Engineers.* Englewood Cliffs, N. J.: Prentice-Hall, 1961.

Wulff, John. *Powder Metallurgy.* Cleveland, O.; American Society for Metals, 1942.

Foundry Metallurgy

Chapter

18

18.1 Introduction. Casting of metals to obtain a desired shape is perhaps the oldest process known to man; yet only recently has the casting process begun to be scientifically understood and controlled so that high quality parts may be economically and constantly produced. Casting involves pouring molten metal into a prepared mold cavity of the desired shape and allowing the metal to solidify (Fig.18–1). The resulting casting may be cleaned, trimmed, machined, heat treated, and finished ready for assembly and use. Millions of castings are produced each year and find application as parts of automobiles, tractors, harvesting equipment, household appliances, communication equipment, and many other uses. High precision and high speed casting processes have been developed which can produce parts with extremely close tolerances, requiring very little, if any, machining.

Parts may be produced by casting from almost all metals. Cast parts may range in size from those weighing a fraction of a pound to several tons.

The properties of a given casting are largely controlled by the nature of the metal being cast, the type of mold, size and shape of the casting, melting and pouring procedure, and its cooling rate. The above factors may be divided into two major divisions, (1) the metallurgical principles involved in making the casting, and (2) the casting process, mold materials and techniques, and equipment. Only the metallurgical principles including casting design considerations will be considered here.

18.2 Melting Principles. Melting of a metal requires that the thermal activity of its atoms exceed the strength of the interatomic bonding, allowing the atoms to move contin-

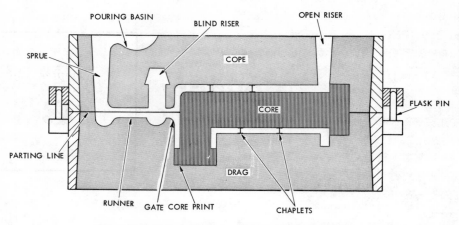

Fig. 18–1. Schematic of a mold cavity for casting.

uously with a random distribution. As heat is applied to a solid metal, causing its temperature to rise, the vibrational activity of its atoms also increases. When the atoms become so active that they are no longer held in a relatively stable crystalline relationship, melting occurs.

An interesting fact about melting is that the energy content of materials in the solid state is lower than the energy content in the liquid state, even when at the same temperature. Thus, when the melting point of a metal is reached, additional heat must be added before melting will proceed, even though there is no temperature change. This additional heat is called the *latent heat of fusion*. It is needed to allow the metal to overcome the attraction force holding the atoms in regular

crystalline positions. Melting results in complete obliteration of the original metallic lattice and allows the atoms random motion that is characteristic of liquids. Even so, with the enormous number of atoms present in even a small quantity of metal it is expected that some atom aggregrates at any given instant are in positions exactly corresponding to their ordered arrangement in the solid state. These chance aggregates will be continuously disbanding and reforming elsewhere; the higher the temperature, the faster the changes.

Another interesting observation about melting is that since metals are not completely homogeneous, and frequently consist of more than one phase, they do not melt simultaneously in all portions of the metal. Melting usually begins at grain boundaries, at regions of eutectic composition, or in regions where lower melting point constituents are found.

18.3 Melting Practices. Two problems that must be taken into consideration when melting metals are: (1) the tendency for metals to absorb and dissolve gases, and (2) oxidation of the metal at the elevated temperatures. Molten aluminum, for example, absorbs and dissolves hydrogen, (Fig. 18–2). Upon solidification the dissolved hydrogen is liberated, causing porosity. Measures should be taken to minimize both oxidation of the molten metal and absorption of atmospheric gases; temperature control is especially important here as both tendencies are more pronounced at higher temperatures.

One other problem of melting is that of grain growth in the solidifying casting when excessively high pouring temperatures are used. Accurate temperature measurement and control is very important here also. Typical melting furnaces are shown in Fig. 18–3. The cupola furnace, crucible furnace, and induction furnace are widely used for melting of non-ferrous alloys. The cupola furnace is used mainly for cast iron melting. The electric arc, and the induction melting furnaces are often used for melting of high quality stainless steels, tool steels, and specialty metals. Many other types of furnaces are in use in addition to those illustrated. One recent development is the vacuum furnace (usually electric induction heated) which is used in the production of the highest quality metals and alloys.

The foundry melting operation usually involves much more than simply adding enough heat to a solid metal to melt it. Often the metal must be *alloyed* during melting by

Fig. 18–2. Macrophotograph showing porosity in aluminum ingot due to entrapped gases, 2X.

the addition of other metals. Undesirable impurities in the melt must be removed by *refining* or *degasing* operations. Finally, metals are sometimes inoculated immediately prior to pouring for grain refinement of the metal or to alter metallurgical structure in some other way, such as in the production of ductile iron.

18.4 Alloying. Melting stock for non-ferrous alloys is nearly always *primary* and *secondary* ingot and foundry *returns* (gates, risers, and

scrapped castings from previous operations). Primary ingot is prepared from commercially pure metal by refineries known as smelters. Secondary ingot is a remelt of sorted, melted, and alloyed scrap metals. Secondary ingot is less expensive than primary ingot and sometimes less desirable.

Cast iron is made from pig iron and iron and steel scrap. Alloy steels are usually made from carefully selected scrap steel to which the alloying elements are added. When

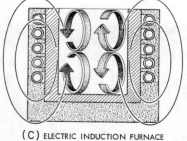

(A) CRUCIBLE FURNACE (GAS OR ELECTRIC HEAT)

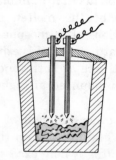

(B) DIRECT ARC ELECTRIC FURNACE

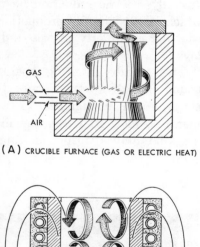

(C) ELECTRIC INDUCTION FURNACE

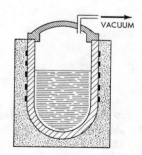

(D) ELECTRIC RESISTANCE HEATED VACUUM MELTING FURNACE

Fig. 18–3. Schematic showing melting furnaces. A, crucible (gas or electric), B, direct arc electric, C, electric induction, D, electric resistance vacuum.

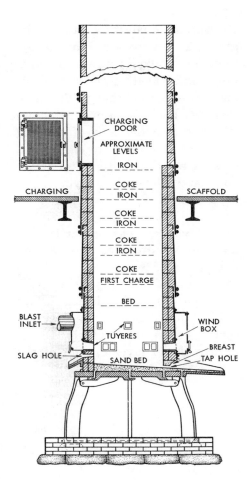

Fig. 18–3. (Cont.) Schematic melting furnace. E, cupola.

one of the alloying elements has a low melting point in comparison with the major element or when it oxidizes easily, it may partially vaporize before the alloying is completed. An example of this is when zinc oxide is produced during the alloying of zinc with copper to produce brass. This causes noxious metal vapors and a variation in the alloy composition from that intended for the final alloy. In order to overcome this problem of alloying, an intermediate alloy may be prepared. The intermediate alloy containing proportional amounts of the alloying elements is prepared then analyzed and added to the final charge in the appropriate quantity based on its analysis. The intermediate alloy, sometimes called a master alloy, is prepared by melting and superheating the lower melting point metal then dissolving the second metal into it. Additional methods of alloying involving the reaction of purified ores, or reactions with metal oxides and metal salts of the alloying metal are also used.

18.5 Fluxes for Refining and Degasing. Fluxes are chemicals used in melting to refine the alloy by (1) removing undesirable impurities, gases, and oxides, and (2) reducing surface oxidation by forming protective coatings. Cover fluxes, Fig. 18–4A form a liquid layer on top of the

melt and are used to prevent oxidation. Thus, the problem of zinc oxidation in brass melting, which was mentioned previously, may be appreciably reduced by a thin cover flux mixture of glass, soda ash, or borax.

Cover fluxes are used on aluminum, magnesium, and other highly oxidizable metals or alloys. Care must be exercised during pouring to see that the fluxes are not entrapped in the casting. The flux used in magnesium casting, for example, contains reactive chlorides and must not be allowed to remain in contact with the cast metal as it will attack the casting.

Other types of flux used are those designed for eliminating impurities,

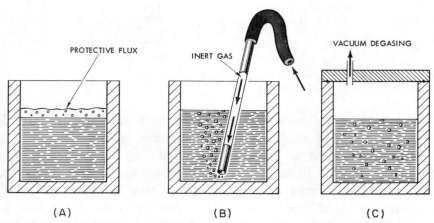

Fig. 18–4. Methods of refining and degasing molten metal. A, protective flux coating, B, bubbling insoluble gas through melt to float out impurities, C, vacuum degasing.

reduction of oxides, and for removal of dissolved gases. The fluxes used for deoxidizing are other metals or compounds which actively combine with the oxygen of the metal being deoxidized, thus freeing the metal. Iron is usually deoxidized with aluminum, silicon, or manganese. Copper is deoxidized by calcium or phosphorus additions, while nickel is deoxidized by the addition of calcium or magnesium. The deoxidizing agents must be used carefully since it is not possible to remove all the deoxidizer from the melt and it may have a deleterious effect on the properties of the cast metal.

Removal of certain impurities and dissolved gases in a melt can often be accomplished by bubbling insoluble gases such as nitrogen, helium, argon, or chlorine through the melt as illustrated in Fig. 18–4B. The small bubbles of gas tend to attach themselves to small particles of oxide and float them to the top. When a dissolved gas such as hydrogen in a melt comes in contact with a bubble of insoluble gas, such as nitrogen, the hydrogen will diffuse into the nitrogen bubble and be carried to the surface of the melt where it is dissipated. Another method of gas removal and of oxide reduction is the vacuum treatment shown in Fig. 18–4C. Reduced pressure over the melt results in evolution of the dissolved gases. Recently developed methods of vacuum melting and

casting show great promise in the reduction and elimination of problems related to surface oxidation and gas absorption. Perhaps prevention is still the best method.

18.6 Inoculants. An inoculant is an addition made to a melt, usually late in the melting operation, to alter the grain size or structure of the cast metal.

In grain refining of aluminum alloys, for example, additions of less than 0.2 percent titanium or 0.02 percent boron are sufficient to reduce grain size from 0.10 inches in diameter to 0.005 inches in diameter. Magnesium alloys are usually grain refined with very small additions of zirconium while magnesium-aluminum alloys may be refined with small additions of carbon to the melt.

A very important inoculation treatment for cast iron is the addition of magnesium or cesium to produce ductile iron. The addition of as little as 0.04 percent residual magnesium alters the graphite flakes into almost perfect spheroids with a tremendous improvement in mechanical properties of the metal.

A treatment similar to that used in producing ductile iron is the sodium treatment of aluminum-silicon alloys. In this case minute quantities of sodium (or phophorus) alter the needle-like Al_3Si precipitate to form a finely divided eutectic structure. This treatment substantially improves the ductility of aluminum-silicon alloys containing more than 8 to 10 percent silicon.

The theory of inoculants is not clearly understood but seems to be due to more subtle causes than simply changes in chemistry. Grain refiners and graphitizers probably promote nucleation by introducing foreign nuclei into the liquid metal while those that produce ductile iron may change the surface tension of solidifying particles thereby altering the nucleation and growth rates.

18.7 Solidification. One of the most important but least appreciated phases of metal casting is that of metal solidification. The techniques for melting, metal handling, and mold preparation, are fairly well understood and controlled. All too often, however, the metal is poured into the mold and solidification is left to proceed as it may. Poor control of the solidification phase in metal casting is responsible for coarse grains, internal porosity, shrinkage cavities, hot tearing or rupture of the metal, and general unsoundness of the casting.

Many factors which govern the mechanism of metal solidification are not completely understood but enough is known to permit treatment of the practical aspects of solidification.

Shrinkage of metal is a volume reduction accompanying the temperature drop from pouring temperature to room temperature. Cast metals undergo *liquid contraction* as they cool from the pouring temperature to the solidification temperature, *solidification contraction* as they freeze, and *solid contraction* as the solid casting cools to room temperature.

Fig. 18–5 is a graph showing the shrinkage encountered in solidifying and cooling steel. The total volume change exceeds 10 percent in the example shown. Of the first two contractions, liquid and solidification contraction, solidification is by far the more important. It requires the use of special techniques to hold the metal liquid and allow it to feed into the solidifying and shrinking casting to prevent internal pipe and shrinkage cavities. The final solid contraction results in castings which are smaller than the mold cavity and in some situations results in internal stresses which may cause warping or hot tearing.

When a molten pure metal or some eutectic alloys are poured into a cold mold, the metal freezes by forming a solid skin of equiaxed grains in what is termed the chill zone, Fig. 18–6. Following the initial thermal shock the solid skin grows progressively inward with a relatively uniform advancing front until the entire casting is solid. The growth is perpendicular to the mold wall and results in the formation of

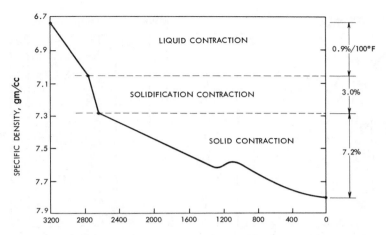

Fig. 18–5. Shrinkage in the three stages of solidifying and cooling low carbon steel.

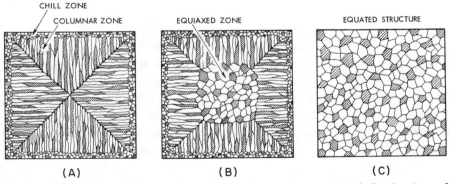

Fig. 18–6. Cast structures. A, grain structure of a nearly pure metal, B, structure of alloy which freezes over wide temperature range, note nucleation and new grains near center, the equiaxed zone, C, structure of metals which solidify over wide range. (By permission: *Foundry Engineering*, John Wiley & Son)

elongated or columnar crystals. Pure metals and those that freeze with a very narrow solidification temperature range may have only chill and columnar grains. As successive layers of solid are built up, the liquid level drops because of solidification contraction as illustrated in Fig. 18–7, with the development of pipe and centerline shrinkage.

The time required for solidification of a casting is affected by the casting volume and surface area and has been generalized into the form

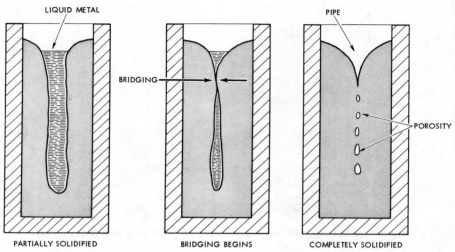

Fig. 18–7. Pipe cavity and centerline shrinkage in solidifying ingot.

known as *Chvorinov's Rule.* This rule which is shown below may be used to approximate the time for castings of different shapes to solidify when poured into the same mold material.

Solidification Time,

$$= K \left(\frac{\text{casting volume}}{\text{casting surface area}^2} \right)$$

Where:

K is a constant dependent upon the thermal conductivity of the metal and mold.

Thus a 2 inch steel cube will freeze in one-fourth the time required for a 4 inch steel cube when poured at the same temperature and into the same mold material. The rate of solid skin formation is affected by the shape of the casting and builds up on flat surfaces roughly as the square root of the elapsed solidification time. Fig. 18–8 shows drawings of steel castings which were allowed to partially solidify then inverted and the remaining liquid poured out, leaving a shell. It should be observed that solidification is more rapid at external angles because the heat is more easily dissipated at these points; conversely, the solidification is slower at internal angles. It should also be observed that solidification progresses from the thinner sections toward the heavier sections, since the heat is removed more easily from the thin sections.

Metal alloys solidify in a somewhat different manner than pure

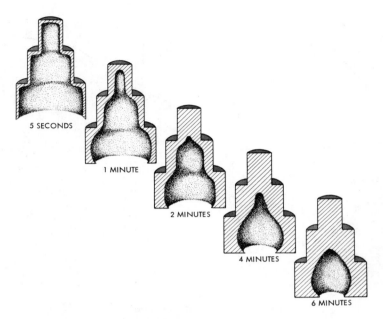

Fig. 18–8. Progressive solidification of steel casting. Liquid metal was poured out after time shown leaving shell.

metals. Essentially pure metals freeze by forming a solid skin which grows progressively inward. Fig. 18–9A. Only in extremely pure metals, however, is the interface between the liquid and solid skin a smooth wall. Even small amounts of impurities or alloying elements in the melt tend to be rejected by the solidifying metal at the solid-liquid interface. These elements lower the freezing point of the remaining liquid forming a somewhat mushy zone. During the freezing of an alloy, which is schematically illustrated in Fig. 18–9B, freezing progresses by growth of dendrites into the liquid melt. The result is a jagged solid-liquid interface with minute pine-tree-like dendrites protruding into the liquid, shown in Fig. 18–10, for a cast monel alloy.

As solidification progresses, the original dendritic branches thicken and interlock with other dendrites. As the amount of liquid remaining becomes smaller and as the interdendritic passages become smaller, the liquid remaining experiences difficulty in trying to feed the shrink-

575

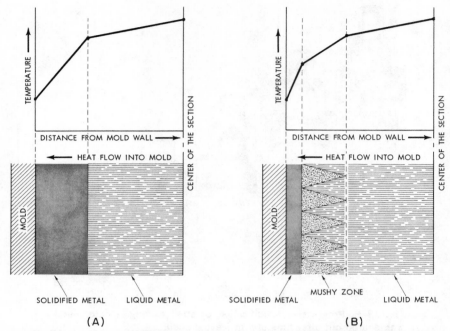

Fig. 18–9. Temperature distribution and shape of solid interface during solidification of casting. A, pure metals and some eutectic alloys, B, alloys with broad solidification range.

ing casting and inter-dendritic microporosity or microshrinkage results.

The width of the mushy zone of a solidifying casting depends upon the temperature range between the liquidus and solidus of the alloy. Alloys such as medium carbon steels, that freeze over a large temperature range tend to have a wide mushy zone. These metals nucleate new grains late in the solidification process causing the development of a region

termed the *equiaxed zone*, Fig. 18–6B. This casting shows all three types of cast structures, chill, columnar, and equiaxed.

In cast irons and most non-ferrous alloys, the temperature range of the mushy zone is sufficiently wide that nucleation occurs throughout the melt, rather than only at the mold interface. This results in rather fine, equiaxed grains forming throughout the casting as shown in Fig. 18–6C. The addition of grain refiners tends

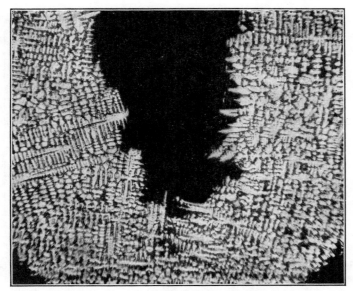

Fig. 18–10. Dendritic growth typical of solidification of alloys. Monel metal shown. Liquid metal drained away from dendrites by inadequate feeding, 25X. (Malcolm S. Burton, Cornell University)

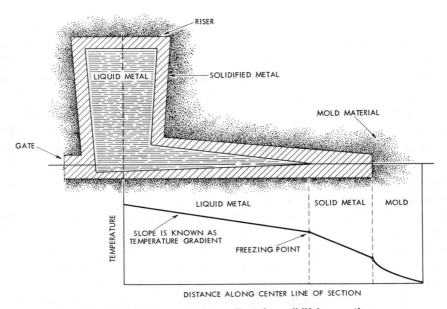

Fig. 18–11. Temperature gradient for solidifying casting.

577

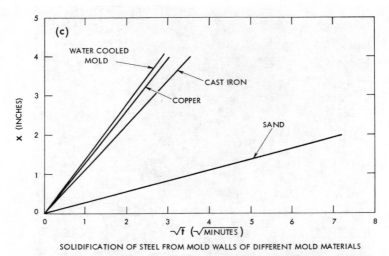

SOLIDIFICATION OF STEEL FROM MOLD WALLS OF DIFFERENT MOLD MATERIALS

Fig. 18–12. Solidification rates for molds having different thermal capacities. (By permission: *Foundry Engineering*, John Wiley & Son)

to reduce the growth of columnar grains and promotes a uniform, equiaxed structure.

Cooling rate and temperature gradient of a solidifying casting plays an important role in determining internal microshrinkage, segregation, and grain size. Fig. 18–11 is the temperature gradient for a solidifying metal casting. Castings cool primarily by giving up heat to the mold material and in the conduction of this heat away from the casting. A mold with high thermal capacity, such as a metal mold, Fig. 18–12, is able to quickly and continuously remove heat from the casting surface. A high thermal gradient is thereby

established between the mold wall and the casting itself. This results in a small distance between the liquidus and solidus regions of the freezing casting and, consequently, small dendrites. With small dendrites, the liquid metal has only to flow a short distance in order to properly feed the shrinkage of a solidifying metal casting and microporosity is minimized.

In contrast to the above, mold materials of low thermal capacity and low heat conductivity result in the formation of large dendrites. This increases the susceptibility for microshrinkage as the liquid experiences difficulty in feeding the shrinking casting through the maze of small interdendritic spaces. Although shrinkage microporosity may not be visible its effect upon strength and pressure tightness of a casting may be considerable.

18.8 Casting Problems. Metal casting, with its many processes, ramifications, and parameters, has been largely an art. Difficulties and problems encountered in metal casting have been frequently solved by trial-and-error methods. Even today, as foundry work is slowly moving from the art stage into a science, there is no substitute for a skilled and experienced foundryman. However, the problem, as with all rapidly evolving technologies, arises that there is insufficient time to give prospective foundry workers many years of experience. This has precipitated the keen need for passing on information from one generation to the next by educational training programs, and for exploration of the basic parameters of metal casting. One starting point for aiding the student to become more knowledgeable about foundry work might be in the investigation of casting problems. As long as the castings satisfy the requirements of the designer and can be economically and satisfactorily cast, there are few problems. However as soon as an anomaly occurs, there is need for investigation as to its causes and further preven-

tion. Most casting problems may be classified as one or more of the following:

1. Shrinkage,
2. Porosity,
3. Misruns and cold shuts,
4. Segregation and coring,
5. Excessive grain growth,
6. Inclusions, and
7. Poor design.

The first six of these problems may be generally alleviated by choosing the correct mold materials, pouring the metal at the appropriate temperature and rate, use of gates and risers capable of setting up favorable temperature gradients, use of chills, and padding to promote directional solidification and use of proper fluxes and deoxidants. The problems associated with casting design are also associated with many of the problems of powder metallurgy design, forging design, and design of machined parts and will be discussed in some detail in the next section. Fig. 18–13 shows common casting defects along with recommendations for their remedy.

18.9 Casting Design Considerations. Functional requirements and appearance are primary considerations when designing a new casting. But just as important is the selection of a casting process that can fulfill both of the above needs and do it economically. Many features of the design will be influenced by the casting

DEFECT	PROBABLE CAUSES	REMEDY
1. SHRINKAGE 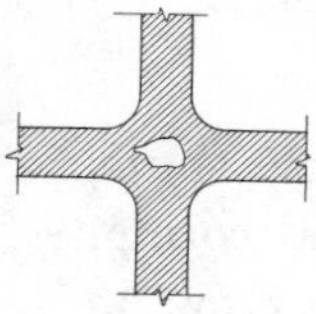	1. IMPROPER FEEDING 2. LACK OF PROGRESSIVE SOLIDIFICATION	1. USE ADDITIONAL RISERS OR A COMBINATION OF CHILLS AND RISERS 2. MODIFY GATING TO SUPPLY COOLER METAL TO AREAS WHERE SHRINKAGE IS 3. MODIFY DESIGN FOR MORE UNIFORM WALL THICKNESS
2. POROSITY 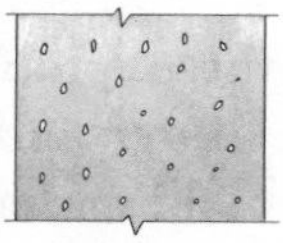	1. DISSOLVED GASES 2. EVOLVED GAS FROM DAMP SAND OR VOLATILE BINDERS 3. MECHANICALLY ENTRAPPED GASES 4. SHRINKAGE MICROPOROSITY	1. USE VACUUM OR FLUX DEGASING 2. USE SAND WITH PROPER PERMEABILITY, PROPER BAKING, AND LOWER MOISTURE 3. CHANGE GATING TO REDUCE MECHANICAL AGITATION 4. REDUCE POURING TEMPERATURE 5. USE CHILLS TO ACCELERATE SOLIDIFICA-TION
3. MISRUNS AND COLDSHUTS 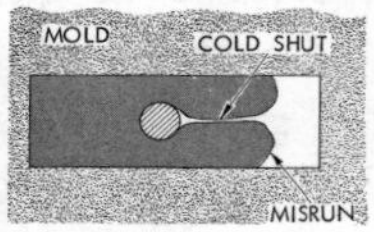	1. LOW FLUIDITY OF MOLTEN METAL	1. INCREASE POURING TEMPERATURE 2. ALTER GATING SYSTEM TO PROVIDE EASIER ENTRY TO MOLTEN METAL INTO MOLD CAVITY 3. CHANGE TO AN ALLOY HAVING GREATER FLUIDITY
4. SEGREGATION, CORING & DENDRITIC GROWTH 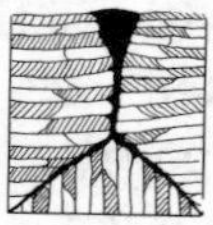	1. POOR AGITATION IN MELTING OR HOLDING FURNACES 2. EXCESSIVELY SLOW COOLING	1. PROVIDE PROPER AGITATION AND MIXING, i.e. INDUCTION HEATING 2. MODIFY GATING AND CHILLING TO IMPROVE SOLIDIFICATION CONDITIONS
5. EXCESSIVE GRAIN SIZE 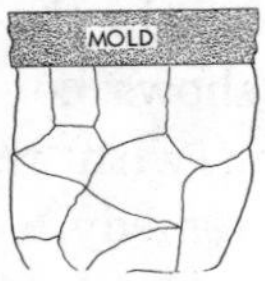	1. EXCESSIVE POURING TEMP. 2. PROLONGED SOLIDIFICATION TIME	1. REDUCE POURING TEMPERATURE 2. USE MOLD MATERIAL WITH HIGHER HEAT CONDUCTIVITY 3. USE INOCULANTS
6. CRACKING (HOT TEARS) 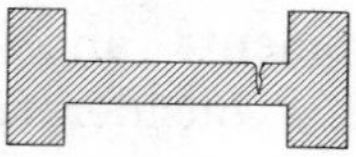	1. HOT SHORTNESS 2. EXCESSIVE RESTRICTION OF CASTING DURING CONTRACTION	1. MINIMIZE RESISTANCE TO SHRINKAGE BY USE OF CORES HAVING GREATER COLLAPSBILITY 2. STRENGTHEN SECTIONS SUBJECT TO CRACKING 3. CHANGE CHILLING, GATING, AND RISER-ING TO REDUCE HOT SPOTS
7. INCLUSIONS 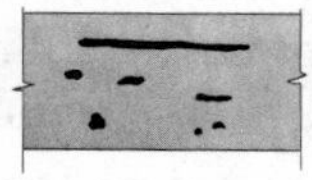	1. REFRACTORIES 2. SAND FROM MOLD OR CORE 3. OXIDE FILMS AND DROSS	1. CLEAN EQUIPMENT 2. HANDLE MOLDS AND CORES CAREFULLY 3. HOLD MELT WITHOUT AGITATION FOR A SHORT TIME PRIOR TO POURING 4. AVOID EXCESSIVE AGITATION OF THE MELT

Fig. 18–13. Common defects with causes and remedies.

process chosen. Functional design factors which should affect the choice of a casting process include: size and shape of the casting, minimum section thickness, dimensional tolerance, and surface finish. The economic factors which should also affect selection of a casting process are the number of pieces to be cast and relative machining and finishing costs.

The advantages of using castings for engineering parts are well known by designers. Parts can be produced with great complexity and in almost any size. A large variety of metals may also be selected by the designer giving him a choice of physical and mechanical properties for the cast metal part.

In an effort to make known to the designer those foundry practices which affect casting design, a number of foundry organizations have published casting design rules. A summary of these general casting design rules are given here.

Rule 1. *Consult a foundryman or patternmaker before releasing a final casting drawing.*

All too often a design may suit the engineering department but it is one which cannot be produced with the strength and functional properties intended by the designer. Consultation between the designer and foundryman will permit consideration of problems such as type of pattern required, shrinkage allowance, type of pattern to be used, allowances for machined surfaces, locating and holding points, dimensional limitations, and manufacturing costs.

Rule 2. *Construct a small model or pictorial drawing of the casting.*

A scale model such as that in Fig. 18–14 will help the foundryman decide how to mold the casting, detect potential casting weaknesses, determine the placement of gates and risers, and in general, help him determine how the metal will enter the mold, how solidification will proceed, and which parts will have to be fed to assure casting soundness.

Rule 3. *Design for uniform section thickness.*

Whenever possible section thickness should be uniform throughout the casting. Fig. 18–15. Uniform section thickness simplifies gating and feeding problems. It also minimizes the danger of solidification shrinkage and cracking due to unequal thickness. If additional strength is needed sections may be stiffened by reinforcing ribs. Large flat areas are sometimes difficult to feed properly and may become warped. Slightly curved or ribbed construction will minimize these difficulties.

Minimum section thickness of a casting is dependent upon the casting process used, the type of gating used, and upon the fluidity of the metal. As mentioned previously, the fluidity of a given alloy is dependent upon the pouring temperature. The

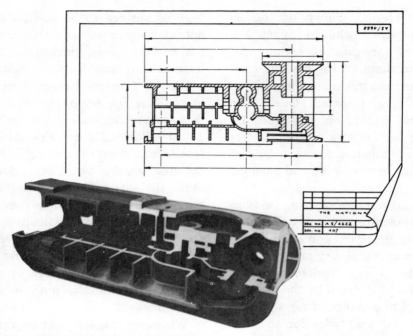

Fig. 18–14. Scale model of casting made by designer helps patternmaker plan gating, risering, and foresee potential weaknesses. (Steel Founder's Society of America)

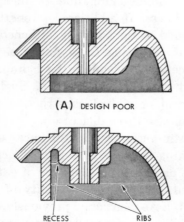

(A) DESIGN POOR

(B) DESIGN GOOD

RECESS   RIBS

Fig. 18–15. Design for uniform section thickness. A, is poor because of heavy section joined to thin one. B, recessing and ribs maintain strength and remove shrinkage danger in heavy section.

Table 18–1. Recommended minimum section thickness, inches.

METAL	CASTING PROCESS *				
	GREEN OR DRY SAND	SHELL MOLD	PERMANENT MOLD CASTINGS	INVESTMENT OR PLASTER CASTING	DIE CASTING
ALUMINUM ALLOYS	.125-.140	.062-.125	.100-.160	.025-.060	.030-.080
COPPER ALLOYS	.090-.140	.090-.125	.100-.200	.040-.060	.030-.080
GRAY IRON	.125-.200	.125-.150	.250-.400	-----	-----
MAGNESIUM	.125	.125-.150	.100-.160	-----	-----
STEEL	.250-.250	.180-.250	.250-.250	.030-.060	.050-.080
ZINC	.125-.250	-----	.100-.200	-----	.015-.050

*THE DESIGN OF THIN-WALL CASTINGS MAY INVOLVE THE CONSIDERATION OF A NUMBER OF FACTORS INCLUDING THE NECESSITY OF INDIVIDUALLY GATING ANY ISOLATED MASSES, CONSIDERATION OF THE SIZE OF CASTING, CAREFUL SELECTION AND CONTROL OF METAL COMPOSITION, AND SELECTION OF CHEMICALLY BONDED MOLDING MATERIALS WITH LOW MOISTURE CONTENT AND CONSEQUENTLY REDUCED CHILLING ACTION. FIGURES PRESENTED ARE FOR USUAL CONDITIONS ON SMALL CASTINGS OF 1"-6" LENGTH. MINIMUM SECTION THICKNESS MAY VARY CONSIDERABLY FROM THE VALUES PRESENTED WHEN SPECIAL TECHNIQUES ARE EMPLOYED.

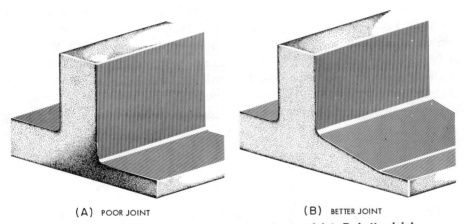

(A) POOR JOINT   (B) BETTER JOINT

Fig. 18–16. Joining different thicknesses. A, poor joint, B, better joint.

pouring temperature in turn has a significant influence upon physical properties of the cast metal. Thus, a compromise is required between these somewhat conflicting factors.

The minimum section thicknesses for a number of metals and processes is given in Table 18–1.

When sections of different thicknesses in the same casting cannot

583

be avoided, the section thickness should be increased gradually as illustrated in Fig. 18–16. When joints are to be made the increased mass can be estimated by the inscribed circle method shown in Fig. 18–17. This increased mass should be kept as small as possible to reduce heat concentration and shrinkage. In Fig. 18–18 are some ways of designing joints to minimize increased mass at joints.

Rule 4. *Avoid sharp corners and angles.*

Solidification of molten metal proceeds at right angles to the mold face or the plane of the cooling surface. When two or more sections meet as in Fig. 18–19A, mechanical weakness is induced at the junction and free cooling is interrupted creating a "hot-spot" and a potential shrinkage cavity. Fig. 18–19B shows the metal structures which occur

with smooth blended fillets and rounds. One caution is that the radii must not be excessively large or Rule 4 will be violated causing possible shrink defects.

With the casting process the designer has the ability to produce small or large radii as desired. He should thus choose radii which will best satisfy the design requirements. Some corner designs were checked by photoelastic models for stress concentration, then later cast in steel and stress concentration factors determined with SR-4 strain gages. Fig. 18–20. The agreement between the two methods was within about 5 percent. A stress concentration factor of 2.03 shown for the sharp corner means that the maximum value of stress at the corner will be over twice as great as that found in the adjacent sections. Inclusions or flaws in the material or

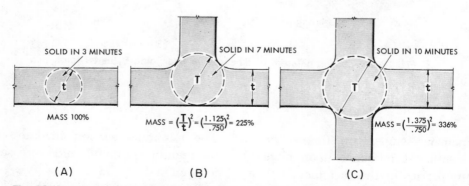

SOLID IN 3 MINUTES

SOLID IN 7 MINUTES

SOLID IN 10 MINUTES

MASS 100%

$$MASS = \left(\frac{T}{t}\right)^2 = \left(\frac{1.125}{.750}\right)^2 = 225\%$$

$$MASS = \left(\frac{1.375}{.750}\right)^2 = 336\%$$

(A)

(B)

(C)

Fig. 18–17. Inscribed circle method for estimating mass at joint. Solidification time is determined by mass within circle.

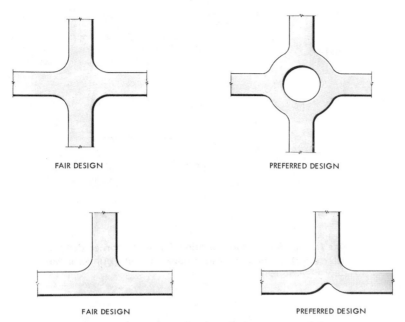

FAIR DESIGN PREFERRED DESIGN

FAIR DESIGN PREFERRED DESIGN

Fig. 18–18. Two methods for reducing mass at joints.

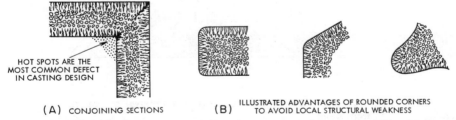

HOT SPOTS ARE THE
MOST COMMON DEFECT
IN CASTING DESIGN

(A) CONJOINING SECTIONS

(B) ILLUSTRATED ADVANTAGES OF ROUNDED CORNERS
TO AVOID LOCAL STRUCTURAL WEAKNESS

Fig. 18–19. Avoid sharp corners and angles. A, hot spot and weak corner due to stress concentration, B, rounded corners avoid local weakness. (Meehanite Metals Corp.)

surface defects also serve as stress raisers. It is at these areas of high stress that failures begin.

Rule 5. *Provide draft and shrinkage allowances.*

In casting design, sufficient draft or taper, must be provided on all vertical faces to permit the pattern to be withdrawn from the mold. Draft allowance is shown graphi-

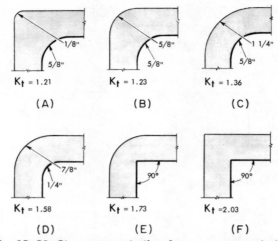

Fig. 18–20. Stress concentration for some corner designs determined by photoelastic models. (Steel Founder's Society of America)

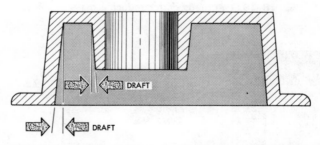

Fig. 18–21. Draft allowance to permit removal of pattern from mold.

cally in Fig. 18–21. In the case of permanent mold castings draft must be provided to permit the casting to be removed from the metal mold. Sand castings usually require a draft of at least $\frac{1}{16}$" per foot. Permanent mold castings may require 2 degrees on each side for outer surfaces and 5 degrees for inner surfaces. To compensate for linear shrinkage of the molten metal during solidification and cooling the pattern is made larger than the desired casting. For aluminum the pattern may be $\frac{5}{32}$ in./foot larger and for cast steel $\frac{1}{4}$ in./foot larger than the final casting size.

Rule 6. *Design to minimize residual casting stresses.*

All castings contain residual stresses. These may be of such magnitude as to cause distortion in machining, or weakened, or cracked castings. Whenever full strength properties are needed, the casting should be stress relief annealed. In Fig. 18–22A is a pattern for simple shape consisting of three parallel members of equal length, joined at their ends by rigid cross members. During cooling, the outer members cool faster than the center member causing distortion indicated in Fig. 18–22B.*

The student should be aware that compression stresses are developed in the first part of a casting to solidify and tension stresses are in the last part to solidify. For this reason, sections of varying sizes should be carefully blended. For example, in a spoked wheel Fig. 18–23A, the rim and hub solidify while the spoke is still partially liq-

*Metals which have a phase change (Fig. 18–5, 1200°F) during solid state cooling may have a reversal of residual stresses normally expected.

 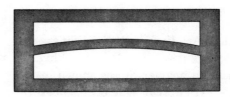

(A) ORIGINAL PATTERN   (B) DISTORTED CAST PART

Fig. 18–22. Distortion of simple shape due to rapid cooling of outer members. A, original pattern, B, distorted cast part.

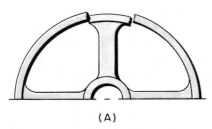 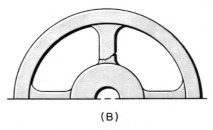

(A)   (B)

Fig. 18–23. Cast wheel incorrect designs. A, heavy spoke solidifies after hub and rim causing tension and breakage at rim, B, large hub causes stress and breakage at hub.

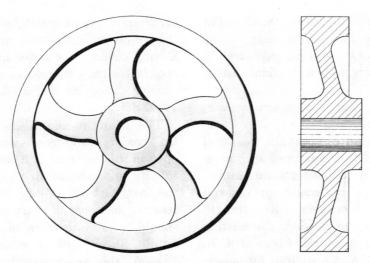

Fig. 18–24. Modifications to allow for shrinkage without causing stress. Note odd number curved inclined spokes and blended sections.

uid. As the spoke solidifies and shrinks it will result in cracking. By using a heavier rim as shown in Fig. 18–23B a tear will occur near the hub. Fig. 18–24 shows two ways to minimize these problems. In the first instance, an odd number of curved spokes are used. The curved spoke will tend to straighten slightly, thereby offsetting the dangers of cracking. The odd number of spokes will have less direct tensile stress along the arms than one having an even number; in addition, it will have greater resiliency to accommodate residual internal stresses. In the second instance, the cross section of the wheel shows the spokes

at a slight incline to the hub. This allows for blending the sections to avoid excessive variations in thickness and also allows the spokes to straighten slightly during cooling to reduce residual internal stresses.

18.10 Inspection and Testing of Castings. Regardless of the casting method used and the care exercised in handling and processing the metal in the foundry it is extremely difficult to produce castings without at least some minor imperfections. Fortunately, these minor imperfections do not impair the usefulness of the castings for the majority of applications, and therefore should not be the cause for rejection of the casting. Proper interpretation of inspection results requires good judgment in distinguishing between castings that will satisfactorily meet the operating requirements and those that will not.

Castings may, for convenience, be grouped into three classes of quality: (1) "perfect" castings such as those required for highly stressed structural parts for military and civilian aircraft or for high pressure or high temperature service. These castings require 100 percent x-ray examination; (2) moderately stressed castings subject to low pressure service or to normal fatigue or impact stresses where failure would not involve loss of life or serious work stoppage. These castings would be initially produced under x-ray inspection control but after standards had been established and production begun, would require only periodic x-ray inspection checking; (3) normal commercial castings for low-stress, non-structural applications. These castings do not require x-ray inspection, but are satisfactory if they meet visual inspection standards and are adequate for subsequent machining or processing operations.

Some general methods of casting inspection in common use are:

1. X-ray examination,
2. Pressure testing,
3. Visual inspection.

X-ray examination may be used to reveal blow holes, sand spots, and inclusions, internal shrinkage, hot tears, cracks, unfused chaplets, and internal chills.

Any discontinuities within the metal affect the intensity of the radiation reaching the x-ray film and produce variations in the density of the photographic image. Cavities are registered as darker areas on the film while heavy inclusions show up as lighter regions. Comparison of the x-ray film with radiographic standards such as those available from the American Society for Testing and Materials may be made as a means of determining the seriousness of certain internal flaws.

Pressure testing is required of all castings which must be leak-proof or pressure-tight. This test is usually made by sealing external parts in a special fixture after which air or liquid pressure is applied. Leaks are revealed by either immersing the entire assembly into a water tank or by applying a water soap solution to the exterior. In either case leaks will be revealed by bubbles appearing on the surface of the casting. Liquid pressure tests may involve examination of a hydraulic gage mounted in the pressure line to reveal leakage.

Visual examination, as the name implies, is simply inspection of the external surfaces of the casting by eye. The more obvious surface imperfections are revealed by this inspection. Dimensional inspection using measuring instruments, and inspection fixtures may be part of the visual inspection.

Many castings which do not initially pass inspection may be salvaged by welding, impregnation with chemicals to promote pressure tight-ness, peening to close leaks, and abrasive machining or blending to remove objectional surface irregularities. The best treatment, however, for reducing defects, is good casting design, proper metal selection, melting and handling, and well applied foundry practices.

References

Avnet-Shaw Corp. *The Shaw Process*. Plainview, N. Y.

Burton, Malcolm S. *Applied Metallurgy for Engineers*. New York: McGraw-Hill, 1956.

Flinn, Richard A. *Fundamentals of Metal Casting*. Reading, Mass.: Addison-Wesley Pubg., 1963.

Gray Iron Founders' Society. *Gray Iron Castings Handbook*. Cleveland, O., 1957.

Malleable Founders' Society. *Malleable Iron Castings*. Cleveland, O. 1960.

Meehanite Metals Corp. *Casting Design as Influenced by Foundry Practice, Bulletin #44*. White Plains, N. Y. 1965.

Steel Founders' Society of America. *Steel Casting Rules and Data. Section 1*. Cleveland, O. 1965.

Taylor, H. F., Flemmings, M. C., and Wulff, J. *Foundry Engineering*. New York: John Wiley & Son, 1959.

Welding Metallurgy

19.1 Introduction. Joining of metals by soldering, brazing, and welding has progressed empirically from ancient times. Early Romans used solders, referred to by Pliny as "argentarium" and "tertarium." Argentarium was described as containing equal parts of tin and lead; tertarium consisted of two parts lead to one of tin. Filler metals for soldering with these same compositions are still in use today. The skill of early craftsmen can be deduced from their work. For example: Etruscan ornaments were embellished by gluing tiny gold grains onto their surface using a mixture of copper salt and organic gum. Firing the assembly to a temperature in excess of 1650°F carbonized the gum and reduced the copper salt to copper. The copper then alloyed with the gold to braze the minute gold grains in place. Welding, like soldering and brazing, is also an ancient art. Prehistoric gold bracelets have been found which show that ends were forge welded together. It was not until the latter half of the 19th century, however, that discoveries were made which paved the way for modern welding practices. Moissan used the carbon arc for melting metals in 1881, and in 1887 Bernandos of Russia was applying this arc to welding. Experiments with consumable electrodes were shortly thereafter conducted by Slavianoff. The oxyacetylene blowpipe was introduced in 1895 by LeChatelier. Resistance welding probably had its conception by J. Joule in 1856 when he electrically heated two wires, then forged them together. However, it appears to have been Elihu Thompson, an American engineer, who first employed contact resistance welding in the year 1877.

Progress in welding developments were slow at first. Craftsmen guarded their secrets jealously. Welded joints were not always trust-

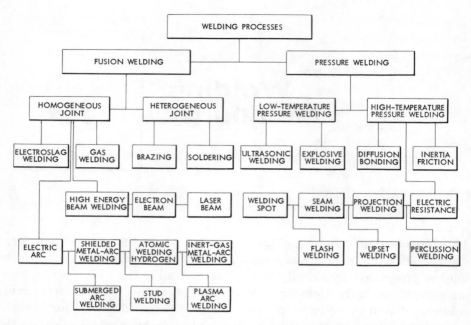

Fig. 19–1. Welding processes with major groups of fusion and pressure welding.

worthy and prior to World War I, welding was avoided rather than required or encouraged. During the 1930's, industrial welding gradually gained acceptance and during World War II was extensively used for joining in tanks, ships, and aircraft. Within the past 15 to 20 years there has been a virtual cascade of technical and scientific knowledge about welding and an increasing tendency for specialization. The use of welding in nuclear and space age applications has been a strong impetus in changing welding from an art to a science. Fig. 19–1 illustrates the major welding processes in use today.

Welding is here thought of in its broad generic sense to include pressure welding, fusion welding, brazing, and soldering. Welding is basically a metallurgical process and will therefore be considered in metallurgical terms. Weldability, or the capacity of metals to be joined satisfactorily, may be thought of as encompassing three parts: (1) the metallurgical compatibility for a specific process, (2) the ability to produce mechanical soundness, and (3) serviceability under special requirements.

Metallurgical compatibility means that the parent metal will join with the weld metal, and in the degree of dilution encountered in the process utilized, without the formation of unwanted or deleterious constituents or alloy phases. Mechanical soundness indicates that the joint will conform to engineering and industrial standards in freedom from discontinuities, gas porosity, shrinkage, incomplete penetration, slag, and cracks as revealed by radiographic, ultrasonic, or other inspection methods. Serviceability under special requirements indicate that, in addition to meeting the first two requirements, the weldment may be required to perform under such conditions as low temperature impact, high temperature loading, and corrosive environments. Weldability of metals varies considerably with the welding process utilized; factors

Fig. 19–2. Sulfur embrittlement in a root bend weld joint in nickel sheet. Left side cleaned with solvent and clean cloth, right side with solvent and dirty cloth. Note cracking. (Huntington Alloy Products Div., The International Nickel Co.)

593

which play a role in determining weldability by both pressure welding and fusion welding will be covered.

Pressure Welding

19.2 Metal Contact. Pressure welding is a metallurgical bonding operation in which the weld metals are brought into intimate contact under pressure. In some pressure welding processes as in resistance welding, the metals being joined may be heated to a high temperature, while in other processes, such as with cold pressure welding, the welding may occur at room temperature. In either the low temperature or the high temperature pressure welding processes good metal contact is essential for sound welds. This means that the surfaces to be joined must be free from dirt, oil, grease, paint, coatings, and other foreign material. These materials may contain sulfur, lead, phosphorus, or other elements which may lead to weld embrittlement. The methods used for surface cleaning have been discussed in Chapter 8 and may include vapor degreasing, ultrasonic cleaning, caustic bath degreasing, and other methods. Solvent solutions should be removed from the weld area by thorough water washing to prevent weld contamination. Fig. 19–2 shows the effect of sulfur embrittlement from cleaning nickel with solvent and a dirty cloth.

Oxide surface film can be a barrier to both high temperature and low temperature pressure welding. It must be removed because of its higher melting point than that of the base metal. In low temperature pressure welding the oxide film prevents intimate contact on an atomic scale of the surfaces to be joined. The oxide film should be removed by grinding, abrasive grit blasting, machining, or chemical etching.

19.3 Bonding Pressure. In cold pressure welding localized coalescence of the metals to be welded is achieved through the application of pressure and without external heating. The faying surfaces must be cleaned immediately by wire brushing prior to welding. The pieces are then placed in contact and sufficient pressure is applied to produce about 30 to 80 percent cold working. Pressure required for aluminum is 25,000 to 35,000 psi, and may be applied either by impact or by a slow squeezing action, both methods being equally effective. Pressure is applied over a narrow strip so the metal can flow away from the weld zone and rupture films of gas, moisture, and oxides, and obtain close enough atomic proximity to obtain a weld union, shown in Fig. 19–3. Metals which have been successfully cold pressure welded include aluminum, copper, lead, nickel, zinc, and monel metal. Cold pressure welds in duc-

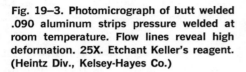

Fig. 19–3. Photomicrograph of butt welded .090 aluminum strips pressure welded at room temperature. Flow lines reveal high deformation. 25X. Etchant Keller's reagent. (Heintz Div., Kelsey-Hayes Co.)

tile metals yield welds whose strength is 90 to 100 percent of that of the parent metal. The highly deformed metal at this weld joint is subject to recrystallization and the formation of new, stress-free grains during thermal treatments such as stress-relieving operations.

One of the greatest advantages of cold welding lies in the simplicity of the tools required, the speed with which it can be effected, and the lack of a heat affected zone. Recent experiences with aerospace vehicles show that under the high vacuum conditions of outer space, undesirable cold welding may occur between adjacent parts. Efforts are being made to better understand and control the cold pressure welding process.

In high temperature pressure welding processes, notably resist-

ance, inertia, and friction welding, heat is obtained in the weld area from the resistance of the work to the passage of an electrical current or from the heat due to rubbing friction, Fig. 19–4 and 19–5. Coalescence in the weld area is achieved by heating and application of pressure. In most of these processes welds are made without the workpiece being melted, or at least with very little melting. The pressure applied makes possible coalescence at lower temperatures; and the forging action results in a finer grain structure in the weld. The pressure required for electric resistance welding usually varies from 4000 psi to 8000 psi although it may require forces as high as 25,000 psi for flash welding. Too little pressure gives high resistance between the electrode and the workpiece with result-

595

ant surface burning and electrode pitting. Too high pressure, on the other hand, may cause molten metal to be squirted from the faying surfaces, or indentation of the workpiece by the electrodes. The best condition exists when moderate pressure is applied prior to and during the initial welding stages in order to insure proper contact resistance; and then the welding pressure increased just as the proper welding temperature is reached. Friction welding requires axial pressure of 1500 psi for welding a 1 inch diameter carbon steel bar and speed of 1500 rpm, while the same size stainless steel bar requires 12,000 psi at 3000 rpm.

19.4 Bonding Temperature. In pressure welding, the bonding temperature may be above the recrystallization temperature as in electrical resistance welding, or below the recrystallization temperature as in

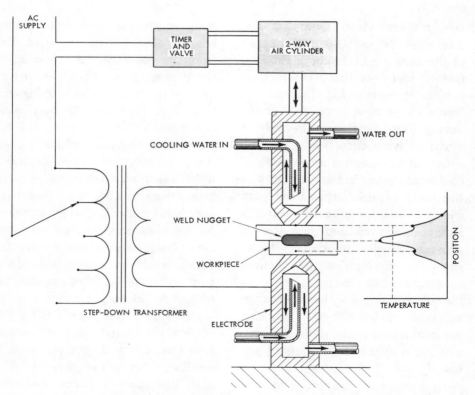

Fig. 19–4. Resistance welding. A, schematic of equipment. (Sciaky Bros.)

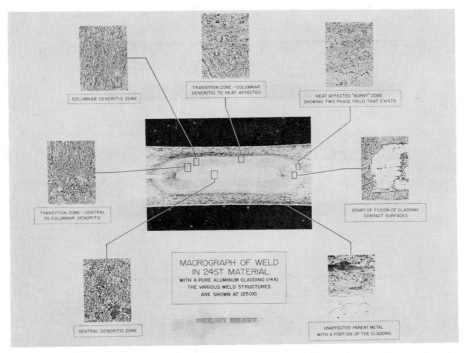

Fig. 19–4. (Cont.) Resistance welding. B, weld nugget showing microstructure at various positions. (Sciaky Bros.)

cold butt or lap welding. Diffusion bonding operations such as friction welding and explosive welding usually employ bonding temperatures which are ½ to ⅔ of the melting temperatures of the metals being joined. In ultrasonic welding the temperatures may be 35 to 50 percent of the melting temperature. The elevated temperature accelerates diffusion movements of the atoms at the joint interface and also softens the metal to permit more intimate contact at the joint interface. In Fig. 19–6A is the microstructure of dissimilar metals joined by ultrasonic welding. A schematic drawing of the device used for inducing high-frequency vibratory energy into the weld area is shown in Fig. 19–6B. When the bonding temperature is sufficiently high recrystallization along the weld interface may occur, yielding a fine grain size.

19.5 Metal Bond Strength. In pressure welding there is no customary weld fusion zone since the liquid phase is not usually present; however, the material in the weld may

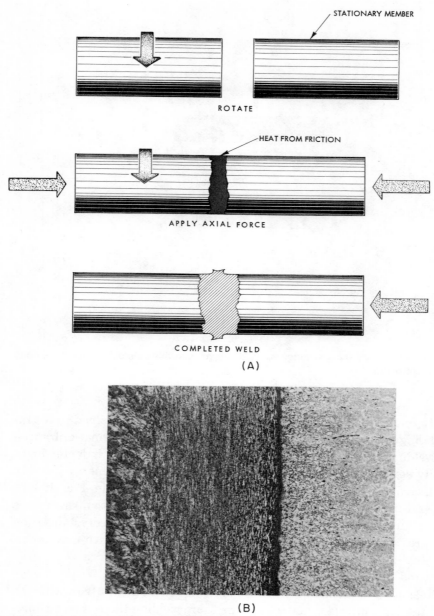

Fig. 19–5. Inertia welding. A, schematic of principle involved. B, photomicrograph of aluminum bronze welded to carbon steel. 75X. Note area of working between the metals. Interface temperatures were below the melting points of either face, avoiding shrinkage, gas porosity, and voids.

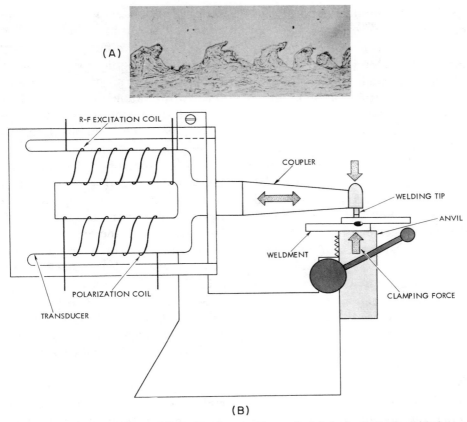

Fig. 19–6. Ultrasonic welding. A, microstructure of nickel sheet (top) welded to molybdenum. 200X. Interface ripples show plastic flow has occurred locally. (A. L. Phillips, American Welding Society) B, schematic of setup. Transducer moves welding tip at 20 to 40 kilocycles. Moving tip disturbs oxide film placing metals in intimate contact and causing a weld.

undergo some changes. Microstructural characteristics of pressure welds may include one or more of the following:

1. *Interfacial effects* such as surface film disruption and dispersion,

2. *Working effects* such as plastic flow, grain distortion, and edge extrusion,

3. *Heat effects* such as recrystallization, precipitation, phase transforming and diffusion.

599

Solid-state bonding is the bonding of metals by application of pressure alone. This involves the formation of a metallurgical joint between similar or dissimilar metals by causing adjoining atoms at joint interfaces to combine by interatomic attraction in the solid state. Factors which govern the compatibility and strength of dissimilar joint metals include the atomic size factor, lattice structure type, and surface cleanliness. Solid state bonding is difficult to achieve since adjoining surfaces must be atomically clean and brought within atomic distances before interatomic forces can establish a bond. This process differs from diffusion bonding in that no atomic diffusion is required.

Diffusion bonding involves the application of heat and pressure in the formation of a metallurgical joint between similar or dissimilar metals by the interdiffusion of interfacial atoms across the joint interface. The term is usually applied to solid-state diffusion but may also involve a liquid phase. Pressure is applied to bring the joining surfaces within atomic distances. Heat is applied to increase atom movement across the joint interface. Although many dissimilar metals are now being joined by diffusion bonding techniques that were thought to be unweldable there are certain limitations. For example: In addition to atomic size factor and lattice structure type mentioned previously, there is also the solid state solubility factor.

Some metals have extremely low solid state solubility, thus limiting atom movement, especially in the short times available for solid state diffusion.

Since diffusion bonds are generally void free a major problem is to distinguish between high strength

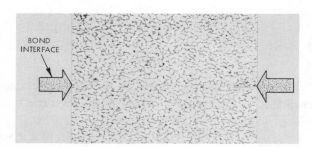

Fig. 19–7. Photomicrograph of diffusion bonded joint in titanium. Joint is barely discernable under microscope. 250X. Kroll's etch. (North American Rockwell Corp.)

and low strength bonds. Diffusion bonding can lead to joints as strong as the parent metal; it can also lead to bonds with brittle intermetallic compounds or difficult to detect bond separations at the joint interface. X-ray examination is not very effective in showing bond quality since the molten fusion zone is absent. The joint may not even be discernible by microscopic examination in some instances (Fig. 19–7). Ultrasonic and eddy current techniques are also quite limited in determining bonding quality except where gross non-bonding is present. High resolution die penetrants have shown some promise in detecting surface cracks but systematic destructive test sampling seems to be the best measure of weld quality.

19.6 Advantages and Limitations of Pressure Welding. The main advantages of pressure welding are primarily due to the lower temperatures employed and the simplicity of the equipment required. These advantages include:

1. High efficiency in most processes,

2. Joining of dissimilar metals,

3. Very limited oxidation or gas absorption,

4. Minimized shrinkage and cracking problems,

5. Recrystallization across the boundary (if above the recrystallization temperature), and

6. Complete metallurgical joint continuity.

Other advantages may be added to this list depending upon the specific pressure welding process employed.

Limitations of pressure welding may include:

1. Extreme surface cleanliness required in some processes,

2. Size or thickness limitations on some processes,

3. Fit up tolerances are usually critical, and

4. Careful control of pressure, temperature, and time.

In spite of the limitations listed, diffusion bonding is finding important applications in aerospace, nuclear, and commercial production.

Fusion Welding

19.7 Basic Principles of Fusion Welding. In fusion welding processes, such as those in Fig. 19–1, the base metals are always melted, and in many cases, filler metal is added. Heat for melting the base metal and the filler rod may be supplied by means of an oxyacetylene gas flame, Fig. 19–8, an electric arc, Fig. 19–9, or from high energy beams such as electron beam, Fig. 19–10, laser beam, Fig. 19–11 or plasma arc, Fig. 19–12.

It has been said that fusion welding involves all of the principles of metallurgy. In fusion welding, for

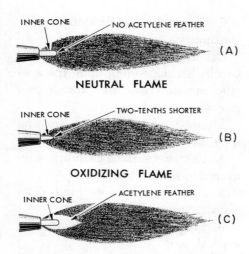

Fig. 19–8. Oxyacetylene flames. A, oxidizing. B, neutral. C, reducing.

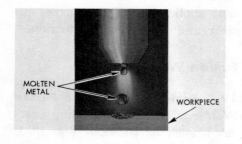

Fig. 19–9. High speed photograph shows metal deposited from argon-shielded arc to weld puddle. (Air Reduction Co.)

Fig. 19–10. Electron beam welding a heat treated gear cluster. (Sciaky Bros.)

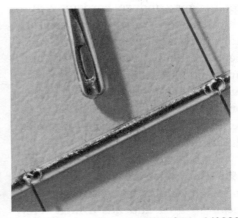

Fig. 19–11. A, Laser beam vaporizing a 0.020″ dia hole in tungsten at 10,700°F in 1/1000 sec. B, Laser beam welds of 0.0003″ tungsten and 0.020″ nickel wires in microminiature electronic manufacturing. (Hughes Aircraft Co.)

Fig. 19–12. Low current plasma arc used for edge welding on stainless steel. Note needle-like arc. (Linde Div., Union Carbide Corp.)

example, the metal is melted, refined by fluxes, alloyed, and resolidified. All of the defects common to castings may be found in welds including blowholes, large columnar dendrites, segregation of constituents, hot tears, and so on. The base metal near the weld zone is subjected to high temperatures. These high temperatures with subsequent rapid

603

Metallurgy Theory and Practice

cooling may induce hardening and embrittlement, or with slow cooling, annealing and weakening, all dependent upon the alloy content and upon previous cold working. Thus, fusion welding is a rather complex process and must be interpreted in terms of the metallurgical principles discussed in earlier chapters.

19.8 Heat Distribution. The heat in fusion welding is applied in a very localized area and has a high tem-

perature gradient as illustrated in Fig. 19–13. In the case of a metallic arc some electrode and some base metal are melted, producing liquid metal known as *weld metal*. The temperature distribution is important since it influences the complex metallurgical changes which take place in the weld region. If both the thermal cycles involved and the heat treating response of the alloy being welded are known, it is theoretically possible to predict the resulting mi-

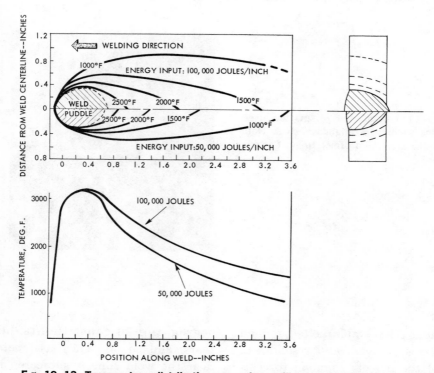

Fig. 19–13. Temperature distribution around metallic arc weld. A, maximum temperature with heating and cooling rates along weld. B, weld area temperature isotherms. Skewing is due to motion of arc in direction indicated.

604

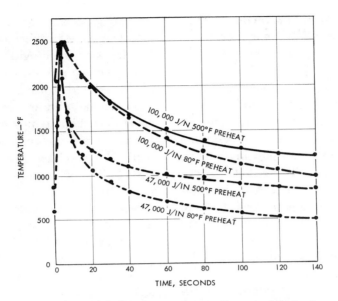

Fig. 19–14. Initial plate temperature and energy input rate on weld metal temperature and cooling rate.

crostructure and mechanical properties of the weld. Unfortunately, this phase of welding is rather complex and has not been developed to the point where weld properties are easily calculated. There are, however, a number of variables which have been investigated and which have been found to influence the temperature distribution in arc welding. These variables include the energy input, measured in joules* per lineal inch of weld, the preheat temperature, the weld geometry, and the thermal characteristics of the material. In Fig. 19–14 is shown the effect of preheating and changing the energy input rate. It should be noted that approximately the same weld metal temperature is achieved with the low input as with the high energy input. However, increasing the energy input does cause an increase in the time of exposure to temperature near the peak temperature, and it does decrease the cooling rate. Many of the problems of producing metallurgically suitable welds arises from the heat effects of welding on the parent metal and upon the control of the cooling rate following the welding operation.

*The energy input in joules per inch for arc welding is calculated as follows:

$$\text{joules/inch} = \frac{60 \times \text{arc voltage} \times \text{arc current}}{\text{arc travel speed, inches/minute}}$$

19.9 Weld Metal Protection. During fusion welding the molten metal in the weld "puddle," Fig. 19–15, is very susceptible to oxidation and must be protected from the atmosphere. Customary methods of protection include the use of fluxes, inert gases, or vacuum. Welding fluxes generally contain ingredients such as SiO_2, TiO_2, FeO, MgO, Al_2O_3, and other materials. In addition to providing a gaseous shield to prevent atmospheric contamination, welding fluxes may act as scavengers to reduce oxides, add alloying elements to the weld, control surface tension in the weld puddle and influence the shape of the weld bead during solidification, form a slag to carry off impurities, protect the hot metal, and slow the cooling rate.

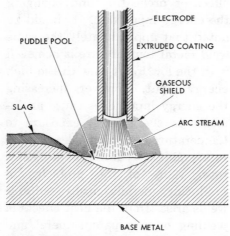

Fig. 19–15. Weld puddle and flux covering of electric arc.

ELECTRODE

PUDDLE POOL

EXTRUDED COATING

GASEOUS SHIELD

SLAG

ARC STREAM

BASE METAL

Inert gases such as argon, helium, nitrogen, or sometimes carbon dioxide are used for providing a protective envelope around the weld area in the shielded tungsten arc and shielded metallic arc welding processes. The high temperature of the arc in these processes vaporizes any oxide film which may be present and the blanket of inert gas prevents it from reforming.

Vacuum conditions are frequently used for electron beam welding. A "hard vacuum" of 0.1 micron of mercury allows efficient welding of even highly reactive vacuum-melted material as titanium, zirconium, and hafnium with gaseous impurities totaling less than 1 part per million. This compares very favorably to the 20 parts per million as the best gaseous impurity level attainable with inert gas-shielded methods. Although vacuum conditions are expensive to attain in terms of "pumpdown" time and equipment cost, they do permit extremely high quality welds, free from oxides, nitrides, and carbides.

19.10 Weld Metal Solidification. Fusion welding has been likened to a casting process in which the molten metal is contained in a solid metal mold as shown in Fig. 19–16A. The liquid metal is nucleated and begins to solidify at the area of contact with the cooler parent metal. During cooling, heat flows outward

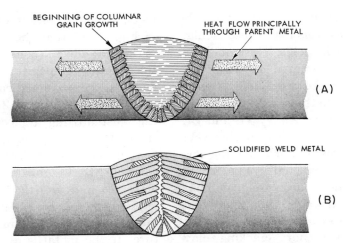

BEGINNING OF COLUMNAR
GRAIN GROWTH

HEAT FLOW PRINCIPALLY
THROUGH PARENT METAL

(A)

SOLIDIFIED WELD METAL

(B)

Fig. 19–16. Solidification in single pass arc weld. A, nucleation and columnar grain growth. B, solidified weld metal.

from the weld zone, causing columnar dendritic grains to grow inwardly toward the center of the weld puddle as shown in Fig. 19–16B. Unless adequately protected molten weld metals tend to dissolve atmospheric gases such as hydrogen, oxygen, or nitrogen. During solidification, these gases will be evolved, forming blow holes within the weld. This situation is similar to that previously discussed in connection with blowhole formation in castings but differs in two respects. First, the heating and cooling rates are much more rapid in welding than in casting, thus minimizing the time for gases to be dissolved. Second, the welding atmosphere and flux cover

can be controlled, also reducing the amount of dissolved gases.

Other problems which may be encountered during weld metal solidification in addition to that of gas blow holes previously cited are coring segregation, and hot cracking. Coring is the microscopical separation or segregation of elements in an alloy; it is not usually troublesome in carbon steels, but is particularly evident in non-ferrous and stainless steel welds. Coring in welds may be minimized by rapid cooling. Post heat treatments will usually eliminate coring developed during welding.

Segregation is a major cause of weakness in cast metal ingots where columnar dendrites meet at nearly right angles; this may cause similar weaknesses in welds. This condition is greatly alleviated in longitudinal welds where the solidification is along the length of the weld. Grain refiners such as Al, V, Ti, and Zr, are sometimes added to the weld metal to provide nuclei for the formation of many small grains. These grain refiners decrease columnar grain size in welds, thus improving ductility and minimizing hot-short cracking. Hot-short cracking may also be caused by excessive shrinkage during the last stages of solidification. Cracking can often be minimized by preheating the weld area prior to actual welding and by reducing the restraining or clamping force during weld metal solidification.

19.11 Heat Affected Zone. Metal adjacent to the weld area is heated and cooled during welding and thus undergoes a heat treatment. This heat treatment may cause the metal to become hardened and embrittled and crack when cooled, or it may anneal and soften the metal. In determining how the metal will be affected in this *heat affected zone* (Fig. 19–17) during welding no new metallurgical principles are involved. The metal goes through a heating and cooling treatment, the effect of which can be predicted from the fundamental principles of heat treat-

ment. For example, it can be seen in Fig. 19–17 that the temperature in the heat affected zone ranges from the melting temperature (Point 1) down to the lower critical temperature, (Point 3). Point 2 represents the austenitizing temperature for steel. Metal adjacent the weld is heated to near its melting point and consequently is subject to grain coalescence and growth. This coarse grained metal has less desirable properties than a fine-grained structure. If the metal is quench hardenable, the coarse grained regions have higher hardenability and, consequently, a greater tendency to form a hard, brittle martensitic structure which may crack after welding. During the welding of hardenable steels, every effort must be made to avoid the formation of martensite and to ensure the formation of softer transformation products such as ferrite and pearlite.

Fig. 19–18 is a typical isothermal transformation diagram for a quench hardenable steel. When the steel is heated above the A_e temperature, austenitized, and then rapidly cooled, martensite will form. Slower cooling tends to form the softer structures of pearlite and ferrite. The hardness of the martensite is dependent upon the carbon content of the alloy; low carbon martensite being softer than high carbon martensite. Alloys with high hardenability tend to form martensite even upon

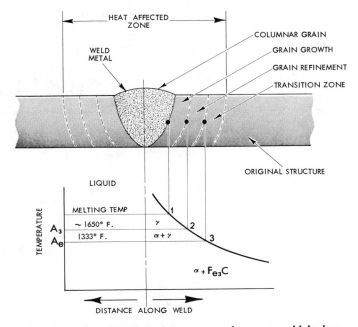

Fig. 19–17. Heat affected zone around an arc weld in low carbon steel.

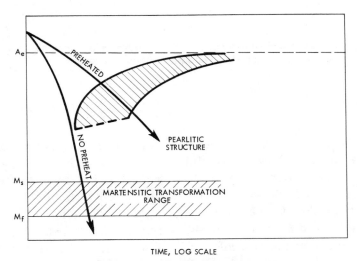

Fig. 19–18. Isothermal diagram showing effect of preheating on martensitic hardening in weld region.

609

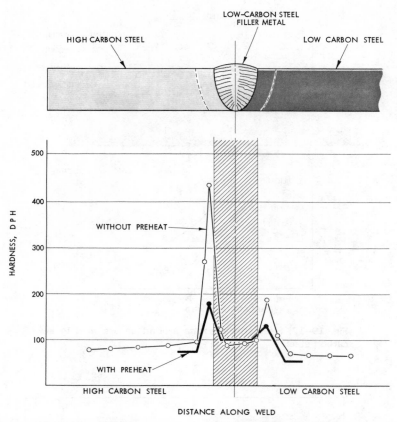

Fig. 19–19. Hardness cross-section through weld joining high carbon to low carbon steel, with and without preheating.

slow cooling and require special considerations to achieve successful welds. By examination of Fig. 19–18 it can be seen that martensite forms at a relatively low temperature between the M_s and M_f temperatures. When the metal in the heat affected zone is cooled slowly enough to avoid the formation of martensite cracking is avoided. One method of inducing slow cooling is by preheating the parent metal to a temperature of about 200°F. to 400°F just prior to welding. The effect of preheating to reduce weld hardness is shown in Fig. 19–19. Other factors which can be controlled during the welding process are the rate of heat input,

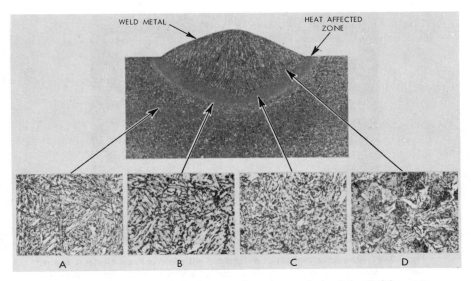

Fig. 19–20. Weld nugget and heat affected zone in heat treated low carbon manganese steel. Macrograph, 8X. Photomicrographs, 500X. A, parent metal (tempered martensite). B, tempered martensite. C, fine-grained ferrite, carbide, and fine pearlite. D, adjacent to weld showing ferrite and pearlite-Widmanstaten Structure. (H. F. Ebling, A. O. Smith Corp.)

plate thickness, thermal conductivity of the plate and geometry of the weld joint. Cooling rate decreases with high heat input, with thinner plate, with lower plate conductivity, and with simple butt welds rather than fillet welds. Fillet welds surround the weld region with more cool parent metal to conduct heat away from the weld region and thus tend to cool the weld more rapidly.

It should be apparent from the previous discussion that the heat affected zone of quench hardening metals contains a variety of microstructures, ranging from very hard martensite to coarse pearlite. These structures are clearly shown in Fig. 19–20. This heat affected zone usually represents the weakest area in a weld, since the desirable properties imparted by hot working or cold working have been lost. Much of the variation in microstructures which may exist in welds can be reduced by suitable heat treatment, such as normalizing, following welding. It is obvious, however, that heat treating cannot restore all of the benefits of previous hot or cold working, consequently steps should be taken to minimize or control the width of the

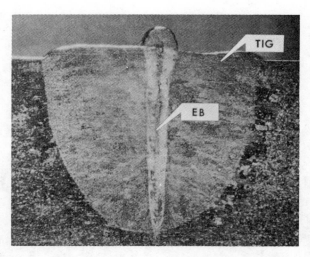

Fig. 19–21. Photomicrograph of a Tig weld with low voltage electron beam weld in center of Tig nugget to show relative width of heat affected zone in each process. (Airesearch Mfg. Co.)

heat affected zone. One step may involve the use of increased energy input. For example, a high energy electron beam weld is pictured in Fig. 19-21 compared with a convenional tungsten-arc weld. The electron beam weld has a much narrower heat affected zone because of its extremely high energy input rate. Other methods which may be employed to control the width of the heat affected zone are to increase the welding speed, use lower preheat temperatures, or use multiple pass welds.

19.12 Parent Weld Metal. Weld metal composition can generally be adjusted for welding nearly any an-alysis of metal. The parent weld metal, however, is usually chosen for its intended service and may not be ideally suited for welding. Thus, the limiting feature in most welds is not the weld filler metal but rather the parent metal.

The parent metal may be either a pure metal or solid solution alloy, with or without allotropic transformation; a eutectic alloy; a quench hardenable alloy; an age hardenable alloy; or it may consist of two dissimilar metals. Each of these combinations presents special problems which must be considered. In addition to the problems of metallurgical compatibility and the possible development of new structures and

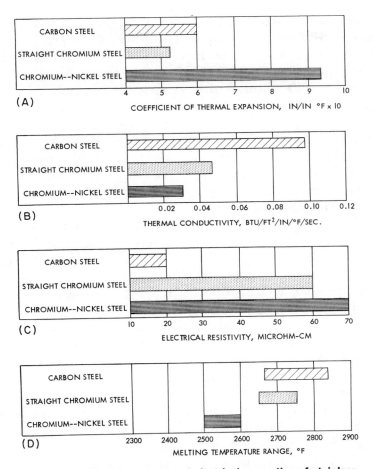

(A)

COEFFICIENT OF THERMAL EXPANSION, IN/IN °F x 10

(B)

THERMAL CONDUCTIVITY, BTU/FT^2/IN/°F/SEC.

(C)

ELECTRICAL RESISTIVITY, MICROHM-CM

(D)

MELTING TEMPERATURE RANGE, °F

Fig. 19–22. Relative thermal and electrical properties of stainless and carbon steels helps explain different behavior during welding.

phases during welding, there are often differences in coefficients of thermal expansion, thermal conductivity, electrical resistance, and melting points which must be dealt with in order to produce successful welds. For example, in Fig. 19–22A through D are shown the relative thermal and electrical properties of stainless and carbon steels. This comparison helps to explain why these materials behave differently during welding. For a given energy input, stainless steel heats up faster and melts sooner than carbon steels. Once a stainless steel weldment is heated, it remains hot longer than a carbon steel weldment. Perhaps one of the most important differences is that the expansion rate of austenitic stainless steel is approximately 1.5 times that of carbon steel. Thus, the problems which are encountered in welding stainless steel to low carbon steel should be apparent.

When welding dissimilar metals, the weld is composed not only of a filler metal, but also contains substantial quantities of both the parent metals. During the joining process the filler metal composition is changed or diluted by the two joined metals. Thus, the goal in welding of dissimilar metals is to achieve a high degree of metallurgical compatibility between the filler metal and the parent metals. For a number of years the general rule was that the filler metal should match the composition of the more highly alloyed of the two metals being joined. Further research into this problem has shown that the ideal filler metal for dissimilar metal joining should have the following properties:

1. High tolerance for dilution by elements in the parent metals,

2. Thermal expansion characteristics compatible with those of the metals to be joined,

3. High strength and ductility at both high and low temperatures,

4. Resistance to oxidation and corrosion,

5. Ability to withstand extreme service temperatures over long periods of time, and

6. Melting point near that of the parent metals.

Modern developments have provided the materials and techniques to weld dissimilar metals formerly thought impossible to join by fusion.

19.13 Residual Stresses in Welded Joints. Residual stresses in welded joints and resulting distortion arise primarily from two sources: (1) thermal stresses and (2) reaction stresses.

Thermal welding stresses are the result of resistance to thermal expansion and contraction offered by the pieces being welded. These thermal stresses may exist even if a piece is not joined to another structure or restrained in any manner. Fig. 19–23A shows a welded metal plate. As the weld metal and the metal adjacent to the joint are heated, expansion occurs. Freedom to expand in the direction parallel to the weld is largely prevented, however, by the very much larger rigid plate. As a result, the material in the weld zone is not free to expand lengthwise and consequently becomes thickened or *upset*. Fig. 19–23B illustrates what would happen if the weld metal in the weld zone (the upset metal plus filler metal) could be separated upon

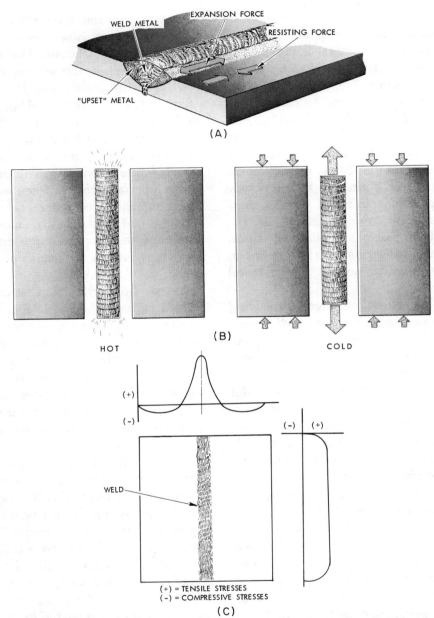

Fig. 19–23. Internal forces in fusion welding. A, expansion and upsetting of weld metal due to resisting forces of cooler plates. B, shrinkage of weld metal if separated from parent metal showing internal tensile stresses. C, stresses parallel to and perpendicular to weld in steel plate.

cooling. The cold weld would be shorter than the plate. However, since the weld metal is attached to the plates, it is stretched in tension upon cooling. The plates are placed under a corresponding compressive stress as shown in Fig. 19–23C. Also Fig. 19–23C shows the stress pattern perpendicular to the weld. This stress results from restriction to transverse shrinkage by previously deposited weld metal, causing some transverse residual stress. If the welded plates are relatively thick, additional stress will be present in the direction normal to the metal surface.

The effect of thermal welding stresses have not been well understood, and there is still no complete agreement regarding them. There does not seem to be evidence that thermal stresses have any harmful effect on the strength of weldments, except in the case where notches are present, or where the metal has become embrittled by the heat effects of welding. Normally, when a static load is applied to a part having high residual stress plastic flow occurs as soon as the sum of the residual stress and the load stress exceed the yield strength of the metal; thus, the load is uniformly distributed and the part is able to withstand further loading.

Reaction stresses are those due to the resistance to thermal expansion and contraction by the structure to which a piece is joined. Thus reaction stresses exist only when two pieces which are joined by welding will form a rigid structure. The magnitude of reaction stresses never exceed the yield strength of the material and seldom exceed 15,000 psi. Perhaps the most important effect of reaction stresses is their tendency to cause cracking during or immediately following welding. This problem is especially critical when welds are being made where there is great rigidity and restraint to normal shrinkage that occurs in the direction transverse to the length of the weld.

Types of distortion that occur in welded joints are shown in Fig. 19–24. The weld becomes shorter because of longitudinal stresses and is distorted angularly because of an unsymmetrical weld. During machining residual stresses can cause warpage particularly when weldments are machined so as to unbalance the stress system.

Stress reduction in weldments may be accomplished by a number of methods such as preheating prior to welding, use of multiple pass welding operations, inducing plastic flow by shot peening, or by stress relieving heat treatments. Of the above, stress can be more safely and more completely relieved by heat treatment. As the weld is heated its yield strength is reduced to a value lower than the residual stress value; this allows plastic flow to occur and the

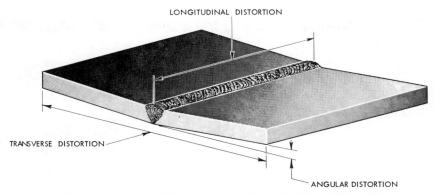

LONGITUDINAL DISTORTION

TRANSVERSE DISTORTION

ANGULAR DISTORTION

Fig. 19–24. Distortion encountered in butt welded plates.

residual stress is relieved. Ductility of the metal is also improved during this treatment. Uniform cooling from the stress relieving temperature reduces possibilities of introducing further stresses during the stress relieving treatment.

Brazing and Soldering

19.14 Description. Brazing is defined as the joining of metals with a non-ferrous filler metal melting at a temperature of above 800°F but below the melting point of the metal being joined. Soldering, on the other hand, is similar to brazing but is done below 800°F. The molten filler metal wets the parent metal and flows by capillary action between the heated but unmelted joint members. Typical brazed and solder joint designs are shown in Fig. 19–25. Both soldered and brazed joints are gen-

erally of relatively large area and very small thickness. There are a number of significant requirements of brazing and soldering including the following: (1) the filler metal is always different in composition from the parent metal, (2) the joint clearance must be small (of the order of 0.001 — 0.005 in.) for capillary action to exist, (3) bonding occurs between the parent metal and the filler metal and not between the parent metals, (4) surfaces must be cleaned to allow the filler material to wet them, and then protected by flux to reduce oxidation, (5) the 800°F temperature is an arbitrary one, set to distinguish brazing from soldering.

19.15 Bonding Mechanism. The bond between a soldering or brazing metal and the parent metal is generally due to some diffusion of the

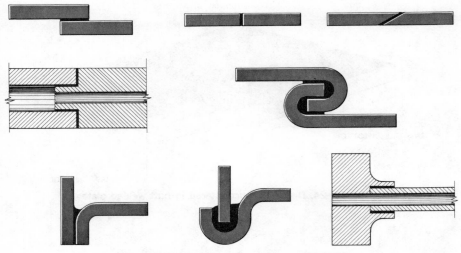

Fig. 19–25. Configurations of brazed joints.

filler metal into the hot base metal and to some surface alloying of the metals. Fig. 19–26A is a photomicrograph of a typical brazed joint and Fig. 19–26B is a soldered joint. Normally, alloying during soldering and brazing is a surface effect and extends only a few thousandths of an inch into the surfaces being joined.

In some instances, however, alloying and diffusion can be so extensive as to cause problems.

For example, some aluminum and magnesium filler metals will alloy and diffuse completely through thin aluminum sheets if held at the brazing temperature longer than is necessary to fill the braze joint.

In general, soldering and brazing alloys will wet a metal surface provided that: (a) they form an inter-metallic compound with the solid, or (b) the solid metal can take the filler metal into solution.

Solders and brazing alloys do not usually wet metals with oxide surface coatings. It is, therefore, the purpose of fluxing agents to remove the bulk of the oxide and expose clean metal. Surface texture also plays a role in determining how well the filler metal will spread. Spreading is benefited if the surface of the joint consists of a series of interconnecting scratches along which the filler metal may be drawn by capillary forces. It should be pointed out that wetting is not *absolutely* essential to the formation of a bond. This is demonstrated by the fact that, although lead does not wet steel, if molten lead is allowed to solidify

618

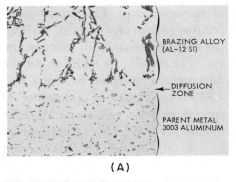

(A)

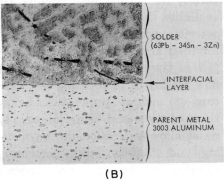

(B)

Fig. 19–26. Photomicrographs of heterogeneous joints. A, brazed joint. B, soldered joint.

and brazing. Bond strength in soldered joints may reach 8,000 psi while in brazed stainless steel joints strengths of 130,000 psi have been reported.

Soldering is generally used where excessive heat is detrimental to the parts to be joined and where high strength is not required. Brazing on the other hand is used where stronger joints are required and where the operating temperature of the parts may reach 450-500°F or higher.

19.16 Soldering Alloys. The most generally useful soldering and brazing alloys are those close to a eutectic composition and which therefore have a narrow melting range. Soldering alloys are principally alloys of lead and tin, although many other combinations of low melting point metals are in use.

19.17 Soldering Fluxes. Soldering often requires very active fluxes to clean the metal surface and to assist the flow of solder. Such fluxes invariably leave corrosive residues which must be removed. In those instances which preclude the use of corrosive flux because of difficulties of removal, as in electronic work, non-corrosive fluxes are used whose residue may remain on the joint without causing corrosion. Such mild fluxes do not have a vigorous cleaning action and cannot be expected to dissolve thick layers of

in contact with a clean, oxide-free steel surface, the two metals will be strongly bonded. Even though there is no intersolubility between lead and steel, the metallic bond is formed whenever the distance between the two metal surfaces approach the normal lattice spacing. It is apparent that the probability of achieving the required intimate contact is very much greater when interdiffusion takes place during normal soldering

metallic oxides. The parent metals must be well cleaned prior to soldering and must be protected from oxidation during heating of the joint. Many theories have been proposed to explain the mechanism of flux action. The most widely held view is that the flux removes the oxide film from the base metal by dissolving or loosening the film and floating it off into the main body of the flux. The molten flux then forms a protective blanket to prevent the film from reforming until it is displaced by the liquid solder which reacts with the parent metal to form an intermetallic bond.

19.18 Brazing Alloys. In principle, any metal or alloy melting above 800°F but below the melting point of the metals to be joined could be used as a brazing filler metal. The most common brazing alloys, however, have been copper and silver, or their alloys. The emergence of new engineering materials in recent years and the stringent demands placed upon them has led to the development of a number of new brazing filler metals other than those based on silver or copper. The American Welding Society and the American Society for Testing and Materials jointly have issued specifications for brazing filler metal (AWS A5.8, ASTM B260). This specification lists brazing filler metals under the following seven classifications: (1)

aluminum-silicon, (2) copper-phosphorus, (3) silver, (4) gold, (5) copper and copper-zinc, (6) magnesium, and (7) nickel. In addition to the filler metals classified by the AWS-ASTM specifications a number of special purpose non-standard filler metals are available for application in jet and rocket propulsion and nuclear energy. Although a wide variety of brazing filler metals are available, they all have certain characteristic properties. They are noted for their ability to (1) wet the surface of the metals being joined, (2) flow readily at the brazing temperature, (3) melt at temperatures below those of the metal being joined, (4) resist alloy separation during melting, and (5) form joints possessing suitable mechanical and physical properties.

Most of the brazing filler metals are ductile and can be supplied in rolled strip or drawn wire forms.

19.19 Brazing Fluxes. Borax has been used for centuries as a brazing flux and remains to this day as one of the most useful components of commercial brazing flux. Both borax and boric acid are reduced by chemically active metals to form low melting point borides. They also dissolve non-refractory oxide films at the brazing temperature used for copper alloy filler metals. Additions of fluorides and fluoroborates to borax-boric acid mixture lowers the melt-

Fig. 19–27. Induction brazing operation. (L. C. Miller Co.)

ing point so it can be used with silver solders. The most effective fluxes for aluminum, magnesium, titanium, and zirconium are mixtures of chlorides and fluorides. Fluxes are available as pastes, powders, or liquids, and should be selected on the suitability and economy for each intended application. Most fluxes are corrosive and should be removed after brazing is completed. Borax flux residues after brazing are often glass-like and require thermal shock (quenching) or abrasive or chemical action to remove them. Chloride-

fluoride flux residues are water soluble and may be removed by immersion in hot water. In some cases brazing may be done in a protective atmosphere or under vacuum without the use of flux.

19.20 Heating Methods. A number of heating methods are used in commercial soldering and brazing. Torch heating using ordinary gas welding equipment is used for both soldering and brazing. It is now used primarily for repair work. With manually operated torches it is difficult

Metallurgy Theory and Practice

to obtain a uniform temperature throughout the joint and much depends upon the skill of the operator. Furnace soldering and brazing are used for high production runs; generally on complicated assemblies which can be heated without damage to any of its components.

Induction heating has many advantages in soldering and brazing and is widely used because of its high speed, localized heating and capabilities for automatic operation. A typical induction brazing operation is shown in Fig. 19–27. Another heating method is that of dip soldering and brazing in which the assembly is immersed in a bath of molten filler metal. The bath provides the

required heat and filler metal for the joint. A number of other heating methods are used, each with its own advantages and limitations.

In summary, the big advantages of brazing and soldering are that they can be used to join a large variety of dissimilar metals. Since the parent metals are not melted, they do not need to have similar metallurgical characteristics nor similar melting temperatures. It is only necessary that each parent metal be able to bond with the filler metal. Thus, the problem of brazing and soldering resolves itself into the selection of filler metal, flux, and a heating method suitable for the assembly to be joined.

References

American Welding Society. *Brazing Manual.* New York: Reinhold Publishing, 1955.

———. *Fundamentals of Welding Metallurgy.* New York, 1963.

———. *Soldering Manual.* New York, 1959.

———. *Welding Handbook. 5th ed. sec. 3.* New York, 1964.

Burton, Malcolm S. *Applied Metallurgy for Engineers.* New York: McGraw-Hill, 1956.

Lancaster, J. F. *The Metallurgy of Welding, Brazing, and Soldering.* New York: American Elsevier Publg., 1965.

Lyman, Taylor, ed. *Metals Handbook, 1948 ed.* Cleveland, O.: American Society for Metals.

Patton, W. J. *The Science and Practice of Welding.* Englewood Cliffs, N. J.: Prentice-Hall, 1967.

622

Metallurgy of Machining

20.1 Introduction. Efficient metal cutting processes are an essential element in economical, competitive manufacturing. The cost of metal cutting enters into the cost of every manufactured article, either in producing the article itself, or in making the machines used in its production. Metal cutting is big business. Every year over 15 million tons of metal are whittled away into chips in the United States. The cost of this metal cutting has been estimated to be as high as 34 billion dollars a year. A large part of every dollar spent in manufacturing goes for the cutting of metal. Therefore it is important, from an economic consideration, to study metal cutting processes in an attempt to make them more efficient. Just imagine the annual savings if metal cutting could be made even 1 percent more efficient—a savings of $340,000,000!

Metal cutting is an extremely complex process involving many variables and complicated physical, chemical, thermal, and metallurgical relationships. The metal cutting process has been studied in many countries since the beginning of the century and still has many avenues needing scientific investigation. This chapter is intended only to acquaint the student with the fundamental principles of metal cutting physics and to dwell more specifically on the metallurgical aspects of conventional and new metal removal processes.

20.2 Machinability. The measure of the ease with which metals may be cut is termed machinability. A more exact definition of machinability is difficult since the ease with which a given material may be worked depends on machine variables such as cutting speed, depth of cut, tool material, tool geometry, nature of cut, and environment; it also changes with work material

623

variables such as hardness, tensile strength, alloy composition, microstructure, previous cold work, strain hardenability, shape and dimension, and rigidity of the workpiece. Thus, a material which may have a high machinability rating under one set of conditions, may have a lower machinability rating under different conditions. It should be observed that much of the published data on machinability rating of metals does not sufficiently specify the conditions under which the original test was conducted to permit accurate duplication of the original test conditions. Furthermore, caution must be observed in trying to apply information gained from turning tests to drilling or milling operations. Thus, much of the published data should be used only as a first approximation to setting machine speeds, feeds, depth of cut, tool angles, and other items.

An example of published machinability ratings is given in Table 20-1. These ratings are based on the 100 percent rating of AISI B-1112 cold drawn steel. A cutting speed of 180 surface feet per minute (with a coolant) was used in establishing the rating which yielded a 60 minute tool life when using a high speed

Table 20-1. Machinability ratings for various steels.

AISI CLASSIFICATION	RELATIVE MACHINABILITY RATING PER CENT	BRINELL HARDNESS
B1113	135	179-220
B1112	100	179-229
C1109	85	137-166
C1118	80	143-179
C1022	70	187-229
C1020	65	137-174
C1040*	60	179-229
A3140	55	187-229
A8650*	50	183-241
A2340	45	187-241
E9315	40	179-229
E52100	30	183-229
HIGH CARBON, HIGH CHROMIUM TOOL STEEL	25	

*SPHEROIDZED ANNEAL

steel cutting tool. A material with a 50 percent machinability rating is supposed to be cut at one half of the speed of B-1112. In establishing machinability ratings a number of different tests have been devised. The most popular test is the turning test. Turning tests are frequently based on measuring tool life or tool wear at higher than normal cutting speeds. The test is conducted by turning the material under investigation at different cutting speeds and measuring the tool life for each speed. Next a curve of tool life versus cutting speed is plotted. From this curve can be read the cutting speed for any desired tool life under the conditions established for the cut. Tool life is normally the elapsed cutting time in minutes between tool sharpenings or tool changes.

In actual shop conditions, tool life may also be based on the number of pieces machined between tool changes. Other criteria for establishing tool life based on horsepower requirements, surface finish, and size failure have also been used.

A simple drilling test has been developed to measure the drilling time or penetration in test specimens. The set-up for the drilling test is shown in Fig. 20–1. This test has been correlated with machinability data obtained with some conventional turning tests, but there are also some instances where the data does not correlate well. This is a sim-

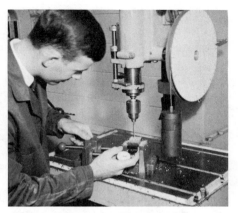

Fig. 20–1. Drilling penetration test set-up for machinability. Time to drill a ½" deep hole is used as a measure of machinability.

ple, quick test. It requires a minimum of material and because of these features, is gaining wider acceptance.

Recently a number of studies have been made in an attempt to predict machinability from physical properties of the material to be machined. Properties such as hardness, microstructure, tensile strength, elevated temperature yield strength, thermal conductivity, lattice structure, and other factors have been used in these predictions. Some results are quite promising and show good correlation between predicted and "actual" machinability ratings. In spite of the great amount of work that has been done, the effect of many variables such as cutting fluids, tool materials, tool geometry, and other factors have not been quantitatively evaluated.

20.3 Metal Cutting Physics. The shape, dimensions, and color of chips obtained from a metal cutting operation throw considerable light on the nature and cutting conditions of that operation. Irrespective of the type of machining operation (turning, milling, drilling, grinding, etc.) three basic types of chips or combinations of them are found: (1) discontinuous or segmental, (2) continuous, and (3) continuous with built-up edge (BUE). These three types of chips are illustrated by photomicrographs in Fig. 20–2.

Discontinuous chips consist of individual metal segments which in some cases adhere loosely to each other after the chip has been formed. In other cases the segments come from the cutting tool as distinct and unconnected segments produced by actual fracture of the metal ahead of the cutting edge. This type of chip is frequently found in the machining of brittle materials. Discontinuous chips are also found when cutting ductile materials at very low cutting speeds with large chip thickness and small rake angles. When the discontinuous chip is associated with brittle materials it ordinarily affords fair surface finish, low power consumption, and reasonably good tool life. However, when the discontinuous chip is associated with ductile materials, the surface finish is often poor, cutting temperatures

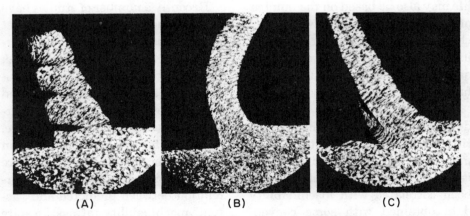

Fig. 20–2. Three types of chips from metal cutting. A, Type I, discontinuous or segmental, B, Type II, continuous, C, Type III, continuous with built up edge (BUE). (Cincinnati Milling Machine Co.)

rather high, and tool wear excessive.

Continuous chips are formed by continuous deformation of the metal ahead of the tool without fracturing followed by a smooth flow of the chip up the tool face. Continuous chips are ordinarily obtained when cutting ductile materials at cutting speeds above 200 surface feet per minute. This type of chip is associated with low friction between the tool and chip, and large positive rake angles. The conditions which produce the continuous chip lead to good surface finish on the workpiece, good tool life, and minimum heat. One problem introduced by the continuous chip is that of chip disposal, which is ordinarily handled by the use of chip breakers on the cutting tools.

Continuous chips/built-up edge (BUE) are similar to the continuous chips except that a built-up edge is present on the nose of the tool. The built-up edge is an unstable mass of metal which adheres to the face of the tool while the chips shear past it up the tool face. Fragments of the built-up edge are continuously forming and then shedding off onto the finished surface as the tool progresses through the workpiece. This type of chip is frequently formed when cutting ductile materials at ordinary cutting speeds with high speed steel tools. The conditions which favor the BUE chip are (1) ductile materials without friction

reducing alloy additives, (2) high speed steel tool bits, (3) low cutting speeds, (4) lack of cutting fluids, (5) rough ground tool face, and (6) thick chips. The BUE type of chip is often associated with poor surface finish resulting from fragments continually shedding on the finished surface from the built-up edge. Wear occurs by cratering on the tool face at the point of contact with the chip and by abrasion of the tool flank by contact with escaping severely work hardened fragments of the built-up edge.

Chip color is a rough indicator of cutting temperature and can be used as a guide to selecting the best cutting speed range. Chip color is an important but often overlooked guide for establishing optimum cutting speeds.

Each type of cutting tool has a heat sensitive range in which the temperature from cutting causes the tool to soften and rapidly fail. Tools which can maintain their hardness at elevated temperature ranges can be operated at higher cutting speeds than those which soften at lower temperatures. An example of the steels which soften readily upon heating are the carbon tool steels. From the relationship between cutting speed and tool temperature has come the term high speed steel which resulted from the high temperature hardness of tungsten and molybdenum base tool steels. The

Table 20–2. Temperature sensitive range for typical cutting tool materials.

TOOL MATERIALS	APPROX. TEMP. SENSITIVE RANGE + − 50°	TYPICAL CUTTING SPEED SFPM
. CARBON TOOL STEEL	400°F	40
. HIGH SPEED STEEL	1100°F	100
. CAST ALLOY	1500°F	180
. TUNGSTEN CARBIDE	1700°F	500
. TITANIUM CARBIDE	---	1200
. CERAMICS	2100°F	2000

WHEN CUTTING MACHINERY STEEL, TOOL LIFE APPROX. 1 HOUR

temperature sensitive range for a number of cutting tool materials is given in Table 20–2. Tools which are run too slowly are inefficient and result in low production. Tools which are operated at excessive cutting speeds overheat and dull quickly, requiring frequent tool changes.

Geometry of chip formation is important. While the basic mechanism of chip formation is actually the same for all types of chips resulting from metal cutting operations, the continuous chip resulting from orthogonal cutting is the simplest and therefore lends itself very nicely to analysis. In orthogonal or two dimensional cutting the tool is set with its cutting edge perpendicular to the direction of tool travel as shown in Fig. 20–3. When cutting a tubular workpiece as shown in this arrangement the chip is assumed to be essentially two dimensional with cutting forces in a single plane. The width of the chip is assumed to remain constant. The geometry of the cut is shown in Fig. 20–4.

In the figure, t_1 represents the depth of cut and t_2 the thickness of the chip as it passes off the tool face. The tool rake angle is (α), and the angle (ϕ) is known as the "shear plane" angle. The base mechanism of metal cutting is that of deformation of the metal ahead of the cutting edge. This deformation proceeds by a process of shear in a narrow zone extending from the cutting edge to the work surface. This shear zone which is actually a narrow band may, for purposes of mathematical analysis, be treated

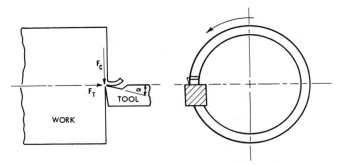

Fig. 20–3. Orthogonal cutting using hollow tube to eliminate any radial cutting force. (Cincinnati Milling Machine Co.)

Fig. 20–4. Chip geometry in orthogonal cutting with Type II continuous chip. (Cincinnati Milling Machine Co.)

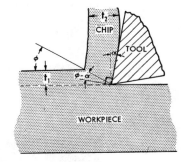

as a single plane commonly known as the shear plane. As the metal lying ahead of the tool reaches this plane it is displaced by shear to form the chip which then slides up the face of the tool. This shearing action may be seen in the photomicrographs of the chips in Fig. 20–2. The formation of a chip by shear deformation and its motion upward along the tool face has been likened to the sliding and movement of a pile of inclined playing cards when pushed. This analogy is illustrated in Fig.

20–5A. The slope of the inclined cards corresponds to the shear zone angle of Fig. 20–4, and movement of the cards corresponds to movement of thin laminae of metal along crystallographic shear planes, Fig. 20–5B. With very thin laminae the chip flows over the tool face like a continuous ribbon.

In Fig. 20–6 is shown what happens as the shear zone angle is changed. A small shear zone angle results in a thick chip and a long shear path requiring large shearing

629

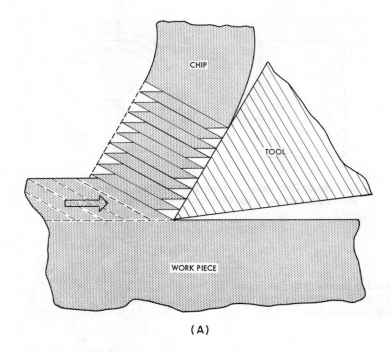

(A)

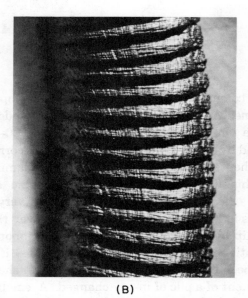

(B)

Fig. 20–5. Chip formation. A, Piispanen's card analogy, B, chip corresponding to card analogy, front of chip shown 5X.

630

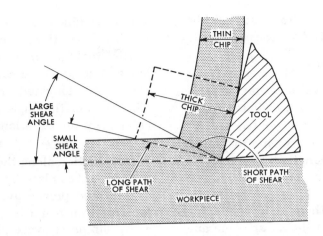

Fig. 20—6. Diagram showing effect of shear angles on chip thickness and length of shear plane. (Cincinnati Milling Machine Co.)

forces. On the other hand, a large shear angle results in a thin chip and a short path of shear with minimum shearing forces required. In practice the shear angle may be changed depending upon the cutting conditions and the material cut. The shear angle is an important geometrical quantity in the cutting of metal, and one over which we have some control.

A number of important quantities can be calculated from the geometrical relationships shown in Fig. 20–4. For example: The velocity of shear, V_s, of the chip with respect to the workpiece is given by the following expression:

$$V_s = V \frac{\cos \alpha}{\cos (\alpha - \phi)},$$

where:

V is the cutting speed of the tool relative to the workpiece, α is the true rake angle, and ϕ is the shear zone angle.

A second geometrical quantity of interest is the velocity of chip flow, V_f, as it passes up the cutting tool. The expression which has been derived is:

$$V_f = V \frac{\sin \phi}{\cos (\phi - \alpha)}$$

but since t_1/t_2 is equal to $\sin \phi/\cos (\phi - \alpha)$ the same expression may be given by:

$$V_f = \frac{t_1 V}{t_2}$$

The ratio t_1/t_2 is referred to as the chip thickness ratio, r_c. This ratio is not difficult to measure and may be used to calculate the shear angle directly from the equation:

$$\tan \phi = \frac{r_c \cos \alpha}{1 - r_c \sin \alpha}$$

Another value, that of shear strain, e, may be calculated from:

$$\epsilon = \cot \phi + \tan (\phi - \alpha)$$

This value represents the amount of deformation that the metal undergoes in the process of chip formation.

Actually there is a small difference between the shear plane angle (ϕ) and the direction or grain elongation as illustrated by the model in Fig. 20–7. It may be seen that as the "tool" was moved toward the left and sheared the cards, the circles were elongated into ellipses. The value of the angle may be calculated by the equation:

$$\cot \psi = \cot \phi + \tan (\phi - \alpha)$$

Deformation along the shear zone not only takes place by plane shear as indicated previously, but by deformation on a much smaller scale as by slip on cleavage planes, twinning of crystals, rotation of slip

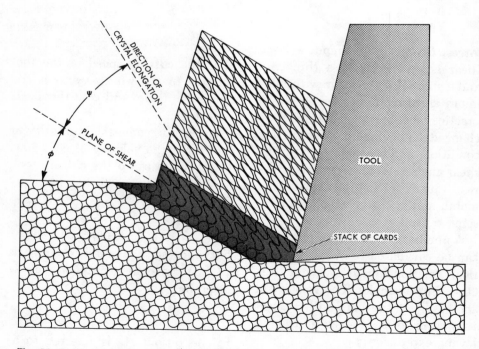

Fig. 20–7. Model using stack of cards with circles drawn on the edge to represent undeformed crystal structure. Model shows difference between direction of shear and direction of crystal elongation. (Cincinnati Milling Machine Co.)

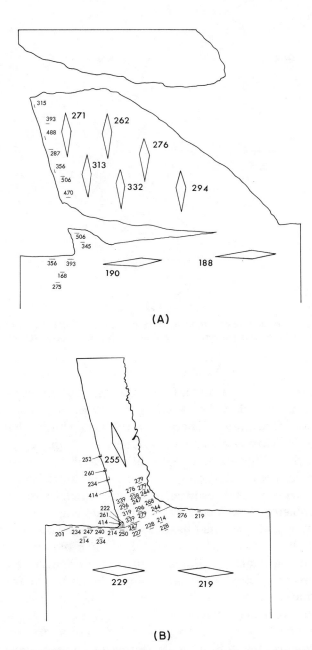

Fig. 20–8. Hardness distribution in chip, workpiece, and built-up edge. Traced from photo. Hardness is in Knoop values. A, discontinuous chip, B, continuous chip. (Cincinnati Milling Machine Co.)

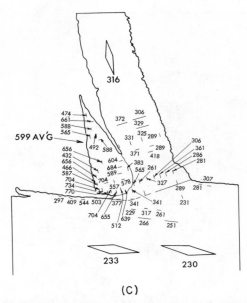

Fig. 20–8. (Cont.) Hardness distribution in chip, workpiece, and built-up edge. Traced from photo. Hardness is in Knoop values. C, continuous chip with BUE. (Cincinnati Milling Machine Co.)

planes, subdivision of grains, and adjustment of grain boundaries to accommodate elongated grains. There is much yet to be learned about deformation of the chip, workpiece, and tool point in metal cutting.

Hardening from metal cutting may create problems. The geometry of chip formation just discussed has a direct effect on the amount of deformation and hardening in and around the area of actual cutting. It is known that both the chip and surface of the workpiece may be severely hardened by machining processes. Evidence for this statement is clearly shown in Fig. 20–8. Work hardening here becomes a

problem. In all cases hardness is greatest at the tool chip interface with a maximum Knoop value of 770 at the built-up edge whereas the hardness of the material being machined is approximately 240. Fragments which rub across the tool with the chip promote rapid tool wear. Fragments of the built-up edge which remain on the machined surface evidently provide a very poor bearing surface. In other words, the quality of a surface obtained from a BUE chip is very poor from the point of view of surface deformation and fragmentation. Continuous chips and discontinuous chips both offer much less hardened

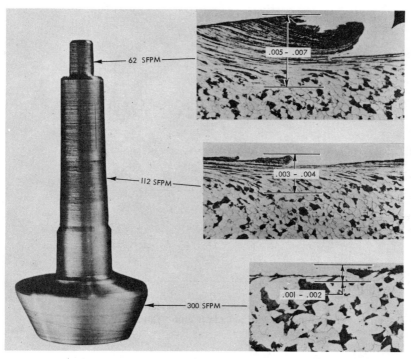

Fig. 20–9. Photomicrograph of depth of deformation in AISI 4140 steel at various cutting speeds. Minimum deformation occurs above 600 fpm. (Jones & Lamson Machine Co.)

and distorted work surfaces. In Fig. 20–9 is shown the depth of surface deformation at various cutting speeds. Minimum surface deformation occurs at speeds above 600 surface feet per minute. At slower speeds surface deformation may extend as much as .005-.007 inches into the workpiece.

Forces acting on a cutting tool in general are three dimensional. In the case of orthogonal cutting which we are considering for the sake of simplicity, the entire force system lies in a single plane as shown in Fig. 20–10. Cutting force components are measured by special devices called metal cutting dynamometers. The horizontal force component, F_c, measures the total amount of force required to move the tool through the workpiece. The force component, F_t, is a force tending to push the tool out of the workpiece. Both

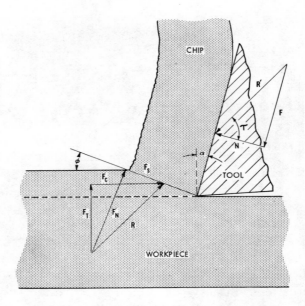

Fig. 20–10. Diagram of force system for orthogonal cutting with a Type II chip. (Cincinnati Milling Machine Co.)

forces F_c and F_t cause deflection of the tool relative to the workpiece. The force system in Fig. 20–10 is arrived at by assuming that the chip is a body in equilibrium under the action of the forces exerted upon it at the tool face and at the shear plane. The force components at the tool face are F and N. These forces act on the chip. Their resultant is equal in magnitude but opposite in direction to the resultant of forces F_t and F_c.

The force F is known as the frictional force and represents the frictional resistance met by the chip in sliding over the tool face. The force N is the normal force. The ratio F/N is the coefficient of friction, (μ) between the chip and tool. Since the chip is in equilibrium a reduction of the coefficient of friction between the tool and chip can affect a reduction in the overall cutting force.

Now that the basic geometrical and force relationship of metal cutting has been established we shall consider how the metallurgical factors of the workpiece affect cutting, then tool properties and tool life and basic problems in metal cutting, and finally, new metal removal processes including electrical discharge machining and electrochemical machining.

20.4 Metallurgical Factors. There are a number of metallurgical factors which play a significant role in determining the machinability of any given workpiece. Among these factors are: chemical composition, previous processing treatments, and physical characteristics.

Chemical composition and additives of a steel have major influences on machinability. The composition affects the structure, mechanical properties, and heat treatment re-

sponse. However, because the effect of composition is somewhat obscured by other factors, it is difficult to assess the exact influence of each element. Carbide forming elements such as chromium, tungsten, molybdenum, and vanadium tend to decrease machinability by increasing the metal hardness. Their detrimental effects may be overcome by annealing treatments. Elements which dissolve in ferrite, including nickel, and manganese, ordinarily reduce machinability because of the increased toughness and hardness they impart to the steel. Elements which form hard abrasive inclusions (such as alumina or silica) are also detri-

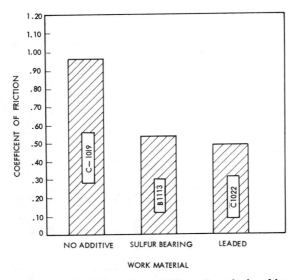

Fig. 20–11. Effect of alloy additions in reducing friction in low carbon steels.

mental to machining. Elements which form soft inclusions appear to have a beneficial effect. Chemical additives which form soft inclusions in metals include sulfur, lead, phosphorus, selenium, and tellurium. These additives have been used to improve the machinability of steel, stainless steel, tool steel, copper alloys, aluminum alloys, and other metals. The effect of alloy additions on coefficient of friction when machining low carbon steels with sintered carbide tools is shown in Fig. 20–11. Approximately a 50 percent reduction in friction is achieved in carbon steel by means of sulfur or lead additives, the lead being slightly better in this respect. The

effect of these additives is illustrated in Fig. 20–12. The additives are either insoluble in the metal, as is the case of lead additions, or else appear as inclusions as sulfur which forms manganese sulfide. These additives are distributed randomly throughout the workpiece and smear over the tool face forming a low shear strength film during the cutting action. The additives have good lubricity and reduce friction and welding tendencies at the tool-chip interface.

Processing treatments which affect machining characteristics include hot working, cold working, and heat treatment. Experience indicates that large grain size is or-

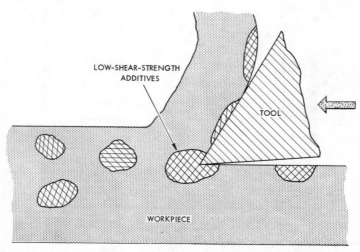

LOW-SHEAR-STRENGTH ADDITIVES

TOOL

WORKPIECE

Fig. 20–12. Additives smear over tool face forming a low strength film reducing friction and welding tendencies.

dinarily preferred for most types of machining operations on steel. Small grain size is preferred if extra fine surface finishes must be achieved. The desired structural grain size can usually be obtained by controlling the finishing temperature during *hot working*. For carbon steels the finishing temperatures are held at about 1700°F when high machinability is desired and when the steel is to be machined prior to further heat treatment. Alloy steels, on the other hand, may be finished at slightly lower temperatures to avoid large grains with their attendant higher hardenability and hardness upon cooling. The effect of hot working is minor when compared with the effects of chemical composition or annealing heat treatments.

Machining characteristics of most steels and many soft non-ferrous metals can be greatly improved by *cold working*. Cold working decreases the ductility of metals thus promoting clean shearing and chip breakage. Cold working may be used to increase hardness of soft, gummy alloys to the range most desirable for good machining. For steels this hardness range is between 190-220 BHN. Cold working is most beneficial on metals which lose ductility rapidly and which exhibit low strain hardening characteristics. Care should be exercised in the machining of austenitic stainless steels since these alloys tend to harden

considerably by cold working. Excessive roughing cuts should be minimized to prevent work hardening. The tool should not be allowed to rub on the work or else glazing will occur and the tool will be unable to enter the workpiece. Free machining grades should be used whenever high production rates are required.

The machining characteristics of most metals may be markedly affected by *heat treatment*. For example, steels containing less than 0.10 percent carbon machine better after quenching in water from above the A_3 temperature. The improvement is due to lowered ductility in the quenched steel with a reduced tendency to drag during cutting. Annealing and normalizing heat treatments for medium carbon steels should be designed to coarsen the grain size and break up the continuity of the ferrite matrix as much as possible. Higher carbon and high alloy steels should receive heat treatment to break up any carbide network and to spheroidize the carbides. In Table 20–3 are listed the recommended microstructures for production turning of carbon and alloy steels. In Fig. 20–13 are shown typical microstructures for plain carbon and alloy steels. The structures are arranged in order of decreasing machinability with the low carbon steels containing sulfur additives rated highest and the block pearlite rated lowest.

Cast iron parts that require ex-

Table 20–3. Recommended microstructure for production turning of carbon and alloy steels.

| TYPES OF STEEL | MICROSTRUCTURE | |
	DESIRABLE	UNDESIRABLE
LOW CARBON STEELS 0.08 - 0.25% C	1. COLD DRAWN 2. LAMELLAR PEARLITE 3. FREEDOM FROM FERRITE SEGREGATION 4. UNIFORM GRAIN SIZE	1. SPHEROIDIZED 2. BANDED 3. ABNORMAL PEARLITE
MEDIUM CARBON ALLOY STEELS 0.30 - 0.50% C	1. LAMELLAR PEARLITE OR 2. FINE PEARLITE 3. UNIFORM GRAIN SIZE	1. COARSE SPHEROIDIZED STRUCTURE WITH FREE FERRITE 2. SORBITIC BANDING 3. COARSE GRAINS 4. SEGREGATION
HIGH CARBON ALLOY STEELS 0.50 - 0.70% C	1. FINE OR COARSE SPHEROIDITE	1. PARTIALLY SPHEROIDIZED PEARLITIC STRUCTURE 2. FINE PEARLITE (SORBITE)

tensive machining may be given annealing treatments prior to machining to decompose any massive carbides present in the as-cast matrix and thus improve machinability.

Physical characteristics of metals are significant in determining machinability ratings. These characteristics include: grain size, hardness, microstructure, tensile properties, atomic structure, and thermal conductivity. *Grain size* has already been mentioned with regard to specific alloys. Although there is a divergence of opinion about the relationship of grain size to machining quality, it is generally agreed that some improvement in machining performance is achieved with large

structural grain size. However, the improvement is not proportional to the degree of coarsening.

Another factor which affects machinability is the effect of alloy additions. Steels which have coarse grains as a result of high temperature hot working and which normally have good machining properties have poorer machining properties when deoxidized with aluminum because of the abrasive action of the Al_2O_3 on the tool.

Many investigators have tried to closely correlate Brinell *hardness* numbers and machinability but without much success. The reason that much of this work has been unsuccessful is because composition,

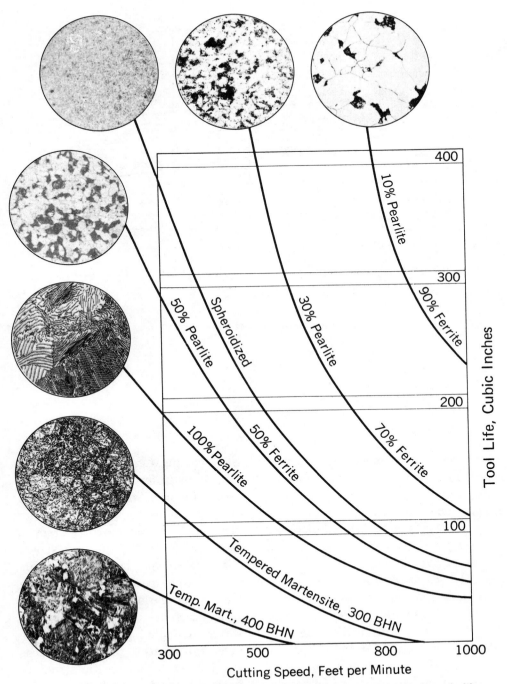

Fig. 20–13. Microstructures of carbon and alloy steels in order of decreasing machinability. (By permission: *Metal Machining and Forming Technology,* **Ronald Press)**

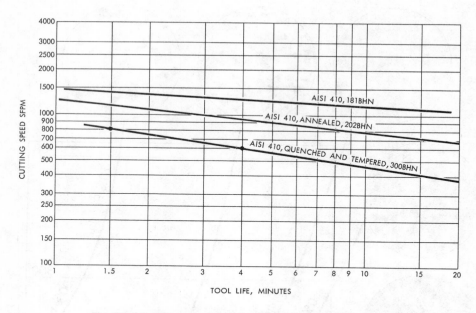

Fig. 20—14. Effect of microstructure on tool life.

structure, and mechanical factors are not taken into account by a simple Brinell hardness test. Nevertheless, some consider Brinell hardness an indication of machinability and specify a range of 190–220 BHn for ease of machining.

The *microstructure* of metals has an important influence on machinability. It should be kept in mind that the final structure of many alloys is not only a function of the original chemical composition, but is also dependent upon cooling rate.

Slow cooling tends to produce soft structures in ferrous metals such as steel and cast iron whereas rapid cooling tends to produce harder structures of fine pearlite, or even free carbide. In Fig. 20–14 is shown the relationship between tool life and cutting speed when machining steel with various microstructures using carbide cutting tools. The softer laminated pearlitic structure gives the greatest tool life whereas the tempered martensitic structure gives the shortest tool life.

Unfortunately the microstructure that is best for one kind of machining operation may not be the best for another operation. These differences are shown in Table 20–3. A spheroidized structure, for example, may be best for turning whereas a tempered martensitic structure in the same steel is best for broaching.

The *tensile properties* in metals which affect machining include elastic modulus, yield strength, ultimate strength, work hardening exponent, elongation, and reduction of area. The elastic modulus is the only one of the tensile properties which is relatively independent of the others. The elastic modulus determines rigidity of the workpiece. High modulus materials require less support to prevent deflection away from the cutting tool. Low modulus materials require a large amount of strain before plastic flow and chip removal are initiated.

The other tensile properties listed, including hardness, are inter-related in their influence on metal cutting. Tensile properties are an index of machinability only in giving comparative values of strength and ductility for different steel compositions. An optimum combination of hardness, strength, and ductility are required to yield maximum cutting efficiency.

The work hardening exponent, sometimes called the *Meyer* exponent, is derived from the slope of the true stress-strain curve beyond the yield-stress point. The greater the slope in this plastic region of the curve, the more rapidly the material hardens due to the component of tool force normal to the workpiece surface. The normal force can be reduced with sharp tools and with high rake angles. When a material work hardens during cutting the difficulty of taking subsequent cuts on the surface increases as does the magnitude of residual stresses in the workpiece during the last cut. Care must be taken to see that the cutting edge of the tool is below the work hardened layer of the previous cut. This applies to feed rate as well as to depth of cut. Likewise the tool must never be allowed to dwell and rub on the workpiece. In summary, it may be stated that as tensile properties increase, the ease of machining will decrease.

20.5 Cutting Tools. Cutting tool selection is an important factor in improving metal cutting efficiency. Cutting tools must be able to withstand high unit forces, high temperature, and various kinds of tool wear.

Cutting forces are related to the strength of the material being cut. In some cases the unit stress may be very high. For example, assume the cutting force, F_c, is 400 lbs when cutting a high strength alloy with a depth of cut of .100 in. and a feed

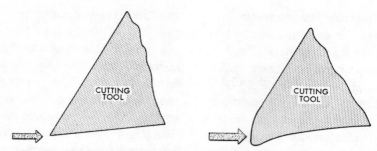

Fig. 20–15. Probable tool tip deflection during machining. Left, low cutting forces. Right, high cutting forces.

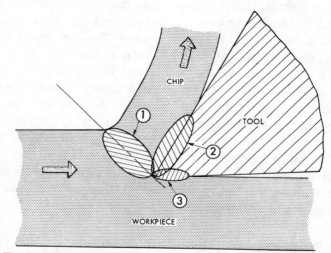

Fig. 20–16. Sources of heat in cutting. 1, heat of shearing. Produced by plastic deformation along shear zone. 2, sliding friction at tool face. 3, rubbing friction between tool and workpiece.

rate of .010 in./rev. The area of the tool on which the chip load is applied is equal to the feed rate × the depth of cut, or in this case, .001 sq in. Thus the stress on the cutting tool tip is as follows:

$$\text{Unit stress} = \frac{\text{cutting force, } F_c}{\text{Area}} =$$

$$\frac{400 \text{ lbs.}}{.001 \text{ sq in.}} = 400{,}000 \text{ psi}$$

This is indeed a very high stress and probably near the maximum force that would be encountered in metal cutting. The need for high strength tool materials can be seen from the above example. In Fig. 20–15 is il-

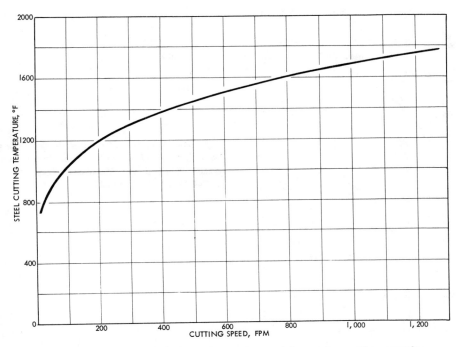

Fig. 20–17. Tool-chip interface temperatures at increasing cutting speeds.
(After K. J. Trigger)

lustrated the probable tool deflection resulting from high cutting forces.

Cutting temperatures are usually responsible for the greatest problems in most machining operations. Fig. 20–16 shows the three sources of heat in metal cutting; (1) *heat of shearing* during plastic deformation of metal along the shear plane, (2) *sliding friction* at the tool chip interface due to rubbing of the chip over the tool, and (3) *rubbing friction* between the tool flank and workpiece.

Most of the mechanical energy expended during cutting is transformed into heat, 75 percent or more of which is carried away in the chip, the balance going into the tool and workpiece. The heat of shearing may be reduced by increasing the shear plane angle, or by reducing the strength of the material by an annealing or tempering operation. Sliding friction can be reduced by use of cutting fluids, increased rake angles, or higher cutting speeds. Rubbing friction can be minimized by

645

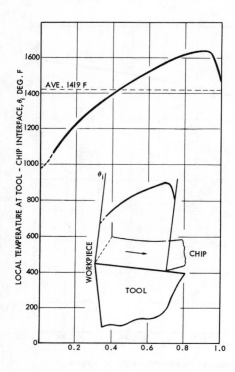

Fig. 20–18. Temperature distribution along tool-chip interface when machining AISI 4150 steel with carbide tool at 456 sfpm. (After Trigger and Chao)

proper clearance angles, and by replacing tools when they become dull. In Fig. 20–17 is shown typical tool chip interface temperatures when cutting steel at increasing cutting speeds.

From the graph it may be seen that a cutting speed of 100 sfpm will give a cutting temperature of about 1000°F, 400 sfpm, a temperature of about 1400°F, and 1300 sfpm, a temperature of about 1700°F. Not all tool materials can withstand the higher temperatures and must consequently be operated at slower cut-

ting speeds. The highest cutting temperatures are not at the extreme tool tip but occur near the point where the chip curls away from the tool as shown in Fig. 20–18.

Tool wear may be attributed to two fundamental causes: (1) metal transfer and (2) plowing wear. At low cutting speeds and high pressures metal transfer occurs by welding of the chip to the tool face. The weld interface is stronger than the two metals involved and when breaking occurs in the chip, it results in transfer of metal from the chip to the tool face, forming the built-up edge. The built-up edge forms and breaks down hundreds of times per second. The hardened particles breaking from the built-up edge abrade the tool, resulting in accelerated wear. At higher cutting speeds and at temperatures above the recrystallization temperatures welding occurs without strain hardening.

(A) FLANK WEAR

Fig. 20–19. Types of tool wear.

(B) CRATER WEAR

(C) CHIPPING

When this weld is broken the surfaces separate in the same plane in which they were formed. A built-up edge does not form but as the surfaces separate, small particles of the tool matrix may be transferred to the chip, causing tool wear. Metal transfer is affected mainly by the hardness of the two surfaces, the surface finish, cutting pressures, and conditions of lubrication.

Plowing wear in metal cutting results when hard particles embedded in or attached to the surface of the chip slide over the tool face. Particles on the chip may result from tiny hardened fragments from the built-up edge or from abrasive carbides in the material matrix.

Tool wear generally is evidenced by one or more of the following in Fig. 20–19: flank wear is shown at A, cratering B, or chipping C. The spike occurring on the region of flank wear is frequently caused by work hardening of layers previously cut. It may also result when machining materials with an abrasive outer

Fig. 20–20. Parts produced by electrical discharge machining. (EDM). (Elox Corp. of Michigan)

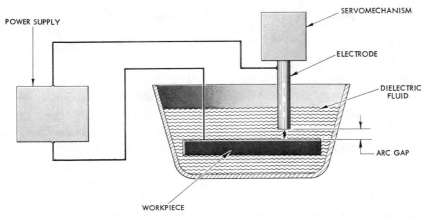

POWER SUPPLY

SERVOMECHANISM

ELECTRODE

DIELECTRIC FLUID

ARC GAP

WORKPIECE

Fig. 20—21. Electrical discharge machining circuit. Arc gap about 0.001″ maintained by machine.

surface. These two types of wear—metal transfer and plowing—are the main causes of tool wear.

When flank wear becomes excessive, around .030-in. for carbide tools, the tool is regarded as "worn-out." Continued use will result in excessive frictional heat at the cutting point; it will rapidly lose its hardness and catastrophic tool failure is imminent.

20.6 Electrical Discharge Machining. Machining metal by means of a high frequency electric discharge is a relatively new process that is gaining wide acceptance. Electrical discharge machining (EDM) is being used to machine complex shapes (Fig. 20–20) in extrusion and form-ing dies, machine hardened metals or carbides, slice thin sections from metallic semiconductors, and shape space age alloys. The basic EDM circuit is shown in Fig. 20–21. Repeated discharge of current between the tool (electrode) and metal workpiece results in vaporizing the workpiece, producing a cavity which is a duplicate of the electrode shape. A dielectric fluid is used to maintain a constant electrical resistance across the spark gap and to flush away products of the erosion. Metal removal rate and surface finish are controlled by the frequency and intensity of the sparks. High frequency, low current settings give low metal removal rates and fine finishes. Low frequency, high current discharges give relatively rapid metal removal rates but coarse surface finishes.

Because of the uniformity of craters formed by electric discharge machining, EDM surfaces require much less polishing than conventionally machined surfaces to achieve surfaces suitable for molding dies, extrusion dies, and the like. It is generally accepted that EDM surfaces are only about 80 percent as rough as those produced by conventional methods with comparable surface roughness readings. Typical EDM surfaces are shown in Fig. 20–22.

The EDM process which works equally well on hardened or annealed materials, has been widely accepted by toolmakers. Use of EDM in the aerospace industries was thought to be the answer to machining of difficult-to-machine high strength alloys but its use has been somewhat hampered by the fact

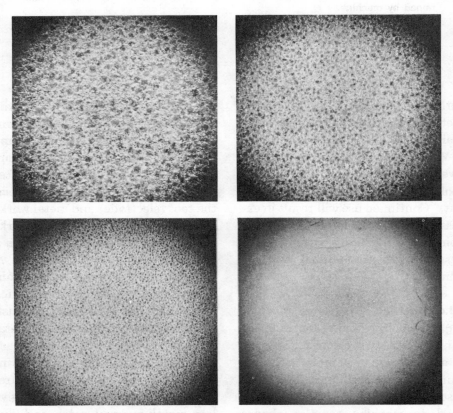

Fig. 20–22. EDM surface finishes. 3X. Upper left, 500 rms. Upper right, 250 rms. Lower left, 125 rms. Lower right, 32 rms.

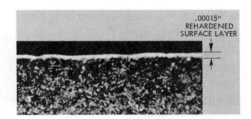

Fig. 20–23. Rehardened layer 0.00015″ deep on AISI 01 tool steel produced by electrical discharge machining. White layer has R C64 hardness compared to R C57 for base metal. (American Iron and Steel Institute)

that parts produced by EDM frequently have lower fatigue strength than conventionally machined parts. This fact, coupled with results of a study of failures of tools produced by EDM has led to a better understanding of the metallurgical effects of electrical discharge machining.

Laboratory studies have shown a shallow white surface layer, observable only by microscopic means, to exist on the surface of certain tools machined by EDM techniques. When present, the layer varied in depth from a few "tenths" to several thousandths from tool to tool. The

condition was evident on materials cut in the hardened condition but not on those cut in the annealed condition. Laboratory studies confirmed the white layer to be a rehardened zone resulting from high temperatures of the spark discharge and the subsequent quenching action of the dielectric fluid.

For example: the base hardness of the AISI 01 mold steel shown in Fig. 20–23 was Rockwell C57 while the hardness of the white layer was Rockwell C64. Sufficient heat had been generated to cause complete rehardening of that surface. It

Fig. 20–24. Rehardened layer and tempered zone beneath it from electrical discharge machining of tool steel. Rehardened layer frequently causes tool failure. Polishing and honing will remove it. (American Iron and Steel Institute)

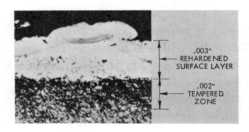

is safe to say that the surface temperature was close to 1500°F. It is thought that electrical discharge machining produces a hardened layer on all materials containing sufficient carbon to produce hardening but that subsequent heat treatment of annealed steel masks its presence. Depth of the rehardened zone has been found to vary from a minimum of .00015″ to a maximum of .005″. The shallowest layer was produced by high frequency, low current discharges. Conversely, the deepest rehardened zone was produced during rapid metal removal with low frequency, high amperage sparks. Under some conditions a tempered zone exists immediately beneath the rehardened surface layer as shown in Fig. 20–24.

It has been observed that the hard white layer is considerably richer in carbon than the base material. The fact that a hydrocarbon dielectric fluid was present in the high temperature discharge zone has led to the conclusion that increased carbon is derived from the dielectric fluid. Also pickup of minute particles from the electrode material have also been observed in this surface layer.

A recent Air Force sponsored research program conducted at Cincinnati Milling Machine Co. has conclusively shown that surface damage and surface roughness can be accurately controlled by the judicious selection of tool material and dielectric fluid, and if necessary, surface damage and contamination can be reduced to a negligible minimum.

Most EDM produced surfaces require a light honing or lapping operation to achieve the desired smoothness. This same finishing operation is excellent for removing the undesirable rehardened zone.

20.7 Electrochemical Machining. In the process of electrochemical machining material is removed from the workpiece by controlled electrochemical dissolution similar to reverse electroplating. A schematic illustration of the electrochemical machining process is shown in Fig. 20–25. The electrolyte is pumped rapidly through the tool to flush away the dissolved metal. A gap between the tool and workpiece of less than 0.010 is maintained. A D.C. power supply of about 5-15 volts maintains very high current density between the tool and workpiece. In many instances 100 to 1500 amp per square inch is used; however, the tendency is toward higher currents.

The rate of metal removal with electrochemical machining is proportional to current density and is normally much higher than with electrical discharge machining. Penetration rate is proportional to current density for a given workpiece. The penetration rate for selected materials is shown in Fig. 20–26.

Advantages of electrochemical

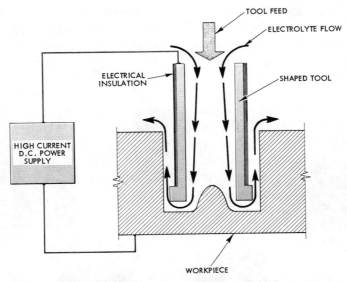

Fig. 20–25. Schematic of electrochemical machining process.

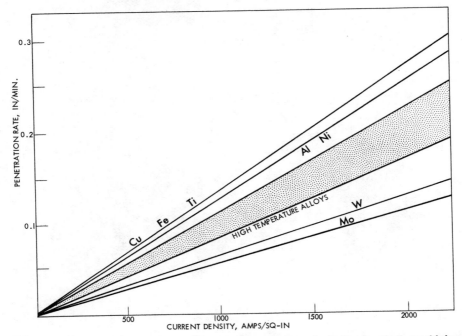

Fig. 20–26. Tool penetration rate and current density of electrochemical machining of various materials.

machining (ECM) include the formation of burr-free surfaces, no tool wear, absence of thermal damage, and very good surface finishes. The ECM process easily machines materials harder than 400 BHN, and

Fig. 20–27. Parts produced by electrochemical machining.

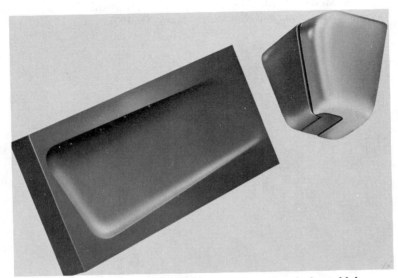

Fig. 20–27. (Cont.) Parts produced by electrochemical machining.

gives relatively high metal removal rates. There are a few drawbacks with ECM, including difficulty of maintaining uniform hydraulic electrolyte flow, inability to produce sharp square corners and flat bottoms, and an apparent reduction of fatigue strength on some metals. The apparent 10-15 percent reduction of fatigue strength is due to the fact that ECM removes the stressed surface layers, leaving the surface in a stress-free condition. Mechanical finishing frequently imparts compressive stresses to the surface raising the fatigue strength of a metal above its intrinsic value. The apparent loss of fatigue strength on ECM produced parts can be restored by mild surface treatments such as hand honing and shot peening. In Fig. 20–27 are shown typical parts produced by electrochemical machining.

References

Black, Paul H. *Theory of Metal Cutting*. New York: McGraw-Hill, 1961.

Brierley, Robert G. and Siekmann, H. J. *Machining Principles and Cost Control*. New York: McGraw-Hill, 1964.

Ernst, Hans. *Physics of Metal Cutting*. Cincinnati, O.: The Cincinnati Milling Machine Co.

Ernst, Hans and Merchant, M. Eugene. *Chip Formation, Friction, and Finish*. Cincinnati, O.: The Cincinnati Milling Machine Co.

Merchant, M. Eugene. *Machining—Theory and Practice*. Cleveland, O.: American Society for Metals, 1950.

Merchant, M. Eugene. *Metal Cutting Research—Theory and Application*. Cincinnati, O.: The Cincinnati Milling Machine Co., 0000.

Monarch Machine Tool Co. *Speeds and Feeds for Better Turning Results*. Sidney, O., 1957.

Index